Lecture Notes in Artificial Intelligence   11198

Subseries of Lecture Notes in Computer Science

LNAI Series Editors

Randy Goebel
   University of Alberta, Edmonton, Canada
Yuzuru Tanaka
   Hokkaido University, Sapporo, Japan
Wolfgang Wahlster
   DFKI and Saarland University, Saarbrücken, Germany

LNAI Founding Series Editor

Joerg Siekmann
   DFKI and Saarland University, Saarbrücken, Germany

More information about this series at http://www.springer.com/series/1244

Larisa Soldatova · Joaquin Vanschoren
George Papadopoulos · Michelangelo Ceci (Eds.)

Discovery Science

21st International Conference, DS 2018
Limassol, Cyprus, October 29–31, 2018
Proceedings

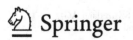
Springer

Editors
Larisa Soldatova (iD)
Goldsmiths University of London
London, UK

George Papadopoulos (iD)
University of Cyprus
Nicosia, Cyprus

Joaquin Vanschoren (iD)
Eindhoven University of Technology
Eindhoven, The Netherlands

Michelangelo Ceci (iD)
Università degli Studi di Bari Aldo Moro
Bari, Italy

ISSN 0302-9743   ISSN 1611-3349 (electronic)
Lecture Notes in Artificial Intelligence
ISBN 978-3-030-01770-5   ISBN 978-3-030-01771-2 (eBook)
https://doi.org/10.1007/978-3-030-01771-2

Library of Congress Control Number: 2018956724

LNCS Sublibrary: SL7 – Artificial Intelligence

This Springer imprint is published by the registered company Springer Nature Switzerland AG
The registered company address is: Gewerbestrasse 11, 6330 Cham, Switzerland

Preface

The 21st International Conference on Discovery Science (DS 2018) was held in Limassol, Cyprus, during October 29–31, 2018. The conference was co-located with the International Symposium on Methodologies for Intelligent Systems (ISMIS 2018), which was already in its 24th year. This volume contains the papers presented at the 21st International Conference on Discovery Science, which received 71 international submissions. Each submission was reviewed by at least three committee members. The committee decided to accept 30 papers. This resulted in an acceptance rate of 42%. Invited talks were shared between the two meetings. The invited talks for DS 2018 were "Automating Predictive Modeling and Knowledge Discovery" by Ioannis Tsamardinos from the University of Crete, and "Emojis, Sentiment, and Stance in Social Media" by Petra Kralj Novak from the Jožef Stefan Institute. The invited talks for ISMIS 2018 were "Artificial Intelligence and the Industrial Knowledge Graph" by Michael May from Siemens, Germany, "Mining Big and Complex Data" by Sašo Džeroski from the Jozef Stefan Institute, Slovenia, and "Bridging the Gap Between Data Diversity and Data Dependencies" by Jean-Marc Petit from INSA Lyon and Université de Lyon, France. Abstracts of all five invited talks are included in these proceedings.

We would like to thank all the authors of submitted papers, the Program Committee members, and the additional reviewers for their efforts in evaluating the submitted papers, as well as the invited speakers. We are grateful to Nathalie Japkowicz and Jiming Liu, ISMIS program co-chairs (together with Michelangelo Ceci), for ensuring the smooth coordination with ISMIS and myriad other organizational aspects. We would also like to thank the members of the extended DS Steering Committee (consisting of past organizers of the DS conference) for supporting the decision to organize DS jointly with ISMIS this year, and in particular Sašo Džeroski for supporting us in bringing this decision to life. We are grateful to the people behind EasyChair for making the system available free of charge. It was an essential tool in the paper submission and evaluation process, as well as in the preparation of the Springer proceedings. We thank Springer for their continuing support for Discovery Science. The joint event DS/ISMIS 2018 was organized under the auspices of the University of Cyprus. Financial support was generously provided by the Cyprus Tourism Organization and Austrian Airlines. Finally, we are indebted to all conference participants, who contributed to making this momentous event a worthwhile endeavor for all involved.

October 2018

Larisa Soldatova
Joaquin Vanschoren
Michelangelo Ceci
George Papadopoulos

Organization

Symposium Chair

George Papadopoulos   University of Cyprus, Cyprus

Program Committee Co-chairs

Larisa Soldatova   Goldsmiths University London, UK
Joaquin Vanschoren   Eindhoven University of Technology, The Netherlands

Program Committee

Mihalis A. Nicolaou	Imperial College London, UK
Ana Aguiar	University of Porto, Portugal
Fabrizio Angiulli	DEIS, University of Calabria
Annalisa Appice	University of Bari Aldo Moro, Italy
Martin Atzmueller	Tilburg University, The Netherlands
Gustavo Batista	University of São Paulo, Brazil
Albert Bifet	LTCI, Telecom ParisTech, France
Hendrik Blockeel	Katholieke Universiteit Leuven, Belgium
Paula Brito	University of Porto, Portugal
Fabrizio Costa	University of Exeter, UK
Bruno Cremilleux	Université de Caen, France
Andre de Carvalho	University of São Paulo, Brazil
Ivica Dimitrovski	Macedonia
Kurt Driessens	Maastricht University, The Netherlands
Saso Dzeroski	Jozef Stefan Institute, Slovenia
Floriana Esposito	Università degli Studi di Bari Aldo Moro, Italy
Peter Flach	University of Bristol, UK
Johannes Fürnkranz	TU Darmstadt, Germany
Mohamed Gaber	Birmingham City University, UK
Joao Gama	University of Porto, Portugal
Dragan Gamberger	Rudjer Boskovic Institute, Croatia
Crina Grosan	Brunel University, UK
Makoto Haraguchi	Hokkaido University, Japan
Kouichi Hirata	Kyushu Institute of Technology, Japan
Jaakko Hollmén	Aalto University, Finland
Geoffrey Holmes	University of Waikato, New Zealand
Bo Kang	Data Science Lab, Ghent University, Belgium
Kristian Kersting	TU Darmstadt, Germany
Takuya Kida	Hokkaido University, Japan
Masahiro Kimura	Ryukoku University, Japan

Ross King	University of Manchester, UK
Dragi Kocev	Jozef Stefan Institute, Slovenia
Petra Kralj Novak	Jozef Stefan Institute, Slovenia
Stefan Kramer	Johannes Gutenberg University Mainz, Germany
Tetsuji Kuboyama	Gakushuin University, Japan
Niklas Lavesson	Jönköping University, Sweden
Nada Lavrač	Jozef Stefan Institute, Slovenia
Philippe Lenca	IMT Atlantique, France
Gjorgji Madjarov	Ss. Cyril and Methodius University, Republic of Macedonia
Giuseppe Manco	ICAR-CNR, Italy
Elio Masciari	ICAR-CNR, Italy
Vlado Menkovski	Eindhoven University of Technology, The Netherlands
Robert Mercer	University of Western Ontario, Canada
Vera Miguéis	FEUP, Portugal
Anna Monreale	University of Pisa, Italy
Andreas Nuernberger	Otto von Guericke University of Magdeburg, Germany
Pance Panov	Jozef Stefan Institute, Slovenia
George Papadopoulos	University of Cyprus, Cyprus
Panagiotis Papapetrou	Stockholm University, Sweden
Ruggero G. Pensa	University of Turin, Italy
Bernhard Pfahringer	University of Waikato, New Zealand
Gianvito Pio	University of Bari Aldo Moro, Italy
Ronaldo Prati	Universidade Federal do ABC, Brazil
Jan Ramon	Inria, France
Chedy Raïssi	Inria, France
Stefan Rueping	Fraunhofer, Germany
Noureddin Sadawi	Imperial College London, UK
Kazumi Saito	Univesity of Shizuoka, Japan
Tomislav Smuc	Rudjer Boskovic Institute, Croatia
Larisa Soldatova	Goldsmiths, University of London, UK
Jerzy Stefanowski	Poznan University of Technology, Poland
Allan Tucker	Brunel University, UK
Peter van der Putten	LIACS, Leiden University and Pegasystems, The Netherlands
Jan N. van Rijn	Leiden University, The Netherlands
Joaquin Vanschoren	Eindhoven University of Technology, The Netherlands
Celine Vens	KU Leuven Kulak, Belgium
Herna Viktor	University of Ottawa, Canada
Veronica Vinciotti	Brunel University, UK
Akihiro Yamamoto	Kyoto University, Japan
Leishi Zhang	Middlesex University, UK
Albrecht Zimmermann	Université Caen Normandie, France
Blaz Zupan	University of Ljubljana, Slovenia

Additional Reviewers

Adedoyin-Olowe, Mariam
Amna, Dridi
Ardito, Carmelo
Barella, Victor H.
Bioglio, Livio
Cunha, Tiago
De Carolis, Berardina Nadja
Gu, Qianqian
Guarascio, Massimo

Loglisci, Corrado
Nogueira, Rita
Orhobor, Oghenejokpeme Orhobor
Panada, Dario Panada
Pedroto, Maria
Pirrò, Giuseppe
Pisani, Paulo Henrique
Pliakos, Konstantinos
Ritacco, Ettore

Invited Talks

Emojis, Sentiment and Stance in Social Media

Petra Kralj Novak

Jozef Stefan Institute

Abstract. Social media are computer-based technologies that provide means of information and idea sharing, as well as entertainment and engagement handly available as mobile applications and websites to both private users and businesses. As social media communication is mostly informal, it is an ideal environment for the use of emoji. We have collected Twitter data and engaged 83 human annotators to label over 1.6 million tweets in 13 European languages with sentiment polarity (negative, neutral, or positive). About 4% of the annotated tweets contain emojis. We have computed the sentiment of the emojis from the sentiment of the tweets in which they occur. We observe no significant differences in the emoji rankings between the 13 languages. Consequently, we propose our Emoji Sentiment Ranking as a European language-independent resource for automated sentiment analysis. In this talk, several emoji, sentiment and stance analysis applications will be presented, varying in data source, topics, language, and approaches used.

Automating Predictive Modeling and Knowledge Discovery

Ioannis Tsamardinosi

University of Crete

Abstract. There is an enormous, constantly increasing need for data analytics (collectively meaning machine learning, statistical modeling, pattern recognition, and data mining applications) in a vast plethora of applications and including biological, biomedical, and business applications. The primary bottleneck in the application of machine learning is the lack of human analyst expert time and thus, a pressing need to automate machine learning, and specifically, predictive and diagnostic modeling. In this talk, we present the scientific and algorithmics problems arising from trying to automate this process, such as appropriate choice of the combination of algorithms for preprocessing, transformations, imputation of missing values, and predictive modeling, tuning of the hyper-parameter values of the algorithms, and estimating the predictive performance and producing confidence intervals. In addition, we present the problem of feature selection and how it fits within an automated analysis pipeline, arguing that feature selection is the main tool for knowledge discovery in this context.

Mining Big and Complex Data

Saso Dzeroski

Jozef Stefan Institute and Jozef Stefan International Postgraduate School,
Slovenia

Abstract. Increasingly often, data mining has to learn predictive models from big data, which may have many examples or many input/output dimensions and may be streaming at very high rates. Contemporary predictive modeling problems may also be complex in a number of other ways: they may involve (a) structured data, both as input and output of the prediction process, (b) incompletely labelled data, and (c) data placed in a spatio-temporal or network context.

The talk will first give an introduction to the different tasks encountered when learning from big and complex data. It will then present some methods for solving such tasks, focusing on structured-output prediction, semi-supervised learning (from incompletely annotated data), and learning from data streams. Finally, some illustrative applications of these methods will be described, ranging from genomics and medicine to image annotation and space exploration.

Artificial Intelligence and the Industrial Knowledge Graph

Michael May

Siemens, Munich, Germany

Abstract. In the context of digitalization Siemens is leveraging various technologies from artificial intelligence and data analytics connecting the virtual and physical world to improve the entire customer value chain. The internet of things has made it possible to collect vast amount of data about the operation of physical assets in real time, as well as storing them in cloud-based data lakes. This rich set of data from heterogeneous sources allows addressing use cases that have been impossible only a few years ago. Using data analytics e.g. for monitoring and predictive maintenance is nowadays in wide-spread use.

We also find an increasing number of use cases based on Deep Learning, especially for imaging applications. In my talk I will argue that these techniques should be complemented by AI-based approaches that have originated in the knowledge representation & reasoning communities.

Especially industrial knowledge graphs play an important role in structuring and connecting all the data necessary to make our digital twins smarter and more effective. The talk gives an overview of existing and planned application scenarios incorporating AI technologies, data analytics and knowledge graphs within Siemens, e.g. building digital companions for product design and configuration or capturing the domain knowledge of engineering experts from service reports using Natural Language Processing.

Bridging the Gap between Data Diversity and Data Dependencies

Jean-Marc Petit

INSA Lyon and Universit de Lyon, France

Abstract. Data dependencies are declarative statements allowing to express constraints. They turn out to be useful in many applications, for example from database design (functional, inclusion, multi-valued, dependencies) to data quality (conditional functional dependencies, matching dependencies, denial dependencies,). Their practical impacts in many commercial tools acknowledge their importance and utility. Specific data dependencies have been proposed to take into account data diversity encountered in practice, i.e. inconsistency, uncertainty, heterogeneity. In this talk, I will introduce the main ingredients required to unify most of data dependencies proposed in the literature. Two approaches will be presented: The first one is a declarative query language, called RQL, which is a user-friendly SQL-like query language devoted to data dependencies. The second one is to study structural properties on data domains to define data dependencies through a lattice point of view.

Contents

Reinforcement Learning

Streams and Time Series

Subgroup and Subgraph Discovery

Text Mining

Applications

Classification

Addressing Local Class Imbalance in Balanced Datasets with Dynamic Impurity Decision Trees

Andriy Mulyar and Bartosz Krawczyk[(✉)]

Department of Computer Science, Virginia Commonwealth University,
401 West Main Street, Richmond, VA 23284, USA
{mulyaray,bkrawczyk}@vcu.edu

Abstract. Decision trees are among the most popular machine learning algorithms, due to their simplicity, versatility, and interpretability. Their underlying principle revolves around the recursive partitioning of the feature space into disjoint subsets, each of which should ideally contain only a single class. This is achieved by selecting features and conditions that allow for the most effective split of the tree structure. Traditionally, impurity metrics are used to measure the effectiveness of a split, as ideally in a given subset only instances from a single class should be present. In this paper, we discuss the underlying shortcoming of such an assumption and introduce the notion of local class imbalance. We show that traditional splitting criteria induce the emergence of increasing class imbalances as the tree structure grows. Therefore, even when dealing with initially balanced datasets, class imbalance will become a problem during decision tree induction. At the same time, we show that existing skew-insensitive split criteria return inferior performance when data is roughly balanced. To address this, we propose a simple, yet effective hybrid decision tree architecture that is capable of dynamically switching between standard and skew-insensitive splitting criterion during decision tree induction. Our experimental study depicts that local class imbalance is embedded in most standard classification problems and that the proposed hybrid approach is capable of alleviating its influence.

Keywords: Machine learning · Decision trees · Splitting criteria
Class imbalance

1 Introduction

Among a plethora of existing machine learning and pattern classification algorithms, decision trees have emerged as one of the most popular and widely-used. While not being the algorithm with the highest predictive power (decision trees are often considered weak classifiers), they offer a number of unique benefits. They are interpretable, allowing for an explanation of the decision process that

© Springer Nature Switzerland AG 2018
L. Soldatova et al. (Eds.): DS 2018, LNAI 11198, pp. 3–17, 2018.
https://doi.org/10.1007/978-3-030-01771-2_1

leads to a given classification and the gleaning of valuable insights from the analyzed data [19]. They can handle various types of data, as well as missing values. They are characterized by low computational complexity, making them a perfect choice for constrained or dynamic environments [12,20]. They are simple to understand and implement in distributed environments [5], which allows them to succeed in real-life industrial applications. Finally, they have gained extended attention as an excellent base component in ensemble learning approaches [21].

The general idea behind decision tree induction lies in a sequential partitioning of the feature space, until the best possible separation among classes is achieved. Ideally, each final disjunct (leaf) should contain instances coming only from a single class. This is achieved by using impurity metrics that measure how well each potential split separates instances from distinct classes. While a perfect separation may potentially be obtained, it may lead to the creation of small disjuncts (e.g., containing only a single instance) and thus to overfitting. Therefore, stopping criteria and tree post-pruning are used to improve the generalization capabilities of decision trees [10]. However, one must note that these mechanisms are independent from the used splitting criterion and thus will not alleviate other negative effects that may be produced when using standard impurity criteria.

In this paper, we focus on the issue that while each split created during decision tree induction aims at maximizing the purity of instances in a given disjunct (i.e., ensuring they belong to the same class), the underlying class distributions will change at each level. Therefore, we will be faced with a problem of local class imbalance. We state that this is a *challenge inherent to the learning process*. Contrary to the well-known problem of global class imbalance [14], we do not know the class proportions a priori. They will evolve during decision tree induction and thus global approaches to alleviate class imbalance, such as sampling or cost-sensitive learning, cannot be used before tree induction. This also prevents us from using any existing splitting criteria that are skew-insensitive [2,7,16], as they do not work as well as standard ones when classes are roughly balanced.

We propose to analyze in-depth the problem of local class imbalance during decision tree induction and show that it affects most balanced datasets at some point of classifier training. This affects the tree structure, leading to the creation of bias in tree nodes towards the class that is better represented. In order to show that this issue impacts the generalization capabilities of decision trees, we propose a simple, yet effective hybrid decision tree architecture. Our proposed classifier is capable of switching between standard and skew-insensitive splitting metrics, based on the local class imbalance ratio in each given node. The main contributions of this paper include:

- Analysis of the local class imbalance phenomenon that occurs during decision tree induction.
- Critical assessment of standard and skew-insensitive splitting criteria.
- A hybrid decision tree architecture that is capable of dynamically switching between splitting criteria based on local data characteristics.

– A experimental study that showcases the need for analyzing emergence of local class imbalance and its impact on the predictive power of decision trees.

2 Decision Tree Induction

In our work we adopt the following notation [3] for facilitation of discussion: for a given node, n, in a binary decision tree there corresponds a class of splits $\{s\}$ defined on the instances in n. To keep decision tree induction tractable, this class of splits is limited to only splits corresponding to axis-parallel hyperplanes bi-partitioning the local decision space of n; we will discuss $\{s\}$ under this restriction. A goodness-of-split function, $\theta(s, n)$, is defined and the best split taken as the s that maximizes $\theta(s, n)$. Denote the class proportions, or probabilities, (we will use the terms interchangeably) $\mathbf{p} = (p_0, p_1)$ where p_0 represents the proportion of majority class instances in n and and p_1 the minority - notice $p_0 \geq p_1$ with equality when the class distribution of n is balanced. Finally, define the local imbalance ratio $n_I = \dfrac{p_1}{p_0}$ as the proportion of minority class to majority class instances in n. Notice, it is irrelevant what class a given instance belongs to as we are only concerned with the distribution of classes amongst all instances in n.

With this notation, we arrive at the characterization of a split s as sending a proportion of instances P_L to the left child resulting in class probabilities $\mathbf{p}_L = (p_{0,L}, p_{1,L})$ and likewise a proportion of instances $P_R = 1 - P_L$ to the right child resulting in class probabilities $\mathbf{p}_R = (p_{0,R}, p_{1,R})$. This lends to the definition [4] of goodness-of-split as:

$$\theta(s, n) = \phi(\mathbf{p}) - P_L \phi(\mathbf{p}_L) - P_R \phi(\mathbf{p}_R)$$

where $\phi(\mathbf{p})$ is an impurity function.

Table 1. Definitions of impurity functions/criterion

Criterion	Definition
Gini	$1 - \sum\limits_{p_i \in \mathbf{p}} p_i^2$
Entropy	$-\sum\limits_{p_i \in \mathbf{p}} p_i \log(p_i)$
Hellinger distance	$\dfrac{1}{\sqrt{2}} \sqrt{\sum\limits_{p_i, p_j \in \mathbf{p}} \left(\sqrt{p_i} - \sqrt{p_j}\right)^2}$

2.1 Impurity Criterion

An impurity function $\phi(\mathbf{p}) : [0,1]^2 \to [0,1]$ measures the homogeneity of the distribution of classes in the region defining $\mathbf{p}$. When the subset of the decision space defining $\mathbf{p}$ is completely homogeneous $\phi(\mathbf{p}) = \phi(1,0) = \phi(0,1) = 0$, and when completely heterogeneous $\phi(.5,.5) = 1$. $\phi(\mathbf{p})$ takes on all values between these extrema and can, in the general multi-class case, be visualized as a strictly convex function over the unit hyper-cube taking on zero's at the vertices's in its range.

Common impurity functions found in decision tree learning include Gini and Information Gain (Entropy) [4]. In recent literature, new criterion for impurity have been formulated that exhibit properties favorable when addressing modern challenges in Machine Learning; DKM [13] and Hellinger distance [7], in particular, have been shown to be impurity criterion highly robust to class imbalances in $\mathbf{p}$ - the latter more-so than the former. Table 1 gives definitions for Gini, Entropy, and normalized Hellinger distance but is by no measure an exhaustive listing of impurity criterion present in literature.

3 Impurity Criteria and Class Imbalance

With the wide selection of impurity metrics present in literature, the question arises: which metric is best? This no-free-lunch-esque query produces a response not unlike that of the theorem - it depends. More precisely, the impurity metric that best characterizes the homogeneity, or lack thereof, of the decision space modeled by $\mathbf{p}$ is a function of the class distribution [8] of $\mathbf{p}$; that is, we can use the class imbalance statistic n_I to determine the behavior of $\phi(\mathbf{p})$. We will now describe a method developed in [8] for measuring the performance of various incarnations of $\phi(\mathbf{p})$ under class imbalance and analyze two interesting cases: Hellinger distance and Gini impurity.

Table 2. ROC surfaces of splitting criteria

Criterion	TP node ROC surface eqn.
Gini	$\dfrac{1+c}{tpr + c \cdot fpr} \cdot tpr \cdot fpr$
Hellinger	$\sqrt{\left(\sqrt{tpr} - \sqrt{fpr}\right)^2 + \left(\sqrt{1-tpr} - \sqrt{1-fpr}\right)^2}$

3.1 ROC Surfaces

In [8], Flach proposes an extension to Receiver-Operator-Characteristic (ROC) curve analysis that allows for the projection of decision tree node impurity criteria into ROC space. With this tool, it is possible to directly compare the

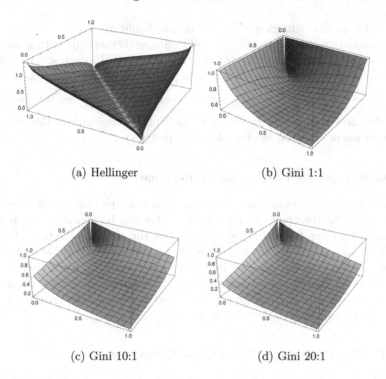

(a) Hellinger   (b) Gini 1:1

(c) Gini 10:1   (d) Gini 20:1

Fig. 1. Skew surfaces of Hellinger distance and Gini impurity. Hellinger distance retains a constant skew surface over all levels of class imbalance. (a) depicts a skew surface of Hellinger distance, (b) depicts the Gini skew surface with $c = 1$ (completely class balanced). (c) depicts the Gini skew surface with $c = \frac{1}{10}$ (10:1 class imbalance). (d) depicts the Gini skew surface with $c = \frac{1}{20}$. The iso-line diagonal of the above isosurfaces corresponds to splits that result in complete class separation hence having an impurity of zero (ie., completely pure).

theoretical performance of different impurity criteria under varying levels of class imbalance in both an analytical and geometrical sense. Flach's model is constructed as follows: a given split s can be seen a sending a set of positive predictions (true positive and false positive instances) to one child and a set of negative predictions (true negative and false negative) instances to the other child of a given node in a binary decision tree. In this manner, we can represent the split with entries in a 2×2 contingency table (confusion matrix) and utilize previously formulated methods [8] for projecting classifier evaluation metrics into ROC space accordingly. The generalized split isosurface becomes:

$$m = Imp(pos, neg) - (tp + fp) \cdot Imp\left(\frac{tp}{tp + fp}, \frac{fp}{tp + fp}\right)$$

$$- (tn + fn) \cdot Imp\left(\frac{tn}{tn + fn}, \frac{fn}{tn + fn}\right)$$

as derived from the definition of the goodness-of-split $\theta(s,n)$.

This model, as applied to Gini [8] and Hellinger distance [6], with normalization yields Table 2 of functions (geometrically interpreted as surfaces in 3-space) mapping the true positive rate (tpr), false positive rate (fpr), and class imbalance $c = \dfrac{tn + fn}{tp + fp} = \dfrac{1}{n_I}$ to $\phi(\mathbf{p})$ for all possible splits of a decision node. We direct the interested reader to [8] for an in-depth excursion into the geometric interpretation of common machine learning metrics.

3.2 Gini Impurity Under Class Imbalance

The Gini impurity demonstrates a strong sensitivity to class skews when projected under Flach's node impurity model. This can be visualized by plotting the Gini isosurfaces for $c = \{1, \frac{1}{10}, \frac{1}{20}\}$. Notice the striking difference between Figs. 1b and c where isosurfaces for Gini are generated utilizing class imbalances of 1 and $\frac{1}{10}$ respectively. As class imbalance escalates, Gini isosurfaces become flatter meaning that in a decision node that is highly homogeneous (appearing closer to the edges of the isosurface) class skew forces the Gini criterion to give a more biased estimate of node impurity [8].

3.3 Hellinger Distance Under Class Imbalance

The Hellinger distance, a skew insensitive impurity metric capturing the divergence of the class distribution modeled by $\mathbf{p}$ [7], under Flach's model produces an identical isosurface under all values of n_I. This is demonstrated analytically by observing the corresponding equation for Hellinger distance in Table 2 is not a function of c. Consequently, this means that the performance of Hellinger distance as a decision tree impurity metric does not degrade in the presence of increased class skew.

4 Local Class Imbalance in Balanced Datasets

When combating class imbalance, state-of-the-art approaches largely consider only the global imbalance present across an entire dataset. In the context of decision trees, this a priori statistic does not take into account the new decision problems emerging during learning. During decision tree induction, a super-decision problem is recursively bi-partitioned into sub-decision problems by means of finding the partitions maximizing θ until a termination condition is reached. The subsets of the decision space encapsulated by these sub-problems are not guaranteed to have similar class distributions as that of their super-spaces in-fact, it is easy to visualize class imbalance exacerbation in induced sub-spaces as a consequence of the very goal of $\phi(\mathbf{p})$. The concept of local class imbalance, the imbalance ratio at a given tree node captured by the statistic n_I, reconciles the discussion of class imbalance in the context of decision tree learning. In general, a viable statistic or measure determining the difficulty of accurate tree induction

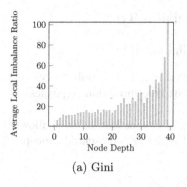
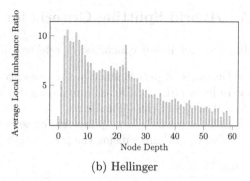

(a) Gini   (b) Hellinger

Fig. 2. Average imbalance ratio at node depth in decision trees induced over benchmark datasets. (a) depicts decision trees induced using Gini impurity, (b) depicts decision trees induced using Hellinger distance.

over a dataset must take into account the ever changing class distributions at each child node (sub-space), not only the root (entire decision space).

State-of-the-art approaches to dealing with imbalanced data would traditionally not be applied to a dataset without a global class imbalance [11]. A consequence of the above observation concludes that, in the context of decision tree learning, considering only global class imbalance is misguided as local class imbalances can arise throughout the learning process.

4.1   Gini Impurity on Local Class Imbalance

It is known that splits maximizing θ when $\phi(\mathbf{p})$ is the Gini impurity have the property of sending solely instances belonging to the majority class, p_0, to one child n_L and all other instances to the other child n_R [3]. This property, while working towards the end goal of pure leaf nodes, drastically amplifies the class imbalance present throughout the induction process. Breiman's theoretical proof of this property is shown empirically in Fig. 2a. Acknowledging scale, we observe Gini decision trees initially induced on global class balanced datasets being forced to calculate optimal splits on harshly local class imbalanced regions of the decision space.

4.2   Hellinger Distance on Local Class Imbalance

Decision Tree induction utilizing the Hellinger distance as $\phi(\mathbf{p})$ has been shown to select splits independent from class skews in $\mathbf{p}$ [7]. We observe empirically in Fig. 2b that this insensitivity to skew results in a stable local class imbalance ratio throughout all levels of tree depth. We do note that in exchange for this convenience, trees induced using Hellinger distance require many more levels to arrive at pure child nodes.

5   Hybrid Splitting Criteria for Decision Trees

In summary, it was concluded or referred that Gini impurity:

- Degrades in performance in the presence of even minor class imbalances.
- Induces large local class imbalances throughout tree induction regardless of initial global class imbalance.
- Inclines towards large jumps in the separation of classes leading to shallower trees but in turn causing the possible oversight of under-represented concepts.

while Hellinger distance:

- Does not degrade in performance in the presence of high levels of class imbalance.
- Does not exacerbate local class imbalance throughout tree induction.
- Incorporates fine grain details of the decision space into split selection allowing for the learning of under-represented concepts at the cost of deeper trees, and intuitively speculating, the possible learning of noise and lack of generalization.

We propose a hybrid decision tree that dynamically selects the impurity criterion best suited for giving an earnest impurity measurement based on the local class distribution induced by a potential split of the decision space local to a node. By coupling the skew insensitivity and local class balancing nature of the Hellinger distance with Gini impurities local class imbalancing but excellent large-scale class separating ability, we achieve a dynamic splitting criterion that increases classification performance over the utilization of a single impurity metric whilst having no effect on induction complexity.

Formally, when considering a possible split s defining two prospective children n_L and n_R over the decision space of a node n we define the dynamic impurity criterion:

$$\phi_D\left(\mathbf{p}, \alpha\right) = \begin{cases} \phi_H\left(\mathbf{p}\right) & n_{c,I} \geq \alpha \\ \phi_G\left(\mathbf{p}\right) & otherwise \end{cases}$$

where $n_c \in \{n_L, n_R\}$, $n_{c,I} = \dfrac{\mathbf{p}_{c,0}}{\mathbf{p}_{c,1}}$ (the proportion of majority to minority class instances), α is the imbalance ratio threshold, and $\phi_H\left(\mathbf{p}\right), \phi_G\left(\mathbf{p}\right)$ are the Hellinger distance and Gini impurity respectively. A description of the algorithm is given below. A ready-to-apply implementation is available as a fork of the official sci-kit learn python machine learning repository [17].

This formulation takes advantage of the Gini impurities excellent class separating ability while diminishing it's under performance in the presence of class skews by allowing the Hellinger distance to quantify the impurity of a region when necessary. The parameter α is to be tuned dependent on local class imbalance severity during tree induction.

Algorithm 1 Dynamic Splitting Criterion Algorithm

1: **procedure** IMP($\mathbf{p}, \alpha$)   ▷ Computes impurity of decision space characterized by $\mathbf{p}$
2:     $n_I \leftarrow \frac{p_0}{p_1}$
3:     **if** $n_I \geq \alpha$ **then**
4:         **return** $\phi_H(\mathbf{p})$
5:     **else**
6:         **return** $\phi_G(\mathbf{p})$
7:     **end if**
8: **end procedure**
9:

---

6   Experimental Study

We designed the following experimental study in order to answer the following two research questions posed in this manuscript:

- Does local class imbalance in balanced datasets impact the predictive power of inducted decision trees?
- Is the proposed hybrid architecture for dynamic splitting criteria selection capable of improving decision tree performance?

6.1   Datasets

Data benchmarks employed for empirical analysis consist of a subset of the *Standard datasets* found in the *KEEL-dataset* repository [1] . The *KEEL-dataset* repository contains a well categorized collection of datasets that span over various application domains. Table 3 comprises the data benchmarks utilized in our empirical study of hybrid splitting criterion decision trees. The column IR refers to the global imbalance (root imbalance) ratio of the dataset. This study considers solely initially balanced datasets to demonstrate the emergence of local class imbalances during tree induction.

6.2   Set-up

Decision trees with dynamic impurity selection are evaluated against decision trees utilizing traditional Gini and Entropy splitting criterion. Multiple trials with increasing α thresholds are utilized on each respective benchmark to illustrate the need for tuning of α based on the severity of local class imbalance introduced during tree induction. Experimentation is conducted on CART Decision Trees [4] as implemented in the scikit-learn [17] repository with the modification of our dynamic splitting criterion. All default parameters are left unchanged as provided in the DecisionTreeClassifier documentation. Evaluation is conducted with use of stratified 5×2 cross validation as already performed in KEEL [1] recording accuracy, recall (sensitivity) on the target class, F-measure, G-Mean, and area under the ROC curve (AUC). Additionally, we conduct a Shafer post-hoc statistical analysis over multiple datasets [9] with significance level $\alpha = 0.05$.

Table 3. Characteristics of benchmark datasets.

Dataset	#Inst	#Feat	#Class	IR	Dataset	#Inst	#Feat	#Class	IR
ring	7400	20	2	1.01	Australian	690	14	2	1.24
banana	5,300	2	2	1.23	bupa	345	6	2	1.37
saheart	462	9	2	1.88	haberman	306	3	2	2.81
heart	270	13	2	1.25	ionosphere	351	33	2	1.80
magic	19,020	10	2	1.84	tic-tac-toe	958	9	2	1.89
pima	768	8	2	1.86	appendicitis	106	7	2	4.00
sonar	208	60	2	1.15	spambase	4,597	57	2	1.53
spectfheart	267	44	2	3.84	titanic	2,201	3	2	2.09
twonorm	7,400	20	2	1.00	wdbc	569	30	2	1.69

6.3   Results and Discussion

Table 4 presents results of evaluations across our twenty data benchmarks. The corresponding row labeled dynamic(α) corresponds to a decision tree induced with dynamic impurity selection using α as the imbalance threshold. An exhaustive search $[2 \leq \alpha \leq 200]$ is conducted to find the α threshold that maximizes accuracy on each respective dataset. Due to space constraints enumeration of evaluation performance of every tree in the exhaustive search is not feasible but we note that alpha thresholds with in the vicinity of the one showcased in Table 4 had similar, superior performance to trees induced using gini, hellinger, and entropy. Additionally, Table 5 presents the outcomes of a post-hoc statistical analysis of results. We give a broad summary of general performance below.

We summarize the performance of dynamic impurity selection over thresholds $[2 \leq \alpha \leq 200]$ on our benchmarks with assistance of Figs. 3a–e. In Figs. 3a–e, each point (α, E) on the variable curve corresponds to the mean value of the referenced evaluation metric, E, when utilizing an imbalance ratio threshold α over all data benchmarks. The solid and dashed constant lines correspond to the mean value of E when utilizing Gini impurity and Hellinger distance respectively. We conclude primarily the need for tuning of α as clearly visible from increased average performance at certain thresholds values across all evaluation metrics metrics. We would like to note an interesting observation concerning the average limiting behavior ($\alpha > 100$) where the predominant majority of splits are chosen utilizing the Gini impurity - the performance of dynamic impurity selection becomes constrained between the performance of Gini impurity and Hellinger distance implying that it may be best to restrict a search for an optimal α threshold to $\alpha \leq 100$.

Dynamic impurity selection featured across-the-board performance improvement or matching on every benchmark except appendicitis when compared to the widely used Gini impurity in regards to accuracy or AUC. This is due to the fact that when using a large α threshold the majority of splits chosen will be splits determined by the Gini impurity. It is interesting to consider datasets such as bupa, ionosphere, magic, sonar, spectfheart, twonorm, wdbc, and tic-

Table 4. Results obtained by examined decision tree induction approaches with respect to five performance metrics. In the column titled *Impurity*, dynamic(α) corresponds to a tree induced using dynamic impurity selection with α as the imbalance threshold for selecting the splitting criterion to utilize at a potential split. The parameter α showcased corresponds to the α value that maximized AUC under an internal 5x2 cross validated search of trees induced at thresholds spanning $[2 \leq \alpha \leq 200]$. Un-pruned trees are grown that differ only by the splitting criterion used during induction.

Dataset	Impurity	Accuracy	Recall	F-Measure	G-Mean	AUC	Dataset	Impurity	Accuracy	Recall	F-Measure	G-Mean	AUC
appendicitis							australian						
	gini	**0.8961**	**0.9765**	**0.9384**	0.7288	0.7732		gini	0.8464	0.8826	0.8636	0.8406	0.8419
	hellinger	**0.8961**	**0.9765**	**0.9384**	0.7288	0.7732		hellinger	0.8522	0.8851	0.8689	0.8472	0.8481
	entropy	0.8866	0.9647	0.9324	0.7246	0.7674		entropy	0.8522	0.8852	0.8688	0.8462	0.8481
	dynamic(14)	0.8771	0.9412	0.9249	**0.7462**	**0.7806**		dynamic(30)	**0.8609**	**0.8981**	**0.8773**	**0.8549**	**0.8562**
saheart							banana						
	gini	0.6795	0.8111	0.7674	0.5896	0.6212		gini	0.8908	0.9196	0.9028	0.8868	0.8874
	hellinger	0.6558	**0.8179**	0.7566	0.5329	0.5840		hellinger	**0.8928**	**0.9224**	**0.9048**	**0.8887**	**0.8894**
	entropy	0.6687	0.8078	0.7601	0.5716	0.6070		entropy	0.8891	**0.9224**	0.9018	0.8843	0.8852
	dynamic(68)	**0.6860**	0.8177	**0.7725**	**0.5964**	**0.6276**		dynamic(82)	0.8915	0.9189	0.9034	0.8878	0.8883
bupa							haberman						
	gini	0.7101	0.6828	0.6640	0.7048	0.7064		gini	0.6698	0.8178	0.7833	0.4411	0.5390
	hellinger	0.6957	0.6414	0.6407	0.6824	0.6882		hellinger	0.6861	**0.8400**	0.7965	0.4652	0.5494
	entropy	0.7043	0.6621	0.6510	0.6955	0.6985		entropy	0.6861	0.8356	0.7958	0.4599	0.5534
	dynamic(28)	**0.7217**	**0.6759**	**0.6708**	**0.7126**	**0.7154**		dynamic(26)	**0.6896**	0.8356	**0.7980**	**0.4692**	**0.5597**
heart							ionosphere						
	gini	0.8074	**0.8667**	0.8336	0.7942	0.8000		gini	0.9318	0.8815	0.9028	0.9197	0.9208
	hellinger	0.8074	0.8933	0.8393	0.7858	0.7967		hellinger	0.9374	0.8892	0.9108	0.9259	0.9268
	entropy	0.8111	**0.8667**	0.8381	0.7987	0.8042		entropy	0.9345	0.8892	0.9072	0.9238	0.9246
	dynamic(62)	**0.8259**	**0.8867**	**0.8512**	**0.8121**	**0.8183**		dynamic(44)	**0.9460**	**0.9129**	**0.9242**	**0.9383**	**0.9387**
magic							pima						
	gini	0.8677	0.9411	0.9022	0.8301	0.8367		gini	0.7370	0.8760	0.8126	0.6458	0.6767
	hellinger	0.8693	**0.9448**	0.9036	0.8305	0.8374		hellinger	**0.7616**	0.8680	0.8257	**0.6980**	**0.7156**
	entropy	0.8686	0.9432	0.9030	0.8303	0.8370		entropy	0.7500	**0.8820**	0.8209	0.6654	0.6927
	dynamic(118)	**0.8716**	0.9442	**0.9051**	**0.8345**	**0.8409**		dynamic(14)	0.7591	**0.8820**	**0.8267**	0.6829	0.7058
ring							sonar						
	gini	0.9343	**0.9749**	0.9363	0.9338	0.9347		gini	0.8071	0.8917	0.8340	0.7899	0.8006
	hellinger	0.9370	0.9599	0.9379	0.9369	0.9372		hellinger	0.8028	**0.9103**	0.8353	0.7762	0.7943
	entropy	**0.9378**	0.9645	0.9389	**0.9377**	**0.9381**		entropy	0.8077	0.9008	0.8349	0.7858	0.7999
	dynamic(102)	0.9374	0.9724	**0.9390**	0.9371	0.9378		dynamic(42)	**0.8266**	0.9008	**0.8493**	**0.8111**	**0.8207**
spambase							spectfheart						
	gini	0.9476	0.9709	0.9573	0.9408	0.9413		gini	0.8126	0.4182	0.4852	0.6168	0.6665
	hellinger	0.9484	0.9720	0.9581	0.9416	0.9421		hellinger	0.7862	0.3455	0.4135	0.5566	0.6230
	entropy	0.9450	**0.9752**	0.9555	0.9360	0.9368		entropy	**0.8275**	0.4545	0.5189	0.6410	0.6892
	dynamic(104)	**0.9493**	0.9741	**0.9588**	**0.9421**	**0.9426**		dynamic(26)	0.8240	**0.4909**	**0.5303**	**0.6630**	**0.7005**
titanic							twonorm						
	gini	**0.7887**	**0.9832**	**0.8631**	0.6118	0.6822		gini	0.9439	**0.9622**	0.9450	0.9437	0.9439
	hellinger	**0.7887**	**0.9832**	**0.8631**	0.6118	0.6822		hellinger	0.9441	0.9598	0.9450	0.9439	0.9440
	entropy	**0.7887**	**0.9832**	**0.8631**	0.6118	0.6822		entropy	0.9427	0.9600	0.9437	0.9425	0.9427
	dynamic(8)	0.7828	0.9235	0.8521	**0.6708**	**0.7058**		dynamic(84)	**0.9508**	0.9679	**0.9517**	**0.9506**	**0.9508**
wdbc							tic-tac-toe						
	gini	0.9578	0.9832	0.9671	0.9480	0.9490		gini	0.9321	0.8310	0.8934	0.9044	0.9083
	hellinger	0.9614	0.9860	0.9698	0.9523	0.9530		hellinger	0.9185	0.8431	0.8774	0.8986	0.9008
	entropy	0.9508	0.9749	0.9614	0.9417	0.9427		entropy	0.9217	0.8341	0.8805	0.8984	0.9011
	dynamic(12)	**0.9649**	**0.9888**	**0.9725**	**0.9561**	**0.9567**		dynamic(22)	**0.9436**	**0.8583**	**0.9124**	**0.9207**	**0.9236**

tac-toe in which the Hellinger distances performs worse or equivalent to the gini impurity. On these benchmarks, dynamic impurity selection outperformed gini, hellinger, and entropy in AUC. We believe this to be a manifestation of the conclusions summarized in Sect. 5. The Hellinger distance's faculty to over fit regions of the decision space resulting in low generalization performance becomes alleviated when coupled with Gini impurities blissful ignorance of difficult to capture, minority regions if combined appropriately.

Table 5. Results of Shafer post-hoc test with p-values for proposed dynamic approach vs. reference approaches. Symbol > stands for situation when the dynamic approach is found to be statistically better and symbol = stands for a situation when there is no significant differences.

	Accuracy	Recall	F-Measure	G-Mean	AUC
vs. gini	= (0.8931)	= (0.0531)	> (0.0374)	> (0.0279)	> (0.0185)
vs. hellinger	= (0.5062)	= (0.0726)	> (0.0402)	> (0.0400)	> (0.0247)
vs. entropy	= (0.9301)	= (0.0519)	> (0.0328)	> (0.0281)	> (0.0173)

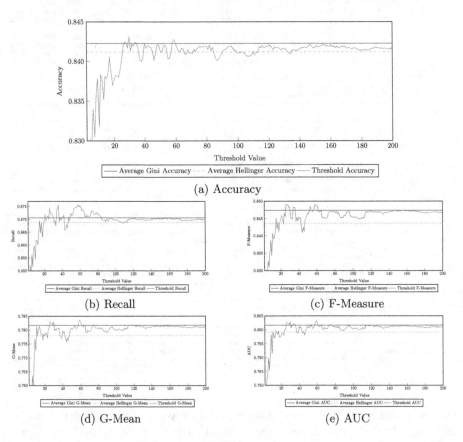

(a) Accuracy

(b) Recall

(c) F-Measure

(d) G-Mean

(e) AUC

Fig. 3. Averages of evaluation metrics over all data benchmarks at given α thresholds plotted against average evaluation metric value of Gini impurity and Hellinger distance over all benchmarks.

7 A Limitation of Dynamic Impurity Selection

We note the following possible limitation of dynamic impurity selection in application. We found that dynamic impurity selection necessitates an appropriate imbalance threshold. Depending on domain, it may be infeasible to perform a cross-validated grid search to find the optimal α threshold due to memory constraints. This limitation appears in any methodology that utilizes a tunable parameter to increase classification performance. However, the low induction complexity of decision trees compared to other learners such as Support Vector Machines aids in the relative feasibility of any parameter search.

8 Conclusions and Future Works

In this paper, we have critically assessed existing splitting criteria for decision tree induction and discussed the potential difficulties they may impose on the learning process. We have showed that while the minimization of impurity seems a viable approach, it may also result in a phenomenon we named as local class imbalance. We stated that class imbalance must be considered as not just a characteristic of data but as a *property inherent to the learning process*. This allowed us to propose a simple, yet effective hybrid decision tree architecture. It was based on dynamic selection of splitting criteria at each node - on local data properties. Our approach produced results outperforming or matching state-of-the-art decision tree performance on benchmarks of various sizes and spanning multiple domains. Results obtained during our work encourage us to pursue this topic further. We envision the following directions for our future research:

Using a more diverse base of splitting criteria. At this point, we alternate between two possible splitting criteria in our hybrid approach. It seems interesting to explore the unique strengths of other splitting metrics in order to further improve the robustness of our framework.

Considering advanced data characteristics. Currently, we base the selection of splitting criteria on the class imbalance ratio in a given node. However, disproportion between classes is not the sole indicator of the learning difficulty. We plan to incorporate instance-level difficulty metrics [18] and analyze more in-depth properties of minority class [15] to offer a better selection mechanism.

Extend the architecture to multi-class problems. So far, we have analyzed only binary datasets. Their multi-class counterparts will offer even a greater challenge, as local class imbalance with multiple classes will be much more difficult to analyze and handle in an efficient manner.

Acknowledgements. This work is supported by the VCU College of Engineering Deans Undergraduate Research Initiative (DURI) program.

References

1. Alcalá-Fdez, J., Fernández, A., Luengo, J., Derrac, J., García, S.: KEEL data-mining software tool: data set repository, integration of algorithms and experimental analysis framework. Mult.-Valued Log. Soft Comput. **17**(2–3), 255–287 (2011)
2. Boonchuay, K., Sinapiromsaran, K., Lursinsap, C.: Decision tree induction based on minority entropy for the class imbalance problem. Pattern Anal. Appl. **20**(3), 769–782 (2017)
3. Breiman, L.: Technical note: some properties of splitting criteria. Mach. Learn. **24**(1), 41–47 (1996)
4. Breiman, L., Friedman, J.H., Olshen, R.A., Stone, C.J.: Classification and Regression Trees. Wadsworth (1984)
5. Cano, A.: A survey on graphic processing unit computing for large-scale data mining. Wiley Interdisc. Rew. Data Min. Knowl. Discov. **8**(1) (2018)
6. Cieslak, D.A., Chawla, N.V.: Learning decision trees for unbalanced data. In: Daelemans, W., Goethals, B., Morik, K. (eds.) ECML PKDD 2008. LNCS (LNAI), vol. 5211, pp. 241–256. Springer, Heidelberg (2008). https://doi.org/10.1007/978-3-540-87479-9_34
7. Cieslak, D.A., Hoens, T.R., Chawla, N.V., Kegelmeyer, W.P.: Hellinger distance decision trees are robust and skew-insensitive. Data Min. Knowl. Discov. **24**(1), 136–158 (2012)
8. Flach, P.A.: The geometry of roc space: understanding machine learning metrics through roc isometrics. In: Proceedings of the Twentieth International Conference on International Conference on Machine Learning, pp. 194–201. ICML'03, AAAI Press (2003). http://dl.acm.org/citation.cfm?id=3041838.3041863
9. García, S., Fernández, A., Luengo, J., Herrera, F.: Advanced nonparametric tests for multiple comparisons in the design of experiments in computational intelligence and data mining: experimental analysis of power. Inf. Sci. **180**(10), 2044–2064 (2010)
10. Hapfelmeier, A., Pfahringer, B., Kramer, S.: Pruning incremental linear model trees with approximate lookahead. IEEE Trans. Knowl. Data Eng. **26**(8), 2072–2076 (2014)
11. He, H., Garcia, E.A.: Learning from imbalanced data. IEEE Trans. Knowl. Data Eng. **21**(9), 1263–1284 (2009). https://doi.org/10.1109/TKDE.2008.239
12. Jaworski, M., Duda, P., Rutkowski, L.: New splitting criteria for decision trees in stationary data streams. IEEE Trans. Neural Netw. Learn. Syst. **29**(6), 2516–2529 (2018)
13. Kearns, M.J., Mansour, Y.: On the boosting ability of top-down decision tree learning algorithms. In: STOC, pp. 459–468. ACM (1996)
14. Krawczyk, B.: Learning from imbalanced data: open challenges and future directions. Prog. AI **5**(4), 221–232 (2016)
15. Lango, M., Brzezinski, D., Firlik, S., Stefanowski, J.: Discovering minority sub-clusters and local difficulty factors from imbalanced data. In: Yamamoto, A., Kida, T., Uno, T., Kuboyama, T. (eds.) DS 2017. LNCS (LNAI), vol. 10558, pp. 324–339. Springer, Cham (2017). https://doi.org/10.1007/978-3-319-67786-6_23
16. Li, F., Zhang, X., Zhang, X., Du, C., Xu, Y., Tian, Y.: Cost-sensitive and hybrid-attribute measure multi-decision tree over imbalanced data sets. Inf. Sci. **422**, 242–256 (2018)

17. Pedregosa, F., et al.: Scikit-learn: machine learning in python. J. Mach. Learn. Res. **12**, 2825–2830 (2011)
18. Smith, M.R., Martinez, T.R., Giraud-Carrier, C.G.: An instance level analysis of data complexity. Mach. Learn. **95**(2), 225–256 (2014)
19. Weinberg, A.I., Last, M.: Interpretable decision-tree induction in a big data parallel framework. Appl. Math. Comput. Sci. **27**(4), 737–748 (2017)
20. Woźniak, M.: A hybrid decision tree training method using data streams. Knowl. Inf. Syst. **29**(2), 335–347 (2011)
21. Woźniak, M., Graña, M., Corchado, E.: A survey of multiple classifier systems as hybrid systems. Inf. Fusion **16**, 3–17 (2014)

Barricaded Boundary Minority Oversampling LS-SVM for a Biased Binary Classification

Hmayag Partamian[(⊠)], Yara Rizk, and Mariette Awad

Department of Electrical and Computer Engineering, American University of Beirut, Beirut, Lebanon
{hkp00,yar01,mariette.awad}@aub.edu.lb

Abstract. Classifying biased datasets with linearly non-separable features has been a challenge in pattern recognition because traditional classifiers, usually biased and skewed towards the majority class, often produce sub-optimal results. However, if biased or unbalanced data is not processed appropriately, any information extracted from such data risks being compromised. Least Squares Support Vector Machines (LS-SVM) is known for its computational advantage over SVM, however, it suffers from the lack of sparsity of the support vectors: it learns the separating hyper-plane based on the whole dataset and often produces biased hyper-planes with imbalanced datasets. Motivated to contribute a novel approach for the supervised classification of imbalanced datasets, we propose Barricaded Boundary Minority Oversampling (BBMO) that oversamples the minority samples at the boundary in the direction of the closest majority samples to remove LS-SVM's bias due to data imbalance. Two variations of BBMO are studied: BBMO1 for the linearly separable case which uses the Lagrange multipliers to extract boundary samples from both classes, and the generalized BBMO2 for the non-linear case which uses the kernel matrix to extract the closest majority samples to each minority sample. In either case, BBMO computes the weighted means as new synthetic minority samples and appends them to the dataset. Experiments on different synthetic and real-world datasets show that BBMO with LS-SVM improved on other methods in the literature and motivates follow on research.

Keywords: Biased datasets · Linearly separable features
Weighted means · Barricaded boundary minority oversampling
Kernel matrix

1 Introduction

Advancement in communication, the emergence of Internet of Things and wireless sensor networks have allowed the widespread collection of data from various sources. These diverse sources of data result in noisy, unstructured and often

© Springer Nature Switzerland AG 2018
L. Soldatova et al. (Eds.): DS 2018, LNAI 11198, pp. 18–32, 2018.
https://doi.org/10.1007/978-3-030-01771-2_2

biased data that make processing it difficult. Specifically, classification of large and biased datasets is one of the leading challenges in data analytics since traditional machine learning algorithms are usually biased towards the majority class [17]. In many real life applications such as text classification [12], handwriting recognition [13], seismic data analysis [28], fraud detection [11], medical data [4,42], and spam filtering [5,29] to name a few, the data is imbalanced or biased, i.e. the important class has significantly less instances than the other class.

Motivated to contribute to the problem of classifying biased data using LS-SVM, we propose Barricaded Boundary Minority Oversampling (BBMO), a novel approach that adds synthetically-created data to the minority class based on the demographic data distribution at the boundary of the two classes, in an attempt to improve classification. While borderline-SMOTE-2 oversamples the data at the boundary by considering the nearest majority neighbor along with the k-nearest minority neighbors, BBMO extracts the closest majority samples to each minority sample and computes their respective weighted means. The calculated weighted means are then added to the minority samples as synthetic data which form a kind of "barricade" around the minority boundary samples in the direction of the closest majority samples. This procedure ensures the removal of the bias and produces a better defined boundary by altering the distribution of the minority at the boundary differently from other proposed techniques. Experimental results on multiple datasets motivate follow on research.

The rest of the paper is organized such that Sect. 2 briefly describes existing work. Section 3 describes the proposed variations of the BBMO formulations in detail. Section 4 demonstrates the experimental results and evaluates the performance on publicly available databases while Sect. 5 concludes the paper.

2 Literature Review

The numerous efforts that have aimed to learn from biased datasets can be divided into either data, algorithmic, or kernel based. At a data level, resampling methods such as minority class over-sampling by replication to balance the class distribution and under-sampling of the majority samples which randomly eliminates samples from the majority class [15,19,24], were proposed to resample the data prior to training. Because under-sampling methods proved to be inefficient due to the loss of important information [2], the Synthetic Minority Over-Sampling Technique (SMOTE) [8] added "synthetic" data to the minority class using k-nearest neighbor. Other extensions of the SMOTE algorithm have been developed: one based on the distance space [18], SMOTE-RSB which uses the rough set theory [25], the Safe Level-SMOTE [6] where safe level coefficients are computed by considering the majority class in the neighborhood, and the Borderline-SMOTE [14] which oversamples the minority class at the borderline by considering the nearest minority neighbors. An extension of Borderline-SMOTE, borderline-SMOTE-2, oversamples the minority class by considering the nearest neighbor from the majority class in addition to the k-nearest minority neighbors. Also, SPIDER [31] locally oversamples the minority class while filtering difficult examples from the majority class.

Extensions to Support Vector Machines (SVM) [35] are among techniques that have been proposed at an algorithmic level to handle imbalanced datasets [27]. Veropoulos et al. [36] assigned different error costs for each class while Tang et al. [33] merged a cost sensitive learning approach that extracted a smaller number of the support vectors (SV), with different error costs. Another popular technique is the one-class SVM which estimates the probability density function and gives a positive value for the elements in the minority class and a negative value for everything else [26,34] as if the cost function of the majority class samples is taken to be zero [20]. Scholkopf et al. [30] proposed matching the data into a new feature space using a kernel function and separating the new samples from the origin with a maximum margin. Support Vector Data Description (SVDD) finds a sphere that encloses the minority class while separating the outliers. Since kernel parameters influence the size of the region, correct tuning is essential for satisfactory accuracy [43]. zSVM modified the decision boundary in such a way to remove the minority class's bias towards the majority class [16]. Other techniques have also been proposed to solve the problem of biased datasets that include the combination of two or more different algorithmic approaches [1,2,21,23,24,37]. Kernel modification methods among which are Class Boundary Alignment (CBA) and Kernel Boundary Alignment (KBA), transform the kernel function to enlarge the region around the minority class in an attempt to overcome the imbalance problem [38–40]. Wu and Chang point out that since positive samples lie further from the ideal boundary, SVMs trained on imbalanced data produce skewed hyper-planes.

3   BBMO

LS-SVM's original formulation [32] introduced two major changes to SVM. First, the error term in SVM was changed to a least square error. Second, the inequality constraint was changed into equality. Thus, the hyper-plane's orientation is controlled by the data instead of the SV. This leads to LS-SVM's lack of sparsity since all the dataset is considered to behave as SV [41]. However, it is computationally much faster when compared to quadratic programming algorithms and can be written and solved as a system of linear equations which allows incremental and distributed extensions. Taking LS-SVM drawbacks into consideration, BBMO creates synthetic minority samples at the hyper-plane boundary separating both classes, in the direction of the majority boundary samples to push the hyper-plane away from the minority to remove the bias.

Two variations of BBMO are represented: BBMO1 handles the linearly separable case by adding synthetic data in the direction of all majority boundary samples and BBMO2 handles the more general linearly non-separable case that uses the kernel matrix values and adds synthetic data around each boundary minority sample in the direction of all close majority boundary samples if there are any. In this paper, we adopt the following nomenclature:

- X: Training samples
- X_1: Minority class samples

- X_2: Majority class samples
- m: Number of minority samples
- n: Total number of samples
- d: Dimension of input space
- B: Boundary samples
- X_{B1}: Boundary minority samples
- X_{B2}: Boundary majority samples
- n_{B1}: Number of boundary minority samples
- n_{B2}: Number of boundary majority samples
- n_z: Number of synthetic weighted means
- α: Lagrangian multipliers of LS-SVM
- α_{max}: Maximum Lagrangian multiplier value of the minority samples
- α_{min}: Minimum Lagrangian multiplier value of the majority samples
- θ_1: Threshold value for the linearly separable case
- θ_2: Threshold value for the linearly non-separable case
- $K(.,.)$: Kernel matrix
- γ: Weight of the weighted means
- IR: Imbalance ratio of the dataset, $IR = \frac{m}{n-m}$
- Z: Weighted mean, the "barricade"
- X_{BZ}: Boundary samples with "barricade"

3.1 BBMO1

First, let us consider the linearly separable case. LS-SVM computes the Lagrange multipliers of all the samples and produces positive Lagrange multipliers for all the minority samples within the boundary points that form band B and negative values for the majority samples (labels are taken to be 1 for the minority and -1 for the majority), as shown in Fig. 1. To find the samples closest to the boundary, we first find the absolute maximum Lagrange multipliers of both classes, then we select the samples based on a threshold θ_1. Next, we compute the inter-class weighted means of the selected boundary samples by considering all the combinations, forming a "barricade" Z in front of the minority boundary samples in the direction of the majority boundary samples. The set Z represents the weighted means which employ a weight γ that varies between 0.5 and 1 (since the synthetic samples are closer to the minority boundary samples). We assume the weight varies with the imbalance ratio (IR) of the dataset according to: $\gamma = 0.5 + \frac{1}{2IR}$. Table 1 summarized the pseudo code of the proposed algorithm.

For the linearly separable case, BBMO1 uses LS-SVM Lagrange multipliers whose computation is mainly a matrix inversion with complexity $\Theta(n^3)$. The selection of the boundary samples has a complexity of $\Theta(2n)$ while the computation of the weighted means has a complexity of $\Theta(4n_z)$. The weighted means are included in the LS-SVM formulation resulting in a larger matrix inversion with complexity $\Theta((n + n_z)^3)$. The overall complexity of the BBMO1-LS-SVM algorithm is thus: $\Theta(n^3 + (n+n_z)^3 + 2n + 4n_z) < \Theta(3n^3)$ knowing that $n_z << n$ and n is large.

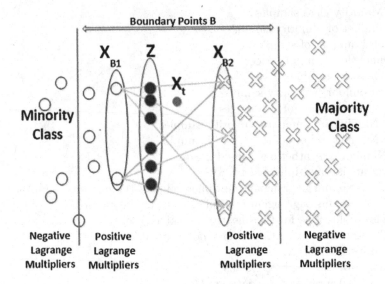

Fig. 1. Illustrative extraction of the boundary samples of the linearly separable case and the formation of the "barricade" Z by calculating the weighted means of all the boundary samples.

Table 1. BBMO1 pseudo code

---

1. $\forall x \in X$:
       Compute α using LS-SVM
2. $\forall x_i \in X_1,\ i = 1, 2, ..., m$:
       $\alpha_{max} = max(\alpha_i)$
       $\forall x_j \in X_2,\ j = 1, 2, ..., n - m$:
       $\alpha_{min} = min(\alpha_j)$
3. Let $0.8 \leq \theta_1 \leq 1$
       $\forall x_i \in X_1$:
       If $\alpha_i > \theta_1.\alpha_{max}$,
          Add x_i to X_{B1}
       $\forall x_j \in X_2$:
       If $\alpha_j < \theta_1.\alpha_{min}$,
          Add x_j to X_{B2}
4. $\forall x_k \in X_{B1}$, where $k = 1, 2, ..., n_{B1}$
       $\forall x_l \in X_{B2}$, where $l = 1, 2, ..., n_{B2}$
       Compute $z_p = \gamma x_k + (1 - \gamma)x_l$, where $p - 1, 2, ..., n_z$
       $n_z = n_{B1}.n_{B2}$, $z_p \in Z$, $\gamma = 0.5 + \frac{1}{2IR}$, $IR = \frac{m}{n-m}$
5. Add Z to X_1 and train using LS-SVM

Table 2. BBMO2 pseudo code

1. $\forall x_i \in X_1,\ i = 1, 2, ..., m,$
   $\forall x_j \in X_2,\ j = 1, 2, ..., n - m$:
       Find $K(x_i, x_j)$
2. $\forall x_i \in X_1$:
       Compute $\mu_i = \max_j\{K(x_i, x_j)\}$
3. Calculate $M = \max_i\{\mu_i\}$
4. Let $0.8 \le \theta_2 \le 1$
   $\forall x_i \in X_1,$
   $\forall x_j \in X_2$:
       If $K(x_i, X_j) > \theta_2.M,$
           Add x_i to X_{B1} and x_j to $X_{B2}(q)$ where $q = 1, 2, ..., n_{B1}$
5. $\forall x_k \in X_{B1}$ where $k = 1, 2, ..., n_{B1},$
   $\forall x_r \in X_{B2}(q)$ where $r = 1, 2, ..., Card(X_{B2}(q))$:
       Compute $z_p = \gamma x_k + (1 - \gamma)x_r$ where $p = 1, 2, ..., Card(X_{B2}(q)),$
       $z_p \in Z,\ \gamma = 0.5 + \frac{1}{2IR},\ IR = \frac{m}{n-m}$
6. Add Z to X_1 and train using LS-SVM

3.2   BBMO2

When data becomes linearly non-separable, kernels are typically used in classification problems. With BBMO2, the Lagrange multipliers were not used to choose the boundary samples, as in BBMO1. Instead, the RBF kernel matrix is used to extract the boundary samples because it represents the distances between the samples in the kernel space. Therefore, we propose retaining the maximum of these kernel values (since the higher their values, the more similar the samples are). More specifically, we select the boundary samples of the minority class with the corresponding closest majority boundary samples according to a threshold θ_2. For each boundary minority sample and its corresponding "near" majority boundary samples, the weighted means are computed and added to the minority class. The BBMO2 pseudo code in Table 2 describes the details of this algorithm.

Figure 2 shows another example with BBMO2 and the RBF kernel on a synthetic 2D dataset. The minority sample "b" finds majority samples "1" and "2" to be closest within the threshold value and generates two synthetic samples in the direction of these majority samples. Sample "d" finds only sample "5" and generates a weighted mean in that direction. Other samples are not oversampled as there are no close majority samples at the boundary. The minority samples' distribution is increased in the direction of the majority as indicated by the green band. The created "barricade" around the minority boundary samples allows a wider region for the minority samples to populate in. Other oversampling techniques have also aimed at providing more general data distribution for the minority such as SMOTE but assumptions vary.

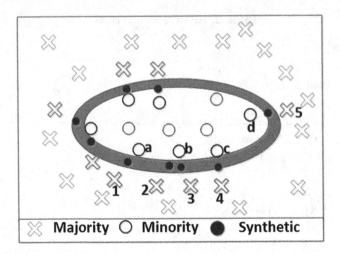

Fig. 2. BBMO2 oversamples in the direction of the closest majority for all the minority samples at the boundary.

To understand the difference between SMOTE and BBMO, consider the toy example in Fig. 3. SMOTE oversamples the minority samples by introducing synthetic samples along the line joining any or all "k" minority class nearest neighbors depending on the amount of over-sampling needed [8]. As shown in Fig. 3(left), samples "a", "b", and "c" are nearest neighbors, synthetic samples are generated along their lines. The three synthetic samples in the circle now occupy most of this region although the region contains the majority sample "1". On the other hand, BBMO2 adds synthetic data around the boundary minority samples in the direction of all near (not nearest) majority samples. Sample "a" has two close majority samples close to it, thus two synthetic samples are added in their direction, as shown in Fig. 3(right). Similarly, synthetic samples are added for samples "b" and "c". As we can observe, the circular region is not totally occupied by the minority but only a portion of it.

The way oversampling techniques perturb the distribution of the minority sample plays an important role in classification results, especially the samples at the boundary. While SMOTE reserves larger areas for the minority samples in the direction of the nearest neighbors, BBMO reserves only the regions surrounding the minority samples at the boundary in the direction of the close majority samples. Thus, BBMO leaves unknown regions unoccupied until new samples arrive and help construct the ideal boundary.

BBMO2 uses kernel matrix values to find the boundary samples; the complexity of computing the kernel matrix is $\Theta(n^2 d)$ [9]. The operation to select the boundary samples has a complexity of $\Theta(m(n - m))$. Similar to BBMO1, the computation of the weighted means has a complexity $\Theta(4n_z)$ while the weighted means causing a larger matrix inversion in the LS-SVM formulation has a complexity of $\Theta((n + n_z)^3)$. The overall complexity of the BBMO2-LS-SVM algorithm is thus: $\Theta(n^2 d + m(n - m) + (n + n_z)^3 + 4n_z) < \Theta(3n^3)$ with $n_z << n$ and n is large.

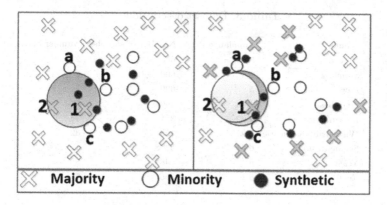

Fig. 3. Comparing how SMOTE (left) and BBMO2 (right) modify the minority class distribution.

4   Experimental Results

4.1   Experimental Setup

Experiments were run on a Windows machine with an Intel Core i7 CPU. The libSVM library was adopted to train SVM models [7] and the publicly available SMOTEBoost [10] implementation of the SMOTE algorithm was used to benchmark our proposed algorithms. In all our experiments, the values of the trade-off constant C of LS-SVM were adjusted using a grid search. A 5-fold cross validation was performed; the test data samples were chosen carefully to include minority and majority samples. Then, the results of the 5-folds were averaged and reported in the tables. In the linearly non-separable cases, the RBF kernel was adopted and a grid search was performed to obtain the values of sigma and the penalty term C.

Multiple publicly available datasets, described in Table 3, were used to benchmark our proposed algorithms. They were retrieved from KEEL data repository [3] and UCI machine learning repository [22]. These datasets were selected based on their imbalance ratios, number of samples, number of attributes and best reported accuracy in literature. In all the tables below, the datasets are sorted in ascending imbalance ratio order.

To evaluate the performance of the proposed algorithms, multiple performance metrics were considered. Since we consider imbalanced datasets, the accuracy of a classifier's performance, the proportion of the total number of predictions that were correctly classified as computed by (1), is a misleading measure. A classifier might produce high overall accuracy but still misclassify most of the important minority class due to the bias.

$$\text{Accuracy} = \frac{TP + TN}{TP + TN + FP + FN} \tag{1}$$

Instead, other measures are used to evaluate the performance of classifiers with imbalanced datasets. For binary classification, true positive (TP), true neg-

Table 3. Dataset description

Type	Name	Imbalance ratio	Number of samples	Number of attributes
Synthetic	Clover	5	600	2
	Subclass	5	600	2
	Paw	5	600	2
Real world	Spambase	1.54	4601	57
	Ionosphere	2.02	351	34
	Wisconsin Diagnostic	3.21	198	32
	Segmentation	6.01	2308	19
	Yeast	8.11	1484	8
	Ecoli	8.19	336	7

ative (TN), false positive (FP), and false negative (FN), which combine to form a confusion matrix shown in Table 4, provide a better idea of the classifiers performance on each class. From the confusion matrix, multiple measures, suitable for imbalanced dataset classification, can be computed. Recall or true positive rate, computed in (2), represents the percentage of the correctly classified positive samples. It is a measure of completeness or the number of examples of the positive class that were labeled correctly, whereas the precision, computed in (3), defines a measure of exactness or of the examples labeled as positive, how many are actually labeled correctly. The geometric mean (G-mean), computed in (4), is a measure that takes the square root of the product of the precision and recall and describes the degree of inductive bias in terms of a ratio of positive accuracy and negative accuracy. The F-measure, computed in (5), is another metric that relates precision and recall with a parameter β and is referred to as F1-score when $\beta = 1$.

Table 4. Confusion matrix

		Predicted class	
		Yes	No
Actual Class	Yes	TP	FN
	No	FP	TN

$$\text{Recall} = \text{TP} - \text{rate} = \frac{TP}{TP + FN} \tag{2}$$

$$\text{Precision} = \frac{TP}{TP + FP} \tag{3}$$

$$\text{G} - \text{mean} = \sqrt{Recall \times Precision} \tag{4}$$

$$\text{F} - \text{measure} = \frac{(1 + \beta)^2 \times Recall \times Precision}{\beta^2 \times Recall + Precision} \tag{5}$$

Although these metrics give an insight into the performance of the classifiers on imbalanced datasets, they often do not describe the performance of different classifiers over a range sample distributions [15]. On the other hand, the receiver operating characteristic (ROC) curve analysis can provide a better understanding of classifier performance and makes use of TP rate and FP rate, which are defined in (2) and (6), respectively. Plotting TP rate versus FP rate with a threshold value between [0,1] produces the ROC curve. The area under the ROC curve (AUC) measure describes the performance of the classifier over a range of sample distributions and provides a better measure for evaluating performance of classifiers [15].

$$FP - rate = \frac{FP}{FP + TN} \tag{6}$$

4.2   Performance Results

Table 5 reports the performance of BBMO1 on the linearly separable datasets. We observe that BBMO1 with LS-SVM performs better than LS-SVM in all performance metrics except precision. BBBMO1 trades precision to increase other performance metric values. For example, BBMO1's AUC increases by an average of 1% compared to LS-SVM on the three datasets. Therefore, the boundary minority samples that were not classified correctly before have been correctly classified with minor trade-off with the majority boundary samples. Thus, the bias is removed safely in the linear case, for the considered datasets.

Table 5. Linearly separable case

Data	Accuracy	Recall	Precision	G-mean	F-score	AUC
SVM						
Spambase	0.9126	0.8758	0.8996	0.8876	0.8875	0.9590
Wisconsin	0.9561	0.9339	0.9476	0.9408	0.9408	0.9796
Ionosphere	0.8690	0.7301	0.8909	0.8065	0.8025	0.9180
LS-SVM						
Spambase	0.8858	0.7842	**0.9131**	0.8462	0.8438	0.9538
Wisconsin	0.9543	0.8860	**0.9892**	0.9362	0.9348	0.9741
Ionosphere	0.8578	0.6523	**0.9253**	0.7769	0.7652	0.9107
BBMO1+LS-SVM						
Spambase	**0.9127**	**0.8896**	0.8890	**0.8893**	**0.8893**	**0.9660**
Wisconsin	**0.9701**	**0.9483**	0.9706	**0.9594**	**0.9593**	**0.9819**
Ionosphere	**0.883 0**	**0.8005**	0.8724	**0.8357**	**0.8349**	**0.9207**

BBMO2 performs well on most of the linearly separable and non-linearly separable datasets with the exception of the Subclass, Ecoli, and Ionosphere datasets, as shown in Table 6. BBMO2 with LS-SVM outperforms LS-SVM and records better or comparable results when compared with SVM on many of the datasets, proving its effectiveness on non-linearly separable data.

Table 6. Linearly non-separable case

Data	Algorithm	Accuracy	Recall	Precision	G-mean	F-measure	AUC
Paw	SVM	97.160	0.930	**0.903**	0.917	0.916	**0.992**
	LS-SVM	96.970	0.900	0.902	0.901	0.901	0.990
	BBMO2+LSSVM	**97.670**	**0.980**	0.893	**0.935**	**0.930**	0.991
Subclass	SVM	94.500	**0.860**	0.821	**0.840**	**0.840**	0.984
	LS-SVM	**94.670**	0.760	**0.902**	0.828	0.825	**0.986**
	BBMO2+LSSVM	93.500	0.810	0.805	0.808	0.806	0.976
Clover	SVM	96.560	0.885	**0.995**	**0.938**	**0.937**	0.967
	LS-SVM	96.940	0.900	0.871	0.885	0.885	0.962
	BBMO2+LSSVM	**97.650**	**0.938**	0.889	0.913	0.908	**0.976**
Yeast	SVM	95.350	0.881	0.798	0.838	0.837	0.939
	LS-SVM	93.740	0.711	**0.869**	0.786	0.782	0.926
	BBMO2+LSSVM	**96.130**	**0.905**	0.859	**0.882**	**0.873**	**0.946**
Ecoli	SVM	**95.210**	0.807	**0.884**	**0.845**	**0.844**	**0.948**
	LS-SVM	92.260	0.615	0.840	0.718	0.710	0.923
	BBMO2+LSSVM	94.650	**0.867**	0.807	0.837	0.830	0.941
Segment	SVM	99.490	0.975	0.990	0.982	0.982	0.982
	LS-SVM	99.420	0.970	0.990	0.980	0.980	0.981
	BBMO2+LSSVM	99.490	0.975	0.990	0.982	0.982	0.982
Ionosphere	SVM	**95.183**	0.906	**0.958**	**0.932**	**0.931**	**0.994**
	LS-SVM	93.730	0.936	0.895	0.915	0.915	0.979
	BBMO2+LSSVM	94.590	**0.960**	0.899	0.929	0.919	0.981
Spambase	SVM	91.560	**0.885**	0.891	0.888	**0.888**	0.932
	LS-SVM	91.340	0.860	**0.915**	0.887	0.887	0.930
	BBMO2+LSSVM	**91.820**	0.883	0.907	**0.895**	0.887	**0.942**
Wisconsin	SVM	95.420	0.915	**0.961**	**0.938**	**0.937**	**0.974**
	LS-SVM	94.900	0.906	0.956	0.930	0.930	0.970
	BBMO2+LSSVM	**95.430**	**0.929**	0.947	**0.938**	0.932	0.972

Next, we consider the influence of the inducer on BBMO and compare it to SMOTE. BBMO2 outperformed SMOTE regardless of the inducer, except on Ecoli, as shown in Table 7. Furthermore, BBMO2 was not affected by its inducer since it performed equally well with SVM and LS-SVM.

Figure 4 shows how the accuracy, recall and the G-mean vary with the variation of the number of boundary samples chosen. Since the RBF kernel's sigma parameter choice affects the kernel matrix, θ_2 needs to be tuned. The threshold θ_2 defines the number of samples chosen at the boundary and hence the number of oversampled data forming the barrier. In our experiments, θ_2 was chosen

Table 7. Comparing BBMO2 and SMOTE

Data	Accuracy	F1-score	AUC	Accuracy	F1-score	AUC
	SMOTE+SVM			BBMO2+SVM		
Paw	97.50	0.916	**0.992**	97.65	**0.930**	**0.992**
Subclass	92.40	**0.821**	0.957	93.35	0.806	0.962
Clover	95.06	**0.933**	0.962	**97.65**	0.908	0.975
Yeast	95.35	0.837	0.939	**96.13**	**0.873**	0.943
Ecoli	**96.94**	**0.844**	**0.948**	94.65	0.830	0.937
Segment	99.49	0.982	0.982	99.49	0.982	0.982
Spambase	90.75	0.881	0.932	91.52	0.887	0.936
	SMOTE+LS-SVM			BBMO2+LS-SVM		
Paw	97.10	0.908	0.990	**97.67**	**0.930**	0.991
Subclass	91.90	0.810	0.933	**93.50**	0.806	**0.976**
Clover	95.06	0.933	0.954	97.35	0.908	**0.976**
Yeast	95.15	0.833	0.938	95.76	0.872	**0.946**
Ecoli	96.55	0.835	0.947	94.25	0.829	0.941
Segment	99.49	0.982	0.981	99.49	0.982	0.982
Spambase	90.98	0.886	0.935	**91.82**	**0.888**	0.942

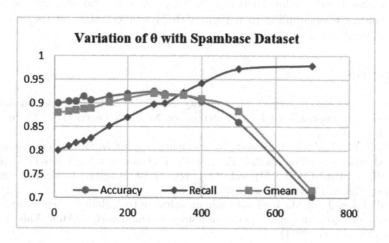

Fig. 4. Accuracy, recall, and G-mean variation as the number of selected boundary minority samples increases by varying the value of the threshold θ.

$0.8 < \theta_2 < 1$ (which can be understood as 80% of the maximum value) and sigma was in the order of tens. This ensured the selection of the closest samples. For the Spambase dataset, θ_2 was set to 0.998 which resulted in oversampling by approximately 300 samples. As shown in Fig. 4, when the number of synthetic samples varied between 300 and 400 samples a slight trade-off between the different metrics was produced. Furthermore, when the number of oversampling of the minority using BBMO2 increased, accuracy and G-mean decreased while the recall increased which means that the majority samples were being misclassified at the expense of the minority class. Thus, θ_2 needs to be tuned correctly to select a good representative amount of boundary samples and subsequently a representative number of oversampled minority boundary samples.

5 Conclusion

We presented in this paper a new approach of oversampling the boundary minority samples to improve the performance of the LS-SVM in the presence of bias in data. This is particularly important when building a model to extract information from data. BBMO adds synthetic minority samples by computing the weighted means of selected boundary samples. BBMO2 is a generalization of BBMO1 to linearly non-separable data. BBMO2 uses kernel matrix values of each minority sample while BBMO1 uses Lagrange multipliers to determine whether it is a boundary sample or not and to extract the closest majority boundary samples based on a threshold value. The weighted means were computed and added to the dataset. Experiments validated our expectations where BBMO2 with LS-SVM removed LS-SVM's bias and performed better than LS-SVM and SMOTE on most of the datasets. BBMO only introduced synthetic samples near the boundary between existing minority and majority samples while avoiding oversampling in regions with high uncertainty in the minority class distribution.

References

1. Ajeeb, N., Nayal, A., Awad, M.: Minority svm for linearly separable imbalanced datasets. In: International Joint Conference on Neural Networks (IJCNN), pp. 1–5. IEEE (2013)
2. Akbani, R., Kwek, S., Japkowicz, N.: Applying support vector machines to imbalanced datasets. In: Boulicaut, J.-F., Esposito, F., Giannotti, F., Pedreschi, D. (eds.) ECML 2004. LNCS (LNAI), vol. 3201, pp. 39–50. Springer, Heidelberg (2004). https://doi.org/10.1007/978-3-540-30115-8_7
3. Alcalá-Fdez, J., et al.: Keel data-mining software tool: data set repository, integration of algorithms and experimental analysis framework. J. Mult.-Valued Log. Soft Comput. 17 (2011)
4. Awad, M., Motai, Y., Näppi, J., Yoshida, H.: A clinical decision support framework for incremental polyps classification in virtual colonoscopy. Algorithms 3(1), 1–20 (2010)
5. Blanzieri, E., Bryl, A.: A survey of learning-based techniques of email spam filtering. Artif. Intell. Rev. 29(1), 63–92 (2008)
6. Bunkhumpornpat, C., Sinapiromsaran, K., Lursinsap, C.: Safe-level-SMOTE: safe-level-synthetic minority over-sampling technique for handling the class imbalanced problem. In: Theeramunkong, T., Kijsirikul, B., Cercone, N., Ho, T.-B. (eds.) PAKDD 2009. LNCS (LNAI), vol. 5476, pp. 475–482. Springer, Heidelberg (2009). https://doi.org/10.1007/978-3-642-01307-2_43

7. Chang, C.C., Lin, C.J.: Libsvm: a library for support vector machines. ACM Trans. Intell. Syst. Technol. **2**(3), 27 (2011)
8. Chawla, N.V., Bowyer, K.W., Hall, L.O., Kegelmeyer, W.P.: Smote: synthetic minority over-sampling technique. J. Artif. Intell. Res. **16**, 321–357 (2002)
9. Cristianini, N., Shawe-Taylor, J.: An introduction to support vector machines (2000)
10. Das, B.: Implementation of smoteboost algorithm used to handle class imbalance problem in data (2012). https://www.mathworks.com/matlabcentral/fileexchange/37311-smoteboost
11. Di Martino, M., Decia, F., Molinelli, J., Fernández, A.: Improving electric fraud detection using class imbalance strategies. In: ICPRAM (2), pp. 135–141 (2012)
12. Dumais, S., Platt, J., Heckerman, D., Sahami, M.: Inductive learning algorithms and representations for text categorization. In: Proceedings of the 7th International Conference on Information and Knowledge Management, pp. 148–155. ACM (1998)
13. Hajj, N., Awad, M.: Isolated handwriting recognition via multi-stage support vector machines. In: 6th IEEE International Conference on Intelligent Systems, pp. 152–157. IEEE (2012)
14. Han, H., Wang, W.-Y., Mao, B.-H.: Borderline-SMOTE: a new over-sampling method in imbalanced data sets learning. In: Huang, D.-S., Zhang, X.-P., Huang, G.-B. (eds.) ICIC 2005. LNCS, vol. 3644, pp. 878–887. Springer, Heidelberg (2005). https://doi.org/10.1007/11538059_91
15. He, H., Garcia, E.A.: Learning from imbalanced data. IEEE Trans. Knowl. Data Eng. **21**(9), 1263–1284 (2009)
16. Imam, T., Ting, K.M., Kamruzzaman, J.: z-SVM: An SVM for improved classification of imbalanced data. In: Sattar, A., Kang, B. (eds.) AI 2006. LNCS (LNAI), vol. 4304, pp. 264–273. Springer, Heidelberg (2006). https://doi.org/10.1007/11941439_30
17. Khanna, R., Awad, M.: Efficient learning machines: theories, concepts, and applications for engineers and system designers. Apress (2015)
18. Köknar-Tezel, S., Latecki, L.J.: Improving svm classification on imbalanced data sets in distance spaces. In: 9th International Conference on Data Mining, pp. 259–267. IEEE (2009)
19. Kotsiantis, S., Kanellopoulos, D., Pintelas, P., et al.: Handling imbalanced datasets: a review. GESTS Int. Trans. Comput. Sci. Eng. **30**(1), 25–36 (2006)
20. Kowalczyk, A., Raskutti, B.: One class svm for yeast regulation prediction. ACM SIGKDD Explor. Newsl. **4**(2), 99–100 (2002)
21. Li, P., Chan, K.L., Fang, W.: Hybrid kernel machine ensemble for imbalanced data sets. In: 18th International Conference on Pattern Recognition, vol. 1, pp. 1108–1111. IEEE (2006)
22. Lichman, M.: UCI machine learning repository (2013)
23. Nayal, A., Jomaa, H., Awad, M.: Kerminsvm for imbalanced datasets with a case study on arabic comics classification. Eng. Appl. Artif. Intell. **59**, 159–169 (2017)
24. Ou, Y.Y., Hung, H.G., Oyang, Y.J.: A study of supervised learning with multivariate analysis on unbalanced datasets. In: International Joint Conference on Neural Networks, pp. 2201–2205. IEEE (2006)
25. Ramentol, E., Caballero, Y., Bello, R., Herrera, F.: Smote-rsb*: a hybrid preprocessing approach based on oversampling and undersampling for high imbalanced data-sets using smote and rough sets theory. Knowl. Inf. Syst. **33**(2), 245–265 (2012)
26. Raskutti, B., Kowalczyk, A.: Extreme re-balancing for SVMS: a case study. ACM Sigkdd Explor. Newsl. **6**(1), 60–69 (2004)

27. Rizk, Y., Mitri, N., Awad, M.: An ordinal kernel trick for a computationally efficient support vector machine. In: 2014 International Joint Conference on Neural Networks (IJCNN), pp. 3930–3937. IEEE (2014)
28. Rizk, Y., Partamian, H., Awad, M.: Toward real-time seismic feature analysis for bright spot detection: a distributed approach. IEEE J. Sel. Top. Appl. Earth Obs. Remote. Sens. (2017)
29. Saab, S.A., Mitri, N., Awad, M.: Ham or spam? a comparative study for some content-based classification algorithms for email filtering. In: 17th IEEE Mediterranean Electrotechnical Conference, pp. 339–343 (2014)
30. Schölkopf, B., Platt, J.C., Shawe-Taylor, J., Smola, A.J., Williamson, R.C.: Estimating the support of a high-dimensional distribution. Neural Comput. 13(7), 1443–1471 (2001)
31. Stefanowski, J., Wilk, S.: Improving rule based classifiers induced by modlem by selective pre-processing of imbalanced data. In: Proceedings of the RSKD Workshop at ECML/PKDD, Warsaw, pp. 54–65. Citeseer (2007)
32. Suykens, J.A., Vandewalle, J.: Least squares support vector machine classifiers. Neural Process. Lett. 9(3), 293–300 (1999)
33. Tang, Y., Zhang, Y.Q., Chawla, N.V., Krasser, S.: SVMS modeling for highly imbalanced classification. IEEE Trans. Syst. Man Cybern. Part B (Cybern.) 39(1), 281–288 (2009)
34. Tax, D.M., Duin, R.P.: Support vector domain description. Pattern Recognit. Lett. 20(11), 1191–1199 (1999)
35. Vapnik, V.: The Nature of Statistical Learning Theory. Springer science & business media, Berlin (2013)
36. Veropoulos, K., Campbell, C., Cristianini, N., et al.: Controlling the sensitivity of support vector machines. In: Proceedings of the International Joint Conference on Artificial Intelligence, pp. 55–60 (1999)
37. Wang, X., Matwin, S., Japkowicz, N., Liu, X.: Cost-sensitive boosting algorithms for imbalanced multi-instance datasets. In: Zaïane, O.R., Zilles, S. (eds.) AI 2013. LNCS (LNAI), vol. 7884, pp. 174–186. Springer, Heidelberg (2013). https://doi.org/10.1007/978-3-642-38457-8_15
38. Wu, G., Chang, E.Y.: Adaptive feature-space conformal transformation for imbalanced-data learning. In: International Conference on Machine Learning, pp. 816–823 (2003)
39. Wu, G., Chang, E.Y.: Class-boundary alignment for imbalanced dataset learning. In: ICML 2003 workshop on learning from imbalanced data sets II, pp. 49–56. Washington (2003)
40. Wu, G., Chang, E.Y.: KBA: Kernel boundary alignment considering imbalanced data distribution. IEEE Trans. Knowl. Data Eng. 17(6), 786–795 (2005)
41. Yang, J., Bouzerdoum, A., Phung, S.L.: A training algorithm for sparse LS-SVM using compressive sampling. In: IEEE International Conference on Acoustics Speech and Signal Processing, pp. 2054–2057. IEEE (2010)
42. Yang, P., Xu, L., Zhou, B.B., Zhang, Z., Zomaya, A.Y.: A particle swarm based hybrid system for imbalanced medical data sampling. BMC Genomics 10(3), S34 (2009)
43. Zhuang, L., Dai, H.: Parameter optimization of kernel-based one-class classifier on imbalance learning. J. Comput. 1(7), 32–40 (2006)

Dynamic Classifier Chain with Random Decision Trees

Moritz Kulessa and Eneldo Loza Mencía[✉]

Knowledge Engineering Group, Technische Universtität Darmstadt,
Darmstadt, Germany
{mkulessa,eneldo}@ke.tu-darmstadt.de

Abstract. Classifiers chains (CC) is an effective approach in order to exploit label dependencies in multi-label data. However, it has the disadvantages that the chain is chosen at total random or relies on a prespecified ordering of the labels which is expensive to compute. Moreover, the same ordering is used for every test instance, ignoring the fact that different orderings might be best suited for different test instances. We propose a new approach based on random decision trees (RDT) which can choose the label ordering for each prediction dynamically depending on the respective test instance. RDT are not adapted to a specific learning task, but in contrast allow to define a prediction objective on the fly during test time, thus offering a perfect test bed for directly comparing different prediction schemes. Indeed, we show that dynamically selecting the next label improves over using a static ordering of the labels under an otherwise unchanged RDT model and experimental environment.

Keywords: Multi-label classification · Random decision trees
Classifier chains

1 Introduction

Contrary to multi-class classification, where only one class label is expected to be associated to an example, multi-label classification (MLC) is the task of assigning a subset of all possible labels to an example. In this task, it is considered crucial to take the dependencies between labels into account. Classifier chains (CC) [18] and their extensions (cf. Sect. 2) have proven to be a simple but powerful method for exploiting label dependencies in MLC. Similarly to the binary relevance decomposition method these methods train a binary predictor for each of the labels. However, they are organized in a chain so that successive classifiers can make use of the predictions of the previous ones. This enables CC to capture dependencies between labels.

Nevertheless, this simple technique has several shortcomings, especially regarding the chain. Firstly, the ordering in which the labels are predicted in the chain has to be fixed beforehand. To find a sequence which best allows to consider dependencies between labels is a non-trivial task [12] and methods which try to explore different orderings are usually computationally much

© Springer Nature Switzerland AG 2018
L. Soldatova et al. (Eds.): DS 2018, LNAI 11198, pp. 33–50, 2018.
https://doi.org/10.1007/978-3-030-01771-2_3

more expensive than the often taken option of just choosing a random ordering. Secondly, the assumption that there is one single prediction ordering which works best always for every possible single test instance might hold only in very restricted scenarios. Instead, our assumption in this work is that the ordering in which labels should be predicted in order to obtain the best performance highly depends on the specific context, namely the test instance at hand. The question of how to dynamically choose an appropriate ordering for individual instances instead of the entire datasets has been little researched so far. Da Silva et al. [20] made a first attempt by letting a nearest neighbor classifier decide which ordering to use for a given instance. However, the dynamic selection was restricted to a pre-determined set of static label orderings. The approach of Nam et al. [15] predicts the positive labels at the beginning of the chain, but the ordering in which these are predicted is pre-determined.

In this work, we propose to use random decision trees (RDT) for the purpose of constructing dynamic chains, since these trees have a series of convenient and appealing properties (Sect. 3).

Foremost, the construction of the model is independent of the specific leaning task. This has the advantage that the objective can easily be changed during prediction without the need for modifying the trees. Our dynamic classifier chains extension of RDT is strongly relying on this property. Instead of choosing the next label to predict from the pre-determined ordering, our proposed method predicts the label for which the RDT is most confident given the current context (Sect. 4). Our experiments on a series of datasets confirm that it is advantageous to predict the labels in such a dynamic way w.r.t. predictive performance (Sect. 5).

Moreover, we propose to take advantage of the flexibility of RDT to build a controlled experimental setup where not only the training hyper parameters can be fixed, but also the respective models (Subsect. 4.1). This allows us to directly measure the impact of certain modifications, as well as to compare conceptually different approaches on a fair basis. For instance, we use this possibility in our experimental evaluation to analyze the specific utility of considering previous predictions, or to compare CC to our dynamic CC using the same actual ensemble of trees.

2   Multi-label Classification and Classifier Chains

Multi-label classification is the task of learning a mapping from instances $X \in \mathcal{X}$ to subsets $Y \subset \mathcal{Y}$ of a finite set of non-exclusive class labels $\mathcal{Y} = \{y_0, \ldots, y_n\}$. For convenience, Y is often represented as binary vector $Y = (y_0, \ldots, y_n)$ where y_i is 1 if the label is relevant (positive), otherwise 0 for irrelevant (negative) labels. An extensive overview over MLC is provided by Tsoumakas et al. [22].

The simplest method for solving MLC tasks is the binary relevance method (BR) where each label is handled as a single classification task for which a classifier is trained. Formally, we learn a function $h_i : \mathcal{X} \rightarrow \{0, 1\}$ for each y_i. According to this, each classification of a label is independent of the values of the other labels.

Another method is the label power-set method (LP) which reduces the problem of MLC to a single multi-class classification task by representing each possible combination of labels as one separate and exclusive class. This approach naturally considers to predict labels in dependence of the remaining labels, hence focusing on predicting correct label combinations. However, in addition to the obvious limitations due to the exponential growth of label combinations, LP does not allow to predict label combinations which have not been seen in the training data.

A more flexible approach of considering label dependencies was proposed by [18] by using classifier chains. CC enhances the idea of BR and executes the binary classifiers in a chain which has the advantage that subsequent classifiers can use the information of the already predicted labels. More formally, each $h_i : \mathcal{X} \times \{0,1\}^{i-1} \to \{0,1\}$ uses the real labels $y_1, \ldots, y_{i-1}$ for training and the corresponding predictions $\hat{y}_1, \ldots, \hat{y}_{i-1}$ produced by previous classifiers in the chain during testing.

Further analysis revealed that the ordering of the classifiers has an effect on the predictive performance [18,20]. Usually this ordering is chosen randomly or different random orderings have to be evaluated to find a good chain order. A straight-forward solution is to use ensembles of classifier chains. Nevertheless it turned out that these ensembles are often unnecessarily large for which reason Li and Zhou [11] proposed a method to composite the ensemble. By doing so a subset of CC is selected while keeping or improving the predictive performance of the ensemble.

However, creating and maintaining an ensemble of CC is not always feasible [8]. Another way to handle the label ordering problem is to determine a good chain sequence in advance. For this purpose methods such as genetic algorithms [8], Bayesian networks [21] or double Monte Carlo optimization technique [17] have been used. On the other hand, the classification sequence can be determined during the classification process by finding similar instances in the training set and using the label ordering which works well on these instances [20]. However, this method is not appealing in terms of time complexity since a new CC model has to be build on the fly.

A further improvement of CC could be achieved with probabilistic classifier chains (PCC) [2]. While the training process of both methods is the same, PCC modifies the classification procedure by considering the joint probability of each possible label assignment. According to this, Bayes optimal predictions can be created which makes PCC superior to CC [10]. However, this process has a much higher time complexity and is only feasible for datasets with not more than 15 labels [2]. To tackle this problem beam search [10] or A* search [13] can be used to perform the inferences which speeds up the process. In [14] an overview of inference methods for PCC is given. Nevertheless, PCC also relies on a predefined chain ordering for which reason ensembles of PCC have been introduced [2].

The research on CC and PCC contributed to the understanding and formalization of label dependencies in MLC. For instance, Dembczyński et al. [3] found that these methods are able to exploit so called unconditional dependencies which exist globally on the whole dataset, but also conditional dependencies which only appear locally in the instance space. Moreover, they also discovered that certain multi-label evaluation measures can be orthogonal to each other and optimizing them requires different approaches. For instance, methods such as LP and CC are tailored towards finding the correct label combination, which corresponds to the mode of the joint label distribution, whereas for correctly predicting each label individually (measured by the Hamming loss) it might be sufficient to use approaches such as BR.

3   Random Decision Trees for Multi-label Classification

Introduced by Fan et al. [6], the approach of RDT is an ensemble of randomly created decision trees. More precisely, the tests at the inner nodes are chosen completely at random. This is the major difference compared to classical decision tree algorithms [26], but also to the well known algorithm family of Random Forest [1], where only the subset of features which each tree learner can use is randomly drawn. In contrast, RDT do not optimize any objective function during training, yet they are able to achieve competitive and robust performance [25]. Moreover, by increasing the number of trees in the ensemble the estimation risk can be decreased [25] while we never tend to overfit [4]. Computational complexity is another major advantage because the random selection takes no time compared to computing information gain or similar heuristics [26].

In addition to these guarantees, the characteristics of RDT offers a wide range of possibilities since the random construction is independent of the learning task. For instance, Zhang et al. [26] make use of this property for large scale MLC problems since the computational costs do not depend on the number of labels in their formulation. Zhang et al. [25] propose to abstract RDT with hash functions which is claimed to handle MLC problems in an even more efficient way. RDT were also successfully applied to multi-label stream data and for handling concept drifts with only small modifications to the original algorithm [9]. Depending on the particular needs, RDT can flexibly be constructed before the arrival of the training data [6,9,26] or by taking advantage of it [7,25].

In the following, we describe the general construction and prediction process of RDT as well as the adaptations to be taken for the MLC setting. In particular, we propose an extension to the weighting of the trees in the ensemble based on their individual confidences computed by the Gini-index (Sect. 3).

Training. The construction of the trees for RDT is done recursively, as for most decision tree learners, with the aforementioned difference that the features in the inner nodes are chosen randomly. Hence, starting from the root node, inner nodes are constructed recursively by distributing the training instances according to the test as long as the stopping criterion of maximum depth or minimum number of instances is not fulfilled. Discrete features are chosen without replacement in contrast to continuous features, for which additionally a randomly picked instance determines the threshold [5]. In the case that no further test can be created a leaf will be constructed in which information about the assigned instances will be collected. In MLC, for instance, we might track the number of instances N_k^θ in leaf k of tree θ in relation to the number of positive values $n_k^\theta(i)$ for label y_i. However, any other information could be collected depending on the learning task at hand.

Prediction. During prediction, an instance is forwarded from the root to a leaf node passing the respective tests in the inner nodes. In case of missing features, the function $U = q(\theta, X)$ returns the set $U = \{k | k \in [1, T] \subset \mathbb{N}\}$ of the leaves indices in tree θ to which the instance has been assigned to. Following Fan et al. [5], the posterior probability that the specific label y_i is true given an instance X and a tree θ, or an ensemble of trees Θ, respectively, can be formalized as

$$P(y_j = 1 | X, \theta) = \frac{\sum_{k \in q(\theta, X)} n_k^\theta(i)}{\sum_{k \in q(\theta, X)} N_k^\theta}, \quad P(y_i = 1 | X, \Theta) = \frac{1}{|\Theta|} \sum_{\theta \in \Theta} P(y_i = 1 | X, \theta) \quad (1)$$

An obvious option in order to obtain multi-label predictions from the estimations in Eq. 1 is to use a threshold of 50% so that $\hat{y}_j = I\left[P(y_j = 1 | X, \Theta) \geq 0.5\right]$ with $\mathbb{I}[x] = 1$ if x is true and 0 otherwise, which we refer to as the *probability threshold method* (or shortly *probability method*). However, as Quevedo et al. [16] observed, a threshold of 50% is not always ideal. Note that the tests in the tree are not specifically chosen to obtain a high purity of the distributions in the leaves, and in fact many leaves might contribute only with estimates close to the prior distribution, pulling down the average estimates. Thus, Zhang et al. [26] proposed to estimate the average number of relevant labels

$$r(X, \theta) = \frac{\sum_{k \in q(\theta, X)} \sum_{j=1}^{n} n_k^\theta(j)}{\sum_{k \in q(\theta, X)} N_k^\theta}, \quad R(X, \Theta) = \frac{1}{|\Theta|} \sum_{\theta \in \Theta} r(X, \theta) \quad (2)$$

where $R(X, \Theta)$ is rounded in order to get an integer. This value is then used to cut the ranking of labels induced by the distribution of the marginals $P(y_i | X, \Theta)$. We refer to this method as the *label threshold method* or *label method*.

Weighting the trees. As aforementioned, the randomness make for a large variety of distributions which are aggregated, many of them approaching the prior label distribution. Nevertheless, previous RDT approaches for MLC use equal weighting irrespectively. We propose to distinguish between the quality of the collected statistics and to reward trees with higher confidences in their estimates. The *Gini index* is often used for determining the purity of a distribution, which we use in inverted form as follows

$$w(X, \theta) = 1 - \frac{4}{n} \sum_{y_i \in \mathcal{Y}} P(y_i = 1 | X, \theta)(1 - P(y_i = 1 | X, \theta)) \quad (3)$$

in order to weight the estimates of the individual trees, resulting in the overall prediction

$$P(y_i = 1 | X, \Theta) = \frac{1}{\sum_{\theta \in \Theta} w(X, \theta)} \sum_{\theta \in \Theta} P(y_i = 1 | X, \theta) w(X, \theta) \quad (4)$$

Eq. 2 can be adapted accordingly.

We observed a better performance of using the inverted Gini index in preliminary experiments, so that we adopted it as the default setting for our proposed algorithm.

4 Dynamic Predictions with Random Decision Trees

As already stated, a key disadvantage of classical CC is that their predictive performance may be highly influenced by the pre-selected ordering of the labels. In this section, we propose an extension of RDT referred to as Dynamic Classifier Chains (DCC) where the sequence in which the values for the labels are predicted is chosen dynamically during the process of classification (Subsect. 4.3). Moreover, RDT and their randomized construction provides a very convenient controlled environment for experimentation (Subsect. 4.1).

4.1 Test Bed for Multi-label Classification

In Sect. 2 we reviewed some transformation methods for solving MLC tasks with different desirable properties, respectively. The description of RDT in Sect. 3 applies to binary classification problems as well. For instance, we could use RDT as base learner for a binary relevance decomposition, estimating $P(y_i|X, \Theta_i)$ instead of $P(y_i|X, \Theta)$. However, both ensembles Θ_i and Θ are drawn from the same tree distribution which is independently of any y_i. Hence, both estimations approach the same expected value as the number of constructed trees increases.

This key observation lead to the following advantages of RDT. Firstly, we can collapse BR, and other MLC transformation or decomposition methods [22] such as CC as we will see in the following, to a single RDT ensemble without loss in predictive accuracy, therefore saving memory and computational costs. Secondly, and more importantly, RDT can provide a controlled environment where we can compare alternative decomposition methods, prediction methods and other extensions isolated from any side effects since the model can be fixed beforehand and be the same for every analyzed approach.

4.2 Static Chain Ordering

Similarly to BR, we can collapse a classifier chain to a single RDT in the following way: Instead of augmenting the input space $\mathcal{X}$ by only the previous labels, we add the whole label matrix so that $X' \in \mathcal{X} \times \mathcal{Y}$. The prediction of base classifier h_i for label y_i (more specifically, $P(y_i = 1|X, \hat{y}_1, \ldots, \hat{y}_{i-1}, \Theta_i)$) is obtained by setting $\hat{y}_j$ to the previous predictions of h_j, $j < i$, as for CC, but leaving $\hat{y}_j$, $j \geq i$ as *missing*.[1] Remind that when encountering a missing value at inner nodes all branches are visited and aggregated (cf. Sect. 3). As RDT are completely randomized, we can—similarly to Subsect. 4.1—expect on average the

---

[1] We assume, w.l.o.g., that $y_1, y_2, \ldots$ is the ordering of the predicted labels.

same predictions as for a RDT with one node less. In fact, we control in our experiments the percentage of activated label tests with a parameter σ, which allows us to analyze the effect of using previous predictions on an otherwise unchanged model.

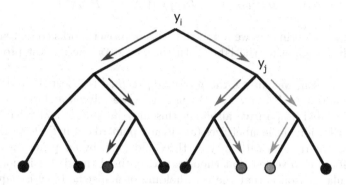

Fig. 1. Example for the refinement of a prediction for a particular instance and decision tree. y_i and y_j indicate tests on labels at the respective inner nodes.

Figure 1 visualizes the prediction process for a label on a single tree: Let us assume that the label to be predicted is y_i, which comes before y_j. In this case neither y_i nor y_j are known, i.e. all three colored branches are followed and the respective leaves are used in order to produce a prediction for y_i. For label y_j the previous label y_i would be known, so that we would skip either the left or right branch, obtaining a label distribution at the leaves which is different and more refined than the previous one. Indeed, we can observe that the number of leaves on which the prediction relies is monotonically decreasing during the classification process. Therefore, the set of leaves to which the instance is assigned in the first iteration will always be a superset of the leaves of the following iterations. This leads to a refinement of the prediction through the iterations.

4.3 Dynamic Chain Ordering

In order to take advantage of the situation that predicting a label before or after another one might be easier depending on the instance at hand, we propose to let the RDT decide which label to predict next. Hence, instead of using the estimated distribution in order to decide whether the i-th label is positive or negative, we use it in order to set the label for which RDT is most confident in its prediction. Labels, which were already predicted, are ignored.

Predicting the next label in the sequence. For convenience, we introduce the following definitions. Let C denote the label candidates which were not yet predicted, P^+ the set of labels which were predicted as relevant, and P^- the irrelevant labels, respectively. Accordingly, we start with $P^+ = P^- = \emptyset, C = \mathcal{Y}$ in the first iteration.

In each iteration, we first decide on the next label to be predicted. We select the label for which the RDT is most confident in the following way and remove it from C:

$$y_i = \arg\max_{y_j \in C} |0.5 - P(y_j = 1|X, P^+, P^-, \Theta)| \tag{5}$$

In preliminary experiments we found that this approach works consistently better than always choosing the label with the lowest or the highest probability, respectively.

With y_i chosen, we can use the *probability method* (cf. Sect. 3) to determine whether to add it to P^+ or P^-. We note that we rely on a threshold of 50% to classify a label as positive although this may be suboptimal regarding the skewed distribution of the label sizes (as already pointed out in Sect. 3). However, preliminary experiments with varying thresholds, e.g. by adapting them to the prior distribution, revealed that choosing the optimal thresholds is non-trivial. One particular reason is that the thresholding of a specific label is required at different stages of the prediction chain, in contrast to using a static ordering, introducing additional dependencies and dynamics of the right threshold. We leave further investigations for future work.

The process of predicting the value with the *label method* also needs further adaption due to the iterative prediction of the labels. The idea is to have predicted exactly $R(X, P^+, P^-, \Theta)$ labels positive after the prediction sequence is completed. Since the prediction changes during the classification process $R(X, P^+, P^-, \Theta)$ has to be re-computed in every iteration. First of all, we can only predict a label positive if the number of already predicted positive labels $|P^+|$ is smaller than $R(X, P^+, P^-, \Theta)$. Moreover, we have to predict a label as positive if we know that all the remaining labels in C need to be predicted positive to ensure that we obtain exactly $R(X, P^+, P^-, \Theta)$ positive labels.

$$y_i = \begin{cases} 1, & \text{if } P(y_i = 1|X, P^+, P^-, \Theta) \geq 0.5 \text{ and } |P^+| < R(X, P^+, P^-, \Theta) \\ 1, & \text{if } n - |P^-| < R(X, P^+, P^-, \Theta) \\ 0, & \text{otherwise} \end{cases} \tag{6}$$

Let us consider again the tree in Fig. 1. The difference to the static chain approach is that the aggregated blue, red and green leaves would be used in order to determine whatever label y_k is most likely given the found distribution, instead of a specific label, in the previous example label y_i. Hence, the RDT could decide to predict y_j instead if they are more confident about it, or any other label with the highest confidence. We believe that this potentially leads to more reliable predictions, both in terms of individual labels as well as label combinations.

In particular Fig. 2 displays an example how the quality of the predictions could differ between a static ordering and a dynamic ordering. Let us assume we want to identify objects in a scene. While the static ordering has to follow a predefined sequence for the classification, it has to classify the label *beach* first. Since

Static Ordering (Classifier Chain)

beach → stairs → bridge ... → city
no   yes   yes   ...   yes

Dynamic Ordering (Dynamic Classifier Chain)

sea → car → cliff ... → beach
yes   no   yes   ...   yes

Fig. 2. Example for different classifications using the static and the dynamic ordering for a picture associated to labels *beach, sea, cliff, bridge, stairs*. See text for explanations.

the beach is not clearly visible on the picture, the label receives a negative value which already introduces an error for future predictions in the chain. Especially for our example, this has the consequence that the scene is classified as a city because stairs and bridges are more correlated with cities than with the seaside. In contrast, the dynamic chain classifies the most obvious targets first for which reason the label *sea* receives a positive classification in the first iteration. This increases the chance to classify the label *cliff* as positive which provides even more evidence for predicting the difficult label *beach* as positive. On other hand, we reduce the probability to incorrectly classify the scene as a city, since we exclude objects which are clearly not identifiable in the picture, such as a car, in early iterations.

Computational Costs. The costs for building the trees and performing the dynamic predictions is conceptually equal to using a static ordering. They mainly depend on the size of the ensemble and the depth of the trees. However, the dynamic approach potentially allows to shorten the prediction process, namely when enough positive (or negative) labels have been already predicted, potentially removing the dependencies on the label size.

5 Evaluation

A key aspect in our experimental evaluation was, of course, to demonstrate that using dynamic, context-dependent predictions improves over using static orderings w.r.t. predictive performance (Subsect. 5.3). A decisive role in this is played by the influence of the previous predictions on the current prediction, which is analyzed in Subsect. 5.2.

Another aspect, we were particularly interested in, was to verify our ideas on the usage of RDT as controlled experimental environment for fair and specific comparisons.

Regarding our proposed dynamic approach, we will mainly distinguish between the two variants using the probability and the label threshold method for determining the value of the next label, respectively. We expected that other hyper parameters would behave quite different with respect to different datasets, both regarding the shapes (and densities) of the input and output spaces. In contrast to other hyper parameters like number of trees or minimum leaf sizes, we decided to consider this aspect separately.

5.1 Setup

For our experiments we have used eight different multi-label datasets from the Mulan repository [23]. An overview of these datasets is provided in Table 1. From the text datasets we have only included *Enron* and *Medical*, which have a relatively small vocabulary, since RDT are known to not perform well on sparse data without further adaptations which we did not want to put in the focus for this work.

Table 1. Dataset statistics: Total number of instances, of nominal and numeric attributes, of labels, average number of labels per instance and distinct label combinations.

Name	Instances	Nominal	Numeric	Labels	Cardinality	Distinct
Flags	194	9	10	7	3.392	54
Emotions	593	0	72	6	1.869	27
Scene	2407	0	294	6	1.074	15
Yeast	2417	0	103	14	4.237	198
Birds	645	2	258	19	1.014	133
Medical	978	1449	0	45	1.245	94
Enron	1702	1001	0	53	3.378	753
CAL500	502	0	68	174	26.044	502

A large variety of evaluation measures exist for MLC. We focus in this work on two of them, namely *subset accuracy* and *micro-averaged F1 measure*. Subset accuracy is a very restrictive evaluation metric since it only measures the percentage of instances for which all labels have been predicted correctly. Especially in the case of predicting a large amount of labels this measure often approaches zero without being able to distinguish. However, the objective of classifier chains is precisely to find exactly the correct label combination (cf. Sect. 2). Hence, we expect the impact of our proposed extensions to be best reflected in the subset accuracy.

Micro-averaged F1 measure is less strict since it also considers partial matches and is therefore often used for providing a general comparison of the predictive quality. However, the measure is to a certain degree orthogonal to subset accuracy [3]. As Dembczyński et al. [3] indicate, it is sufficient to obtain good estimates for the individual labels in order to optimize univariate losses such as F-measure or Hamming loss. Nevertheless, our approach may still benefit from the dependencies captured by the chaining approach with respect to these measure, which is why we include micro-averaged F1 measure (micro F1) in our comparisons.

Given N test instances, corresponding true labels Y_j and predicted labels $\hat{Y}_j$, true positives $tp_j = Y_j \cap \hat{Y}_j$, false positives $fp_j = \hat{Y}_j \setminus Y_j$, false negatives $fn_j = Y_j \setminus \hat{Y}_j$ for the j-th test instance, we obtain the measures as follows:

$$\text{subset accuracy} = \frac{1}{N} \sum_{j=1}^{N} \mathbb{I}\left[Y_j = \hat{Y}_j\right] \qquad \text{micro F1} = \frac{\sum_{j=1}^{N} 2\, tp_j}{\sum_{j=1}^{N} 2\, tp_j + fp_j + fn_j}$$

$$(7)$$

Unless otherwise noted we have chosen to evaluate all combinations of parameter settings of ensembles with 300 decision trees, a maximum depth of 30, a minimum number of instances to create a test of $\{4, 6, 10\}$ and a percentage of label tests of $\{10\%, 20\%, 30\%\}$. Preliminary experiments with RDT revealed reasonable and stable performance for these parameter ranges also on other kind of problems. Furthermore, we compare the results based on the averages of a ten-fold cross validation performed on the whole dataset.

5.2 Independent Predictions vs. Exploiting Previous Predictions

In this experiment we evaluated how the prediction is influenced by the usage of the label tests, i.e., by the usage of the previous predictions in the dynamic chain. At this stage the flexibility of the RDT algorithm pays off since we can choose the ratio σ of activated tests on the labels without the need for adaptations of the model (cf. Subsect. 4.2). Hence, $\sigma = 0$ corresponds to a binary relevance classifier using RDT (or the collapsed version, respectively). Incrementing σ allows to directly observe utility and the effectiveness of exploiting potential label dependencies.

Figure 3 show the benefit for some selected cases w.r.t. subset accuracy but also for improving the univariate micro F1. For instance, we can observe on datasets *Emotions* and *Yeast* a major influence of the label tests on the performance for both prediction and evaluation methods. We can conclude that there is a strong dependency between the labels in the datasets of which we can take advantage. Similar but less pronounced effects can be seen for the remaining datasets except *Enron* and *CAL500*. *Enron* shows that the usage of (possibly wrong) previous predictions can also have a negative impact in some cases, or no impact as for *CAL500*. Moreover, both datasets are also an example for the observation that the best label prediction method is highly dataset dependent.

In general it can be seen that the values for the evaluation measures get better the more label tests are activated. Only on the dataset *CAL500* the label tests seem not to have any influence on the predictive performance. Moreover, on the dataset *Enron* it can be seen that the activation of the label tests have

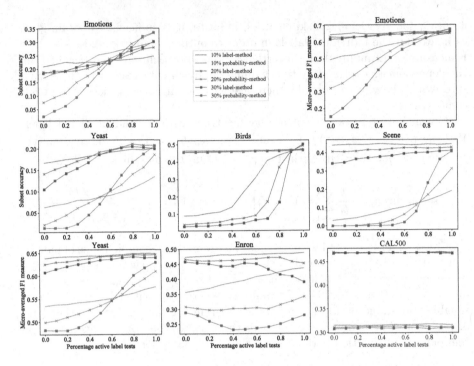

Fig. 3. Influence of label tests on DCC. The y-axis represents the value for the measure and the x-axis represents the percentage σ of activated label tests. The color indicates the prediction method and the style of the line represents the percentage of label tests per tree.

a negative impact on the micro-averaged F1 measure independently of the prediction method in the case that 30% of label tests have been used in the trees. Furthermore, on the dataset *Birds* only the label method benefits from the label tests whereas on the dataset *Scene* only the probability method benefits from the label tests. The values for the corresponding other prediction method stay the same along the activation of the tests.

5.3   Static vs. Dynamic Label Orderings

In this experiment we evaluated the advantage of the dynamic chain ordering in comparison to using a static chain ordering. Taking advantage of our controlled environment, we built for both approaches the same ensemble of trees, respectively, in this case with 20% of label tests. The only difference between the dynamic and the static setup is the ordering of the labels during the prediction process. We compare our proposed dynamic method to the averages over ten different randomly-drawn but fixed orderings used for the static CC approach in Tables 2 and 3.

Table 2. Comparison between dynamic and static chain method for subset accuracy. Bold entries indicate the best results.

	Flags	Emotions	Scene	Yeast	Birds	CAL500	Enron	Medical
DCC LM	**0.1959**	0.2799	**0.4271**	**0.2073**	**0.4930**	0.0000	0.0447	**0.1840**
CC LM	0.1623	0.1098	0.1714	0.0215	0.3673	0.0000	0.0116	0.0034
DCC PM	0.1856	**0.3339**	0.3112	0.1854	0.4698	0.0000	**0.0576**	0.0000
CC PM	0.1835	0.1482	0.0914	0.0407	0.4583	0.0000	0.0286	0.0000

Table 3. Comparison between dynamic and static chain method for micro F1.

	Flags	Emotions	Scene	Yeast	Birds	CAL500	Enron	Medical
DCC LM	**0.7506**	0.6535	**0.4682**	**0.6459**	**0.4484**	**0.4681**	**0.4527**	**0.2696**
CC LM	0.7301	0.4439	0.1935	0.4893	0.1288	0.2945	0.2669	0.0043
DCC PM	0.7448	**0.6774**	0.4344	0.6100	0.0708	0.3114	0.3410	0.0065
CC PM	0.7484	0.4490	0.1656	0.5204	0.0137	0.3093	0.3153	0.0000

The first and foremost observation is that the dynamic chain ordering is clearly superior to the static chain ordering on all datasets. Moreover, the dynamic chain ordering improved the evaluation results of the label method (LM) on all datasets often to a great extent whereas the probability method (PM) does not take major advantage of the dynamic chain ordering on the datasets *Birds*, *CAL500* and *Medical*.

The results suggest that the improvement of DCC over CC relies on high confidence of the classifications in the first iterations. These classifications provide evidence to improve the classification of the difficult labels. This effect is analyzed in more detail in the following experiment.

5.4 Analysis of the Dynamic Sequences

Our approach dynamically produces a different prediction sequence on the labels for each given test instance. We were interested in characterizing and analyzing these sequences, which were selected by the RDT as being most appropriate for producing accurate predictions.

Figure 4 visualizes our results exemplarily for *Yeast*. The heat map on the left shows the average accuracy (color) of predicting the i-th label in the dynamic sequence (y-axis) for the different configurations (x-axis), whereas the right map visualizes the number of labels (color) which were predicted as positive until a certain iteration.

We can observe on *Yeast* as well as on the remaining datasets and independent of the parameter configuration that the predictions of the first iterations are pretty accurate in comparison to the error-prone predictions at the end. One reason is of course that our label selection method chooses the labels where the RDT ensemble is most confident first. On the other hand, this can be the result

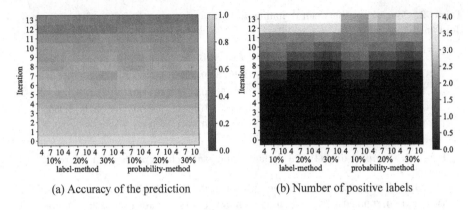

(a) Accuracy of the prediction (b) Number of positive labels

Fig. 4. Heatmaps characterizing the predicted sequences on *Yeast*

of the error propagation since erroneously predicted labels influence the predictions of the following iterations. Furthermore, it can be observed that errors are mostly made on positive labels after predicting almost all the negative labels. This can be again explained by the confidences, which is higher for negative labels due to the sparsity of the label assignments.

5.5   Comparison with Other Classifiers

In order to put the performance of RDT and the different prediction methods in a larger context we present in this section a comparison to a couple of other algorithms. We have evaluated the BR, the LP and the CC method with the J48 WEKA implementation of the C4.5 decision tree learner as the base classifier. This approach represents in our comparison the family of classical decision tree learners which address a learning task by choosing splits at the inner nodes which optimize a certain pre-determined criterion such as the information gain. For the J48 algorithm we have used the default parameter settings which are 0.25 for the confidence threshold for pruning and 2 for the minimum number of instances per leaf. For the CC method we have evaluated ten different random chain orderings and averaged the results. Moreover, we have evaluated the BR and the LP method using RDT. For these methods we have chosen to build ensembles with 300 decision trees, with a maximum depth of 30 and a minimum of four instances to create a test. The dynamic chain methods share the same settings and use 20% of label tests in their ensembles. Hence, the dynamic chain methods use 20% less tests on the features compared to the BR and the LP method because they are replaced by label tests. The ten-fold cross validation results can be seen in Tables 4 and 5.

First of all, it can be seen that the label method outperforms the J48 methods on the datasets *CAL500, Emotions* and *Yeast* in terms of subset accuracy and micro-averaged F1 measure. On the other hand, the J48 methods were able to beat the dynamic chain methods on the datasets *Scene, Enron* and *Medical*.

Table 4. Results for the subset accuracy

	Flags	Emotions	Scene	Yeast	Birds	CAL500	Enron	Medical
DCC LM	0.1959	0.2799	0.4271	0.2073	0.4930	0.0000	0.0447	0.1840
DCC PM	0.1856	0.3339	0.3112	0.1854	0.4698	0.0000	0.0576	0.0000
RDT LP	0.1959	**0.3929**	0.4612	**0.2416**	**0.5008**	0.0000	**0.1363**	0.2566
RDT BR	0.1701	0.2479	0.1379	0.0852	0.4698	0.0000	0.0129	0.0000
J48 BR	0.1443	0.1686	0.3515	0.0633	0.4930	0.0000	0.0593	0.6708
J48 CC	0.2211	0.2169	0.4548	0.1327	0.4998	0.0000	0.0977	**0.6906**
J48 LP	**0.2474**	0.1939	**0.4902**	0.1419	0.4729	0.0000	0.0823	0.6585

Table 5. Results for the micro-averaged F1 measure

	Flags	Emotions	Scene	Yeast	Birds	CAL500	Enron	Medical
DCC LM	0.7506	0.6535	0.4682	**0.6459**	0.4484	**0.4681**	0.4527	0.2696
DCC PM	0.7448	0.6774	0.4344	0.6100	0.0708	0.3114	0.3410	0.0065
RDT LP	0.7310	**0.7168**	0.4953	0.6353	0.3390	0.3349	0.3652	0.3520
RDT BR	**0.7545**	0.5997	0.2433	0.5598	0.1063	0.3178	0.3842	0.0000
J48 BR	0.7416	0.5791	**0.5563**	0.5774	**0.4675**	0.3553	**0.5096**	0.8198
J48 CC	0.7248	0.5806	0.5420	0.5516	0.4576	0.3501	0.4996	**0.8225**
J48 LP	0.7045	0.5668	0.5374	0.5397	0.4259	0.3309	0.3818	0.7527

Especially the results on the dataset *Medical* are conspicuous, where RDT generally performs very poorly. As aforementioned, sparse input data is particularly challenging for RDT-like approaches.

Furthermore, it can be seen that the dynamic chain methods are superior to the RDT-BR method on almost all datasets, as anticipated by the results in Subsect. 5.2.

Of particular interest is the comparison to the RDT-LP method. In terms of subset accuracy this method could outperform the dynamic chain methods on almost all datasets. Especially the results on the datasets *Emotions*, *Yeast* and *Enron* are much better than the results of the other methods. However, a closer examination reveals that the results for the micro-averaged F1 measure of the RDT-LP method are not always that good. The label method could achieve a much higher score for micro-averaged F1 measure on the datasets *Birds*, *CAL500* and *Enron*. Senge et al. [19] observed that LP can benefit from the restricted set of label combinations it can choose from, especially when the number of distinct combinations is relatively low, as it is the case for the used datasets. The other approaches, instead, have to make up valid combinations by concatenating single decisions. Whereas these single decisions might be better than for LP, as seen in terms of micro-averaged F1 measure, the complete combination might still be wrong especially if the cardinality is high.

6   Conclusions and Future Work

In this paper we have proposed a new approach on multi-label classification based on random decision trees and the idea of classifier chains. With our proposed algorithm we have been able to overcome the major problem of the label ordering by dynamically selecting the next label in the sequence depending on the context, namely the instance at hand and the previously predicted labels for it. In comparison to other approaches for CC, which try to pre-compute appropriate sequences, our approach comes at no additional cost, since the framework of RDT allows to perform the necessary inferences completely during prediction time.

In several experiments the dynamic label ordering has been analyzed in depth and compared with other baseline methods. Even though we cannot achieve state-of-the-art results with RDT in some cases and domains, they have appealing properties that allow a fundamental analysis of the advantages and disadvantages of certain approaches. For instance, our experiments revealed the importance of the dynamic label ordering on different datasets, as well as the impact of using the previous predictions. These observations could be made with the guarantee that they were independent of any other factors like the usage of more sophisticated methods or more powerful models.

However, to improve the predictive capabilities of RDT still remains a goal for future work. For instance, the proposed Gini index considers the skew of the counts, but not the number of instances these counts are based on, which could be used as further indicator for the confidence. Efficiency could also be improved if we consider that labels are usually sparse in MLC problems. Therefore, it could be enough to focus on positive labels only, which would considerably reduce the length of the prediction sequence. In addition, as we have seen, RDT have still clear disadvantages on data which is sparse in the feature values, such as text. New types of tests in the inner nodes, which for instance consider disjunctions of several features, could solve this problem. Furthermore, we plan to transfer our ideas on dynamic chains to other kinds of algorithms as well. A first step will be to adapt predictive clustering trees [24]. The construction of these trees does also not necessarily depend on a specific target. However, its clustering may allow for more discriminative distributions at the leaves.

References

1. Breiman, L.: Random forests. Mach. Learn. **45**(1), 5–32 (2001)
2. Dembczyński, K., Cheng, W., Hüllermeier, E.: Bayes optimal multilabel classification via probabilistic classifier chains. In: Proceedings of the 27th International Conference on International Conference on Machine Learning, pp. 279–286 (2010)
3. Dembczyński, K., Waegeman, W., Cheng, W., Hüllermeier, E.: On label dependence and loss minimization in multi-label classification. Mach. Learn. **88**(1–2), 5–45 (2012)
4. Fan, W.: On the Optimality of probability estimation by random decision trees. In: Proceedings of the 19th National Conference on Artificial Intelligence, pp. 336–341 (2004)

5. Fan, W., Greengrass, E., McCloskey, J., Yu, P.S., Drammey, K.: Effective estimation of posterior probabilities: explaining the accuracy of randomized decision tree approaches. In: Proceedings of the 5th International Conference on Data Mining, pp. 154–161 (2005)
6. Fan, W., Wang, H., Yu, P.S., Ma, S.: Is random model better? On its accuracy and efficiency. In: Proceedings of the 3rd IEEE International Conference on Data Mining, pp. 51–58 (2003)
7. Geurts, P., Ernst, D., Wehenkel, L.: Extremely randomized trees. Mach. Learn. **63**(1), 3–42 (2006)
8. Goncalves, E.C., Plastino, A., Freitas, A.A.: A genetic algorithm for optimizing the label ordering in multi-label classifier chains. In: Proceedings of the IEEE 25th International Conference on Tools with Artificial Intelligence, pp. 469–476 (2013)
9. Kong, X., Yu, P.S.: An ensemble-based approach to fast classification of multi-label data streams. In: Proceedings of the 7th International Conference on Collaborative Computing: Networking, Applications and Worksharing, pp. 95–104 (October 2011)
10. Kumar, A., Vembu, S., Menon, A.K., Elkan, C.: Beam search algorithms for multilabel learning. Mach. Learn. **92**(1), 65–89 (2013)
11. Li, N., Zhou, Z.-H.: Selective ensemble of classifier chains. In: Zhou, Z.-H., Roli, F., Kittler, J. (eds.) MCS 2013. LNCS, vol. 7872, pp. 146–156. Springer, Heidelberg (2013). https://doi.org/10.1007/978-3-642-38067-9_13
12. Malerba, D., Semeraro, G., Esposito, F.: A multistrategy approach to learning multiple dependent concepts. Mach. Learn. Stat. Interface chap. 4, 87–106 (1997)
13. Mena, D., Montañés, E., Quevedo, J.R., Coz, J.J.d.: Using A* for inference in probabilistic classifier chains. In: Proceedings of the 24th International Conference on Artificial Intelligence, pp. 3707–3713 (2015)
14. Mena, D., Montañés, E., Quevedo, J.R., Coz, J.J.: An overview of inference methods in probabilistic classifier chains for multilabel classification. Wiley Interdiscip. Rev. Data Min. Knowl. Discov. **6**(6), 215–230 (2016)
15. Nam, J., Loza Mencía, E., Kim, H.J., Fürnkranz, J.: Maximizing subset accuracy with recurrent neural networks in multi-label classification. In: Advances in Neural Information Processing Systems 30 (NIPS-17). pp. 5419–5429 (2017)
16. Quevedo, J.R., Luaces, O., Bahamonde, A.: Multilabel classifiers with a probabilistic thresholding strategy. Pattern Recognit. **45**(2), 876–883 (2012)
17. Read, J., Martino, L., Luengo, D.: Efficient Monte Carlo methods for multidimensional learning with classifier chains. Pattern Recognit. **47**(3), 1535–1546 (2014)
18. Read, J., Pfahringer, B., Holmes, G., Frank, E.: Classifier chains for multi-label classification. Mach. Learn. **85**(3), 333–359 (2011)
19. Senge, R., del Coz, J.J., Hüllermeier, E.: On the problem of error propagation in classifier chains for multi-label classification. In: Spiliopoulou, M., Schmidt-Thieme, L., Janning, R. (eds.) Data Analysis, Machine Learning and Knowledge Discovery. SCDAKO, pp. 163–170. Springer, Cham (2014). https://doi.org/10.1007/978-3-319-01595-8_18
20. da Silva, P.N., Gonçalves, E.C., Plastino, A., Freitas, A.A.: Distinct chains for different instances: an effective strategy for multi-label classifier chains. In: Calders, T., Esposito, F., Hüllermeier, E., Meo, R. (eds.) ECML PKDD 2014. LNCS (LNAI), vol. 8725, pp. 453–468. Springer, Heidelberg (2014). https://doi.org/10.1007/978-3-662-44851-9_29

21. Sucar, L.E., Bielza, C., Morales, E.F., Hernandez-Leal, P., Zaragoza, J.H., Larrañaga, P.: Multi-label classification with Bayesian network-based chain classifiers. Pattern Recognit. Lett. **41**, 14–22 (2014)
22. Tsoumakas, G., Katakis, I., Vlahavas, I.: Mining Multi-label data. Data Mining and Knowledge Discovery Handbook, pp. 667–685 (2010)
23. Tsoumakas, G., Spyromitros-Xioufis, E., Vilcek, J., Vlahavas, I.: MULAN: a java library for multi-label learning. J. Mach. Learn. Res. **12**, 2411–2414 (2011)
24. Vens, C., Struyf, J., Schietgat, L., Džeroski, S., Blockeel, H.: Decision trees for hierarchical multi-label classification. Mach. Learn. **73**(2), 185 (2008)
25. Zhang, X., Fan, W., Du, N.: Random decision hashing for massive data learning. In: Proceedings of the 4th International Workshop on Big Data, Streams and Heterogeneous Source Mining: Algorithms, Systems, Programming Models and Applications, pp. 65–80 (2015)
26. Zhang, X., Yuan, Q., Zhao, S., Fan, W., Zheng, W., Wang, Z.: Multi-label classification without the multi-label cost. In: Proceedings of the Society for Industrial and Applied Mathematics International Conference on Data Mining, pp. 778–789 (2010)

Feature Ranking with Relief for Multi-label Classification: Does Distance Matter?

Matej Petković[1,2]([⊠]), Dragi Kocev[1,2], and Sašo Džeroski[1,2]

[1] Jožef Stefan Institute, Jamova 39, 1000 Ljubljana, Slovenia
[2] Jožef Stefan Postgraduate School, Jamova 39, 1000 Ljubljana, Slovenia
{matej.petkovic,dragi.kocev,saso.dzeroski}@ijs.si

Abstract. In this work, we address the task of feature ranking for multi-label classification (MLC). The task of MLC is to predict which labels from a maximal predefined label set are relevant for a given example. We focus on the Relief family of feature ranking algorithms and empirically show that the definition of the distances in the target space used within Relief should depend on the evaluation measure used to assess the performance of MLC algorithms. By considering different such measures, we improve over the currently available MLC Relief algorithm. We extensively evaluate the resulting MLC ranking approaches on 24 benchmark MLC datasets, using different evaluation measures of MLC performance. The results additionally identify the mechanisms of influence of the parameters of Relief on the quality of the rankings.

Keywords: Feature ranking · Multi-label classification · Relief

1 Introduction

Classification is a task in predictive modelling, where the goal is to learn a model that takes as the input a vector x of descriptive variables (features) x_i, and predicts the class value y that a given example belongs to. If y can take two different values, the task at hand is referred to as binary classification. Otherwise (y can take more than two values), the task at hand is multi-class classification. In both cases, every example is assigned precisely one value. For example, one can predict whether a person has survived a shipwreck where $y \in \{\text{yes}, \text{no}\}$ (binary), or what is the blood type of a person where $y \in \{\text{A}, \text{B}, \text{AB}, \text{0}\}$ (multi-class). In both cases, class values are mutually exclusive.

A related task is multi-label classification (MLC). As opposed to the standard classification, a MLC predictive model predicts which labels from a predefined

We acknowledge the financial supported of the Slovenian Research Agency via the grants P2-0103, J4-7362, L2-7509, N2-0056, and a young researcher grant to MP, the European Commission, through the grants HBP (The Human Brain Project), SGA2, and LANDMARK, and ARVALIS (project BIODIV).

L. Soldatova et al. (Eds.): DS 2018, LNAI 11198, pp. 51–65, 2018.
https://doi.org/10.1007/978-3-030-01771-2_4

set $\mathscr{L}$ are *relevant* for a given example. For example, one can predict which of the genres from the set $\mathscr{L} = \{\texttt{romance}, \texttt{drama}, \texttt{comedy}\}$ are relevant for a given film. Clearly, a film can be $\texttt{drama}$ and $\texttt{comedy}$ at the same time.

There are two main approaches to MLC: problem transformation and algorithm adaptation. From the problem transformation group of methods most widely known are binary relevance and label power set. Binary relevance is a simple method that converts a MLC task to several binary classification tasks with $y \in \{\texttt{yes}, \texttt{no}\}$ where we predict the relevance of each label separately. This approach is often criticized for it cannot make use of the interactions among the labels. In the label power set approach [24], the task of predicting a subset of $\mathscr{L}$ is converted to the task of predicting an element of the power set $2^{\mathscr{L}}$, and thus converting a MLC task to multi-class classification task. However, the number of classes can be as high as $2^{|\mathscr{L}|}$, which results in a very sparse dataset.

The second group of methods are method transformation techniques where an existing method is adapted to a new problem. A prominent member of this group are predictive clustering trees which generalize decision trees, so that they can handle MLC [15] and other structure output prediction tasks [12].

Another important task in machine learning is feature ranking, where the goal is to asses the importance of every descriptive attribute (feature) by using some scoring function. The output of a feature ranking algorithm is a list of features that is sorted with respect to the scores.

Feature ranking is typically considered a part of data preprocessing, since it can be used to reduce the dimensionality of the input space, so that only the features that contain the most information about labels (or target(s) in general) are kept in the dataset. By doing this, we decrease the computational cost of building a predictive model, while the performance of the model is not degraded. Another reason to compute a feature ranking is that dimensionality reduction typically results in models that are easier to understand, which is useful when a machine learning expert works in collaboration with a domain expert. Predictive models, such as decision trees, are easier to interpret when a small number of the most relevant features are used to learn them.

There is a plethora of feature ranking methods for the task of classification [22]. A possible approach to MLC feature ranking is to adapt the binary relevance approach from predictive modelling, where at the first stage, feature importances are computed for every label $\ell \in \mathscr{L}$ separately as in the classification case. After that, the feature importances are averaged over the different labels and a single ranking is returned. In this work, we focus on the RELIEF family of feature ranking algorithms, which are distance based approaches and thus widely applicable. They are part of the *filter methods* which compute the ranking without any additional predictive model [9]. The filters are typically fast, i.e., linear in the number of features, but myopic at the same time, i.e., cannot capture the feature interaction. RELIEF family of the feature ranking algorithms, however, overcomes this, and can successfully discover, e.g., XOR-relation [14].

The rest of the paper is organized as follows. In Sect. 2, the overview of related work is given. In Sect. 3, the proposed feature ranking algorithms are described and analyzed. In Sect. 4, the detailed description of the experimental is given. In Sect. 5, the results of the experiments are presented. In Sect. 6, conclusions and direction for further work are given.

2   Related Work

We start the overview of the related work with the extensions of the RELIEF family to MLC setting that are presented in [20]. There, the binary relevance and label power set approach were applied to the feature ranking scenario. More precisely, in the case of binary relevance approach, feature ranking was computed for every label $\ell \in \mathscr{L}$ separately. This was done by using the Relief algorithm for the standard binary classification [11]. After that, the feature importances were averaged to a single score. In the case of the label power set approach the multi-class extension of the Relief (ReliefF) was used [14].

As mentioned before, these two approaches have some drawbacks. Binary relevance approach does not take the label interactions into account and can be expensive to run when the number of labels is high: we have to solve $|\mathscr{L}|$ feature ranking problems which results in high time or space complexity. High space complexity is also a drawback of label power set approach if the number of different relevant label subsets is high. In that case, the data may also become too sparse for the ranking to be relevant.

Both procedures were evaluated on a rather small subset of ten datasets presented in this study (see Sect. 4.2), in a manner similar to our evaluation procedure, which uses k nearest neighbours classifier. No statistical tests were done and the feature rankings were not compared to any baseline.

Another data transformation approach was presented in [13] where the MLC problem is transformed into $|\mathscr{L}|(|\mathscr{L}| - 1)/2$ binary classification problems - one for each of the label pairs (ℓ_1, ℓ_2) where $\ell_1 \neq \ell_2$. For each binary problem, only the examples for which either ℓ_1 or ℓ_2 is relevant (but not both) are retained in the corresponding dataset. The exclusion of the examples for which both labels are relevant is necessary to avoid ill-defined terms in the equations for importance update. The authors motivate this by claiming that the number of the examples for which both labels are relevant, is small in comparison to the number of examples for which precisely one of the two labels is relevant. However, this may not be the case in some data sets, as observed in [18]. The main drawback of this approach is the computational complexity, since the number of feature ranking problems to solve grows quadratically with the number of labels.

A member of the RELIEF family ReliefF-ML [18] does solve the multi-label ranking problem directly, yet its space complexity is still considerable. The algorithm RReliefF-ML [18] overcomes this issue since it is an extension of the RRelief-efF version that is suitable for regression tasks [14]. In contrast to the extensions, RReliefF(-ML) computes only one group of the nearest neighbors per example which results in significantly smaller space complexity.

The method was empirically shown to yield relevant feature rankings [18] since it statistically significantly outperformed the baseline. For showing statistical significance, Friedman test was used. However, we need to point out that the very basic assumption of the independence of *data samples* (datasets in this case) was not met, since 10 out of 34 are basically different versions of the same data (Corel16k datasets). In our experiments, we also show that the seemingly ad-hoc choice of the target distance may not lead to the best rankings if we want to optimize for a particular evaluation measure.

Regarding pure predictive modelling setting, the authors in [4] show that in general, different evaluation measures result in different optimal classifiers. However, the authors also show that, e.g., *Hamming Loss* and *Subset Accuracy* have the same optimal classifier under some rather strict conditions.

3   MLC-Relief

RELIEF family of feature ranking algorithms calculates the feature importance scores by considering differences in the feature values between pairs of examples (an example and its nearest neighbor). More specifically, if the values of features of a pair of examples from the same class are different then the features' importance decreases. Conversely, if the feature values are different for examples from different classes then the features' importance increases.

In the following, we first introduce the distance measures used within the algorithm. Then, the algorithm is described and its computational complexity (including the complexity of computing different distances) is analyzed. Throughout the paper, F and L always denote the number of features and labels respectively.

3.1   Distances: Why and Which

All methods of the RELIEF family assign feature x_i a weight w_i that is a measure of feature importance in these algorithms. The expected value of the w_i has a nice probability interpretation in the case when both the target and x_i are nominal [14]: simplified to some extent, we have a relation

$$\mathbb{E}[w_i] = \frac{P_{\text{diffAttr, diffTarget}}}{P_{\text{diffTarget}}} - \frac{P_{\text{diffAttr}} - P_{\text{diffAttr, diffTarget}}}{1 - P_{\text{diffTarget}}}, \tag{1}$$

where we define the probabilities $P_{\text{ev}} = P(\text{ev})$ and $P_{\text{ev1, ev2}} = P(\text{ev1} \wedge \text{ev2})$ that base on the events diff/sameAttr (two instances have different/same value of x_i) and diff/sameTarget (two instances have different/same target value). The probabilities from the right hand side of Eq. (1) are modeled as the distances in the corresponding spaces: P_{diffAttr} is modeled by the distance d_i on the domain of feature x_i, $P_{\text{diffTarget}}$ is modeled by the distance $d_{\mathscr{L}}$ on the label set $\mathscr{L}$, and $P_{\text{diffAttr, diffTarget}}$ is modeled as their product $d_i d_{\mathscr{L}}$.

First, the distance on the whole descriptive domain $\mathcal{X}$ is defined via the distances d_i on the domains $\mathcal{X}_i$ of features x_i as

$$d_i(\boldsymbol{x}^1, \boldsymbol{x}^2) = \begin{cases} 1[\boldsymbol{x}_i^1 \neq \boldsymbol{x}_i^2] & : \mathcal{X}_i \nsubseteq \mathbb{R} \\ \frac{|\boldsymbol{x}_i^1 - \boldsymbol{x}_i^2|}{\max\limits_{\boldsymbol{x}} \boldsymbol{x}_i - \min\limits_{\boldsymbol{x}} \boldsymbol{x}_i} & : \mathcal{X}_i \subseteq \mathbb{R} \end{cases} \qquad d_{\mathcal{X}}(\boldsymbol{x}^1, \boldsymbol{x}^2) = \frac{1}{F} \sum_{i=1}^{F} d_i(\boldsymbol{x}^1, \boldsymbol{x}^2) \tag{2}$$

where $\mathbf{1}$ is the indicator function with the values $\mathbf{1}[\texttt{true}] = 1$ and $\mathbf{1}[\texttt{false}] = 0$, and max and min go over the examples $\boldsymbol{x}$ in the training set.

For the distance $d_{\mathscr{L}}$ between two sets of labels S^1 and S^2, we consider four options. The use of the first (Hamming Loss) was proposed in [18].

Hamming Loss. This distance is defined as

$$d_{Hamming}(S^1, S^2) = \left| S^1 \setminus S^2 \cup S^2 \setminus S^1 \right| / L. \tag{3}$$

We observe that this is an analogue of $d_{\mathscr{X}}$ from Eq. (2). Encoding a subset $S \subseteq \mathscr{L}$ as a 0/1 vector $\boldsymbol{s}$, where $\boldsymbol{s}_j = 1 \Leftrightarrow \ell_j \in S$, we have $d_{Hamming}(S^1, S^2) = \frac{1}{L} \sum_{j=1}^{L} d_j(\boldsymbol{s}^1, \boldsymbol{s}^2)$, where the numeric part of d_i in Eq. (2) applies in d_j. We believe that there are more suitable choices for the distance $d_{\mathscr{L}}$ that take into account the set structure.

Accuracy. The similarity between two sets can be also measured by their Jaccard index $|S^1 \cap S^2|/|S^1 \cup S^2|$ which is well defined when at least one of the subsets $S^{1,2}$ is not empty (this is the case in our datasets). We then define

$$d_{Accuracy}(S^1, S^2) = 1 - |S^1 \cap S^2| / |S^1 \cup S^2|. \tag{4}$$

F_1 **distance.** This distance is defined as

$$d_{F1}(S^1, S^2) = 1 - 2|S^1 \cap S^2| / (|S^1| + |S^2|), \tag{5}$$

where the second term can be seen as the harmonic mean of the precision and recall [15]. However, these two measures are not symmetric, thus inappropriate as the distance measures.

Subset Accuracy. This distance is defined as

$$d_{SubsetAcc}(S^1, S^2) = \mathbf{1}\left[S^1 \neq S^2 \right]. \tag{6}$$

It is the strictest, since it does not differentiate between, e.g., almost the same and disjunctive pairs of subsets. This allows for a faster computation of the distance as compared to the other options (Lemma 2).

Except for the d_{F1}, all distances are also metrics. We named them after the measures that they are expected to optimize (defined in Sect. 4.4), and believe that no other standard measures (see [15,27]) allow for a direct derivation of distance definitions.

3.2 Algorithm Description

The calculation of the weights $w_i = importance(x_i)$ using the MLC extension of RReliefF is outlined in Algorithm 1. RReliefF is an iterative procedure. For each of the m iterations, we randomly select an example r from $\mathscr{D}_{\text{TRAIN}}$ (line 4) and find its K nearest neighbors (line 5) using the distance $d_{\mathscr{X}}$ from Eq. (2). After that, we use the neighbors to update the estimates of probabilities that appear in the definition of the weights (1) for all attributes (lines 8–10). The estimates of probabilities are updated with the weighted average of the distances between

---

Algorithm 1 MLC-RReliefF($\mathscr{D}_{\text{TRAIN}}$, m, K, $d_{\mathscr{L}}$)

1: $\boldsymbol{P}_{\text{diffAttr, diffTarget}}$, $\boldsymbol{P}_{\text{diffAttr}}$ = zero lists of length F
2: $P_{\text{diffTarget}} = 0.0$
3: **for** $\iota = 1, 2, \ldots, m$ **do**
4:     $\boldsymbol{r}$ = random example from $\mathscr{D}$
5:     $\boldsymbol{n}_1, \boldsymbol{n}_2, \ldots, \boldsymbol{n}_K = K$ nearest neighbors of $\boldsymbol{r}$
6:     **for** $k = 1, 2, \ldots, K$ **do**
7:         $P_{\text{diffTarget}} += \delta(\ell)d_{\mathscr{L}}\left(\boldsymbol{r}, \boldsymbol{n}_k\right)$
8:         **for** $i = 1, 2, \ldots, F$ **do**
9:             $\boldsymbol{P}_{\text{diffAttr}}[i] += \delta(\ell)d_i\left(\boldsymbol{r}, \boldsymbol{n}_k\right)$
10:             $\boldsymbol{P}_{\text{diffAttr, diffTarget}}[i] += \delta(\ell)d_i\left(\boldsymbol{r}, \boldsymbol{n}_k\right)d_{\mathscr{L}}\left(\boldsymbol{r}, \boldsymbol{n}_k\right)$
11: **for** $i = 1, 2, \ldots, F$ **do**
12:     $w_i = \dfrac{\boldsymbol{P}_{\text{diffAttr, diffTarget}}[i]}{P_{\text{diffTarget}}} - \dfrac{\boldsymbol{P}_{\text{diffAttr}}[i] - \boldsymbol{P}_{\text{diffAttr, diffTarget}}[i]}{1 - P_{\text{diffTarget}}}$

$\boldsymbol{r}$ and its neighbors. Here, the distance $d_{\mathscr{L}}$ from the algorithm input is used. The weight $\delta(k) = 1/(mK)$ ensures that $w_i \in [-1, 1]$ when the algorithms finishes. At the end, the weights w_i are computed (line 12) by using the relation (1).

The default values of the parameters are set as follows. Typically, we iterate over the whole dataset, i.e., $m = |\mathscr{D}_{\text{TRAIN}}|$. By doing this, the estimates of probabilities are expected to be more accurate. The value of K is typically set small enough to capture the local structure in the data. In that way, we implicitly capture the interactions between features [14].

3.3 Computational Complexity

We first analyze the time complexity of a single iteration. Since the space-partitioning data structures, such as kD trees do not perform well when the number of features F is high, we use a brute-force method for finding the nearest neighbors. Hence, the computation of the distances between $\boldsymbol{r}$ and the neighbour candidates takes $\mathcal{O}(MF)$ steps, where $M = |\mathscr{D}_{\text{TRAIN}}|$. In addition to this, the current group of the nearest neighbors must be updated from time to time.

Lemma 1. *The expected number of updates of the group of current nearest neighbors of the instance $\boldsymbol{r}$ is approximately $K \log M$.*

Proof. When we iterate over the neighbors, the group of currently K nearest neighbors is updated if, and only if, at most $K - 1$ better candidates have been found so far. Let $\boldsymbol{n}_k$ be the instances from $\mathscr{D}_{\text{TRAIN}} \setminus \{\boldsymbol{r}\}$, sorted increasingly by the distance to $\boldsymbol{r}$, i.e., $\boldsymbol{n}_1$ is the nearest neighbor and $\boldsymbol{n}_{M-1}$ is the farthest neighbor. Let E_k be the expected number of updates when we find the candidate $\boldsymbol{n}_k$. Then, E_k equals the probability p_k of discovering at most $K - 1$ of the instances $\boldsymbol{n}_1, \ldots, \boldsymbol{n}_{k-1}$ before $\boldsymbol{n}_k$. Probability $p_{k,s}$ of discovering precisely s of them equals the probability that $\boldsymbol{n}_k$ appears in the $(s + 1)$-th position in the random permutation of the instances $\boldsymbol{n}_1, \ldots, \boldsymbol{n}_k$, hence $p_{k,s} = 1/k$, for all $s < k$, and $p_{k,s} = 0$ otherwise. It follows that $p_k = \sum_{s=0}^{K-1} p_{k,s} = \min\{k, K\}/k$.

The total number of expected updates E then equals $E = \sum_{k=1}^{M-1} E_k$, hence $E = \sum_{k=1}^{M-1} p_k = K + K \sum_{k=K+1}^{M-1} \frac{1}{k} = K(1 + H_{M-1} - H_K)$, where k-th harmonic number H_k is defined as $H_k = \sum_{s=1}^{k} 1/s$. Since $\log k < H_k < 1 + \log k$, the leading term in E is indeed $K \log M$.

The overall cost of updating the current nearest neighbours is thus $\mathcal{O}(K \log M \log K)$ if we are using, e.g., the heap structure.

When the neighbours n_k, $1 \le k \le K$, are computed, the distance between their label set and the label set of r are computed. Considering that we store the label sets as 0/1-lists of length L, this takes $\mathcal{O}(KL)$ steps for the distances $d_{Hamming}$, $d_{Accuracy}$ and d_{F1}, since we have to iterate over all labels. In the case of $d_{SubsetAcc}$, we can do much better, knowing that the labels are typically sparse. To be able to obtain a closed form expression, we will assume that all labels have the same probability to be relevant and that they are independent.

Lemma 2. *The expected value of the labels considered in one computation of* $d_{SubsetAcc}$ *is* $\frac{1-p^L(2-p)^L}{(1-p)^2}$, *where p is the probability of a label being relevant.*

Proof. We know that $d_{SubsetAcc}(S^1, S^2) = 1$ as soon as we encounter the label $\ell_l \notin S^1 \cap S^2$. Let X be the number of labels considered. The key observation is that we can easily compute $P(X \ge k) = P(\ell_1, \ldots, \ell_{k-1} \in S^1 \cap S^2) = (1 - p)^{2(k-1)}$. This is useful since $\mathbb{E}[X] = \sum_{x=1}^{L} P(X \ge k)$. We obtained geometric series whose sum equals $\mathbb{E}[X] = \frac{1-p^L(2-p)^L}{(1-p)^2}$.

Table 1 reveals that the dataset `Delicious` has $L = 983$ labels and label cardinality (average number of labels per example) $\ell_c \doteq 19$. Thus, $p \doteq 0.019$ and $\mathbb{E}[X] \doteq 1.04$, which is considerably smaller than L.

After the distances $d_{\mathscr{L}}$ are computed, the probability estimates are updated in $\mathcal{O}(KF)$ steps. After all iterations, the weights are computed in $\mathcal{O}(F)$ steps, thus the final time complexity is $\mathcal{O}(m[MF + K \log M \log K + KL + KF] + F) = \mathcal{O}(m[MF + KL])$ (in the case of $d_{SubsetAcc}$, L the term KL is replaced by $\mathbb{E}[X]$). If the number of labels is high, then the term KL may not be negligible, which was overlooked in [18].

4 Experimental Design

Here, we give the detailed experimental design for evaluating the performance of the proposed distances. We begin by stating the experimental questions and summarizing the MLC datasets used in this study. Then, we present the evaluation procedure and give the specific parameters instantiations of the methods.

4.1 Experimental Questions

The main experimental question is: *Does the choice of the distance $d_{\mathscr{L}}$ matter?*

Furthermore, we investigate (i) whether the knowledge encapsulated in the feature importances leads to better predictive performance of a model, i.e, are the obtained feature rankings relevant, and (ii) how the quality of ranking is influenced by the number of neighbors K and the number of iterations m.

4.2   Datasets

We use 24 MLC benchmark problems. Table 1 presents the basic statistics of the datasets. The number of features ranges from 72 to 52350. The features are numeric and nominal. The label set size L ranges from 6 to 983, while the number of training examples ranges from 322 up to 70000. The average number of labels per example (in $\mathscr{D}_{\text{TRAIN}} \cup \mathscr{D}_{\text{TEST}}$), i.e., *label cardinality* is also given. With the exception of *Delicious* dataset, it ranges between 1.0 and 4.38.

The datasets come from different domains. *Arts, Business, Computers, Education, Entertainment, Health, Recreation, Reference, Science, Social* and *Society* describe the problems of finding relevant subtopics of the given main topic of a web page. *Bibtex* and *Bookmarks* are automatic tag suggestion problems, *Birds* deals with predictions of multiple bird species in a noisy environment. *Corel5k* contains Corel images. *Delicious* contains contextual data about web pages along with their tags. *Emotions* deals with emotions in music. *Enron* contains data about emails. *Genbase* and *Yeast* come from biological domain. *Mediamill* was introduced in a video annotation challenge. *Medical* comes from Medical Natural Language Processing Challenge. *Scene* deals with labelling of natural scenes. *TMC2007-500* is about discovering anomalies in text reports.

4.3   Evaluation Methodology

We adopted the evaluation methodology that has been previously used in MLC context [18] and in the other types of structured output prediction [17].

We use the same train-test split of the datasets as in the Mulan repository http://mulan.sourceforge.net/datasets-mlc.html. A ranking is computed from the training part $\mathscr{D}_{\text{TRAIN}}$ only, and evaluated on the testing part $\mathscr{D}_{\text{TEST}}$.

The quality of the ranking is assessed by using the kNN algorithm where instead of the standard Euclidean distance, its weighted version was used. For two input vectors $\boldsymbol{x}^1$ and $\boldsymbol{x}^2$, the distance between them is defined as

$$d(\boldsymbol{x}^1, \boldsymbol{x}^2) = \sqrt{\sum_{i=1}^{F} w_i d_i^2(\boldsymbol{x}_i^1, \boldsymbol{x}_i^2)}, \tag{7}$$

where d_i is defined by Eq. (2). The weights are set to $w_i = \max\{importance(x_i), 0\}$, since they need to be made non-negative to ensure that d is well defined, and also to ignore the attributes that have smaller values for importance than a randomly generated attribute would have.

The evaluation through a kNN predictive model was chosen because of two main reasons. First, this is a distance based model, hence, it can easily make use of the information contained in the feature importances in the learning phase. The second reason is kNN's simplicity: its only parameter is the number of neighbors, which we set to 15. In the prediction stage, the neighbors' contributions to the predicted value are equally weighted, so we do not introduce additional parameters that would influence the performance.

Table 1. Data characteristics: sizes of train and test part of the dataset, number of features F, labelset size L and label cardinality ℓ_c.

Dataset	$|\mathscr{D}_{TRAIN}|$	$|\mathscr{D}_{TEST}|$	F	L	ℓ_c
Arts [26]	3712	3772	23146	26	1.65
Bibtex [10]	4880	2515	1836	159	2.40
Birds [3]	322	323	260	19	1.01
Bookmarks [10]	70000	17856	2150	208	2.04
Business [26]	5710	5504	21924	30	1.60
Computers [26]	6270	6174	34096	33	1.51
Corel5k [7]	4500	500	499	374	3.52
Delicious[25]	12920	3185	500	983	19.02
Education [26]	6030	6000	27534	33	1.46
Emotions[23]	391	202	72	6	1.87
Enron [1]	1123	579	1001	53	3.38
Entertainment [26]	6356	6374	32001	21	1.41
Genbase [6]	463	199	1185	27	1.25
Health [26]	4557	4648	30605	32	1.64
Mediamill [19]	30993	12914	120	101	4.38
Medical [16]	645	333	1449	45	1.25
Recreation [26]	6471	6357	30324	22	1.43
Reference [26]	4027	4000	39679	33	1.17
Scene [2]	1211	1196	294	6	1.07
Science [26]	3214	3214	37187	40	1.45
Social [26]	6037	6074	52350	39	1.28
Society [26]	7273	7239	31802	27	1.67
TMC2007-500 [21]	21519	7077	500	22	2.22
Yeast [8]	1500	917	103	14	4.24

The second rationale for using kNN as an evaluation model is as follows. If a feature ranking is meaningful, then when the feature importances are used as weights in the calculation of the distances kNN should produce better predictions as compared to kNN without using these weights [28].

4.4 Evaluation Measures

In the following, we denote the sets of true and predicted labels for an example x respectively by $y(x)$ and $\hat{y}(x)$. The measures *Hamming Loss, Accuracy, F_1 Score* and *Subset Accuracy* can be defined in terms of the distances (3)–(6). They are respectively the means (over $\mathscr{D}_{TEST}$) of the values $d_{Hamming}(y(x), \hat{y}(x))$, $1 - d_{Accuracy}(y(x), \hat{y}(x))$, $1 - d_{F1}(y(x), \hat{y}(x))$ and $1 - d_{SubsetAcc}(y(x), \hat{y}(x))$.

Thus, *Hamming Loss* should be minimized while the remaining three should be maximized. We use another four well known measures: *One Error*, *Precision*, *Recall* and area under the pooled precision-recall curve (*pooledAUPRC*). The definitions can be found in [15, 27].

4.5　Statistical Analysis of the Results

For comparing the algorithms, we use the Friedman test. The null hypothesis H_0 is that all considered algorithms have the same performance. If H_0 is rejected by the Friedman's test, we additionally apply Nemenyi or Bonferroni-Dunn post-hoc test. The first is used when we investigate where the statistically significant differences between *any* two algorithms occur, while the second is used when we are interested in the differences between one particular algorithm and the others. A detailed description of all tests is available in [5].

　　The results of the Nemenyi and Bonferroni-Dunn tests are presented on critical distance diagrams. Each diagram shows the average rank of the algorithm over the considered datasets, and the critical distance, i.e., the distance for which average ranks of two considered algorithms must differ to be considered statistically significantly different. Additionally, the groups of algorithms among which no statistically significant differences occur are connected with a line.

　　Before proceeding with the statistical analysis, we round the performances to three decimal points. In the analysis, the significance level was set to $\alpha = 0.05$.

4.6　Parameter Instantiation

Since the sizes of datasets range over different orders of magnitude, the number of iterations m is given as the proportion of the size of $\mathscr{D}_{\text{TRAIN}}$. The considered values are $m \in \{1\%, 5\%, 10\%, 25\%, 50\%, 100\%\}$. On the other hand, since the number of neighbors K controls the level of locality, it is better given in absolute values. Our choice is to consider the following values $K \in \{1, 5, 10, 15, 20, 25, 30, 40\}$.

5　Results

5.1　Does the Distance Matter?

To give every distance as good chance as possible, we compute and evaluate feature rankings for all combinations of the parameters m and K and for every dataset and distance version, the best pair (with respect to the evaluation measure at hand) is chosen.

　　Friedman test rejected the null hypothesis for three of the four evaluation measures that the distance definitions are part of: *Accuracy* ($p = 5.2 \cdot 10^{-4}$), F_1 *Score* ($p = 3.5 \cdot 10^{-4}$) and *Subset Accuracy* ($p = 0.011$). In the case of *Hamming Loss*, the performances are not statistically significantly different ($p = 0.28$). The Bonferroni-Dunn test reveals that $d_{Hamming}$ performs statistically significantly worse than the other three distances, for the evaluation measures *Accuracy* (Fig. 1a) and F_1 *Score* (with qualitatively the same diagram). In the case of

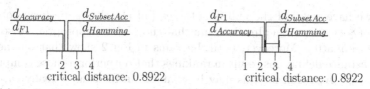

(a) Evaluation measure: *Accuracy* (b) Evaluation measure: *Subset Accuracy*

Fig. 1. Comparison of the four distance functions in terms of (a) *Accuracy*, and (b) *Subset Accuracy*: Critical distance diagrams from Bonferroni-Dunn test with the baseline $d_{Hamming}$.

Subset Accuracy, it has still the worst performance, but it is not statistically significantly worse than $d_{SubsetAcc}$ (Fig. 1b). Interestingly enough, the hypotheses was not rejected for the *Hamming Loss* evaluation measure. Also in this case, the rankings with $d_{Hamming}$ have the worst average rank of 2.9 (as compared to the best average rank of 2.1 that belongs to $d_{Accuracy}$), which leads us to a conclusion that the rankings with $d_{Hamming}$ are indeed to some extent optimized for *Hamming Loss*, but not sufficiently. Average ranks for this four measures are shown on the radar plot in Fig. 2a.

The average ranks of the feature rankings with respect to the other four measures are shown in Fig. 2b. Here, the null hypothesis H_0 is rejected in the case of *Precision* ($p = 0.0011$) and *Recall* ($p = 3.8 \cdot 10^{-4}$). This is not that surprising, since optimizing for F_1 *Score* should directly result in optimized *Precision* and/or *Recall*, as noted after the definition of d_{F1} (Eq. (5)). The results of the follow-up Bonferroni-Dunn tests are similar to those for *Subset Accuracy*: rankings obtained with $d_{Hamming}$ have the worst rank, but are not statistically significantly worse than those obtained with $d_{SubsetAcc}$. Additionally, H_0 is also rejected in the case of *pooledAUPRC* ($p = 0.027$), but in this case, no ranking is statistically significantly different from the one that corresponds to $d_{Hamming}$.

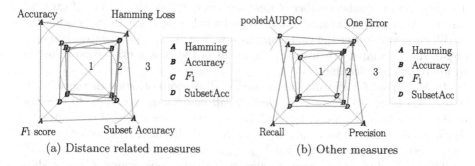

(a) Distance related measures   (b) Other measures

Fig. 2. Average ranks of the rankings computed with the four distance functions (denoted by A, B, C and D), in terms of measures that (a) are, and (b) are not directly related to any of distances.

Since we have rejected the null hypotheses (all algorithms perform equally well) in 6 of 8 cases, we can already claim that choosing an appropriate distance measure does matter. Moreover, both diagrams in Fig. 2 show that our newly proposed distance definitions result in rankings that outperform those computed with $d_{Hamming}$. A reason for this may be that the latter cannot really capture the possible interactions between the labels since it can be decomposed to the per-label distances, as noted in Sect. 3.1. This may also be the reason why the rankings computed with the newly proposed distances are typically closer to each other than to the rankings computed with $d_{Hamming}$.

To detect the differences among the rankings, we also apply Nemenyi post-hoc test. In addition to the relations discovered with Bonferroni-Dunn test, we now know that there is statistically significant difference between d_{F1} and $d_{SubsetAcc}$, when the quality is measured in terms of *pooledAUPRC*.

5.2 Are the Obtained Rankings Relevant?

To answer this question, we partially repeat the analysis from the previous section: in addition to the evaluation of the four ranking types, also the non-weighted 15NN algorithm is evaluated. If we reject the null hypothesis H_0 with Friedman test, the four rankings are compared to the non-weighted 15NN classifier with Bonferroni-Dunn post-hoc test. If there is a statistically significant difference between the weighted 15NN classifier and non-weighted 15NN classifier (in favour of the weighted one), we proclaim the ranking relevant.

H_0 is rejected for all evaluation measures. The corresponding Bonferroni-Dunn tests identifies the following. The distances $d_{Accuracy}$ and d_{F1} always result in relevant rankings. The distance $d_{SubsetAcc}$ fails to result in relevant rankings in the case of *One Error*. The distance $d_{Hamming}$ results in relevant rankings when the quality is measured in terms of *Subset Accuracy* and *pooledAUPRC*.

5.3 Influence of the Parameters m and K

To assess how does the number of iterations m influence the quality of ranking, we choose one of the distance functions and a value for the number of neighbors K. When m varies over the values specified in Sect. 4.6, six different rankings are obtained. We compare their quality in terms of the chosen evaluation measure, by applying the Friedman test.

H_0 is rejected for all values of m and for all versions of target distance in the case of *Accuracy*, *F_1 Score*, *Precision*, *Recall* and *Subset Accuracy*. In the case of *Hamming Loss*, it is never rejected. In the case of *One Error*, it is rejected for d_{F1} when $K \geq 25$ and for $d_{SubsetAcc}$ when $K = 40$. In the case of *pooledAUPRC*, the hypothesis is only rejected for $d_{SubsetAcc}$ when $m = 40$.

The only values of m which are always in the top performing group of algorithms, are 25%, 50% and 100%. A typical critical distance diagram (for $d_{Accuracy}$ and $K = 20$) is shown in Fig. 3a.

To assess the influence of the number of Relief neighbours K, a similar analysis is performed, now with the interchanged roles of m and K: the former is

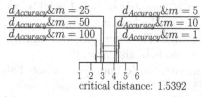

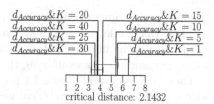

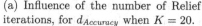

critical distance: 1.5392

critical distance: 2.1432

(a) Influence of the number of Relief iterations, for $d_{Accuracy}$ when $K = 20$.

(b) Influence of the number of Relief neighbors, for $d_{Accuracy}$ when $m = 25\%$.

Fig. 3. Critical distance diagrams from Nemenyi tests that show the influence of the number of (a) iterations, and (b) neighbors, on the quality of the $d_{Accuracy}$ rankings, measured in terms of *Precision*.

fixed and the latter varies. The summary of the results is as follows. Number of neighbors seems to have a lesser influence on the quality, since we do not reject all hypotheses for any of the evaluation measures. However, this is mostly due to the fact that K almost never statistically significantly influence the quality of the $d_{Hamming}$ rankings. For the other distances, the hypothesis is always rejected when the quality is measured in terms of *Accuracy*, F_1 *Score*, *Precision*, *Recall*. This also holds for *Subset Accuracy* with two exceptions for $d_{SubsetAcc}$: $m \in \{1\%, 100\%\}$. Again, no hypothesis is rejected in the case of *Hamming Loss*.

Typically, more is better regarding the number of neighbors and the highest values of K, i.e., $K \in \{30, 40\}$ have often the best average rank. This can be explained by the sparsity of the labels. To properly asses the average label space distance in $\mathscr{D}_{TRAIN}$, one has to consider larger neighborhoods. However, the differences among the algorithms for which $K \geq 15$ are not statistically significant. A typical situation (for $d_{Accuracy}$ and $m = 25\%$) is shown in Fig. 3b.

6   Conclusions and Future Work

In this paper, we propose the use of three distance measures on the target space within an extension of RReliefF approach to feature ranking for MLC tasks. These are the distances that are used within the evaluation measures *Accuracy*, F_1 *Score* and *Subset Accuracy* for predictive performance on MLC tasks. We have shown that using any of these distances always results in rankings of higher quality than the rankings computed with the distance used in the evaluation measure *Hamming Loss* [18]. Additionally, the newly proposed measures outperform the old one in terms of *Precision* and *Recall*, since these two are directly connected to the F_1 *Score*. For more independent measures, such as *pooledAUPRC* and *One Error* we did not observe any differences, so we can conclude that the use of the proposed distance within RReliefF optimizes the corresponding MLC evaluation measures.

We have also shown that all proposed rankings are relevant by comparing the nearest neighbor classifier that uses feature relevance information, to the standard nearest neighbor classifier. Additionally, we measure the influence of the parameters m (number of Relief iterations) and K (number of Relief neighbors) and show that rankings computed from $m = 25\%$ of the training dataset cannot be statistically significantly outperformed on average. The same goes for rankings that were computed by examining the neighborhoods of size $K = 15$.

There are several directions for future work. We plan to find appropriate distance measures for the hierarchical version of the MLC task: hierarchical multi-label classification. Incorporating probabilities in the distances, the RELIEF family can be also extended in the direction of data with missing labels and semi-supervised problems. Once these are solved, we also plan to develop an extension of Relief for seemingly much harder context of unsupervised learning, where there are no target variables and the analogous approach cannot be taken.

References

1. UC Berkeley Enron Email Analysis Project. http://bailando.sims.berkeley.edu/enron_email.html (2018). Accessed 28 June 2018
2. Boutell, M.R., Luo, J., Shen, X., Brown, C.M.: Learning multi-label scene classification. Pattern Recognit. **37**(9), 1757–1771 (2004)
3. Briggs, F., et al.: The 9th annual mlsp competition: new methods for acoustic classification of multiple simultaneous bird species in a noisy environment. In: IEEE International Workshop on Machine Learning for Signal Processing, MLSP 2013, pp. 1–8 (2013)
4. Dembczyński, K., Waegeman, W., Cheng, W., Hüllermeier, E.: On label dependence and loss minimization in multi-label classification. Mach. Learn. **88**(1), 5–45 (2012)
5. Demšar, J.: Statistical comparisons of classifiers over multiple data sets. J. Mach. Learn. Res. **7**, 1–30 (2006)
6. Diplaris, S., Tsoumakas, G., Mitkas, P.A., Vlahavas, I.: Protein classification with multiple algorithms. In: Bozanis, P., Houstis, E.N. (eds.) PCI 2005. LNCS, vol. 3746, pp. 448–456. Springer, Heidelberg (2005). https://doi.org/10.1007/11573036_42
7. Duygulu, P., Barnard, K., de Freitas, J.F.G., Forsyth, D.A.: Object recognition as machine translation: learning a lexicon for a fixed image vocabulary. In: Heyden, A., Sparr, G., Nielsen, M., Johansen, P. (eds.) ECCV 2002. LNCS, vol. 2353, pp. 97–112. Springer, Heidelberg (2002). https://doi.org/10.1007/3-540-47979-1_7
8. Elisseeff, A., Weston, J.: A kernel method for multi-labelled classification. In: Dietterich, T.G., Becker, S., Ghahramani, Z. (eds.) Advances in Neural Information Processing Systems 14. Springer International Publishing (2001)
9. Guyon, I., Elisseeff, A.: An introduction to variable and feature selection. J. Mach. Learn. Res. **3**, 1157–1182 (2003)
10. Katakis, I., Tsoumakas, G., Vlahavas, I.: Multilabel text classification for automated tag suggestion. In: Proceedings of the ECML/PKDD 2008 Discovery Challenge (2008)
11. Kira, K., Rendell, L.A.: The feature selection problem: traditional methods and a new algorithm. In: Proceedings of the Tenth National Conference on Artificial Intelligence, pp. 129–134. AAAI'92, AAAI Press (1992)
12. Kocev, D., Vens, C., Struyf, J., Džeroski, S.: Tree ensembles for predicting structured outputs. Pattern Recognit. **46**(3), 817–833 (2013)

13. Kong, D., Ding, C., Huang, H., Zhao, H.: Multi-label ReliefF and F-statistic feature selections for image annotation. In: 2012 IEEE Conference on Computer Vision and Pattern Recognition (CVPR), pp. 2352–2359 (2012)

14. Kononenko, I., Robnik-Šikonja, M.: Theoretical and empirical analysis of ReliefF and RReliefF. Mach. Learn. J. **55**, 23–69 (2003)

15. Madjarov, G., Kocev, D., Gjorgjevikj, D., Džeroski, S.: An extensive experimental comparison of methods for multi-label learning. Pattern Recognit. **45**, 3084–3104 (2012)

16. Pestian, J.P., et al.: A shared task involving multi-label classification of clinical free text. In: Proceedings of the Workshop on BioNLP 2007: Biological, Translational, and Clinical Language Processing (BioNLP '07), pp. 97–104 (2007)

17. Petković, M., Džeroski, S., Kocev, D.: Feature ranking for multi-target regression with tree ensemble methods. In: Yamamoto, A., Kida, T., Uno, T., Kuboyama, T. (eds.) DS 2017. LNCS (LNAI), vol. 10558, pp. 171–185. Springer, Cham (2017). https://doi.org/10.1007/978-3-319-67786-6_13

18. Reyes, O., Morell, C., Ventura, S.: Scalable extensions of the ReliefF algorithm for weighting and selecting features on the multi-label learning context. Neurocomputing **161**, 168–182 (2015)

19. Snoek, C.G.M., Worring, M., van Gemert, J.C., Geusebroek, J.M., Smeulders, A.W.M.: The challenge problem for automated detection of 101 semantic concepts in multimedia. In: Proceedings of the 14th ACM International Conference on Multimedia, pp. 421–430. ACM, New York (2006)

20. Spolaôr, N., Cherman, E.A., Monard, M.C., Lee, H.D.: A comparison of multi-label feature selection methods using the problem transformation approach. Electron. Notes Theor. Comput. Sci. **292**, 135–151 (2013)

21. Srivastava, A.N., Zane-Ulman, B.: Discovering recurring anomalies in text reports regarding complex space systems. In: 2005 IEEE Aerospace Conference (2005)

22. Stańczyk, U., Jain, L.C. (eds.): Feature selection for data and pattern recognition. Studies in Computational Intelligence. Springer, Berlin (2015)

23. Trochidis, K., Tsoumakas, G., Kalliris, G., Vlahavas, I.: Multilabel classification of music into emotions. In: 2008 International Conference on Music Information Retrieval (ISMIR 2008), pp. 325–330 (2008)

24. Tsoumakas, G., Katakis, I.: Multi-label classification: An overview. Int. J. Data Warehous. Min. pp. 1–13 (2007)

25. Tsoumakas, G., Katakis, I., Vlahavas, I.: Effective and efficient multilabel classification in domains with large number of labels. In: ECML/PKDD 2008 Workshop on Mining Multidimensional Data (MMD'08) (2008)

26. Ueda, N., Saito, K.: Parametric mixture models for multi-labeled text. In: Advances in Neural Information Processing Systems 15, pp. 721–728. MIT Press (2003)

27. Vens, C., Struyf, J., Schietgat, L., Džeroski, S., Blockeel, H.: Decision trees for hierarchical multi-label classification. Mach. Learn. **73**(2), 185–214 (2008)

28. Wettschereck, D.: A study of distance based algorithms. Ph.D. thesis, Oregon State University, USA (1994)

Finding Probabilistic Rule Lists using the Minimum Description Length Principle

John O. R. Aoga[1]([⊠])(iD), Tias Guns[2,3], Siegfried Nijssen[1], and Pierre Schaus[1]

[1] ICTEAM, UCLouvain, Ottignies-Louvain-la-Neuve, Belgium
{john.aoga,siegfried.nijssen,pierre.schaus}@uclouvain.be
[2] VUB, Brussels, Belgium
[3] KU Leuven, Leuven, Belgium
tias.guns@vub.ac.be, tias.guns@cs.kuleuven.be

Abstract. An important task in data mining is that of rule discovery in supervised data. Well-known examples include rule-based classification and subgroup discovery. Motivated by the need to succinctly describe an entire labeled dataset, rather than accurately classify the label, we propose an MDL-based supervised rule discovery task. The task concerns the discovery of a small rule list where each rule captures the probability of the Boolean target attribute being true. Our approach is built on a novel combination of two main building blocks: (i) the use of the Minimum Description Length (MDL) principle to characterize good-and-small sets of probabilistic rules, (ii) the use of branch-and-bound with a best-first search strategy to find better-than-greedy and optimal solutions for the proposed task. We experimentally show the effectiveness of our approach, by providing a comparison with other supervised rule learning algorithms on real-life datasets.

1 Introduction

Rule learning in supervised data is a well-established problem in data mining and machine learning. Compared to many other methods, a clear benefit of rule-based methods is that the rule format is more easy to interpret, and hence is useful in knowledge discovery. Well-known examples of rule learning are

Rule-based classification, in which the aim is to find a set of rules that predicts the class of examples well;

Subgroup discovery, in which the aim is to find a set of rules that describes subgroups of examples in the data; in these subgroups, the distribution of the target attribute is different from the overall population.

The main difference between subgroup discovery and rule-based classification is that rule-based classification aims to find a set of rules that can be applied on

---

J.O.R. Aoga—This author is supported by the FRIA-FNRS (Fonds pour la Formation à la Recherche dans l'Industrie et dans l'Agriculture, Belgium).

L. Soldatova et al. (Eds.): DS 2018, LNAI 11198, pp. 66–82, 2018.
https://doi.org/10.1007/978-3-030-01771-2_5

Table 1. Probabilistic rule lists example

(a) From door opening data

	Rule list	Probability
IF	WEDNESDAY *and* MORNING	0.879
ELSE IF	HOLIDAY *and* THURSDAY	0.011
ELSE IF	THURSDAY *and* AFTERNOON	0.987
ELSE IF	SUNDAY	0.001
ELSE	(Default rule)	0.101

(b) From mushroom dataset

	Rule list	Probability
IF	Gill-spacing is closed *and* No odor	0.95
ELSE IF	Gill-spacing is closed *and* Stalk-shape is tapering	0.0
ELSE IF	Stalk-color-above-ring is white *and* Gill-size is broad	1.0
ELSE IF	Gill-spacing is closed	0.0
ELSE	(Default rule)	0.56

any example to obtain a prediction for that example. Subgroup discovery aims to characterize subgroups of examples, but not necessarily all examples.

Similar to rule-based classification, in this work we are also interested in finding a set of rules that describe a target attribute fully and in an interpretable manner. However, we make a specific assumption that is not common in rule-based classification: we assume that the class attribute has a skewed distribution, and that exact prediction is certainly not possible. The following example illustrates a problem that has these characteristics.

Example 1. Assume that we characterize every minute in a year in terms of the following attributes: the part of the day the minute belongs to (morning, afternoon), the day the minute belongs to (Sunday, Monday, ...), the month the minute belongs to (January, ...) and the minute of the day $(1, 2, ..., 24 \times 60)$; furthermore, over a year we use a sensor to monitor when an individual opens a specific door in his house. Can we use rules to characterize when this individual opens her door?

In this example, the event of "opening a door" is expected to be a rare event; if we use a classification algorithm on the above dataset, we will notice that the class attribute is very unbalanced. Most classification algorithms will either prefer to always predict the default label (the door is closed), or will construct many very specific rules to cover the small number of examples that are the exception. The reason for this is that many rule-based classifiers find *lists* of rules; a rule that makes an error in its prediction, cannot be corrected by a later rule. Hence, most classification rule learning algorithms favor rules with lower recall but high precision.

In this paper, we propose a new algorithm for finding rule lists, designed to work well in this specific setting. It identifies simple probabilistic rule lists, such as in Table 1. Hence, the rule mining setting studied in this work can be characterized by these properties:

- it learns rules with probabilities in the head; these probabilities represent the class distribution for the examples covered by the rule, and should not be understood as class prediction;
- the list of rules is intended to characterize the class distribution over the entire data, in contrast to subgroup discovery;
- it favors smaller rule lists to ease interpretation.

Finding lists of rules that satisfy these requirements is not a straightforward task. To address these challenges, this paper proposes the following contributions.

1. We propose a new optimization criterion based on the *Minimum Description Length* (MDL) principle; this criterion aims to find rule lists that are small, yet characterize the target distribution well.
2. We propose a new search algorithm based on branch-and-bound search; this search algorithm aims to find the global optimum for the proposed optimization criterion under given constraints.

The approach that we take in this work is a *pattern set mining* approach. We first use itemset mining algorithms to find a candidate set of itemsets. From this set, we select a subset that describes the target attribute well. From the pattern set mining perspective, we propose a new supervised optimization criterion for selecting a set of *free* patterns, and a new search algorithm for finding a set of patterns that optimizes the criterion.

In the remainder of the paper, we first present related work in Sect. 2. In Sect. 3 we present the problem of finding probabilistic rule lists. Then, we describe our Minimum Description Length (MDL)-based approach in terms of the formalization and algorithms for solving it. Finally, we show experiments in Sect. 5 before concluding.

2   Related Work

This work builds on a number of areas in the literature.

Rule-based classification. There is a large literature on rule-based classification; a good overview of these algorithms, including classic algorithms such as CN2 and RIPPER, can be found in a textbook by Fürnkranz et al. [4]. Two types of rule-based classifiers can be distinguished: classifiers based on rule sets and on rule lists. In set-based classifiers, all rules that match an example are used to obtain a prediction for that example. In list-based classifiers, the first matching rule is used; we build on this class of methods.

Covering algorithms are the most popular type of rule learning algorithm. These algorithms iteratively search for a rule to add to a rule set or list. Most often, in each iteration a greedy search algorithm is used, which constructs a rule by iteratively adding the condition that improves the quality of the rule the most.

The main challenge faced by pure covering algorithms is that later rules cannot correct errors made by earlier rules in a rule list. Such algorithms hence need to favor precision over recall to obtain accurate classifiers. As a result rule lists may become unnecessarily long. One way to solve this is using *pruning*: the rule set is reduced in a post-processing step.

Pattern-based classification. Compared to traditional rule learning algorithms, pattern-based classifiers use pattern mining algorithms, such as frequent itemset mining algorithms, to identify candidate rules [12]. These frequent item-sets are post-processed to construct rule sets or rule lists. Most of these post-processing approaches use heuristic search algorithms, although the use of exact search has also been studied [6].

Pattern set mining. From a pattern mining perspective, selecting a small set of patterns from a larger set of patterns can be seen as a pattern set mining problem [12]. In contrast to unsupervised methods, supervised methods aim to find a balance between pattern sets that are non-redundant and that are accurate. One popular approach for evaluating the quality of a pattern set is based on the Minimum Description Length principe, as pioneered in the unsupervised setting by the KRIMP algorithm [10]. Exact methods for pattern set mining were studied by Guns et al. [6], among others, but these studies did not consider scoring functions based on MDL or did not exploit freeness, as we do.

Subgroup discovery. Strongly related to both pattern mining and rule-based classification is subgroup discovery. Subgroup discovery differs from classification in that it does not aim to build a predictive model; rather, subgroup discovery algorithms are intended to return small and interpretable sets of *local* patterns; subgroups are not necessarily ordered in a specific manner. For this reason, traditional subgroup discovery algorithms were modifications of covering based rule-learning algorithms to explicitly allow for overlap between patterns [7].

Bayesian rule lists. Most related to this work is recent work by Yang et al. [11] on probabilistic rule lists. This work also finds ordered lists of probabilistic rules. Contrary to our work, however, the aim of the work of Yang et al. is to identify accurate classifiers, and not to identify as small and interpretable representations of the class distribution as possible. Furthermore, Yang et al. use a sampling based algorithm to identify good sets of patterns. We propose an alternative, exact algorithm in this work.

3   The Probabilistic Rule List Mining Problem

This work is motivated by the creation of a probabilistic rule list that summarizes labeled data well. In order to be easily interpretable, the rule list and the individual rules should be concise.

We assume the data is described by a set of discrete attributes. These attributes can be represented as a set of Boolean properties using a one-hot encoding. These properties are referred to as items in the following, in line with the itemset mining literature.

More formally, let $\mathcal{I} = \{1, \cdots, m\}$ represent a set of m possible items and let $\mathcal{F} \subseteq 2^{\mathcal{I}}$ be a set of itemsets built on those items. A probabilistic rule list (PRL) built on $\mathcal{F}$ is a sequence of rules of the form $\mathcal{R} = \langle (I^{(1)}, p^{(1)}), (I^{(2)}, p^{(2)}), \cdots, (I^{(k)}, p^{(k)}) \rangle$ with p^i being a probability and $I^i \in \mathcal{F}, \forall i = 1, \ldots, k-1$ and $I^k = \emptyset$. This latter is the default rule. The sequence of itemsets in the rule list can be expressed as membership to the regular language: $\langle I^1, \ldots, I^k \rangle \in \mathcal{L}(\mathcal{F}^* \cdot \emptyset)$ with

$\mathcal{F}^*$ the Kleene operator on $\mathcal{F}$. Table 1 shows two example rule lists (generated from different data).

The rule list has a sequential interpretation, in that the set of data instances that match the first rule I^1 are assumed to have a positive label with probability p^1. The other data instances, those that do not match I^1, but do match I^2 have a probability of p^2 to be positive, etc. The final empty set $I^k = \emptyset$ hence captures all instances not matched by the other rules.

We now formalize the problem of creating the probabilistic rule list based on $\mathcal{F}$ and a dataset $\mathcal{D}$.

Definition 1. *As input we receive a set of itemsets $\mathcal{F}$ that can be used to compose the rule list, and a database $\mathcal{D}$ of instances, with for each a Boolean target attribute: $\mathcal{D} = \{(t, I_t, a_t) \mid t \in \mathcal{T}, I_t \subseteq \mathcal{I}, a_t \in \{+, -\}\}$, where the set $\mathcal{T}$ contains the instance or transaction identifiers $\mathcal{T} = \{1, \ldots, n\}$. The database can be split into a positive $\mathcal{D}^+$ and negative $\mathcal{D}^-$ database, based on the target attribute value (+ or −).*

The problem of finding a probabilistic rule list is formalized as: $\mathrm{argmin}_{\mathcal{R}} \, \mathrm{score}(\mathcal{R}, \mathcal{F}, \mathcal{D})$ where $\mathcal{R} = \langle (I^{(1)}, p^{(1)}), (I^{(2)}, p^{(2)}), \cdots, (I^{(k)}, p^{(k)}) \rangle$ is a probabilistic rule list such that $\langle I^{(1)}, \ldots, I^{(k)} \rangle \in \mathcal{L}(\mathcal{F}^* \cdot \emptyset)$ and score is an optimization criterion. Various optimization criteria can be defined, including criteria inspired by classification rule learning, subgroup discovery and pattern set mining. Our aim in this work is to develop an optimization criterion that explicitly favors smaller rule lists that describe the entire target distribution well. For this purpose, we will use the Minimum Description Length principle, discussed in the next section.

4   Discovering Probabilistic Rule Lists

4.1   Coverage and Probability of a Rule List

To evaluate the quality of a rule set on a given dataset, we will use a number of concepts taken from the itemset mining literature [1].

Definition 2 (Coverage and support of an itemset). *The set of transactions in a database $\mathcal{D}$ containing an itemset I is called the cover: $\varphi(\mathcal{D}, I) = \{(t, I_t, a_t) \in \mathcal{D} \mid I \subseteq I_t\}$. The size of the cover is called the support $\psi(\mathcal{D}, I) = |\varphi(\mathcal{D}, I)|$.*

Example 2. An example itemset database is given in Fig. 1a. $I = \{A, C\}$ is an example itemset; $\varphi(\mathcal{D}, I)$ contains transaction identifiers $\{1, 2, 5\}$, so $\psi(\mathcal{D}, I) = 3$. The set of frequent itemsets with support at least 4 is $\{\emptyset, \{A\}, \{B\}, \{C\}, \{E\}, \{B, E\}\}$ (Fig. 1b).

In the remainder of this paper, for the sake of simplicity we denote $\varphi(\mathcal{D}, I)$ as $\varphi(I)$ when no ambiguity regarding $\mathcal{D}$ is possible. Similarly, we will use $\varphi^+(I)$ to

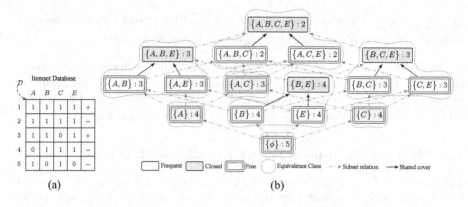

Fig. 1. (a) Itemset Database with positive/negative classes; **(b)** Powerset lattice of $\mathcal{D}$ with equivalence classes.

denote $\varphi(\mathcal{D}^+, I)$ where $\mathcal{D}^+ = \{(t, I_t, a_t) \in \mathcal{D} \mid a_t = +\}$ and likewise for $\varphi^-(I)$ with $a_t = -$.

We are interested in finding a list of rules. Each itemset in the list has a cover that is defined as follows.

Definition 3 (Coverage of an itemset in a sequence). *Assume the sequence of itemsets* $\langle I^{(1)}, \ldots, I^{(k)} \rangle$, *the coverage of an itemset* $I^{(j)}$ *over* $\mathcal{D}$ *is its cover in the database of transactions not covered by the previous itemsets* $I^{(1)}, I^{(2)}, \ldots, I^{(j-1)}$:

$$\Phi(\mathcal{D}, \langle I^{(1)}, \ldots, I^{(k)} \rangle, j) = \varphi\left(\mathcal{D} \setminus \left(\varphi(I^{(1)}) \cup \varphi(I^{(2)}) \cup \cdots \cup \varphi(I^{(j-1)})\right), I^{(j)}\right) \quad (1)$$

with $\Phi(\mathcal{D}, \langle I^{(1)}, \ldots, I^{(k)} \rangle, 1) = \varphi(\mathcal{D}, I^{(1)})$.

Note that in a rule list $\mathcal{R}$, the last itemset is always $I^{(k)} = \emptyset$, which is the *default rule* or final *else-case*. This *empty set* inherently covers all instances not covered by any of the $k - 1$ previous rules since $\varphi(\mathcal{D}, \emptyset) = \{(t, I_t, a_t) \in \mathcal{D} \mid \emptyset \subseteq I_t\} = \mathcal{D}$ for any $\mathcal{D}$.

Given a rule list $\mathcal{R} = \langle (I^{(1)}, p^{(1)}), (I^{(2)}, p^{(2)}), \cdots, (I^{(k)}, p^{(k)}) \rangle$ we will denote by $\Phi(\mathcal{D}, \mathcal{R}, j)$ the cover of the jth itemset in the rule list's sequence of itemsets. If no ambiguity is possible we simply write Φ_j. Similarly $\Phi_j^+ = \Phi(\mathcal{D}^+, \mathcal{R}, j)$ and $\Phi_j^- = \Phi(\mathcal{D}^-, \mathcal{R}, j)$.

When creating a rule list $\mathcal{R}$ from a dataset $\mathcal{D}$ given $\mathcal{F}$, we define the probability $p^{(j)}$ of a rule $I^{(j)}$ as $p^{(j)} = P(a_t = + | (t, I_t, a_t) \in \Phi(\mathcal{D}, \mathcal{R}, j)) = \frac{|\Phi_j^+|}{|\Phi_j^+| + |\Phi_j^-|}$.

Example 3. Assume the running example database (Fig. 1a) and a rule list with corresponding sequence of itemsets $\langle \{A, B, C\}, \{C\}, \emptyset \rangle$. The coverage of $I^{(2)} = \{C\}$ over $\mathcal{D}$ is $\Phi_2 = \{4, 5\}$, instead of $\{1, 2, 4, 5\}$, as the transactions 1 and 2 were already covered by $I^{(1)} = \{A, B, C\}$. Its probability is hence $p^{(2)} =$

$\frac{|\varPhi_2^+|}{|\varPhi_2^+|+|\varPhi_2^-|} = \frac{0}{0+2} = 0$, which indicates that no positive transaction was observed with the condition of the rule, after observing the previous rules.

At this stage, an open question is how to evaluate the quality of a probabilistic rule list $\mathcal{R}$. In this work, we propose to evaluate how well the rule list allows to *compress* the values for the class attribute observed in a training dataset. For this, we will use the Minimum Description Length (MDL) principle.

4.2   Minimum Description Length Encoding of Rule Lists

The Minimum Description Length (MDL) principle [5,8] is a general method for *inductive inference*, based on the idea that '*the more we can compress the data, the more there are regularities in it and the more we learn from it*' [5]. MDL allows making a trade-off between the complexity of rules and their ability to capture the distribution of the class attribute. To do this, we use a two-part code that minimizes the *number of bits* needed to encode the data with a model, as well as the number of bits to encode the model itself. As stated earlier, the focus in this work is on a code that favors simplicity.

Let $\mathcal{M} = M_1, M_2, \ldots$ be a list of model candidates. In two-part MDL, the best model $M \in \mathcal{M}$ to capture information in a given database $\mathcal{D}$ is the one which minimizes the code length $L(M) = L_{model}(M) + L_{data}(\mathcal{D}|M)$, where $L_{model}(M)$ is the length, in bits, of the description of the model itself and $L_{data}(\mathcal{D}|M)$ the length of the data, in bits, when it is encoded with this model.

In our case, models correspond to rule lists of the form $\mathcal{R} = \langle (I^{(1)}, p^{(1)}), (I^{(2)}, p^{(2)}), \cdots, (I^{(k)}, p^{(k)}) \rangle$ with $I^{(j)} \in \mathcal{F}, \forall j \in 1, \ldots, k-1, I^{(k)} = \emptyset$ and $p^{(j)} = \frac{|\varPhi_j^+|}{|\varPhi_j^+|+|\varPhi_j^-|}$. We thus need to define an encoding with $L_{model}(\mathcal{R})$ an encoding of the rule list, and $L_{data}(\cdot|\mathcal{R})$ such that $L_{data}(\mathcal{D}|\mathcal{R})$ can be interpreted as the *coding length* of the distribution of $+/-$'s in $\mathcal{D}$ when it is encoded with $\mathcal{R}$. The best rule list is then the one that minimizes the total length $L(\mathcal{R})$:

$$\mathcal{R}^* = \operatorname*{argmin}_{\mathcal{R} \in \mathcal{L}(\mathcal{F}^* \cdot \emptyset)} L_{data}(\mathcal{D}|\mathcal{R}) + L_{model}(\mathcal{R}), \tag{2}$$

where we identified $\mathcal{R}$ by its sequence of itemsets to ease notation; each itemset has a probability $p^{(j)}$ as defined earlier.

We now first discuss how we encode $\mathcal{R}$ when $k \leq 2$ (i.e. $\mathcal{R} = \langle (\emptyset, p^{(1)}) \rangle$ or $\mathcal{R} = \langle (I^{(1)}, p^{(1)}), (\emptyset, p^{(2)}) \rangle$) and then generalize to the case $k > 2$.

Case $k = 2$: To understand the computation of the coding length of $\mathcal{R}$, we first show how we can encode a target attribute if we have an itemset I and then a *default rule*. Given a rule $(I^{(1)}, p^{(1)})$, we assume that the positive and negative labels in $\varphi(\mathcal{D}, I^{(1)})$ follow a *Bernoulli distribution*, with a probability $p^{(1)}$ for the class label. The probability density of the labels according to I is hence (omitting $\mathcal{D}$ from the notation):

$$Pr\big(a_t = + \mid \varphi(I)\big) = (p^{(1)})^{|\varphi^+(I)|}(1 - p^{(1)})^{|\varphi^-(I)|}. \tag{3}$$

Theorem 1 (Local Coding length of data). *Using Shannons Noiseless Channel Coding Theorem [3] the number of bits needed to encode the class labels of $\mathcal{D}$ using I is at least the logarithm[1] of the probability density of the class labels in $\mathcal{D}$ given I: $L_{local\,data}(\mathcal{D}|I) = -\log P_r(a_t = + \mid \varphi(\mathcal{D}, I))$. Using (3) we can hence encode each positive label at a cost of*

$$L_{local\,data}(\mathcal{D}|I) = \mathcal{Q}(\varphi^+(I), \varphi^-(I)) + \mathcal{Q}(\varphi^-(I), \varphi^+(I)), \qquad (4)$$

with $\mathcal{Q}(a,b) = -a \log \frac{a}{a+b}$.

We will use this bound, which can be approximated closely using arithmetic coding, as the coding length for the class labels. Based on the above theorem and assuming a rule list is $\mathcal{R} = \langle (I^{(1)}, p^{(1)}), (\emptyset, p^{(2)}) \rangle$, the coding length of Φ is

$$L_{data}(\mathcal{D}|\mathcal{R}) = L_{local\,data}(\mathcal{D}|I^{(1)}) + L_{local\,data}(\mathcal{D} \setminus \varphi(I^{(1)})|\emptyset) \qquad (5)$$

Example 4. Assume the rule list is $\mathcal{R} = \langle (\{A, B, C\}, 0.50), (\emptyset, 0.33) \rangle$ and that our database $\mathcal{D}$ (Fig. 1a) is duplicated 256 times. $L_{local\,data}(\mathcal{D}|\{A, B, C\}) = -256 \log 0.5 - 256 \log(1 - 0.5) = 512bits$ and $L_{local\,data}(\mathcal{D} \setminus \varphi(I^1)|\emptyset) = -256 \log 0.33 - 512 \log(1 - 0.33) = 705bits$; then $L_{data}(\mathcal{D}|\mathcal{R}) = 1217bits$.

When we encode the class label using this model, we do not only need to encode the data, but also the model itself.

Definition 4 (Length of the model). *Assume a rule list $\mathcal{R} = \langle (I^{(1)}, p^{(1)}), (\emptyset, p^{(2)}) \rangle$, we represent $(I^{(1)}, p^{(1)})$ as a string "$m_1\, I_1^{(1)}\, \dots\, I_{m_1}^{(1)}\, n_1^+$" where, $m_1 = |I^{(1)}|$ is the number of items in $I^{(1)}$, followed by the identifiers of each item in $I^{(1)}$ and finally the number of positive labels in $\mathcal{D}$: $n_1^+ = |\varphi^+(I^{(1)})|$. The length, in bits, to encode this string is:*

$$L_{local\,model}(I^{(1)}) = \underbrace{\log m}_{|I^{(1)}|} + \underbrace{|I^{(1)}| \log m}_{I_1^{(1)} \dots I_{|I^{(1)}|}^{(1)}} + \underbrace{\log n}_{n_1^+}, \qquad (6)$$

where $\log m$ bits are required to represent m_1, as $m_1 \leq m = |I|$, and also $\log m$ bits for each item identifier plus $\log n$ bits to encode n_1^+. Coding n_1^- is unnecessary as it can be retrieved from the data using the itemset: $n_1^- = |\varphi(\mathcal{D}, I^{(1)})| - n_1^+$. From there, assuming that the itemset database $\mathcal{D}$ and the set of items $\mathcal{I}$ are known, one can easily retrieve the coverage of $I^{(1)}$ and then compute the probability $p^{(1)}$ using the number of positive labels n_1^+. The coding length of the model $\mathcal{R}$ is $L_{model}(\mathcal{R}) = L_{local\,model}(I^{(1)}) + L_{local\,model}(\emptyset)$.

Example 5. We continue on Example 4. To encode the model, the string "3 A B C 256" is encoded: $L_{local\,model}(\{A, B, C\}) = \log 4 + 3 \log 4 + \log 1280 = 19bits$ similarly $L_{local\,model}(\emptyset) = \log 4 + 0 \log 4 + \log 1280 = 13bits$[2] then $L_{model}(\mathcal{R}) = 32bits$. Together with $L_{data}(\mathcal{D}|\mathcal{R}) = 1217bits$ computed in Example 4, the total coding length of $\mathcal{R}$ is $L(\mathcal{R}) = 1217 + 32 = 1249bits$.

---

[1] All logarithms are to base 2 and by convention, we use $0 \log 0 = 0$.
[2] Note that by convention the size of the default rule is $m_2 = 0$.

Case $k > 2$: Assuming now a rule list $\mathcal{R} = \langle (I^{(1)}, p^{(1)}), (I^{(2)}, p^{(2)}), \cdots, (I^{(k)}, p^{(k)}) \rangle$ with $k > 2$. For $k > 1$ we need to modify the definition of $L_{local\,data}$ such that it does not consider parts of the data covered by a previous itemset in the sequence. Hence,

$$L_{local\,data}(\mathcal{D}|I^{(j)}) = \mathcal{Q}(\Phi_j^+, \Phi_j^-) + \mathcal{Q}(\Phi_j^-, \Phi_j^+) \tag{7}$$

and the total coding length is the summation of local lengths:

$$L_{data}(\mathcal{D}|\mathcal{R}) = \sum_{j=1}^{k} L_{local\,data}(\mathcal{D}|I^{(j)}); \tag{8}$$

the coding length of the model is:

$$L_{model}(\mathcal{R}) = \log n + \sum_{j=1}^{k-1} \left(\log m + m_j \log m + \log n \right) \tag{9}$$

To encode the size of $\mathcal{R}$ itself, we need $\log n$ bits. Because all rule list include the *default rule*, we omit these $\log m + \log n$ bits.

Example 6. Fig. 2 shows example rule lists with coding lengths.

4.3   Coding Length Related to Likelihood and Quality of Rule Lists

The coding length of the class labels given a model $\mathcal{R}$ is the number of bits needed to encode the class labels with $\mathcal{R}$. As a consequence of our choice to use Shannon's theorem, this coding length corresponds to the *(-log) likelihood* of the class labels according to the model. In the other words, if we would minimize the coding length of the data only, we would maximize the likelihood of the data under the model. However, as stated earlier, in this work our aim is also to find small and interpretable rule lists. We choose our code such that a relatively large weight is given to the complexity of the model.

Assuming the database of Example 4, the size of the original data is $5 \times 256 = 1280$. Encoding this data with $\mathcal{R}_1 = \langle (\{A, B, C\}, 0.50), (\emptyset, 0.33) \rangle$ we obtained $L_{data}(\mathcal{D}|\mathcal{R}_1) = 1217bits$, $L_{model}(\mathcal{R}_1) = 32bits$ and in total $L(\mathcal{R}_1) = 1249bits$. Instead, when we encode this data with $\mathcal{R}_2 = \langle (\emptyset, 0.40) \rangle$ we obtain $L_{data}(\mathcal{D}|\mathcal{R}_2) = 1243bits$, $L_{model}(\mathcal{R}_2) = 6bits$ and in total $L(\mathcal{R}_2) = 1249bits$. Looking at likelihoods only, one can see that $\mathcal{R}_1$ is a better model for representing this data, as it captures more information than $\mathcal{R}_2$. However, in total, it is not preferable over $\mathcal{R}_2$, since it is more complex to encode. The model coding length penalizes the likelihood and ensures a simple model is preferred.

For our example, the only way to improve $\mathcal{R}_1$ is to add (if possible) a new rule that reduces the error made by $\mathcal{R}_1$ by assuming that the part not covered by $\{A, B, C\}$ is for the *default rule*. Thus, by adding the itemset $\{C\}$ to $\mathcal{R}_1$, which covers all 0s still present, we obtain the best model $\mathcal{R} = \langle (\{A, B, C\}, \frac{1}{2}), (\{C\}, \frac{9}{2}), (\phi, \frac{1}{1}) \rangle$ with $L(\mathcal{R}) = 546bits$ since the *default rule* now covers only remaining 1s.

4.4   A Greedy Algorithm

The probabilistic rule list that minimizes the MDL score (2) can be constructed greedily, extending the list by one rule at each step. Greedy algorithms are known to be efficient and approximate optimal solutions well in other rule learning tasks.

Algorithm 1 shows a greedy algorithm that starts with empty rule list $\mathcal{R}$, and then iteratively finds within a given set of patterns the rule that minimizes the coding length. The local best rule is obtained by considering at each iteration the sub-problem of finding the optimal rule list with $k \leq 2$ on the remaining data. This corresponds to finding the itemset $I^{(1)}$ such that the *coding length* is smallest (Line 3). Once the local best rule is selected the rule list is updated in Line 6 and in Line 7; its coverage is removed from $\mathcal{D}$. The process is then run again until $\mathcal{D}$ is empty or the *default rule* is selected.

Example 7. Assuming our running example, at the first iteration of the greedy algorithm, the minimum code-length $L(\langle\{A, B\}, \emptyset\rangle) = 722bits$ and then it is the greedy solution (See Fig. 2).

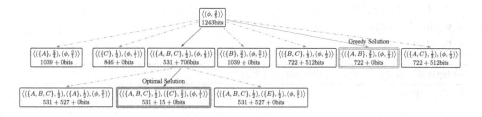

Fig. 2. Finding greedy and optimal solution base on the example of Fig. 1.

The greedy algorithm may be sub-optimal. For instance it fails to discover the $L(\langle\{A, B, C\}, \{C\}, \emptyset\rangle) = 546bits$ on our example.

4.5   Branch-and-Bound Algorithm

For finding solutions that are better than the greedy solution, we propose a best-first branch-and-bound algorithm that can prune away candidates based on a lower-bound on the MDL value. Each node in the search tree is a partial rule list, consisting of a sequence of rules *without* the *default rule*. The children of each node correspond to appending one additional rule from $\mathcal{F}$ to the partial rule list.

Algorithm 2 shows the pseudo-code of this branch-and-bound expansion search. For clarity we omit the probabilities in the rule list representation. The algorithm receives as input a list of rule candidates $\mathcal{F}$ and database $\mathcal{D}$. A priority queue is used to store the set of rule lists not yet expanded, ordered by the code-length obtained when extending the partial rule with the *default rule* (best-first strategy). The initial best rule is the *default rule* (Line 2) and the empty rule list is added as initial search node. As long as the queue is not empty, the priority queue is dequeued and the returned partial rule list is expanded (Line 6). Each new partial rule list is evaluated as if it was completed with the *default rule* ($\emptyset$) and checked whether it is better than the current best rule list (Lines 7, 8).

Before adding the new partial rule list to the queue, a lower-bound on the code length is computed, that is, an optimistic estimate of the code length achievable (see next section). Only if the lower-bound is better than the current best value, the rule list is added to the queue (Lines 9, 10). If not, this part of the search tree is effectively pruned.

Algorithm 1: $Greedy(\mathcal{F}, \mathcal{D})$

---

1 $\mathcal{R} \leftarrow \langle\rangle$
2 **do**
3 $\quad$ $I^* \leftarrow \underset{I \in \mathcal{F}^*}{\operatorname{argmin}} L(\langle (I, p^{(1)}), (\emptyset, p^{(2)}) \rangle)$
4 $\quad$ **if** $L(\langle (I, p^{(1)}), (\emptyset, p^{(2)}) \rangle) \geq L(\langle (\emptyset, p^{(1)}) \rangle)$ **then**
5 $\quad\quad$ $\lfloor\ I^* \leftarrow \emptyset$
6 $\quad$ $\mathcal{R} \leftarrow \mathcal{R} \cup (I^*, p^{(1)})$ $\quad\quad\quad\quad\quad$ ▷ *Add this rule to the rule list*
7 $\quad$ $\mathcal{D} \leftarrow \mathcal{D} \setminus \varphi(I^*)$
8 **while** $I^* \neq \emptyset$;
9 **return** $\mathcal{R}$

Algorithm 2: $Branch\text{-}and\text{-}bound$ $(\mathcal{F}, \mathcal{D})$

---

1 PQ : PriorityQueue $\quad\quad$ ▷ *Partial rule lists ordered by code-length when adding default rule*
2 $best\mathcal{R} \leftarrow \langle\emptyset\rangle,\ best \leftarrow L(best\mathcal{R})$
3 PQ.enqueue-with-priority($\langle\rangle, L(\langle\emptyset\rangle)$)
4 **while** $\mathcal{R} \leftarrow PQ.dequeue()$ **do**
5 $\quad$ **for** *each* $I \in \mathcal{F} \setminus \mathcal{R}$ **do**
6 $\quad\quad$ $\mathcal{R}' \leftarrow \langle \mathcal{R}, I \rangle$
7 $\quad\quad$ **if** $L(\langle \mathcal{R}', \emptyset \rangle) < best$ **then**
8 $\quad\quad\quad$ $\lfloor\ best\mathcal{R} = \langle \mathcal{R}', \emptyset \rangle,\ best \leftarrow L(best\mathcal{R})$
9 $\quad\quad$ **if** *lower-bound*($\mathcal{R}'$) $< best$ **then**
10 $\quad\quad\quad$ $\lfloor\ PQ$.enqueue-with-priority($\mathcal{R}', L(\langle \mathcal{R}', \emptyset \rangle)$)

11 **return** $best\mathcal{R}$

Lower-bound on a partial rule list A good lower-bound is difficult to compute since there is an exponential number of rules that can be added to the list. Because the rule list itself is already evaluated in the algorithm, we are seeking a lower-bound on any expansion of the rule list. The coding length is determined by $L(\mathcal{R}) = L_{model}(\mathcal{R}) + L_{data}(\mathcal{D}|\mathcal{R})$ according to (8) and (9).

The most optimistic expansion is hence achieved with the smallest possible expansion of the rule list yielding the greatest reduction of the coding length for the data. In the best case, this is a rule of length one ($|I^{(j+1)}| = 1$) that perfectly separates the positives from the negatives. In this case, the additional code length of the rule list corresponds to a rule of length one: $L_{local\,model}(I^{j+1}) = \log m + 1 \log m + \log n$ and the addition to the code length of the data is: $L_{local\,data}(\mathcal{D}|I^{j+1}) = \mathcal{Q}(|\varPhi^+_{j+1}|, 0) + \mathcal{Q}(0, |\varPhi^-_{j+1}|) = 0$ with the data coding length of the *default rule* also being 0.

While such a rule expansion may not exist, the resulting value is a valid lower-bound on the code length achievable by any expansion of the partial rule list. This is because any expansion has to be greater than or equal in size to 1, and any expansion will achieve at best a data compression of 0.

Implementation details *Choice of* $\mathcal{F}$. The complexity of Algorithm 2 is $\mathcal{O}(|\mathcal{F}|^d)$ where d is the depth in the best-first search tree. The efficiency of the algorithm strongly depends on $|\mathcal{F}|$ since in the worst case the number of nodes is in $\mathcal{O}(|\mathcal{F}|^{|\mathcal{F}|})$.

To control the size of $\mathcal{F}$ one can consider all frequent itemsets with a given minimum frequency threshold. Because we are interested in a small coding length, we propose to further restrict the set of patterns to the set of *frequent free itemsets* [9]. Known also as generators, a free itemset is the smallest itemset (in size) that does not contain a subset with the same cover: if I is free, $\nexists J \subset I$ s.t. $\varphi(I) = \varphi(J)$. In fact, there may be multiple free itemsets with the same cover and for our purposes just a single one of them is sufficient. In Fig. 1, all the itemsets in a double bordered rectangle are free.

Set representation as bitvectors. Each candidate itemset in $\mathcal{F}$ is represented by the tuple (set of items, set of covered transactions). Operations on sets such as union, intersection, count, ... being at the core of our implementation, they must be implemented very effectively. For this, we represent each set by bitvectors and all the cover computation are bitwise operations on bitvectors. A rule list is represented by an array of *itemset indices* into $\mathcal{F}$. From the index, one can identify the itemset and its coverage. During the search process at each iteration, a new itemset I is added to the partial rule list (Line 6 of Algorithm 2). This operation involves updating the cover of the rule list computed using (1) which depends on all the transactions already covered. To do this effectively, we keep the transactions already covered in a single bitvector $T^{(j)}_{covered} = \varphi(I^{(1)}) \cup \varphi(I^{(2)}) \cup \cdots \cup \varphi(I^{(j)})$. The coverage after the addition of a new itemset $I^{(j+1)}$ is then

$$\varPhi(\mathcal{D}, \mathcal{R} \cup I^{(j+1)}, j + 1) = \neg T^{(j)}_{covered} \cap \varphi(I^{(j+1)}).$$

5 Experiments

We evaluate our approach from three perspectives: (i) the quality of obtained solutions: how expressive and concise are the rule lists; what is the log-likelihood of the data given the lists; (ii) the accuracy and sensibility of our method under various parameters, evaluated using area under ROC curves (AUC), (iii) the predictive power of our method, using AUC as well.

Table 2. Benchmark features

name	anneal	car	australian-cr.	heart-cl.	krvskp	mushroom	primary-tu.	dermatology	gallup	door	soybean
$|\mathcal{D}|$	812	1728	653	296	3196	8124	336	366	15734	3216	630
$|\mathcal{I}|$	89	21	124	95	73	112	31	133	41	11	50
$\frac{|\mathcal{D}^+|}{|\mathcal{D}|}$	0.77	0.7	0.55	0.54	0.52	0.52	0.24	0.2	0.19	0.16	0.15

Table 3. Total code lengths for several datasets (θ is the minimum support for $\mathcal{F}$)

	anneal	car	australian-cr.	heart-cl.	krvskp	mushroom	primary-tu.	dermatology	gallup	door	soybean
θ	20	5	20	20	5	20	20	20	10	1	1
$|\mathcal{F}|$	1361	22	2495	2024	65	1145	214	763	15	35	49
PRLg	587	710	386	262	2594	1978	249	39	10327	1876	356
PRLc	**532**	**628**	**380**	**249**	**845**	**967**	249	39	**10163**	1876	**314**

Note that we add a comparison with other classification methods to properly position our work; our aim is not to build a classification model that is more accurate on commonly used datasets.

Datasets. We use nine annotated datasets publicly available from the $CP4IM$[3] and UCI[4] repositories. We also used the *door* dataset as described in the introduction (Example 1). Furthermore, we used the *Gallup* dataset [2], from a project with the same name on migratory intentions. This data set is not publicly available, but can be purchased. Our objective here is to understand the migratory intentions between two countries by considering the socio-parameters of education, health, security and age. All these datasets have been preprocessed and their characteristics are given in Table 2.

Algorithms. We compare with popular tree-based classification methods such as Random Forests (RF) and decision trees ($CART$) from the *scikit-learn library*, as well as the rule-learning methods $JRIP$ (Weka version of $RIPPER$) and $SBRL$ [11] available in R CRAN (see Sect. 2). We run $SBRL$ with the default setting (number of iterations set to 30.000, number of chains 10 and a lambda parameter of 10).

Protocols. All experiments were run in the JVM with maximum memory set to 8GB on PCs with Intel Core i5 64bits processor (2.7GHz) and 16GB of RAM running MAC OS 10.13.3. Our approach is called PRL (for probabilistic rule lists) and is implemented in Scala. The candidate itemsets $\mathcal{F}$ are the frequent free itemsets. PRL name can be followed by g for greedy or c for complete branch-and-bound. Evaluation of AUC is done using *stratified 10-fold cross-validation*. For the reproducibility of results, all our implementations are open source and available online[5].

---

[3] https://dtai.cs.kuleuven.be/CP4IM/datasets/.

[4] http://archive.ics.uci.edu/ml/datasets.html.

[5] https://projetsJOHN@bitbucket.org/projetsJOHN/mdlrulesets.

Compression power of PRL. Table 3 gives the total code length obtained for the greedy *PRLg* and the complete branch-and-bound *PRLc* approaches. As can be observed, the *compression ratio* (total code length/size of the datasets) is substantial. For instance, it is 10% for the dermatology dataset. For 8/11 instances *PRLc* discovers a probabilistic rule list compressing better than the one obtained with *PRLg*. The gain obtained with *PRLc* is sometimes substantial, for instance on the krvskp and mushroom data sets.

Impact of the parameters. The set of possible itemsets $\mathcal{F}$ to create the rule list is composed of the frequent free itemsets generated with a minimum support threshold θ. Fig. 3a reports the compression ratio for decreasing values of θ. As expected the compression ratio becomes smaller whenever θ decreases. The reason is that the set $\mathcal{F}$ is growing monotonically, allowing more flexibility to discover a probabilistic rule list that compresses well.

Both the greedy and the complete branch-and-bound algorithms can easily limit the size of the probabilistic rule list they produce. This is done by stopping the expansion of the list beyond a given size limit k. Figure 3b reports the compression ratio for increasing values of k. As expected the compression ratio becomes smaller whenever k increases for PRLc and stabilizes at some point when the limit k becomes larger than the length of the optimal rule list. Surprisingly this is not necessarily the case for the greedy approach that is not able to take advantage of longer rule lists on this benchmark.

Regarding the execution time according to the size of the rules, as shown in Fig. 3c, with a time limit of 10 min, we can see that the greedy approach is more scalable. PRLc and SBRL execution time evolves exponentially, PRLc being faster than SBRL though. Note that as soon as the optimal solution is found, in the case of PRLc, the execution time does not increase so much anymore. The reason is that most of the branches are cut-off by the branch-and-bound tree exploration beyond that depth limit.

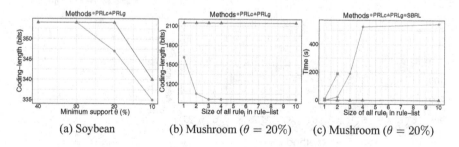

(a) Soybean   (b) Mushroom ($\theta = 20\%$)   (c) Mushroom ($\theta = 20\%$)

Fig. 3. Sensibility of PRL for several settings using mushroom and soybean datasets

Comparison of PRL with existing rule learning algorithms. We compare the rule list produced by our approaches (PRLg and PRLc) and by SBRL [11]. Figure 4a gives the code length for the model and for the data (class labels) for various datasets for the different approaches. Note that the code length for the data corresponds to the *log-likelihood* of the class labels under the rule list. From the rule lists obtained using the training set, the probability (to be positive) of each transaction in the test set is predicted and the coding lengths are computed using the (8) and (9). The reported values are averaged over 10 folds. The model coding length represents the size of the encoding of the initial rule list.

One can see that the PRL approaches are competitive with SBRL. On Fig. 4a, it often obtains the smallest data coding length except for the *mushroom* dataset. The reason is that the test set of *mushroom* is classified perfectly by SBRL. The rule lists produced are arguably shorter with PRLg and PRLc than with SBRL.

The *mushroom* dataset is investigated further in Fig. 4b and 4c. The data coding length and the area under the ROC curve are computed for increasing prefixes of the lists. As we can see, at equal prefix size ($k < 5$) our approach obtains *better likelihood* and is *more accurate* than SBRL. Then beyond $k \geq 5$ SBRL continues to improve on accuracy while PRLg and PRLc stagnates. The lists indeed have reached their optimal length at $k = 5$. This evolution is a clear

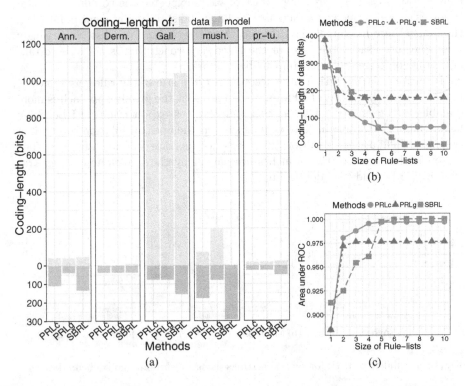

Fig. 4. **(a)** Comparison of coding length in average among *PRL* (g,c) and *SBRL* for different test datasets and **(b and c)** evolution of the coding length of *data only* (top) and the AUC (bottom) for several rule lists size, for *mushroom* dataset, for all 10-folds ($\theta = 10\%$, $|I| = 2$).

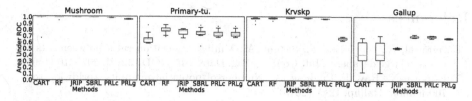

Fig. 5. Comparison of Area under ROC among different methods and four datasets, for all 10-folds ($\theta = 10\%$, $|I| = 1$).

illustration of the difference between the type of rule lists produced by SBRL and our approach. While SBRL lists are more focused on classification, MDL-based lists are a trade-off between the data-coding length (classification) and the complexity of lists (model code length).

Prediction power of PRL and other supervised learning approaches. Although our approach is not designed to generate the best rule list for classification, we evaluate its prediction power in the light of well-known classification methods: *CART*, *RF*, *SBRL* and *JRIP* using 10-fold cross-validation and default settings. For *PRL* the classification is done by associating with each transaction the probability that its label is positive. This probability is that of the first rule of the rule list (obtaining from the training set) that matches with this transaction. The results are shown in Fig. 5.

In general, the AUC of our methods are greater than 0.6 and the optimal solution always has a greater or equal accuracy compared to the greedy approach. The difference becomes significant on databases like *Krvskp* where the difference in compression ratio is also high (Fig. 3).

State-of-the-art methods are often more accurate, except in unbalanced datasets (Gallup, primary-tu.) where our approaches are very competitive. One can see that rule based methods do better on very unbalanced databases like Gallup.

6   Conclusion

This work proposed a supervised rule discovery task focused at finding probabilistic rule lists that can concisely summarize a boolean target attribute, rather than accurately classify it. Our method is in particular applicable when the target attribute corresponds to rare events. Our approach is based on two ingredients, namely, the Minimum Description Length (MDL) principle, and a branch-and-bound search strategy. We have experimentally shown that obtained rule lists are compact and expressive. Future work will investigate the support of multivariate target attributes (> 2 classes) and new types of patterns, such as sequences.

References

1. Agrawal, R., Imieliński, T., Swami, A.: Mining association rules between sets of items in large databases. Int. Conf. Manag. Data (SIGMOD) **22**(2), 207–216 (1993)
2. Esipova, N., Ray, J., Pugliese, A.: Number of potential migrants worldwide tops 700 million. Gallup, USA (2018)
3. Fano, R.M.: The transmission of information. Massachusetts Institute of Technology, Research Laboratory of Electronics Cambridge, Cambridge (1949)
4. Fürnkranz, J., Gamberger, D., Lavrač, N.: Foundations of Rule Learning. Springer Publishing Company, Incorporated (2014)
5. Grünwald, P.D.: The minimum description length principle. MIT press, Cambridge (2007)
6. Guns, T., Nijssen, S., De Raedt, L.: k-Pattern set mining under constraints. IEEE Trans. Knowl. Data Eng. **25**(2), 402–418 (2013)
7. Lavrac, N., Kavsek, B., Flach, P.A., Todorovski, L.: Subgroup discovery with CN2-SD. J. Mach. Learn. Res. **5**, 153–188 (2004)
8. Rissanen, J.: Modeling by shortest data description. Automatica **14**(5), 465–471 (1978)
9. Szathmary, L., Napoli, A., Kuznetsov, S.O.: ZART: a multifunctional itemset mining algorithm. In: Eklund, P.W., Diatta, J., Liquiere, M. (eds.) Proceedings of the 5th International Conference on Concept Lattices and Their Applications, CLA 2007. vol. 331 (2007)
10. Vreeken, J., van Leeuwen, M., Siebes, A.: Krimp: mining itemsets that compress. Data Min. Knowl. Discov. **23**(1), 169–214 (2011)
11. Yang, H., Rudin, C., Seltzer, M.: Scalable bayesian rule lists. In: Precup, D., Teh, Y.W. (eds.) Proceedings of the 34th International Conference on Machine Learning, ICML'17. Proceedings of Machine Learning Research, vol. 70, pp. 3921–3930. PMLR (2017)
12. Zimmermann, A., Nijssen, S.: Supervised pattern mining and applications to classification. In: Aggarwal, C.C., Han, J. (eds.) Frequent Pattern Mining, pp. 425–442. Springer, Cham (2014). https://doi.org/10.1007/978-3-319-07821-2_17

Leveraging Reproduction-Error Representations for Multi-Instance Classification

Sebastian Kauschke[1,2(✉)], Max Mühlhäuser[2], and Johannes Fürnkranz[1]

[1] Knowledge Engineering Group, TU Darmstadt, Darmstadt, Germany
{kauschke,fuernkranz}@ke.tu-darmstadt.de
[2] Telecooperation Group, TU Darmstadt, Darmstadt, Germany
max@tk.tu-darmstadt.de

Abstract. Multi-instance learning deals with the problem of classifying bags of instances, when only the labels of the bags are known for learning, and the instances themselves have no labels. In this work, we propose a method that trains autoencoders for the instances in each class, and recodes each instance into a representation that captures the reproduction error for this instance. The idea behind this approach is that an autoencoder trained on only instances of a single class is unable to reproduce examples from another class properly, which is then reflected in the encoding. The transformed instances are then piped into a propositional classifier that decides the latent instance label. In a second classification layer, the bag label is decided based on the output of the propositional classifier on all the instances in the bag. We show that this reproduction-error encoding creates an advantage compared to the classification of non-encoded data, and that further research into this direction could be beneficial for the cause of multi-instance learning.

Keywords: Multi-instance learning · Denoising autoencoder
Bag classification · Reproduction-error representation

1 Introduction

Multi-instance learning deals with the problem of classifying bags of instances, when only the labels of bags are known for learning, and the instances themselves have no labels. However, it is not known which of the instances are responsible for the bag label, which makes this both an interesting and difficult problem. Dieterich et al. [5] initially mentioned the multi-instance (MI) problem in conjunction with the detection of drug activity based on the molecular structure of proteins. Proteins can rapidly change their shape, so their approach is to gather multiple observed shapes of a protein as a set of observations, and classify this so-called bag instead. In their case, a bag receives a *positive* label when one or more of its instances exhibit the desired behavior. However, this original assumption was later expanded, such that the label can be based on a more generalized composure of the instances in the bag.

L. Soldatova et al. (Eds.): DS 2018, LNAI 11198, pp. 83–95, 2018.
https://doi.org/10.1007/978-3-030-01771-2_6

While MI-learning was initially mainly applied on the drug activity problem, it later became relevant for image and text classification. Although the domains of image and text classification have seen an increased application of deep neural networks, MI-learning still remains relevant, especially in the medical domain, as, e.g., shown in [11,19]. Another example is the domain of image annotation, where labels can be retrieved via a modern search engine, but they may be noisy [18]. Our own motivation for tackling multi-instance problems derives from predictive maintenance scenarios as described in [9,12], where one is often confronted with a not necessarily sequential set of observations, for which no precise label for any observation is given, but it is clear that the set of observations represents a certain state of a machine. Through their non-sequential nature, these bags of observations pose an ideal application for multi-instance classification methods. In general, scenarios with weakly labeled data or potentially noisy input can be considered as multi-instance problems.

There are essentially two ways to approach multi-instance problems: (i) reduce the bags to a single representative instance and classify that instance, and (ii) classify the bag via the distribution of the instances inside the bag. Both techniques have their respective advantages and disadvantages. In this paper, we propose a representation change of the instances and show that it can be beneficial for both approaches. In particular, we aim to create a new representation for instances that captures how well an instance corresponds to the class label of its bag. To that end, we train autoencoders for the individual classes, and encode instances based on their reproduction error. This transformation is simple and efficient to compute, which is ideal for our application. Our results confirm that such a reproduction-error representation improves the results of instance-level and bag-level multi-instance learning.

In Sect. 2 we will give an introduction to multi-instance learning and related work, followed by our definition of the reproduction-error representation in Sect. 3. Section 4 introduces the datasets we have used. Section 5 describes the experiments and the general setup, followed by the results in Sect. 6. We conclude our findings in Sect. 7.

2   Multi-instance Classification

Multi-instance learning is a supervised learning problem, with the goal to predict classes for bags of instances. In the learning phase, a set of bags B is given. Each bag $b \in B$ has a class $c_b \in C$ and contains a set of instances $\mathbf{X}_b$. The instances $\mathbf{x}_{b,i} \in \mathbf{X}_b$ do not have specific labels themselves.

2.1   Multi-instance Assumptions

In the original multi-instance problem formulation by Dietterich [5], it is assumed that for a bag to be positive, it must contain one or more positive instances. This strict assumption was later generalized by other works to cover more complex scenarios [7,17]. For example, Weidman et al. [17] introduced more generalized problem formulations that cover various scenarios. By their definition, the original MI-assumption is the so called *presence-based* MI concept. In addition, they present the *threshold-based* and the *count-based* concept definition. In the *threshold-based* setting, a minimum number or percentage of positive instances in a bag has to be present in order to define a bag as positive. Even more general, in the *count-based* setting there can be both a lower and an upper limit to the amount of positive instances. They proposed their very own method, *TLC* [17], in order to solve these generalized versions of the multi-instance problem.

Whereas Dietterichs and Weidmanns approaches call for methods that determine key instances in the bag and isolate them from the non-key instances, researchers such as Andrews et al. [2] have re-considered this assumption and propose an approach that is based on equal contributions of the instances inside a bag.

2.2 Paradigms of Multi-instance Classification

In general, there are three paradigms w.r.t. handling multi-instance classification [1]: (i) the instance-space paradigm, (ii) the bag-space paradigm, and (iii) the embedded paradigm.

In (i), the main idea is to infer an instance-based classifier first, and make a bag-level prediction based on a meta-decision over the instance-level responses. This is especially difficult, since there are no labels available for the instances at training time. Methods belonging to this category hence have to deal with the problem of how to infer proper labels for the instances. One popular approach, *MiWrapper*, is given by Frank et al. [8], where instances in a bag are considered equally important and therefore will all be assigned the bag label weights proportionate to bag size. For the bag classification, the predicted instance labels are then combined via averaging the class probabilities. In [6], the authors apply an instance-level transformation into a sparse representation via kernels, which helps solve ambiguities between instances. Another example would be the axis-parallel rectangles as proposed by [5], as well as the *Mi-SVM* method [2].

While the instance-space paradigm is concerned with the properties of single instances, the bag-space paradigm (ii) tries to leverage the bag as a whole, and the learning process itself acts in bag-space. This increases the computational complexity, as a bag is not a vector and a comparison between bags must be made. One solution to this problem is to compute a distance function for bags, and use a regular distance-based classifier such as *K-NN* or a SVM. An example for this is *Citation K-NN* [15], which is a modified version of *K-NN*. Another method that belongs to this category is *miGraph* [20], which maps the instances inside a bag into a graph structure, so the dependencies of the instances to each other can be leveraged. This is different to other learners, where independence between instances is assumed. The *miGraph* mapping can be considered as representation transformation on the bag-level.

The final prevalent paradigm is the embedded-space paradigm (iii), where the main idea is to convert a bag into a vectoral form first, and classify afterwards.

One way of doing this is in analyzing statistical properties of bags or their instances, such as the *SimpleMi* method. *SimpleMi* maps the content of a bag into a single instance by calculating the attribute-wise average instance from the bag. This works remarkably well in some scenarios, as we will later show in the experiments (Sect. 6). Extensions thereof, e.g. using max and min in addition to the mean have also been researched [3]. A different approach to the embedded-space paradigm are so called vocabulary-based methods, which first apply some sort of clustering or mapping that can be calculated from the instances in an unsupervised way. In a second layer, a classification decision is then generated based on the distribution of the instances and the clusters. The *TLC* method [17] is an example of such a technique, since it maps the bag into a meta-instance via determining the distribution of its instances in certain parts of the instance-space. The meta-instance can then be classified by a standard propositional classifier.

In our work, we will propose a variant of our method for paradigm (i) and (iii). The main novelty of our approach is the use of an autoencoder to transfer the instances into a different representation, which we will show gives the classifiers built on top of it an advantage compared to their counterpart which does not use that representation. Our approach works very similar to the aforementioned *SimpleMi*, *TLC*, and *MiWrapper*, hence we will build our evaluation based on the comparison with these methods.

3 Reproduction-Error Representations

In this section we will describe our approach and the variants that it incorporates. The general idea of the approaches is to leverage the underlying capabilities of autoencoders. This means we are going to use two features thereof: First, the capability of encoding an instance into a lower-dimensional representation and second, the fact, that an autoencoder can only encode well what it has learned during its training phase. The latter is most important in our case, since we are going to train autoencoders only with a single class of bags, respectively the instances in that bag. We assume the following: an autoencoder trained only on a single class will reproduce instances of bags from other classes worse than the instances from bags of its own class. This is based on the fact that it has seen less or none of those during training time. Especially the instances that differ drastically from the own class might reproduce poorly, which might be exactly the information that is relevant for the classification.

3.1 Autoencoders for Representations

Autoencoders are feed-forward neural networks that are trained to reproduce the input values they are given. While this may sound trivial, it is usually made more complex by adding a constraining hidden layer that has fewer units than the input/output layers, as schematically displayed in Fig. 1. This way, the autoencoder network must be fitted such that it finds a dimension-reduced representation (encoding) of the instance, which it can then decode again to the

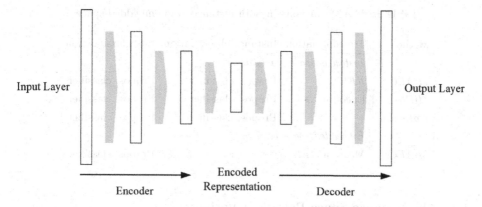

Fig. 1. Example of a deep autoencoder – the central layer provides the encoded representation

initial values. Autoencoders may have one or more hidden layers, and therefore become *deep* autoencoders as well. We will be using both *deep* and *shallow* autoencoders for our experiments. Researchers have found that autoencoders are useful when the task is to detect implicit concepts [14], or even for dimensionality reduction [16], which is in line with our assumption that the label-relevant instances in a bag are different or incorporate a different concept.

3.2   The Reproduction-Error Representation

We first train a set of autoencoders $A = \{A_c\}$, one for each class $c \in C$, based on the instances in all training-bags belonging to that respective class.

The *attribute-wise reproduction error* of an instance $\mathbf{x}$ is defined as $\mathbf{r}_c(\mathbf{x}) = A_c(\mathbf{x}) - \mathbf{x}$, and has the same dimensionality as $\mathbf{x}$. Usually, the reproduction error is calculated as a scalar $e_c(\mathbf{x})$ by calculating the mean error over all attributes.

Our assumption is, that autoencoders can only reproduce well what they have seen during their training. This means, that the reproduction error $e_j(\mathbf{x}), j \in C$ should be lower for instances in bags with class j as opposed to instances from a different bag $k \in C$.

We will use the attribute-wise reproduction errors $\mathbf{r}_c$ to form the *reproduction error representation* (*RER*) of an instance. Since we have $|C|$ autoencoders, we can concatenate the respective $\mathbf{r}_c$ values to a combined vector $\mathbf{R}(\mathbf{x}) = \mathbf{r}_0(\mathbf{x}) \oplus \mathbf{r}_1(\mathbf{x}) \oplus ... \mathbf{r}_{n-1}(\mathbf{x}) \oplus \mathbf{r}_n(\mathbf{x}), n \in C$, which contains all reproductions for $\mathbf{x}$.

Our assumption is that a conversion of the instances to the *reproduction-error representation* will later aid the instance-level classifiers to distinguish between instances of different classes more easily. For testing this assumption, we have implemented four variants of the technique, which use the *RER* as an enhancement to methods that operate in embedded space as well as methods that operate in instance space. The approaches are briefly summarized in Table 1 and explained in detail in the following two sections.

Table 1. *MiAEC* variants in both instance- and embedded-space

Method	Transformation	Instance classification	Bag classification
	Instance-Space Variants		
MiAEC$_i$	*RER*	Propos. classifier	Propos. classifier
MiENC	Encoding only	Propos. classifier	Propos. classifier
NoEncode	—	Propos. classifier	Propos. classifier
	Embedded-space variant		
MiAEC$_e$	Mean of *RER*	—	Propos. classifier

3.3 The Instance-Space Paradigm Variants

In the instance-space paradigm, the goal is to create a classifier for the individual instances, and make a meta-decision over all instances of a bag. We have tested three variants of this scheme, which we now introduce and explain the ideas behind.

MiAEC$_i$: This algorithm is similar to the *MiWrapper*-approach. We first perform the *RER*-transformation, converting the instances into the reproduction-error representation. Based on the *RER*, we train a propositional classifier for instance classification. In order to do so, we must assign class labels to the instances for training, which we do according to the procedure in [8]: Each instance gets the label of the bag it is in. Although this is not optimal w.r.t. the different multi-instance assumptions that exist, Frank et al. [8] bring forth the argument that they assume all instances in a bag are equally relevant for the bag label. However, this leads to the problem that bags with varying numbers of instances will be treated differently. Frank et al. solve this issue by putting a weight on their instances, so that every bag has a combined weight of one. We will also employ this method, and in addition have a meta-layer decide the bag label based upon the frequency of predicted instance labels. This is a simple logistic regression classifier, that is trained on the output of the instance-level classifier.

MiENC: In this variant, instead of the *RER*, we use the encoded form of the instance data as can be retrieved from the autoencoders bottleneck layer. This puts the instances into a lower-dimensional representation. Depending on the data, this representation can be even more useful than the *RER*, as we will see later in the experiments. Except for the representation, *MiENC* uses the same two-level classification approach as *MiAEC$_i$*.

NoEncode: For comparison, we added a third variant that directly works in instance-space: The *NoEncode*-method works exactly like *MiENC*, but it does not use any type of instance transformation, and uses the raw instances instead. Technically, this is very similar to *MiWrapper*, but with the second layer of classification instead of a probabilistic decision.

3.4   The Embedded-Space Paradigm Variant

In addition to the instance-space variants, we investigated a *MiAEC* variant that operates in embedded-space.

MiAEC$_e$ is similar to *SimpleMi*, which aggregates the instances inside a bag via averaging to one single instance with the bag label. Instead of using the bare attributes of the instance, we will be using the *RER*-converted instances. Unlike *MiAEC$_i$* and *MiENC*, this version can not operate on the representation given by the encoder, because the encoder layer has no semantic properties that allow averaging to be applied on it.

4   Datasets

In order to evaluate our efforts, we rely on classical datasets from the domain of multi-instance learning as well as two synthetic datasets that we derived from the popular *MNIST* dataset of handwritten digits. An overview of the datasets is given in Table 2.

4.1   Image Classification – *Elephant*, *Fox* and *Tiger*

Before being overshadowed by deep-learning techniques in recent years, multi-instance learning was popular for tasks of content-based image classification. In the three datasets *Elephant*, *Fox*, and *Tiger* [2], the task is to identify if a bag of image segments contains a certain animal. Each bag represents an image, from which segments have been sampled as instances. The three datasets consist of 200 bags each with an even class distribution, each containing two to 14 instances with 230 attributes.

4.2   Drug Activity Prediction – *Musk*

The *Musk* datasets for drug activity prediction consist of bags that represent molecules. Each instance in the bag is a so called conformation of the molecule. Conformations of molecules are caused by rotating bonds of atoms which make the molecule appear in different shapes, although being chemically equal. In the *Musk* datasets, a feature vector describes these conformations in terms of their surface properties. A molecule should be identified as musk, if it has at least one conformation that represents a conformation that emits a musky smell, hence the name. The *Musk-1* dataset has 92 bags with 476 instances, whereas *Musk-2* has 102 bags with 6598 instances in total. Both have 166 attributes that describe the molecule conformations.

Table 2. Datasets of binary multi-instance classification problems

Dataset	Bags	Instances	Attributes	% pos bags
Elephant	200	1391	230	50.0%
Fox	200	1320	230	50.0%
Tiger	200	1220	230	50.0%
Musk-1	92	476	166	51.1%
Musk-2	102	6598	166	38.2%
Mi-NIST	1500	21740	784	54.0%
Mi-NIST$_2$	1500	21793	784	36.2%

4.3   Handwritten Digits – *Mi-NIST*

In addition to the well known datasets above, we have added two new multi-instance datasets based on the *MNIST*[1] dataset of handwritten digits, which we call *Mi-NIST*.

These datasets consist of 1500 bags of randomly sampled instances from *MNIST*, each bag containing from 9 to 21 instances. A bag is labeled positive, if it contains an instance with the label "8", and negative otherwise. We sampled two datasets this way, such that two class distributions are realized. The dataset *Mi-NIST* consists of 810 positive and 690 negative bags, with a total of 21740 instances. *Mi-NIST$_2$* has fewer positive bags to create a different a-priori distribution. In our comparison these two pose the largest datasets. They are also datasets based on image recognition like *Elephant* etc., but by their properties challenge the classifiers to recognize very specific instances inside the bags, similar to *Musk*.

5   Experiments

For our experiments we chose a usual setup, 10 × 10-fold cross-validation, for all the datasets and algorithms, where the autoencoders are specifically trained only on the training data of the respective fold. We compare our approaches *MiAEC$_i$*, *MiAEC$_e$*, and *MiENC* against the technically similar methods *SimpleMi*, *MiWrapper*, and *TLC*. We used their respective implementations in the WEKA framework, and conducted the evaluation via the WEKA experimenter tool. All the *MiAEC* variants were implemented in python and undergo the very same ten times ten-fold cross validation. As performance measurement we rely on accuracy. In order to demonstrate the added value of the *RER*-transformation, we also include the very same classification strategy as in *MiAEC$_i$* in *NoEncode*, just without the transformation. Comparison of the results of these two variants should clarify, if the transformation yields an advantage. Besides the comparisons of accuracy we will test for statistical significance via the Friedman test with the post-hoc Nemenyi test as described in [4].

---

[1] http://yann.lecun.com/exdb/mnist/.

In the following, we describe the methods we evaluated, and how they were set up.

MiAEC: We trained two type of autoencoders on each dataset. A *shallow* autoencoder with only one hidden layer, which is also the bottleneck, and a *deep* autoencoder with three hidden layers, where the layers before and after the bottleneck consist of an amount of units that is halfway between the input and the bottleneck size. We used the Adam optimizer [10], which uses momentum for learning [13]. Additionally, we included a noise and a dropout layer in order to help with regularization and avoid overfitting. The units in the hidden layers were rectified linear units, whereas the units in the output layer are linear. Since the quality of the reproduction heavily depends on the training of the autoencoder, the parametrization was individually optimized for each dataset: Different noise and dropout values as well as different bottleneck sizes were evaluated via cross-validation with both the *shallow* and *deep* architecture, to get a good reproduction error. The autoencoder was trained with early stopping, the stopping criterium being less than 1% relative accuracy improvement in the training set over the last 4 epochs.
For the classification of *RER*-transformed instances we are using a logistic regression classifier, mainly because it delivers good performance and can be used in the other multi-instance methods as well. Finally, for the classification of bags we are also using logistic regression.

SimpleMi: For *SimpleMi*, two parameters can be set: the classifier and the transformation method. For comparison reasons, we will be using logistic regression as well, and the arithmetic average as transformation.

MiWrapper: *MiWrapper*: [8] has similar options available: a classifier, the transformation method and a weighing method. Likewise, we are choosing logistic regression, arithmetic average and a weighing method that keeps the instance weights such that they sum up to one for each bag.

TLC: *TLC* [17] requires the selection of a classifier and a partition generator. For the partition generator we selected the J48 classifier, a WEKA implementation of C4.5, as the authors did in their original paper. As a classifier, additive logistic regression in the form of LogitBoost is used with decision stumps as weak classifiers.

6   Results

We have applied our and the related methods to seven datasets via 10×10-fold cross-validation, ranked the results and applied the Friedman test with the post-hoc Nemenyi test for critical distances as described in [4]. The results are shown in Table 3 and the critical distance and grouped ranks are displayed in Fig. 2.

The critical distance of the ranks is 3.4, which means that for a significance level of 0.05, the null hypothesis can be rejected. This splits the presented algorithms in two overlapping groups. Essentially, only two algorithms, *MiENC* and *MiAEC_i*, are significantly different to *SimpleMi* (Fig. 2).

Table 3. Experimental results on seven datasets — Accuracy (standard deviation)

Dataset	SimpleMi	MiWrapper	TLC	MiAEC$_i$	MiAEC$_e$	MiENC	NoEncode
Elephant	73.6 (13.2)	84.3 (8.3)	82.5 (8.5)	85.3 (8.1)	81.9 (8.6)	**85.8** (7.5)	83.7 (7.8)
Fox	55.0 (9.5)	59.3 (9.4)	**62.7** (9.6)	59.3 (10.7)	57.8 (10.1)	60.6 (10.4)	58.1 (9.8)
Tiger	76.6 (9.0)	78.6 (8.9)	75.2 (11.2)	**82.1** (8.1)	75.2 (9.7)	80.4 (8.5)	79.5 (8.4)
Musk-1	72.9 (13.0)	79.6 (12.6)	85.2 (12.4)	85.1 (11.5)	**86.7** (10.2)	83.8 (12.1)	83.1 (11.4)
Musk-2	72.3 (13.2)	81.7 (12.7)	78.8 (11.9)	84.8 (10.3)	82.2 (10.9)	**85.5** (10.5)	83.3 (12.1)
Mi-NIST	56.3 (3.9)	57.5 (2.3)	71.5 (4.4)	56.7 (3.7)	**71.7** (2.9)	56.2 (3.6)	54.7 (3.4)
Mi-NIST$_2$	58.1 (4.1)	63.8 (0.3)	76.2 (3.9)	71.3 (4.1)	71.0 (3.7)	**77.9** (3.5)	67.7 (3.5)
Avg. Acc.	66.4	72.1	**76.0**	74.9	75.2	75.7	72.8
Avg. Rank	6.43	4.29	3.50	2.71	4.07	**2.43**	4.57

However, the look at the actual results (Table 3) gives some interesting insights. For a better comparison, we group the results w.r.t. the paradigm that the algorithms were intended for. This means, on the one hand, *SimpleMi*, *TLC* and *MiAEC$_e$*, which operate in embedded-space, and on the other hand, *MiWrapper*, *MiAEC$_i$*, *MiENC*, and *NoEncode* for the instance-space paradigm.

6.1 Embedded Space Results

In this group we look at *SimpleMi*, *TLC* and *MiAEC$_e$*. These algorithms group bags of instances together into one bag-representing instance. *SimpleMi* does this via calculating an arithmetically mean instance over all bag instances, whereas *MiAEC$_e$* does the same, but applies the *RER*-transformation before calculating the mean. As we can see in the results, this leads to advantages in the *Elephant*, both *Musk* and both *Mi-NIST* scenarios, while *Fox* and *Tiger* yield similar results. *TLC* works differently, but shows similar results to *MiAEC$_e$* with only a small advantage. Apparently, the *RER*-transformation is advantageous toward the type of classifier that *MiAEC$_e$* and *SimpleMi* resemble. We assume the cause is, that features which are relevant for distinguishing the classes get alleviated by the transformation, and this remains somewhat intact when the instance-averaging has taken place. The more sophisticated clustering mechanism that *TLC* provides creates a small performance advantage and a better average rank (3.5 vs. 4.07) compared to *MiAEC$_e$*.

6.2 Instance Space Results

In the second group we look at the results of *MiWrapper*, *MiAEC$_i$*, *MiENC*, and *NoEncode*. The best average rank is reached by *MiENC*, but none of the five has a statistically significant advantage regarding the rank. There is also no method that shows an outstanding performance in a specific dataset, either. Since *MiWrapper* and *NoEncode* are practically the same algorithm with only minor differences, we expect the results to be very close to each other. This can

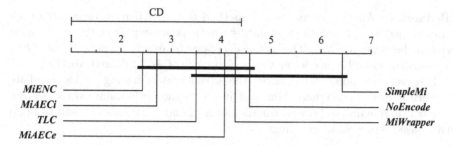

Fig. 2. Comparison of average ranks. The critical distance (CD) is 3.4.

be seen in the results over all the datasets. Both show very similar accuracy, and reach an almost identical average rank (4.29 vs. 4.57).

$MiAEC_i$ uses RER-transformed instances instead of no encoding. Between $MiAEC_i$ and $NoEncode$, we see consistent small improvements over all datasets. On average, the advantage is greater than 2%. $MiENC$ can improve this even further, with a total advantage of 2.9% compared to $NoEncode$. One value stands out: On Mi-$NIST_2$ the improvement of $MiENC$ is 6.6% over $MiAEC_i$. Apparently, in this dataset the encoding has a high impact. This can be explained: in the Mi-$NIST$-datasets we are dealing with hand-written digits in a pixel representation. Since image recognition is one of the main task of neural networks nowadays, the autoencoder can leverage the complex feature-detection abilities that a neural network provides. We suppose a convolutional autoencoder would improve the result even further.

7   Conclusion

In this paper we have presented a method that leverages a special representation of multi-instance data in such a way that classification performance can be enhanced. We have shown that the RER-transformation yields a small but consistent advantage. Learning an autoencoder required to produce this type of representation is a very specific task for each dataset, and can be computationally challenging and time-consuming. However, in case of $MiENC$ the dimension-reduced output benefits the wrapper algorithm. It has to deal with fewer features, in our case a bottleneck layer only 10% of the original size was beneficial. Also, the encoding may enhance certain important aspects of a dataset compared to the original instances, especially when dealing with image-like data. This makes it easier for the wrapper to achieve proper classification.

Therefore, in future work we will look into the area of natural language processing and deep learning in general, where embeddings combined with deep networks are currently proving to be the state-of-the art for some learning problems. For the sake of completeness we would also like to mention that there are other multi-instance classification methods out there, that perform better on the given and other datasets, for example some graph-based or SVM-based

MI-classifiers. Another caveat of our method is, that when all bags contain all types of instances, but with different distributions, we expect the representation error to be generally low. This could decrease the expressiveness of the *RER*-transformation and hence lower the performance of $MiAEC_i$ and $MiAEC_e$.

However, our point is to convey the advantages of having an intermediate instance representation, and show that it affects the performance w.r.t. a given classifier. It remains to be seen whether such methods may also benefit from an error-based representation change.

Acknowledgements. This work has been sponsored by the German Federal Ministry of Education and Research (BMBF) Software Campus project *Effiziente Modellierungstechniken für Predictive Maintenance* [01IS17050]. We also gratefully acknowledge the use of the Lichtenberg high performance computer of the TU Darmstadt for our experiments.

References

1. Amores, J.: Multiple instance classification: review taxonomy and comparative study. Artif. Intell. **201**, 81–105 (2013)
2. Andrews, S., Tsochantaridis, I., Hofmann, T.: Support vector machines for multiple-instance learning. In: Advances in Neural Information Processing Systems - NIPS'03, pp. 561–568 (2003)
3. Bunescu, R.C., Mooney, R.J.: Multiple instance learning for sparse positive bags. In: Proceedings of the 24th International Conference on Machine Learning, pp. 105–112 (2007)
4. Demšar, J.: Statistical comparisons of classifiers over multiple data sets. J. Mach. Learn. Res. **7**, 1–30 (2006)
5. Dietterich, T.G., Lathrop, R.H., Lozano-Pérez, T.: Solving the multiple instance problem with axis-parallel rectangles. Artif. Intell. **89**(1–2), 31–71 (1997)
6. Feng, S., Xiong, W., Li, B., Lang, C., Huang, X.: Hierarchical sparse representation based multi-instance semi-supervised learning with application to image categorization. Signal Process. **94**, 595–607 (2014)
7. Foulds, J., Frank, E.: A review of multi-instance learning assumptions. In: Knowledge Engineering Review, vol. 25, pp. 1–25. Cambridge University Press, Cambridge (2010)
8. Frank, E., Xu, X.: Applying propositional learning algorithms to multi-instance data. (Working paper 06/03). Technical report, University of Waikato, Department of Computer Science (2003)
9. Kauschke, S., Fürnkranz, J., Janssen, F.: Predicting cargo train failures: a machine learning approach for a lightweight prototype. In: Proceedings of the 19th International Conference on Discovery Science - DS'16, pp. 151–166 (2016)
10. Kingma, D.P., Ba, J.: Adam: a method for stochastic optimization. In: CoRR (2014). http://arxiv.org/abs/1412.6980
11. Liu, M., Zhang, J., Adeli, E., Shen, D.: Landmark-based deep multi-instance learning for brain disease diagnosis. Med. Image Anal. **43**, 157–168 (2018)
12. Sipos, R., Fradkin, D., Moerchen, F., Wang, Z.: Log-based predictive maintenance. In: Proceedings of the 20th ACM SIGKDD International Conference on Knowledge Discovery and Data Mining - KDD'14, pp. 1867–1876 (2014)

13. Sutskever, I., Martens, J., Dahl, G., Hinton, G.: On the importance of initialization and momentum in deep learning. In: Proceedings of the International Conference on Machine Learning - ICML'13, pp. 1139–1147 (2013)
14. Vincent, P., Larochelle, H., Manzagol, P.-A.: Stacked denoising autoencoders: learning useful representations in a deep network with a local denoising criterion. J. Mach. Learn. Res. **11**, 3371–3408 (2010)
15. Wang, J., Zucker, J.D.: Solving multiple-instance problem: a lazy learning approach. In: Proceedings of the 17th International Conference on Machine Learning - ICML'00, pp. 1119–1125 (2000)
16. Wang, Y., Yao, H., Zhao, S.: Auto-encoder based dimensionality reduction. Neurocomputing **184**, 232–242 (2016)
17. Weidmann, N., Frank, E., Pfahringer, B.: A two-level learning method for generalized multi-instance problems. In: Lavrač, N., Gamberger, D., Blockeel, H., Todorovski, L. (eds.) ECML 2003. LNCS (LNAI), vol. 2837, pp. 468–479. Springer, Heidelberg (2003). https://doi.org/10.1007/978-3-540-39857-8_42
18. Wu, J., Yu, Y., Huang, C., Yu, K.: Deep multiple instance learning for image classification and auto-annotation. In: Proceedings of the IEEE Conference on Computer Vision and Pattern Recognition - CVPR'15, pp. 3460–3469. IEEE (2015)
19. Yan, Z., Zhan, Y., Zhang, S., Metaxas, D., Zhou, X.S.: Multi-instance multi-stage deep learning for medical image recognition. In: Deep Learning for Medical Image Analysis, pp. 83–104. Academic Press (2017)
20. Zhou, Z.H., Sun, Y.Y., Li, Y.F.: Multi-instance learning by treating instances as non-i.i.d. samples. In: Proceedings of the 26th International Conference on Machine Learning - ICML'09, pp. 1249–1256. ACM (2009)

Meta-Learning

Class Balanced Similarity-Based Instance Transfer Learning for Botnet Family Classification

Basil Alothman[1,2(✉)], Helge Janicke[1,2], and Suleiman Y. Yerima[1,2]

[1] De Montfort University, Leicester LE1 9BH, UK
{heljanic,syerima}@dmu.ac.uk
[2] Faculty of Technology, De Montfort University, Leicester, UK
P14029266@my365.dmu.ac.uk
http://www.dmu.ac.uk/technology

Abstract. The use of Transfer Learning algorithms for enhancing the performance of machine learning algorithms has gained attention over the last decade. In this paper we introduce an extension and evaluation of our novel approach Similarity Based Instance Transfer Learning (SBIT). The extended version is denoted Class Balanced SBIT (or CB-SBIT for short) because it ensures the dataset resulting after instance transfer does not contain class imbalance. We compare the performance of CB-SBIT against the original SBIT algorithm. In addition, we compare its performance against that of the classical Synthetic Minority Over-sampling Technique (SMOTE) using network traffic data. We also compare the performance of CB-SBIT against the performance of the open source transfer learning algorithm TransferBoost using text data. Our results show that CB-SBIT outperforms the original SBIT and SMOTE using varying sizes of network traffic data but falls short when compared to TransferBoost using text data.

Keywords: Similarity-based transfer learning · Botnet detection
SMOTE · TransferBoost

1 Introduction

Transfer learning is one of the active research areas in machine learning [14]. Common machine learning algorithms deal with tasks individually [17], meaning several tasks can only be learnt separately. Transfer learning attempts to learn from one or more tasks (known as source tasks) and use the knowledge learnt to enhance learning in another task (known as the target task). The target and source tasks must be related in one way or another.

Transfer learning is typically employed when there is a limited amount of labelled data in one task (the target task), and sufficient data in another related task (the source task). The idea here is that using only the target data can lead to obtaining models with poor performance since there is insufficient data.

© Springer Nature Switzerland AG 2018
L. Soldatova et al. (Eds.): DS 2018, LNAI 11198, pp. 99–113, 2018.
https://doi.org/10.1007/978-3-030-01771-2_7

Whereas, by transferring knowledge from the source task(s), model quality can be improved.

Transfer learning in network traffic classification was introduced in [19] where feature transfer learning was used, as opposed to our method which is based on instance transfer. The technique is based on projecting the source and target data into a common latent shared feature space and then using this new feature space for model building and making predictions. The technique attempts to preserve the distribution of the data. Although the reported results seem to be reasonably good, there is no freely available tool or code to use for comparison. As this technique is iterative, it can be computationally heavy. The approach we propose in this work is more efficient in terms of speed as it performs instance transfer by performing only one pass over the target as well as source data.

A recent work that applies transfer learning for classification of network traffic can be found in [16]. This work does not propose a new transfer learning method, rather, it only evaluates the performance of an existing open source transfer learning algorithm called TrAdaBoost [5]. Although the results show performance improvement when compared against the base classifier without transfer (referred to as NoTL in the publication), it is noteworthy to mention that TrAdaBoost was extended and enhanced by the introduction of TransferBoost [7] - which is the algorithm that we compare our results against as explained in more detail in [2].

Instance transfer learning has been applied in multiple areas. For example, the recent work in [13] reports an attempt that employs Multiple Instance Learning (MIL) in text classification. This is a two stage method where, in the first stage, the algorithm decides whether the source and target tasks are similar enough to perform transfer which leads to the second stage where transfer is performed.

In this paper, we extend our novel algorithm Similarity Based Instance Transfer Learning (SBIT) and evaluate the performance of the extended version. More detailed explanation of how SBIT works and an evaluation of its performance can be found in our previous work in [2]. We will refer to the extension presented in this work as Class Balanced SBIT (or CB-SBIT) because it ensures that the dataset resulting after instance transfer is class balanced (see Sect. 2 for more details). Our implementation is freely available on Github[1].

The main contributions of this paper are as follows: (1) We introduce an extension to our previous Similarity Based Instance Transfer approach [2] that guarantees class balance in the resulting dataset (avoiding over-fitting) (2) We compare the performance of the extended version of our algorithm against the original version (3) We compare the performance of the extended version of our algorithm against two classical and well known algorithms (4) We show where our algorithm works well and where it does not.

The remainder of this paper is organised as follows: Sect. 2 introduces the SBIT algorithm, highlights one of its current shortcomings, explains the class imbalance problem and provides an overview of the new algorithm CB-SBIT.

---

[1] https://github.com/alothman/CB-SBIT.

Section 3 provides a detailed explanation of experimental setups and results for comparing the performance of CB-SBIT against SBIT, SMOTE and Transfer-Boost using two different types of data. The paper then ends with the conclusions and future work in Sect. 4.

2   Similarity Based Instance Transfer

This section provides a short overview of the original SBIT algorithm [2], class imbalance problem and then introduces the extended version of SBIT (i.e. the CB-SBIT).

2.1   The SBIT Algorithm and its Class Imbalance Problem

The SBIT algorithm is an instance transfer algorithm that scans source datasets one at a time and tries to find similar instances in these datasets to instances in the target dataset. If any similar instance is found, it is transferred to the target dataset which is used later to build a learning model. The pseudo-code of SBIT is provided in Algorithm 1. It is important to bear in mind that SBIT assumes the input target dataset is class balanced (i.e. it contains approximately an equal percentage of classes).

Algorithm 1: The Proposed Transfer Learning Method Algorithm: Similarity-Based Instance Transfer (SBIT)

Input   : Source Datasets $S_1, S_2, \ldots S_n$
Input   : Target Dataset T
Input   : Selected = [ ]
Input   : $thr_1, thr_2, \ldots thr_k$
Output: New Dataset that is the result of $Concatenate(T, Selected)$

1 **for** $S \in [S_1, S_2 \ldots S_n]$: **do**
2   **for** $I_s \in S$: **do**
3   **for** $I_T \in T$: **do**
4           $Sim_1 = ComputeSimilarity1(I_s, I_T)$;
5           $Sim_2 = ComputeSimilarity2(I_s, I_T)$;
6           $\ldots$;
7           $Sim_k = ComputeSimilarity_k(I_s, I_T)$;
8   **if** $Sim_1 > thr_1 \& Sim_2 > thr_2 \ldots \& Sim_k > thr_k$ **then**
9   Add I_s to Selected ;
10

11 $T_{NEW} = Concatenate(T, Selected)$;
12 **Return** T_{NEW};

Careful inspection of Algorithm 1 reveals that SBIT copies an instance from the source data to the target data as soon as it satisfies the similarity criteria (lines 8 and 9). It performs this step without paying attention to the class of that instance. This means it is very possible for instances transferred by SBIT to belong to one class only (or at least for the majority of them to belong to the same class) which leads to creating a new target dataset that is class imbalanced.

2.2   The Class Imbalance Problem

One of the main reasons that cause overfitting is class imbalance [10]. Class imbalance refers to the problem when a classification dataset contains more than one class and number of instances in each class is not approximately the same. For example, there might be a two-class classification dataset that contains 100 instances where the number of instances for one of the classes is 90 and for the other is 10. This dataset is said to be *imbalanced* as the ratio of first class to second class instances is 90:10 (or 9:1). One might train a model that yields 90% accuracy but in reality it could be that the model is predicting the same class for the vast majority of testing data. It is worth mentioning here that evaluation methods such as f1-score, area under the curve, or precision/recall rates provide provide better insight regarding classifier performance when using imbalanced datasets. However, our work focuses on ensuring class balance so an easy to interpret metric such as accuracy can be used.

There are several ways to combat class imbalance [3]. One of these methods is to down sample the majority class (this is sometimes referred to as under sampling). In other words, to randomly select a subset of the instances that belong to the majority class so that the number of instances in each class in the resulting dataset is approximately the same. Another method is to over sample the minority class; which means to randomly duplicate instances from the minority class so the dataset becomes class balanced.

One common technique that falls under this category is the SMOTE algorithm (or the Synthetic Minority Over-sampling Technique [4]) which generates synthetic instances that belong to the minority class rather than generating duplicates.

2.3   The Class Balanced SBIT Algorithm (CB-SBIT)

To avoid class imbalance, the SBIT [2] algorithm discussed in Sect. 2.1 can be modified to ensure the resulting dataset is class balanced.

Recall SBIT assumes that the target dataset is class balanced, the modified version of SBIT makes sure that the new dataset (resulting after selecting instances from source datasets) remains class balanced by using a strict criterion as illustrated in Algorithm 2. This can be achieved in more than one way. For example, it can be done on the fly by keeping track of the ratio of classes of instances transferred from the source datasets and ensuring that whenever an instance is added, the ratio remains almost the same. In other words, it guarantees that approximately the same number of instances from different classes is transferred to the target dataset. Another method is to perform a post-processing step and sub-sample the instances selected for transfer in such a way that the classes are balanced. In our implementation we have both methods although we elected to include the latter in Algorithm 2 (lines 10 and 11).

Algorithm 2: Class Balanced Similarity-Based Instance Transfer (CB-SBIT)

Input : Source Datasets $S_1, S_2, \ldots S_n$
Input : Target Dataset T
Input : Selected = [ ]
Input : $thr_1, thr_2, \ldots thr_k$
Output: New Dataset that is the result of $Concatenate(T, Selected)$

---

1 **for** $S \in [S_1, S_2 \ldots S_n]$: **do**
2   **for** $I_s \in S$: **do**
3   **for** $I_T \in T$: **do**
4          $Sim_1 = ComputeSimilarity1(I_s, I_T)$;
5          $Sim_2 = ComputeSimilarity2(I_s, I_T)$;
6          $\ldots$;
7          $Sim_k = ComputeSimilarity_k(I_s, I_T)$;
8   **if** $Sim_1 > thr_1 \& Sim_2 > thr_2 \ldots \& Sim_k > thr_k$ **then**
9   Add I_s to Selected ;
10

11 $ClassBalancedSelected = SubSample(Selected)$;
12 $T_{NEW} = Concatenate(T, ClassBalancedSelected)$;
13 **Return** T_{NEW};

The *SubSample* function in Algorithm 2 counts the number of instances in each class in the input dataset and randomly removes instances from the majority class(s) until the dataset is class balanced.

3   Experiments and Discussion

In this section we provide a detailed explanation of our experimental setups and discuss the results. We are going to evaluate the performance of some commonly used classifiers on Botnet network traffic data, compare CB-SBIT against the original SBIT and then against two algorithms using data from two different fields.

3.1   Evaluation of Classical Classifiers on Network Traffic Data

In this section we evaluate the performance of several classical classifiers on botnet network traffic data (we use data for the following five botnets: RBot, Smoke_bot, Sogou, TBot and Zeus. In the plot in Fig. 1 these are shown in the x-axis as numbers from one to five. The y-axis in Fig. 1 is the Accuracy.

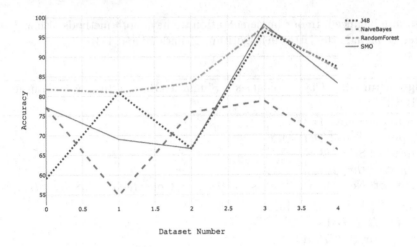

Fig. 1. Performance of classical classifiers on network traffic data

The main purpose of these experiments is to select the best performing algorithm so it can be used for comparison and as the base classifier for SBIT and CB-SBIT.

Figure 1 shows the average accuracy after running a ten-fold cross validation using WEKA's Decision Tree (J48), NaiveBayes, RanfomForest and SMO. It can be noticed that RandomForest scored the highest accuracy in more datasets than any other classifier.

After performing the previous experiments, it becomes clear that Random-Forest should be selected as the base classifier for the transfer learning algorithm developed as part of this work. This is because it performs better than other classifiers on network traffic data.

3.2 CB-SBIT vs SBIT (Using Network Traffic Data)

As explained in Sect. 2, SBIT and its extension CB-SBIT work by selecting instances from source datasets and transferring those instances to the target dataset. Currently the difference between the two algorithms is that CB-SBIT makes sure the new target dataset contains equal percentage of classes. In order to compare the two algorithms against each other, we have created varying sizes of small network traffic datasets. The reason we selected to work on small datasets is that transfer learning is normally applied when data is scarce. These datasets are the same datasets used in [2] (i.e. network traffic data that belong to the following five botnets: *Zeus, TBot, Sogou, RBot* and *Smoke bot*). As explained in detail in [2], each of these botnets has a target dataset and a testing dataset. Datasets that contain network traffic from *Menti, Murlo* and *Neris* botnets were used as source datasets.

The contents of these datasets are derived from the freely available raw Botnet network traffic data which can be found in [15]. As this dataset is in raw format, we used FlowMeter [6] to generate several features that include statistical

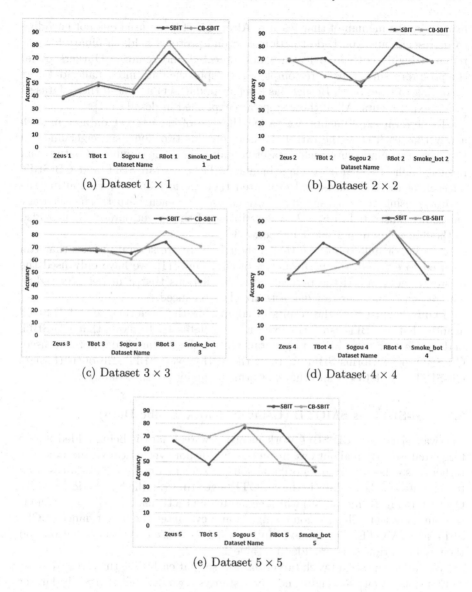

Fig. 2. Accuracy values for CB-SBIT and SBIT

values as well as information such as Source Port, Destination Port and Protocol. Several steps were performed to transform this data into a suitable format for machine learning. The data is in packet capture (PCAP) format and contains traffic data for multiple Botnets as well as Normal traffic. we used FlowMeter to transform it into CSV format. We then followed guidelines provided by the data publisher to assign labels to instances and replaced missing values in each feature by the median of that feature. After this step we used one-hot encoding to represent source port, destination port and protocol fields in binary format, removed highly correlated features and detected and removed Outliers. After the pre-processing steps were completed, we split the data into smaller datasets according to label (each Botnet has a separate dataset) and used these datasets in our experiments. All of these steps are explained in detail in [1].

To perform experiments, we varied the size of each target dataset in such a way that each time the target dataset contains two, four, six, eight and ten instances (we made sure each dataset contains the same number of botnet and normal traffic to guarantee class balance). Then we ran SBIT and CB-SBIT on each of these datasets and evaluated their performance by computing the accuracy using the corresponding test dataset for each botnet. The accuracy values are illustrated in Fig. 2. A description of the target datasets is provided in the first column in Table 1 in Sect. 3.3.

It is important to observe that although there are several metrics that can be used to evaluate the performance of classifiers [11], we have only used the accuracy (accuracy is the percentage of predictions that a model gets right). The reason is that our test datasets are class balanced.

Figure 2 illustrates the results of comparing the performance of CB-SBIT against that of SBIT using the experiment's datasets. It shows that CB-SBIT performs better than SBIT in general. Out of the 25 target datasets we used, CB-SBIT outperforms SBIT in 16 of them. However, SBIT still outperformed CB-SBIT in 6 datasets and they performed equally on three datasets.

3.3   CB-SBIT vs SMOTE (Using Network Traffic Data)

The way SBIT and CB-SBIT work means new real data is being added to the target dataset. By real data we mean the data is not synthetically generated but rather it is collected from its original source. A common algorithm that is used to generate synthetic data is the SMOTE algorithm (or the Synthetic Minority Over-sampling Technique [4]) which generates synthetic instances for a particular class in a dataset. This section compares and evaluates the performance of CB-SBIT and SMOTE. The datasets in Sect. 3.2 were used in this evaluation and their full description is provided in Table 1.

We varied the size of each target dataset so that each time the target dataset contains two, four, six, eight and ten instances - we ensured that each dataset contains the same number of botnet and normal traffic to guarantee class balance. Then we ran CB-SBIT on each of these datasets and saved the resulting target dataset - which now contains the original instances and instances added from source datasets. Using the number of instances of each class in all the resulting datasets, we ran SMOTE to generate new datasets of similar sizes using the original target datasets as the base datasets.

The first column of Table 1 shows the botnet name and the size of the baseline target dataset used (the 1×1 means this dataset contains only two instances, one botnet and one normal, the same concept applies for other sizes). The second column contains the size of the dataset after applying CB-SBIT

Table 1. Datasets resulting after CB-SBIT and SMOTE

Dataset name (size)	Size of dataset generated by CB-SBIT	Size of dataset generated by SMOTE
Zeus 1 (1 × 1)	32 × 32	-
Zeus 2 (2 × 2)	106 × 106	106 × 106
Zeus 3 (3 × 3)	108 × 108	108 × 108
Zeus 4 (4 × 4)	138 × 138	138 × 138
Zeus 5 (5 × 5)	156 × 156	156 × 156
TBot 1 (1 × 1)	42 × 42	-
TBot 2 (2 × 2)	161 × 161	161 × 161
TBot 3 (3 × 3)	211 × 211	211 × 211
TBot 4 (4 × 4)	274 × 274	274 × 274
TBot 5 (5 × 5)	360 × 360	360 × 360
Sogou 1 (1 × 1)	44 × 44	-
Sogou 2 (2 × 2)	67 × 67	67 × 67
Sogou 3 (3 × 3)	147 × 147	147 × 147
Sogou 4 (4 × 4)	170 × 170	170 × 170
Sogou 5 (5 × 5)	252 × 252	252 × 252
RBot 1 (1 × 1)	17 × 17	-
RBot 2 (2 × 2)	34 × 34	34 × 34
RBot 3 (3 × 3)	38 × 38	38 × 38
RBot 4 (4 × 4)	186 × 186	186 × 186
RBot 5 (5 × 5)	212 × 212	212 × 212
Smoke bot 1 (1 × 1)	1 × 1	-
Smoke bot 2 (2 × 2)	52 × 52	52 × 52
Smoke bot 3 (3 × 3)	58 × 58	58 × 58
Smoke bot 4 (4 × 4)	77 × 77	77 × 77
Smoke bot 5 (5 × 5)	96 × 96	96 × 96

using each target dataset as explained above (*number of botnet instances × number of normal instances*). The third column contains the size of the dataset after applying SMOTE using each target dataset. Observe that the cells corresponding to target dataset of size 1 × 1 is empty. This is because SMOTE requires at least two instances of each class to work. Therefore, because SBIT (and CB-SBIT) works normally even when the target dataset contains only one instance of one or more classes, we believe it is fair to conclude that CB-SBIT has a clear advantage over SMOTE when this is the case. In real life there may be cases where only one instance is present for a botnet family - especially when a botnet family is newly discovered.

We have evaluated the performance of RandomForest using each one of them. We have run RandomForest on each dataset and computed the accuracy using the corresponding test dataset for each botnet. The accuracy values are illustrated in Fig. 3.

Inspecting Fig. 3 reveals interesting results. Because SMOTE does not work when the number of instances for any of the classes in the data is less than two, CB-SBIT has a clear advantage in this case. Figure 3a shows a similar behaviour that CB-SBIT performs better when the dataset size is small but

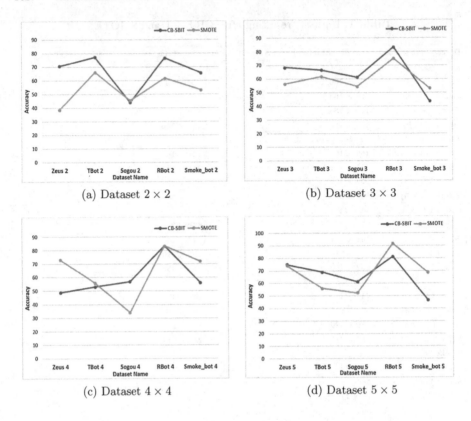

(a) Dataset 2 × 2

(b) Dataset 3 × 3

(c) Dataset 4 × 4

(d) Dataset 5 × 5

Fig. 3. Accuracy values for CB-SBIT and SMOTE

greater than two. When the dataset size is increased gradually, the performance of SMOTE improves and it can be said that it performs equally to CB-SBIT. After using the 25 datasets described in Table 1, CB-SBIT performs better than SMOTE in 17 cases, SMOTE performs better than CB-SBIT in 7 cases and the two of them perform equally in one case. Recall that CB-SBIT (and SBIT) are proposed specifically to address the problem of scarcity of instances in the datasets. Clearly in this scenario CB-SBIT is a better choice than the classical SMOTE.

3.4   CB-SBIT vs TransferBoost (Using Text Data)

For this comparison the popular 20 news groups dataset [12] was used to compare the performance of CB-SBIT against TransferBoost [7] and RandomForest. This dataset consists of 20,000 messages from 20 different netnews newgroups where 1000 messages were collected from each newsgroup. According to the guidelines provided in [12] the 20 groups can be generally categorised into the following six high level categories: computer (contains five sub-categories), miscellaneous (contains only one sub-category), recordings (contains four sub-categories), science (contains four sub-categories), talk (contains three sub-categories) and religion (contains three sub-categories). In order to perform our experiments we have chosen the following six datasets (one from each category): *misc.forsale, comp.graphics, alt.atheism, sci.electronics, rec.autos* and *talk.politics.misc.*

In order to obtain data suitable for machine learning, we used techniques popular in text mining [8]. Text mining involves using several techniques to process (usually unstructured) textual information and generate structured data which can be used to create predictive models and/or to gain some insight into the original textual information. The structured data is usually extracted by analysing the words in the documents and deriving numerical summaries about them.

To be able to use the text documents belonging to the six categories, we created a dataset that has two columns: the first column is the text contained in each document and the second column is the class of that document (which is one of the six categories). After that, we applied the TextToWordVector filter in WEKA [9] with Term Frequency and Inverse Document Frequency [18] (TF-IDF). TF-IDF is a widely used transformation in text mining where terms (or words) in a document are given importance scores based on the frequency of their appearance across documents. A word is important and is assigned a high score if it appears multiple times in a document. However, it is assigned a low score (meaning it is less important) if it appears in several documents.

We used WEKA's default parameters for this filter except for the *number of words to keep*. This parameter is 1000 by default, and we changed it to 10000. In addition to the TextToWordVector, we also used WEKA's NGramTokenizer (with NGramMinSize and NGramMaxSize set to two and three respectively). Not only this, but we also removed Stop Words using a freely available set of stop words. The resulting dataset contained as many as 10530 features and several thousand instances (belonging to the six classes).

The next step was to make sure datasets contained positive and negative examples. We have achieved this by choosing one of the six categories to be our negative class (we randomly chose misc.forsale data). After this, we split the large dataset into smaller datasets according to class and randomly selected a subset of 194 instances from each dataset (except the misc.forsale dataset). Then we randomly selected (without replacement) samples from the misc.forsale dataset and appended them to the other datasets. This was done to ensure that each dataset contains positive and negative instances. At the end of this step we had five datasets as follows: comp.graphics, alt.atheism, sci.electronics, rec.autos and talk.politics.misc (to clarify, the comp.graphics dataset now contains 388 instances, 194 of which are of the comp.graphics class and the remaining 194 are of the misc.forsale class, the same concept applies for the other four datasets).

Since transfer learning requires source and target datasets, we have randomly selected two of the five datasets to be our source datasets (these were the rec.autos and sci.electronics datasets). The remaining three datasets (comp.graphics, alt.atheism and talk.politics.misc) were our target datasets. We have randomly split each of these three datasets into smaller datasets (a target and testing datasets). Each target dataset contained 10 instances (five positive and five negative) and the remaining data was used as our testing datasets. Observe that we made sure we randomly select non-overlapping subsets in all previous steps. Details of these datasets are provided in Table 2.

Table 2. Text dataset details

Dataset name	No of instances	Dataset usage
rec.autos	388 (194 × 194)	Source dataset
sci.elecronics	388 (194 × 194)	Source dataset
alt.atheism_Target	10 (5 × 5)	Target dataset
alt.atheism_Test	378 (189 × 189)	Test dataset
comp.graphics_Target	10 (5 × 5)	Target dataset
comp.graphics_Test	378 (189 × 189)	Test dataset
talk.politics.misc_Target	10 (5 × 5)	Target dataset
talk.politics.misc_Test	378 (189 × 189)	Test dataset

With this setup we have run experiments using RandomForest, TransferBoost and CB-SBIT. When using RandomForest, we have trained it using only the target datasets one at a time. This is because RandomForest only requires one dataset as its input. TransferBoost and CB-SBIT require one Target dataset and one or more Source Datasets, therefore we fixed the source datasets as shown in Table 2 and changed the Target dataset using the Target datasets we have selected. To evaluate, we computed the accuracy of each model using the corresponding test dataset. Our results are illustrated in Table 3.

Table 3. Results using text datasets

Dataset name	CB-SBIT	TransferBoost	RandomForest
alt.atheism	51.06%	89.68%	50.53%
comp.graphics	50.00%	78.84%	50.00%
talk.politics.misc	50.26%	87.56%	52.12%

It is clear from Table 3 that when using textual data, TransferBoost outperforms RandomForest and CB-SBIT. This could be attributed to the nature of the data and how each algorithm works. It can be noticed that the performance of CB-SBIT and RandomForest are almost identical. This is because CB-SBIT uses RandomForest as its base learner and the fact that similarity values between instances in source and target datasets were found to be too small (when compared to the similarity values obtained when using network traffic data).

Table 4 shows computed percentage of similarity values that are greater than 0.5 for two example text and network traffic datasets. The first column of the table shows the two pairs used, while columns two to six show the percentage of similarity results that are greater than 0.5 for the five different types of similarity computation techniques we have used in our work: Tanimoto, Ellenberg, Gleason, Ruzicka and BrayCurtis. Note that the total number of similarity values is the product of the sizes of the pair of families/categories used. Further details on the similarity computation techniques can be found in [2].

Table 4. Percentage of similarity values that are > 0.5 using text and network traffic data

Similarity between	Tanimoto	Ellenberg	Gleason	Ruzicka	BrayCurtis
Graphics - Autos	0.0093%	0.0093%	0.0193%	0.0093%	0.0193%
Politics - Electronics	0.0086%	0.0080%	0.0173%	0.0080%	0.0173%
Zeus - Sogou	12.6311%	91.2733%	97.3485%	7.9254%	14.1463%
TBot - Menti	2.9381%	85.6801%	99.8750%	2.0438%	3.0313%

It is evident that there is much higher similarity in network traffic data than in text data. This means that CB-SBIT could hardly find any instances to transfer from the source to any of the target datasets when using text data. This is an interesting observation especially when it is compared to how CB-SBIT was able to transfer several instances when used with the network traffic data.

4   Conclusions and Future Work

This paper has introduced an extension to a novel transfer learning algorithm that is based on the similarity between instances from the target and source datasets (the SBIT algorithm). The extended version of the algorithm is aware of the percentage of classes in the resulting dataset (resulting after instance transfer) in the sense that it makes sure the classes are balanced. This helps in avoiding several problems such as overfitting and misinterpretation. The paper also included experimental evaluation of the new algorithm (i.e. the CB-SBIT algorithm) against the original SBIT algorithm as well as against two open source commonly used algorithms; the SMOTE and TransferBoost algorithm.

Experimental results show that CB-SBIT outperforms SBIT in majority of the tests; which means CB-SBIT is an improvement over SBIT. When comparing CB-SBIT against SMOTE, several network traffic datasets of various sizes were used and it was evident that CB-SBIT outperforms SMOTE in small datasets (CB-SBIT seems to perform better than SMOTE as the dataset gets smaller). An interesting case was when the dataset contains only one instance of one or more classes. SMOTE does not work in this case whereas CB-SBIT functions normally. On the other hand, text data from the publicly available 20 news groups dataset was used to compare the performance of CB-SBIT against TransferBoost. It was interesting to discover that, even though SBIT outperforms TransferBoost when using network traffic data as it was shown in the original SBIT paper, TransferBoost performs much better than CB-SBIT on text data. This could be due to the nature of the data and the transformations performed in pre-processing it. One interesting observation was made by CB-SBIT is that the similarity values between instances from different topics was very small. This accounts for the poorer performance of CB-SBIT on the text data. Similarity values were observed to be much higher in the network data where CB-SBIT performed very well.

References

1. Alothman, B.: Raw network traffic data preprocessing and preparation for automatic analysis. In: International Conference on Cyber Incident Response, Coordination, Containment & Control (Cyber Incident) - 2018 (2018)
2. Alothman, B.: Similarity based instance transfer learning for botnet detection. Int. J. Intell. Comput. Res. (IJICR) **9**, 880–889 (2018)
3. Chawla, N.V.: Data mining for imbalanced datasets: an overview. Data Mining and Knowledge Discovery Handbook, pp. 875–886. Springer, US (2010). https://doi.org/10.1007/978-0-387-09823-4_45
4. Chawla, N.V., Bowyer, K.W., Hall, L.O., Kegelmeyer, W.P.: Smote: synthetic minority over-sampling technique. J. Artif. Int. Res. **16**(1), 321–357 (2002). http://dl.acm.org/citation.cfm?id=1622407.1622416
5. Dai, W., Yang, Q., Xue, G.R., Yu, Y.: Boosting for transfer learning. In: Proceedings of the 24th International Conference on Machine Learning, pp. 193–200. ICML 2007, ACM, New York, NY, USA (2007). https://doi.org/10.1145/1273496.1273521
6. Draper-Gil, G., Lashkari, A.H., Mamun, M.S.I., Ghorbani, A.A.: Characterization of encrypted and VPN traffic using time-related features. In: ICISSP (2016)
7. Eaton, E., des Jardins, M.: Selective transfer between learning tasks using task-based boosting. In: Proceedings of the 25th AAAI Conference on Artificial Intelligence (AAAI-11), pp. 337–342. AAAI Press (2011). Accessed 7–11 Aug 2011
8. Feldman, R., Sanger, J.: Text Mining Handbook: Advanced Approaches in Analyzing Unstructured Data. Cambridge University Press, New York (2006)
9. Hall, M., Frank, E., Holmes, G., Pfahringer, B., Reutemann, P., Witten, I.H.: The WEKA data mining software: an update. SIGKDD Explor. Newsl. **11**(1), 10–18 (2009). https://doi.org/10.1145/1656274.1656278
10. He, H., Ma, Y.: Imbalanced Learning: Foundations, Algorithms, and Applications, 1st edn. Wiley-IEEE Press (2013)
11. Japkowicz, N., Shah, M.: Evaluating Learning Algorithms: A Classification Perspective. Cambridge University Press, New York (2011)
12. Lang, K.: 20 newsgroups data set. http://www.ai.mit.edu/people/jrennie/20Newsgroups/
13. Liu, B., Xiao, Y., Hao, Z.: A selective multiple instance transfer learning method for text categorization problems. Knowl.-Based Syst. **141**, 178–187 (2018). https://doi.org/10.1016/j.knosys.2017.11.019, http://www.sciencedirect.com/science/article/pii/S0950705117305415

14. Pan, S.J., Yang, Q.: A survey on transfer learning. IEEE Trans. Knowl. Data Eng. **22**(10), 1345–1359 (2010). https://doi.org/10.1109/TKDE.2009.191

15. Samani, E.B.B., Jazi, H.H., Stakhanova, N., Ghorbani, A.A.: Towards effective feature selection in machine learning-based botnet detection approaches. In: 2014 IEEE Conference on Communications and Network Security, pp. 247–255 (2014)

16. Sun, G., Liang, L., Chen, T., Xiao, F., Lang, F.: Network traffic classification based on transfer learning. Comput. Electr. Eng. (2018). https://doi.org/10. 1016/j.compeleceng.2018.03.005, http://www.sciencedirect.com/science/article/ pii/S004579061732829X

17. Torrey, L., Shavlik, J.: Transfer learning. Handbook of Research on Machine Learning Applications, vol. 3, pp. 17–35. IGI Global (2009)

18. Weiss, S., Indurkhya, N., Zhang, T., Damerau, F.: Text Mining: Predictive Methods for Analyzing Unstructured Information. Springer, Berlin (2004)

19. Zhao, J., Shetty, S., Pan, J.W.: Feature-based transfer learning for network security. In: MILCOM 2017–2017 IEEE Military Communications Conference (MILCOM) (2017)

CF4CF-META: Hybrid Collaborative Filtering Algorithm Selection Framework

Tiago Cunha[1]([✉]), Carlos Soares[1], and André C. P. L. F. de Carvalho[2]

[1] Faculdade de Engenharia da Universidade do Porto, Porto, Portugal
{tiagodscunha,csoares}@fe.up.pt
[2] Universidade de São Paulo, ICMC, São Carlos, Brazil
andre@icmc.usp.br

Abstract. The algorithm selection problem refers to the ability to predict the best algorithms for a new problem. This task has been often addressed by Metalearning, which looks for a function able to map problem characteristics to the performance of a set of algorithms. In the context of Collaborative Filtering, a few studies have proposed and validated the merits of different types of problem characteristics for this problem (i.e. dataset-based approach): using systematic metafeatures and performance estimations obtained by subsampling landmarkers. More recently, the problem was tackled using Collaborative Filtering models in a novel framework named CF4CF. This framework leverages the performance estimations as ratings in order to select the best algorithms without using any data characteristics (i.e algorithm-based approach). Given the good results obtained independently using each approach, this paper starts with the hypothesis that the integration of both approaches in a unified algorithm selection framework can improve the predictive performance. Hence, this work introduces CF4CF-META, an hybrid framework which leverages both data and algorithm ratings within a modified Label Ranking model. Furthermore, it takes advantage of CF4CF's internal mechanism to use samples of data at prediction time, which has proven to be effective. This work starts by explaining and formalizing state of the art Collaborative Filtering algorithm selection frameworks (Metalearning, CF4CF and CF4CF-META) and assess their performance via an empirical study. The results show CF4CF-META is able to consistently outperform all other frameworks with statistically significant differences in terms of meta-accuracy and requires fewer landmarkers to do so.

1 Introduction

The task of choosing the best algorithms for a new given problem, the algorithm selection problem, is widely studied in the Machine Learning (ML) literature [4,23,25]. One of the most popular approaches to deal with this problem, Metalearning (MtL), looks for a function able to map characteristics extracted from a problem, named metafeatures, to the performance of a group of algorithms, named metatarget [3]. This function which is learned via a Machine

L. Soldatova et al. (Eds.): DS 2018, LNAI 11198, pp. 114–128, 2018.
https://doi.org/10.1007/978-3-030-01771-2_8

Learning algorithm, named metamodel, can be used later to recommend the best algorithms for new datasets.

MtL has been successfully used to recommend the best algorithm for many tasks [3,15]. This paper is concerned with the use of MtL to recommend the best Collaborative Filtering (CF) for a new recommendation dataset. Several approaches for this task has been proposed by different research groups [1,11,12,17]. Metafeatures proposed in previous studies resulted in a collection of metafeatures for CF recommendation. Two recent studies extended this collection with a set of systematically generated metafeatures [6] and subsampling landmarkers [7], which are performance estimations on samples of the data. More recently, a new strategy named CF4CF was proposed, which, instead of standard MtL approaches, uses CF algorithms to predict rankings of CF algorithms [5]. Instead of metafeatures, algorithm performances are used as training data by CF algorithms to create the metamodel. Using ratings obtained from subsampling landmarkers, CF4CF obtains a predictive performance similar to MtL.

Given the similar predictive performance in MtL and CF4CF, we propose an hybrid approach which combines both: CF4CF-META. The procedure describes each dataset through a combination of problem characteristics and ratings. Then, each dataset will be labeled with the CF algorithms ranked according to their performance. Next, meta-algorithms will be used to train a metamodel. Different from previous MtL-based studies, this work also uses CF4CF's ability to provide recommendations with partial data in a modified Label Ranking approach. To do so, a sampling and regularization procedure is included at prediction time. The predictive performance analysis has shown that this makes the procedure more effective. This work presents several contributions to CF algorithm selection:

- **Frameworks:** this work presents a detailed explanation and formal conceptualization of the state of the art CF algorithm selection frameworks. This is done in order to understand current contributions and frame the proposed CF4CF-META approach.
- **CF4CF-META:** a novel algorithm selection framework is proposed. It leverages simultaneously problem characteristics and rating data as the independent variables of the problem. Furthermore it modifies the standard Label Ranking MtL procedure to deal with partial rankings at prediction time. As far as the authors know, this work is the first of its kind.
- **Empirical comparison:** this work presents an exhaustive experimental study of metalevel and baselevel performance of the existing frameworks. To do so, several combinations of metafeatures for MtL-based algorithm selection frameworks are considered. The goal is to show that despite using the same data and methods as the related work, CF4CF-META performs better than its competitors.

This document is organized as follows: Sect. 2 introduces CF and MtL and describes related work on CF algorithm selection; Sect. 3 presents and formalizes the CF algorithm selection frameworks and introduces CF4CF-META; Sect. 4

describes the experimental setup used, while Sect. 5 presents and discusses the results obtained; Sect. 6 presents the main conclusions and the directions for future work.

2 Related Work

2.1 Collaborative Filtering

CF recommendations are based on the premise that a user will probably like the items favored by a similar user. For such, CF employs the feedback from each individual user to recommend items to similar users [27]. The feedback is a numeric value, proportional to the user's appreciation of an item. Most feedback is based on a rating scale, although other variants, such as like/dislike actions and clickstream, are also suitable. The data structure used in CF, named rating matrix R, is usually described as $R^{U \times I}$, where U is the set of users and I is the set of items. Each element of R is the feedback provided by users to items.

Since R is usually sparse, CF attempts to predict the rating of promising items that were not previously rated by the users. To do so, CF algorithms are used. These algorithms can be organized in memory-based and model-based [2]. Memory-based algorithms apply heuristics to R to produce recommendations, whereas model-based algorithms induce a model from R. While most memory-based algorithms adopt Nearest Neighbor strategies, model-based are usually based on Matrix Factorization [27]. CF algorithms are discussed in [27].

The evaluation is usually performed by procedures that split the dataset into training and test subsets (using sampling strategies, such as k-fold cross-validation [13]) and assesses the performance of the induced model on the test dataset. Different evaluation metrics exist [16]: for rating accuracy, error measures like Root Mean Squared Error (RMSE) and Normalized Mean Absolute Error (NMAE); for classification accuracy, Precision/Recall or Area Under the Curve (AUC) are used; for ranking accuracy, one of the common measure is Normalized Discounted Cumulative Gain (NDCG).

2.2 Metalearning

Metalearning (MtL) attempts to model algorithm performance in terms of problem characteristics [25]. One of the main tasks approached with MtL is the algorithm selection problem, which was first conceptualized by Rice [20]. He defined the following search spaces: problem, feature, algorithm and performance, represented by sets P, F, A and Y. The problem is then described as: for a given instance $x \in P$, with features $f(x) \in F$, find the mapping $S(f(x))$ into A, such that the selected algorithm $\alpha \in A$ maximizes $y(\alpha(x)) \in Y$ [20].

The metadataset in algorithm selection is comprised of several meta-examples, each represented by a dataset. For each meta-example, the predictive features, named metafeatures, are extracted from the corresponding dataset. Each meta-example is associated with the respective target algorithm performance (often, the best algorithm or ranking of algorithms, according to their performance) [3]. Next, a ML algorithm is applied to the metadataset to induce a predictive metamodel, which can be used to recommend the best algorithm(s) for a new dataset. When a new problem arises, one needs just to extract and process the corresponding metafeatures from the new dataset and process them via the MtL model to obtain the predicted best algorithm(s). Thus, MtL has two levels: the baselevel (a conventional ML task applying ML algorithms to problem-related datasets) and the metalevel (apply ML algorithms to meta-datasets).

One of the main challenges in MtL is to define metafeatures able to effectively describe how strongly a dataset matches the bias of ML algorithms [3]. The literature identifies three main groups [3,18,22,25]: statistical and/or information-theoretical (obtain data descriptors using standard formulae); landmarkers (fast estimates of algorithm performance on datasets) and model-based (extraction of properties from fast/simplified models). It is then up to the MtL practitioner to propose, implement and validate suitable characteristics which hopefully will have informative value to the algorithm selection problem.

Since the procedure of designing metafeatures is complex, it is important to highlight recent efforts to help organize and systemically explore the search space. A systematic metafeature framework proposed in [19] leverages on generic elements: object o, function f and post-function pf. To derive a metafeature, the framework applies a function to an object and a post-function to the outcome. Thus, any metafeature can be represented using the notation $\{o.f.pf\}$.

2.3   Collaborative Filtering Algorithm Selection

CF algorithm selection was first addressed using MtL [1,11,12,17]. An overview of their positive and negative aspects can be seen in [9]. These approaches assessed the impact of several statistical and information-theoretical metafeatures on the MtL performance. The characteristics mostly described the user, although a few characteristics related to items and ratings were already available. However, these studies failed in terms of representativity of the broad CF algorithm selection problem, since the experimental setup and nature and diversity of metafeatures were very limited. To address this problem, in [6] the authors proposed systematic metafeatures (which include the metafeatures used in the earlier studies) as well as the use of an extensive experimental setup. In [8], this work was extended to investigate the impact of systematic metafeatures when the goal is to select the best ranking of algorithms, instead of just the best algorithm. For such, the problem was modelled using Label Ranking [14,26].

The algorithm selection problem was also approached using subsampling landmarkers [7], which are metafeatures related to the performance estimation on small samples from the original datasets. Although the problem was modeled in different ways, the authors were unable to find a representation that could be better than the previous systematic metafeatures [6]. In spite of this, these metafeatures were very important for the next CF algorithm selection proposal: CF4CF [5], where the problem of recommending CF algorithms is approached using CF algorithms. Despite the obvious motivation for using recommendation algorithms in a recommendation task, another goal was to provide an alternative to the traditional metafeatures. CF4CF leverages the algorithm performance to create a rating matrix and subsampling landmarkers as initial ratings for prediction. Experimental results showed its ability to be an alternative to standard MtL approaches and the importance of subsampling landmarkers as ratings.

3   Hybrid Algorithm Selection Framework

This work proposes a hybrid framework for the CF algorithm selection problem: CF4CF-META. To explain how it works, we first present and formalize both MtL and CF4CF approaches separately. Table 1 presents the notation used in this document with regards to Rice's framework. Notice that $F = F' \cup F''$, meaning that we use metafeatures from both dataset and algorithm approaches.

Table 1. Mapping between Rice's framework and the CF algorithm selection problem.

Sets	Description	Our setup	Notation
P	Instances	CF datasets	$d_i, i \in \{1, \ldots, |P|\}$
A	Algorithms	CF algorithms	$a_j, j \in \{1, \ldots, |A|\}$
Y	Performance	CF evaluation measures	$y_k, k \in \{1, \ldots, |Y|\}$
F'	Dataset characteristics	Systematic metafeatures	$mf_l, l \in \{1, \ldots, |F'|\}$
F''	Algorithm characteristics	Subsampling landmarkers	$sl_m, m \in \{1, \ldots, |A| \times |Y|\}$

3.1   Metalearning

An overview of the current state of the art MtL for CF approach [8] is presented in Fig. 1. The process combines the systematic metafeatures [6] with standard Label Ranking. Combined, rankings of algorithms can be predicted for new datasets. The problem requires datasets d_i, metafeatures mf_l and algorithms a_j.

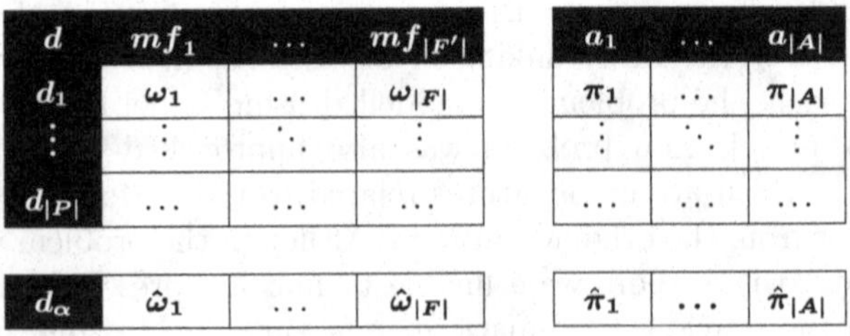

d	mf_1	$\ldots$	$mf_{|F'|}$		a_1	$\ldots$	$a_{|A|}$
d_1	ω_1	$\ldots$	$\omega_{|F|}$		π_1	$\ldots$	$\pi_{|A|}$
$\vdots$	$\vdots$	$\ddots$	$\vdots$		$\vdots$	$\ddots$	$\vdots$
$d_{|P|}$	$\ldots$	$\ldots$	$\ldots$		$\ldots$	$\ldots$	$\ldots$

d_α	$\hat{\omega}_1$	$\ldots$	$\hat{\omega}_{|F|}$		$\hat{\pi}_1$	$\ldots$	$\hat{\pi}_{|A|}$

Fig. 1. Metalearning process overview. Organized into training and prediction stages (top, bottom) and independent and dependent variables (left, right).

The first step in the training stage is to create the metadataset. For such, all datasets d_i are submitted to a systematic characterization process, which yields the metafeatures $\omega = mf(d_i)$. These are now the independent variables of the predictive task. Alternatively, we can use problem characteristics from feature space F'' or even combinations of metafeatures from both feature spaces F' and F''. We will analyse the merits of several approaches in the experimental study.

To create the dependent variables, each dataset d_i is associated with the respective ranking of algorithms π, based on the performance values for a specific evaluation measure y_k. This ranking considers a static ordering of the algorithms a_j (using for instance an alphabetical order) and is composed by a permutation of values $\{1, ..., |A|\}$. These values indicate, for each corresponding position l in the algorithm ordering, the respective ranking position. Modelling the problem this way enables to use Label Ranking algorithms to induce a metamodel. The metamodel can be applied to metafeatures $\hat{\omega} = mf(d_\alpha)$ extracted from a new dataset d_α to predict the best ranking of algorithms $\hat{\pi}$ for this dataset.

3.2 CF4CF

CF4CF [5], illustrated in Fig. 2, is an alternative algorithm selection methodology that uses CF algorithms to predict rankings of CF algorithms. This figure shows the main difference regarding MtL: no metafeatures from F' are used to train the metamodel. Instead, CF4CF uses subsampling landmarkers sl_m.

d	a_1	a_2	a_3	$\dots$	$a_{|A|-1}$	$a_{|A|}$
d_1	ϵ_1	ϵ_2	ϵ_3	$\dots$	$\epsilon_{|A|-1}$	$\epsilon_{|A|}$
$\vdots$	$\vdots$	$\ddots$	$\ddots$	$\ddots$	$\ddots$	$\vdots$
$d_{|P|}$	$\dots$	$\dots$	$\dots$	$\dots$	$\dots$	$\dots$

d_α	$\hat{\epsilon_{sl1}}$	$\dots$	$\hat{\epsilon_{slN}}$	$\hat{\epsilon}_1$	$\dots$	$\hat{\epsilon}_{|A|}$

Fig. 2. CF4CF process overview organized into training and prediction (top, bottom). The prediction stage shows the subsampling landmarkers ϵ_{sl} and predicted ratings $\hat{\epsilon}$.

To create the metadatabase, which in this case is simply a rating matrix, the rankings of algorithms π for every dataset d_i are used. The rankings are converted into ratings by a custom linear transformation rat. The reason is two-fold: to allow any evaluation measure y_k and to enable the usage of any CF algorithm as metamodel. Thus, every dataset d_i is now described as a ratings vector $\epsilon = \left(rat(\pi_n)\right)_{n=1}^{|A|}$. The aggregation of all ratings produces the CF4CF's rating matrix. Next, a CF algorithm is used to train the metamodel.

The prediction stage requires initial ratings to be provided to the CF model. However, it is reasonable to assume that, initially, no performance estimations exist for any algorithm at prediction time. Hence, CF4CF leverages subsampling landmarkers, a performance-based metafeature to obtain initial data. To that end, CF4CF is able to work by providing N subsampling landmarkers and allow CF to predict the remaining $|A| - N$ ratings. Hence, a subset of landmarkers $(sl_m)_{m=1}^{N}$ for dataset d_α are converted into the partial ranking π'. Such ranking is posteriorly converted into ratings also using the linear transformation rat. Thus, the initial ratings are now given by $\hat{\epsilon}_{sl} = \left(rat(\pi'_n)\right)_{n=1}^{N}$. Providing these $\hat{\epsilon}_{sl}$ ratings, the CF metamodel is able to predict the missing $\hat{\epsilon}$ ratings for the remaining algorithms. Considering now the entire set of ratings $r(d_\alpha) = \hat{\epsilon}_{sl} \cup \hat{\epsilon}$, the final predicted ranking $\hat{\pi}$ is created by decreasingly sorting $r(d_\alpha)$ and assigning the ranking positions to the respective algorithms a_j.

3.3   CF4CF-META

The main contribution from this paper is the hybrid framework CF4CF-META, described in Fig. 3. It shows all datasets d_i represented by a union of both types of metafeatures (systematic mf_l and subsampling landmarkers as ratings sl_m) and associated with rankings of algorithms a_j. The process is modeled as a Label Ranking task, similarly to MtL. However, the prediction stage is modified to fit CF4CF's ability to deal with incomplete data. As we will see in the experimental study, this change has great impact on predictive performance.

d	mf_1	$\cdots$	$mf_{|F|}$	sl_1	$\cdots$	$\cdots$	$sl_{|A|}$		a_1	$\cdots$	$a_{|A|}$
d_1	ω_1	$\cdots$	$\omega_{|F|}$	ϵ_1	$\cdots$	$\cdots$	$\epsilon_{|A|}$		π_1	$\cdots$	$\pi_{|A|}$
$\vdots$	$\vdots$	$\ddots$	$\vdots$	$\vdots$	$\ddots$	$\ddots$	$\vdots$		$\vdots$	$\ddots$	$\vdots$
$d_{|P|}$	$\cdots$	$\cdots$	$\cdots$	$\cdots$	$\cdots$	$\cdots$	$\cdots$		$\cdots$	$\cdots$	$\cdots$
d_α	$\hat{\omega}_1$	$\cdots$	$\hat{\omega}_{|F|}$	$\hat{\epsilon}_{sl1}$	$\cdots$	$\hat{\epsilon}_{slN}$	$\varnothing$		$\hat{\pi}_1$	$\cdots$	$\hat{\pi}_{|A|}$

Fig. 3. CF4CF-META process overview. Organized into training and prediction stages (top, bottom) and independent and dependent variables (left, right).

To build the new metadatabase, every dataset d_i is submitted to a metafeature extraction process, yielding a vector of metafeatures $\omega = mf(d_i)$. Next, the subsampling landmarkers sl_m are converted into ratings and leveraged as the remaining metafeatures. Notice, however, that although this characterization is similar to CF4CF's, there is a major difference: while in CF4CF the ratings from the original performance were used as training data, here we are bound to use ratings from subsampling landmarkers. Otherwise, we would be using ratings created from the original algorithm performance to predict the rankings also obtained from the original algorithm performance, which would be an invalid procedure. Thus, the ratings definition consider the ranking of algorithms π' created from all available sl_m to obtain the ratings $\epsilon = \left(rat(\pi'_n)\right)_{n=1}^{|A|}$. The independent variables of the algorithm selection problem are now represented as $F = \omega \cup \epsilon$. To create the dependent variables, each dataset d_i is associated with the respective ranking of algorithms π, similarly to MtL. A standard Label Ranking algorithm is then used to train the metamodel.

In the prediction stage, the new dataset d_α is first submitted to the metafeature extraction process, yielding metafeatures $\hat{\omega} = mf(d_\alpha)$. Next, like in CF4CF, N subsampling landmarkers are used to create the initial data. Although CF4CF-META allows to use all subsampling landmarkers, it is important to provide a procedure that allows to calculate fewer landmarkers. This is mostly due to the significant cost in calculating this type of metafeatures, which we aim to reduce without compromising CF4CF-META's predictive performance. However, since we are working with partial rating data like in CF4CF, this means that the metadata is not exactly the same as it would be if we would use systematic and subsampling landmarkers as metafeatures. This small change, as we will see posteriorly, will greatly influence the predictive performance.

Formally, consider a set of landmarkers $(sl_m)_{m=1}^{N}$ for dataset d_α and its respective partial ranking π'. With it, we are able to obtain the initial ratings $\hat{\epsilon_{sl}} = \left(rat(\pi'_n)\right)_{n=1}^{N}$. Unlike in CF4CF, no ratings are predicted for the missing values. However, this is not a problem, since CF4CF-META is able to work with missing values (these are represented in Fig. 3 by $\varnothing$). Aggregating now the metafeatures $mf(d_\alpha) = \omega \cup \epsilon \cup \varnothing$, we are able to predict $\hat{\pi}$.

4 Experimental Setup

4.1 Baselevel

The baselevel is concerned with the CF problem and consider the following dimensions: datasets, algorithms and evaluation. Table 2 presents all 38 CF datasets used in this work with their main characteristics. For the interested reader, the references for the datasets' origins used can be found in [8].

The CF algorithms used are organized into two CF tasks: Item Recommendation and Rating Prediction. For Item Recommendation, the algorithms used are: BPRMF, WBPRMF, SMRMF, WRMF and the baseline Most Popular. For Rating Prediction, the algorithms used are: MF, BMF, LFLLM, SVD++, three asymmetric algorithms SIAFM, SUAFM and SCAFM; UIB and three baselines: GlobalAverage, ItemAverage and UserAverage. Item Recommendation algorithms are evaluated using NDCG and AUC, while for Rating Prediction NMAE and RMSE measures are used. The experiments are performed using 10-fold cross-validation. No parameter optimization was performed in order to prevent bias towards any algorithm.

4.2 Metalevel

The metalevel has three CF algorithm selection frameworks: MtL [8], CF4CF [5] and CF4CF-META. The metafeatures used in MtL follow several methodologies:

Table 2. Datasets used in the experiments. Values within square brackets indicate lower and upper bounds (k and M stand for thousands and millions, respectively).

Domain	Dataset(s)	#Users	#Items	#Ratings
Amazon	App, Auto, Baby, Beauty, CD, Clothes, Food, Game, Garden, Health, Home, Instrument, Kindle, Movie, Music, Office, Pet, Phone, Sport, Tool, Toy, Video	[7k - 311k]	[2k - 267k]	[11k - 574k]
Bookcrossing	Bookcrossing	8k	29k	40k
Flixter	Flixter	15k	22k	813k
Jester	Jester1, Jester2, Jester3	[2.3k - 2.5k]	[96 - 100]	[61k - 182k]
Movielens	100k, 1m, 10m, 20m, latest	[94 - 23k]	[1k - 17k]	[10k - 2M]
MovieTweetings	RecSys2014, latest	[2.5k - 3.7k]	[4.8k - 7.4k]	[21k - 39k]
Tripadvisor	Tripadvisor	78k	11k	151k
Yahoo!	Movies, Music	[613 - 764]	[4k - 4.6k]	[22k - 31k]
Yelp	Yelp	55k	46k	212k

- **MtL-MF:** a set of systematic metafeatures [6], which consider a combinatorial assignment to a set of objects o (rating matrix R, and its rows U and columns I), a set of functions f (original ratings, number of ratings, mean rating value and sum of ratings) and a set of post-functions pf (maximum, minimum, mean, median, mode, entropy, Gini, skewness and kurtosis).
- **MtL-SL:** a collection of subsampling landmarkers [8]. To calculate these metafeatures, random samples of 10% of each CF dataset are extracted. Next, all CF algorithms are trained on the samples and their performance is assessed using all evaluation measures.
- **MtL-MF+SL:** This strategy combines both systematic metafeatures and subsampling landmarkers in an unified set of metafeatures.

The metatarget is created based on all baselevel evaluation measures (NDCG, AUC, NMAE and RMSE) separately. Since only one evaluation measure can be used at a time, 4 different algorithm selection problems are studied.

Regarding algorithms, this work uses variations of the same algorithm: Nearest Neighbours (i.e. kNN). The goal is to compare in the fairest possible way all frameworks. Hence, both MtL and CF4CF-META are represented by KNN [24], while CF4CF uses user-based CF [21]. The baseline is Average Rankings.

The evaluation in algorithm selection occurs in two tasks: meta-accuracy and impact on the baselevel performance. While the first aims to assess how similar are the predicted and real rankings of algorithms, the second investigates how the algorithms recommended by the metamodels actually perform on average for all datasets. To evaluate the meta-accuracy, this work adopts the Kendall's Tau ranking accuracy measure and leave-one-out cross-validation. The impact on the baselevel is assessed by the average performance for different threshold values, t. These thresholds refer to the number of algorithms used in the predicted ranking. Hence, if $t = 1$, only the first recommended algorithm is used. On the other hand, if $t = 2$, the first and second algorithms are used. In this situation, the performance is the best of both recommended algorithms. All metamodels have their hyperparameters optimized using grid-search.

5   Experimental Results

5.1   Meta-Accuracy

The meta-accuracy regarding Kendall's Tau for all algorithm selection frameworks are presented next: Figs. 4 and 5 present the performance for the Item Recommendation and Rating Prediction scopes, respectively. The results are presented for different N, referring to the amount of landmarkers used as initial ratings (expected to affect only CF4CF and CF4CF-META's performances). The landmarkers are randomly selected and the process repeated 100 times.

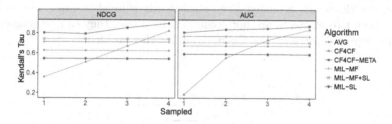

Fig. 4. Ranking accuracy for the item recommendation scope.

The results show CF4CF-META consistently outperforms all other frameworks. They also show its performance increases with N, thus the amount of subsampling landmarkers provided has a positive impact on the framework. It can also be observed that even a single landmarker is enough for CF4CF-META to perform better than the second best framework. Regarding other frameworks, one observes MtL-MF and MtL-MF+SL are always better than AVG, while MtL-SL is always worse. Furthermore, CF4CF is better than AVG only for $N = 3$. However, for $N = 4$ it even surpasses all MtL variations. Also, notice that although MtL-MF and MtL-MF+SL always outperform the baseline, MtL-SL is unable to do the same regardless of the metatarget used.

The results for Rating Prediction are very similar: both CF4CF and CF4CF-META's performances increase with N, MtL-MF and MtL-MF+SL always outperform the baseline, MtL-SL still is worse than the baseline and CF4CF-META performs better than other frameworks for most thresholds. This shows the stability in all CF algorithm selection frameworks, regardless of the evaluation measure used to create the metatarget. However, notice that CF4CF performs better here: it is able to beat the baseline for $N = 4$, which means it needs only 50% of all available landmarkers in Rating Prediction, while it needed 75% in Item Recommendation. Also, for $N = 8$, it is even able to beat CF4CF-META.

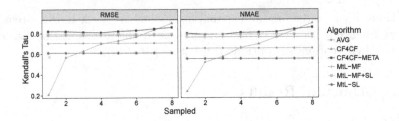

Fig. 5. Ranking accuracy for the rating prediction scope.

In order to validate the observations, we use Critical Difference (CD) diagrams [10]. We represent every framework by the Kendall's Tau performance of the best performing metamodels across all metatargets. Then, we use this technique which applies Friedman test. The resulting diagram represents each framework by its respective ranking position and draws the CD interval. Any two frameworks which are considered statistically equivalent are connected by a line. When two elements are not connected by a line, they can be considered different. Figure 6 presents the CD diagram for this problem.

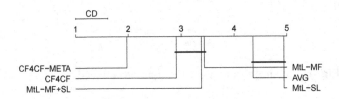

Fig. 6. Critical difference diagram.

These results effectively show that CF4CF-META is better than the remaining frameworks with statistical significance. Furthermore, it presents three frameworks which are better than the baseline, but that hold no statistically significant differences among themselves: CF4CF, MtL-MF and MtL-MF+SL. Lastly, it shows that there is also no statistical significant difference between the baseline and MtL-SL, which has proven to be the worst framework.

5.2 Impact on the Baselevel Performance

Since the algorithm selection task associated with this problem is the prediction of rankings of algorithms, it is important to assess the impact on the baselevel performance considering the rankings predicted by each framework. For such, the frameworks are evaluated considering how many of the first t algorithms in the predicted rankings are used. The goal is to obtain the best performance possible for lower values of t. The results for this analysis, for both the Item Recommendation and Rating Prediction scopes, are presented in Figs. 7 and 8,

respectively. This analysis presents the average baselevel performance for CF4CF
and CF4CF-META for all N subsampling landmarkers used.

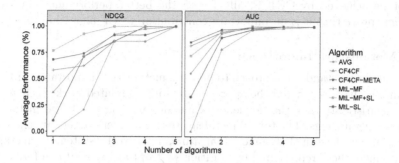

Fig. 7. Impact on the baselevel performance in the item recommendation scope.

According to the results in the Item Recommendation scope, MtL-MF+SL
is the best framework in NDCG (for $N \leq 3$), closely followed by CF4CF-META.
In AUC, CF4CF-META achieves the best performance (for $N \leq 2$), closely
followed by MtL-MF and MtL-MF+SL. Notice that both CF4CF and MtL-SL
perform better than the baseline in NDCG, but fail to do the same in AUC.
This points out the poor stability of CF4CF's predictions.

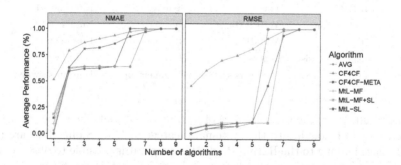

Fig. 8. Impact on the baselevel performance in the rating prediction scope.

In the Rating Prediction scope, CF4CF is consistently better than the
remaining frameworks for the vast majority of thresholds. CF4CF-META is able
to behave better than the remaining competitors in NDCG ($N \leq 5$), although
no significant differences can be observed in RMSE. It is still important to notice
that CF4CF-META is able to beat AVG $t \leq 6$ in NMAE and $t \leq 5$ in RMSE.
Thus, it is a suitable solution for both metatargets in the remaining thresholds.
In this scope all MtL-based frameworks perform quite similarly to the baseline
although usually ranked slightly better.

In summary, although CF4CF-META does not outperform all other frameworks consistently, it is nevertheless the most robust since it is able to always rank at least in second place for all metatargets considered. This stability, which cannot be achieved by CF4CF, allied with the better performances in meta-accuracy, prove that it is an important contribution to CF algorithm selection.

5.3 Metafeature Importance

Since there is no standard approach to perform metafeature importance in Label Ranking, we have replicated the approach used in the related work [8]. It uses a simple heuristic: to rank the frequency of metafeatures on tree-based metamodels for all metatargets. The top 10 metafeatures for all metatargets are shown in Table 3. It can be seen that some metafeatures refer to the systematic metafeatures (namely those represented by notation $\{o.f.pf\}$ ans $sparsity$ and $nusers$), while others refer to ratings (identified by the specific algorithm).

Table 3. Top ranking metafeatures per metatarget in CF4CF-META.

Rank	NDCG	AUC	RMSE	NMAE
1	Most popular	I.count.kurtosis	I.count.kurtosis	I.count.kurtosis
2	WBPRMF	Most Popular	I.mean.entropy	I.mean.entropy
3	I.count.kurtosis	Sparsity	R.ratings.kurtosis	Sparsity
4	I.mean.entropy	I.mean.entropy	U.sum.kurtosis	U.sum.kurtosis
5	U.sum.entropy	R.ratings.kurtosis	I.sum.max	R.ratings.kurtosis
6	BPRMF	I.sum.max	Sparsity	R.ratings.sd
7	U.mean.min	WBPRMF	Nusers	Nusers
8	Nusers	WRMF	R.ratings.sd	I.sum.max
9	U.sum.kurtosis	Nusers	BMF	LFLLM
10	R.ratings.kurtosis	R.ratings.sd	U.sum.entropy	U.mean.skewness

The results show the vast majority of metafeatures available in the top-10 for all metatargets belonging to the systematic category. Only 8 out of 40 are ratings. In fact, looking to the Rating Prediction scope, it is possible to observe how superior is its informativeness: when considering the set of top-8 metafeatures for both RMSE and NMAE metatargets, the sets are exactly the same, although the order shifts. This shows how important are systematic metafeatures for CF algorithm selection in general, and to Rating Prediction in particular. The most important metafeatures of this category are: $I.count.kurtosis$ (best metafeature in 3 metatargets) and $I.mean.entropy$ (top-4 for all metatargets). The results also show ratings can be very effective, when looking, for instance, to the NDCG metatarget: the top-2 metafeatures belong to the ratings category. In fact there is a pattern when it comes to the effectiveness of ratings against systematic metafeatures: in Item Recommendation there are more (and better ranked) ratings in the top-10 metafeatures than in Rating Prediction. The most important metafeatures of this type are ratings for Most Popular and WBPRMF.

6   Conclusions

This work introduced a novel Collaborative Filtering algorithm selection framework. This framework is able to leverage systematic metafeatures and ratings obtained from subsampling landmarkers: CF4CF-META. Based on traditional Metalearning and Collaborative Filtering for algorithm selection (i.e. CF4CF), it incorporates both data and algorithmic approaches to model the problem. The procedure takes advantage of Label Ranking techniques to learn a mapping between both types of metafeatures and the ranking of algorithms, but it introduces a modification at prediction time which is inspired on CF4CF. Several CF algorithm selection frameworks were effectively described and formalized in order to properly present the main contribution from this work. An extensive experimental procedure evaluated all frameworks regarding meta-accuracy and impact on the baselevel performance. The results show that CF4CF-META performs better than the remaining frameworks, with statistically significant differences. Furthermore, CF4CF-META solves a critical problem in CF4CF, by performing better than the remaining frameworks for a reduced amount of subsampling landmarkers used at prediction time. Regarding impact on the baselevel performance, CF4CF-META achieves to be always ranked within the top-2 frameworks for the first positions in the performance ranking for all metatargets. Metafeature importance analysis shows that the data used in this hybrid approach has different impact depending on the CF task addressed: rating data is more important to Item Recommendation, while systematic metafeatures perform better in Rating Prediction. In summary, all of these conclusions allow to understand that CF4CF-META is an important contribution to CF algorithm selection. Directions for future work include the proposal of new metafeatures, to study the impact of context-aware recommendations, to extend the experimental procedure, to study the impact of different algorithms and to assess the CF4CF-META's merits in other domains beyond Collaborative Filtering.

Acknowledgments. This work is financed by the Portuguese funding institution FCT - Fundação para a Ciência e a Tecnologia through the PhD grant SFRH/BD/117531/2016. The work is also financed by European Regional Development Fund (ERDF), through the Incentive System to Research and Technological development, within the Portugal2020 Competitiveness and Internationalization Operational Program within project PushNews (POCI-01- 0247-FEDER-0024257). Lastly, the authors also acknowledge the support from Brazilian funding agencies (CNPq and FAPESP) and IBM Research and Intel.

References

1. Adomavicius, G., Zhang, J.: Impact of data characteristics on recommender systems performance. ACM Manag. Inf. Syst. **3**(1), 1–17 (2012)
2. Bobadilla, J., Ortega, F., Hernando, A., Gutiérrez, A.: Recommender systems survey. Knowl.-Based Syst. **46**, 109–132 (2013)
3. Brazdil, P., Giraud-Carrier, C., Soares, C., Vilalta, R.: Metalearning: Applications to Data Mining, 1st edn. Springer Publishing, Berlin (2009)

4. Brazdil, P., Soares, C., da Costa, J.: Ranking learning algorithms: using IBL and meta-learning on accuracy and time. Mach. Learn. **50**(3), 251–277 (2003)
5. Cunha, T., Soares, C., de Carvalho, A.: CF4CF: recommending collaborative filtering algorithms using collaborative filtering. ArXiv e-prints (2018)
6. Cunha, T., Soares, C., de Carvalho, A.: Selecting collaborative filtering algorithms using metalearning. In: ECML-PKDD, pp. 393–409 (2016)
7. Cunha, T., Soares, C., de Carvalho, A.C.P.L.F.: Recommending collaborative filtering algorithms using subsampling landmarkers. In: Yamamoto, A., Kida, T., Uno, T., Kuboyama, T. (eds.) DS 2017. LNCS (LNAI), vol. 10558, pp. 189–203. Springer, Cham (2017). https://doi.org/10.1007/978-3-319-67786-6_14
8. Cunha, T., Soares, C., de Carvalho, A.C.: A label ranking approach for selecting rankings of collaborative filtering. In: ACM SAC, pp. 1393–1395 (2018)
9. Cunha, T., Soares, C., de Carvalho, A.C.: Metalearning and recommender systems: a literature review and empirical study on the algorithm selection problem for collaborative filtering. Inf. Sci. **423**, 128–144 (2018)
10. Demšar, J.: Statistical comparisons of classifiers over multiple data sets. J. Mach. Learn. Res. **7**, 1–30 (2006)
11. Ekstrand, M., Riedl, J.: When recommenders fail: predicting recommender failure for algorithm selection and combination. In: ACM RecSys, pp. 233–236 (2012)
12. Griffith, J., O'Riordan, C., Sorensen, H.: Investigations into user rating information and accuracy in collaborative filtering. In: ACM SAC, pp. 937–942 (2012)
13. Herlocker, J.L., Konstan, J.a., Terveen, L.G., Riedl, J.T.: Evaluating collaborative filtering recommender systems. ACM Inf. Syst. **22**(1), 5–53 (2004)
14. Hüllermeier, E., Fürnkranz, J., Cheng, W., Brinker, K.: Label ranking by learning pairwise preferences. Artif. Intell. **172**(16–17), 1897–1916 (2008)
15. Lemke, C., Budka, M., Gabrys, B.: Metalearning: a survey of trends and technologies. Artif. Intell. Rev. 1–14 (2013)
16. Lü, L., Medo, M., Yeung, C.H., Zhang, Y.C., Zhang, Z.K., Zhou, T.: Recommender systems. Phys. Rep. **519**(1), 1–49 (2012)
17. Matuszyk, P., Spiliopoulou, M.: Predicting the performance of collaborative filtering algorithms. In: Web Intelligence, Mining and Semantics, pp. 38:1–38:6 (2014)
18. Pfahringer, B., Bensusan, H., Giraud-Carrier, C.: Meta-learning by landmarking various learning algorithms. In: ICML, pp. 743–750 (2000)
19. Pinto, F., Soares, C., Mendes-Moreira, J.: Towards automatic generation of metafeatures. In: PAKDD, pp. 215–226 (2016)
20. Rice, J.: The algorithm selection problem. Adv. Comput. **15**, 65–118 (1976)
21. Sarwar, B., Karypis, G., Konstan, J., Riedl, J.: Analysis of recommendation algorithms for e-commerce. In: ACM Electronic Commerce, pp. 158–167 (2000)
22. Serban, F., Vanschoren, J., Bernstein, A.: A survey of intelligent assistants for data analysis. ACM Comput. Surv. **V**(212), 1–35 (2013)
23. Smith-Miles, K.: Cross-disciplinary perspectives on meta-learning for algorithm selection. ACM Comput. Surv. **41**(1), 6:1–6:25 (2008)
24. Soares, C.: Labelrank: Predicting Rankings of Labels (2015). https://cran.r-project.org/package=labelrank
25. Vanschoren, J.: Understanding machine learning performance with experiment databases. Ph.D. thesis, Katholieke Universiteit Leuven (2010)
26. Vembu, S., Gärtner, T.: Label ranking algorithms: a survey. In: Preference Learning, pp. 45–64 (2010)
27. Yang, X., Guo, Y., Liu, Y., Steck, H.: A survey of collaborative filtering based social recommender systems. Comput. Commun. **41**, 1–10 (2014)

MetaUtil: Meta Learning for Utility Maximization in Regression

Paula Branco[1,2(✉)], Luís Torgo[1,2,3], and Rita P. Ribeiro[1,2]

[1] LIAAD - INESC TEC, Porto, Portugal
{paula.branco,rpribeiro}@dcc.fc.up.pt, ltorgo@dal.ca
[2] DCC - Faculdade de Ciências, Universidade do Porto, Porto, Portugal
[3] Faculty of Computer Science, Dalhousie University, Halifax, Canada

Abstract. Several important real world problems of predictive analytics involve handling different costs of the predictions of the learned models. The research community has developed multiple techniques to deal with these tasks. The utility-based learning framework is a generalization of cost-sensitive tasks that takes into account both costs of errors and benefits of accurate predictions. This framework has important advantages such as allowing to represent more complex settings reflecting the domain knowledge in a more complete and precise way. Most existing work addresses classification tasks with only a few proposals tackling regression problems. In this paper we propose a new method, MetaUtil, for solving utility-based regression problems. The MetaUtil algorithm is versatile allowing the conversion of any out-of-the-box regression algorithm into a utility-based method. We show the advantage of our proposal in a large set of experiments on a diverse set of domains.

1 Introduction

Cost-sensitive learning is important for several practical domains. These methods have been explored thoroughly for classification problems. The study of real world problems and the interest in applications involving the prediction of rare and important phenomena has revealed that these tasks are frequently cost-sensitive [1]. These applications often assume non-uniform costs and benefits that, if disregarded, may result in sub-optimal models and misleading conclusions.

One of the main difficulties associated with these tasks is related with the definition of costs and benefits for the applications, which requires the intervention of domain experts or, at least, need to be provided in an informal way. This happens, for instance, when dealing with tasks known as imbalanced domains where the most important cases are poorly represented. In this setting, we have non-uniform costs and benefits but frequently they are not precisely quantified.

The utility-based learning framework is an extension of cost-sensitive learning that considers both positive benefits for accurate predictions and costs (negative

© Springer Nature Switzerland AG 2018
L. Soldatova et al. (Eds.): DS 2018, LNAI 11198, pp. 129–143, 2018.
https://doi.org/10.1007/978-3-030-01771-2_9

benefits) for misclassifications. It is a more intuitive framework providing information that is easier to understand, and less prone to errors [1,2]. The goal of utility-based learning is the maximization of the predictions' utility, as opposed to cost-sensitive tasks which aim at minimizing the costs.

Although being an important problem with a diversity of applications, most of the research in utility-based learning is still focused on classification. However, many real word applications involve the consideration of costs and benefits in regression tasks. Examples of such applications include the prediction of the concentration of certain particles in the air or forecasting stock returns. In these scenarios we have a continuous target variable with a non-uniform importance over the domain and therefore it is necessary to use utility-based learning solutions. The lack of solutions for tackling utility-based regression problems motivated our work. The main goal of this paper is to propose a new method, MetaUtil, for maximizing the utility of a regression tasks. This new method is inspired by the well-known MetaCost algorithm proposed for cost-sensitive classification tasks. Similarly to MetaCost, the MetaUtil algorithm works as a wrapper method that transforms any standard regression algorithm into a utility-sensitive learner.

This paper is organized as follows. In Sect. 2 the problem definition is presented. Section 3 provides an overview of the related work. Our MetaUtil algorithm is described in Sect. 4 and the results of an extensive experimental evaluation are discussed in Sect. 5. Finally, Sect. 6 presents the main conclusions.

2   Problem Definition

Utility-based learning is framed within predictive tasks, where the goal is to derive a model g that approximates an unknown function $Y = f(\mathbf{x})$. Function f maps a set of p feature variables onto the target variable values. When the target variable is nominal we face a classification task, and when it is numeric we have a regression problem. Model g is obtained using a training set $D = \{\langle \mathbf{x}_i, y_i \rangle\}_{i=1}^{N}$ with N examples. This model can be used to obtain target variable estimates $\hat{y}$ on new data points.

On standard predictive tasks the algorithms are focused on obtaining a model that minimizes a loss function that assigns a uniform importance to all cases, and neither costs nor benefits are taken into account. Still, several real world problems exhibit a non-uniform importance across the domain of the target variable, thus making standard approaches inadequate. Utility-based learning considers a setting where accurate predictions have positive benefits and costs (negative benefits) are assigned to prediction errors. Therefore, it is necessary to adopt a strategy that is able to deal with the information concerning the utility of the predictions. The research community has been mostly concentrated in solving this problem for classification tasks. In this paper we will focus on the less explored problem of utility-based regression.

Utility-based regression assumes the existence of domain knowledge that expresses the benefits and costs for different prediction settings. In classification tasks, this information is typically provided in the form of a cost or cost/benefit (utility) matrix. Torgo and Ribeiro [3] have proposed the concept of utility surface for regression as a continuous version of a utility matrix used in classification tasks. Fully specifying this surface would be too difficult for the end-user given the potentially infinite domain of the continuous target variables. In this context, two alternatives have been put forward to obtain this utility information for regression tasks. The first, and more generally applicable approach, involves using interpolation methods to derive the utility surface using a few user supplied points of this surface. The second alternative, proposed by Ribeiro [1], involves automatically deriving the surface based on some assumptions of the user preferences. More specifically, this automatic method can be used if it is correct to assume that the user preferences involve having accurate predictions for rare extreme values of the target variable. This is a subset of the general problem of utility-based regression. Still, this is an important subset as frequently utility-based regression tasks involve this objective of forecasting rare extreme values. For the data sets used in the experiments carried out in this paper we will assume this goal and thus will use this automatic method of deriving the utility surface. This method is based on the concept of *relevance function*. This function expresses the importance assigned by the user to the different values of the target value.

Definition 1 (Relevance Function). A *relevance function*, which we will denote by $\phi()$, is a function that maps the target variable into a scale of relevance in $[0, 1]$:

$$\phi(y) : \mathcal{Y} \to [0, 1] \tag{1}$$

where 0 represents minimum relevance and 1 represents maximum relevance.

For tasks where the goal is to forecast rare extreme values, Ribeiro [1] has proposed a method to automatically obtain this function using some sampling distribution properties of the target variable in the training set.

Based on this concept of relevance function Torgo and Ribeiro [1,3] have proposed to defined the utility surface as a function of the numeric loss associated with the prediction of $\hat{y}$ for a true value of y, and the respective relevance of these values:

$$U: \mathcal{Y} \times \mathcal{Y} \longrightarrow [-1, 1] \tag{2}$$
$$(y, \hat{y}) \longmapsto U(y, \hat{y}) = g(L(y, \hat{y}), \phi(y), \phi(\hat{y}))$$

where $L()$ is a loss function and $\phi()$ is the relevance function defined for the target variable.

The definition of the utility surface proposed by Torgo and Ribeiro [3] and Ribeiro [1] allows the user to specify which type of errors should be more costly: "false positives" or "false negatives". This is achieved through a parameter $p \in [0, 1]$. Using this parameter it is possible to assign more weight either to false negatives (a relevant case was predicted as non relevant) or to false positives (a non relevant case was predicted as relevant). When $p > 0.5$ the former are considered more serious than the latter, i.e., missing a relevant prediction is considered to have an higher cost than predicting a non relevant case as relevant. On the other hand, when $p < 0.5$ the reverse happens: false negatives are less penalized than false positives. Setting p to 0.5 represents assigning the same cost to both types of errors. Figures 1 and 2 show two utility surfaces obtained automatically for data set `accel`[1] with parameter p set to 0.2 and 0.8, respectively. We can observe that in Fig. 1 the false positive are more costly than false negative, while in Fig. 2 the false negatives have a higher cost.

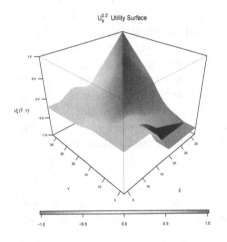

Fig. 1. Utility surface automatically derived for `accel` data set with p set to 0.2.

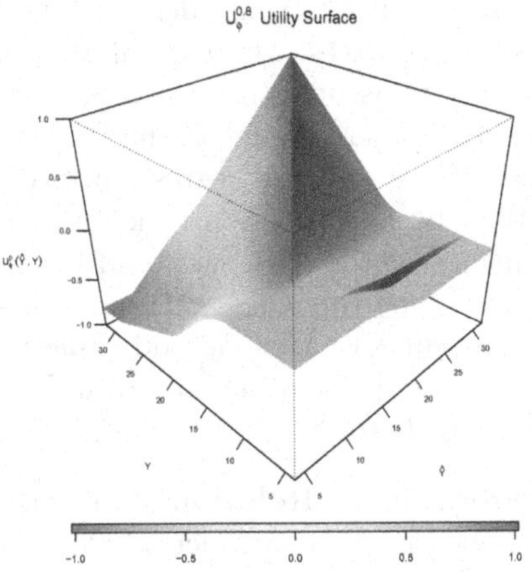

Fig. 2. Utility surface automatically derived for `accel` data set with p set to 0.8.

The goal of utility-based learning is to obtain a model that maximizes the expected utility. In this context, the use of error-based metrics such as the mean squared error or mean absolute deviation is misleading because the user preferences are not taken into account. To evaluate the performance of models in utility-based regression tasks, Ribeiro [1] proposed the use of the normalized mean utility (NMU) measure (cf. Eq. 3). The NMU metric is a normalized version of the mean utility that provides scores in $[0, 1]$, where 1 represents the maximum achievable utility by a model and 0 represents the minimum utility corresponding to the less useful model. We use the NMU metric in our experiments described in Sect. 5.

$$NMU = \frac{\sum_{i=1}^{N} U(y_i, \hat{y}_i) + N}{2N} \tag{3}$$

3   Related Work

As we have mentioned, the research in cost-sensitive and utility-based learning has been mostly focused in classification tasks. In these contexts, a large amount

---

[1] Data set properties described in Sect. 5.

of methods was proposed to tackle these problems that can be categorized into direct methods or meta learning methods [4]. Direct methods change the selected learner internally for making it cost/utility sensitive. Meta learning methods use standard learners and act by changing the data or the used decision threshold. The work presented in this paper was inspired by the seminal work of Domingos [5] where the MetaCost algorithm was proposed for addressing cost-sensitive classification problems.

MetaCost algorithm acts by changing the given training set. An ensemble is generated by applying a classifier to samples with replacement of the training set. The class probabilities of each case are estimated using the votes of the ensemble members. Finally, each training case is relabeled with the Bayes optimal class, i.e., the class that minimizes the conditional risk [6],

$$R(i|\mathbf{x}) = \sum_j P(j|\mathbf{x})C(j,i) \qquad (4)$$

where, $P(j|\mathbf{x})$ is the conditional probability of class j for the example $\mathbf{x}$, and $C(j,i)$ represents the cost of predicting class i when j is the true class.

The new relabeled training set is then simply used to train a new model using a standard classifier.

For utility-based regression only a few works exist in the literature. Some proposals consider special types of utility settings, as it is the case of works that consider different costs for over- and under-predictions (e.g. [7–9]). However, these methods are limited to this particular utility settings. A different approach was recently proposed for maximizing the utility [10] given an arbitrary utility surface. This method adapts the conditional risk minimization defined for classification (cf. Eq. 4), to regression and utility. In this method, the optimal prediction y^* for a case $\mathbf{x}$ is given by an approximation of the following equation:

$$y^* = \underset{z \in Y}{\mathrm{argmax}} \int f_{Y|X}(y|X = \mathbf{x}).U(y,z) \, dy \qquad (5)$$

where $f_{Y|X}(y|X = \mathbf{x})$ represents the conditional probability density function, and $U(y,z)$ is the utility value of predicting z for a true target variable value of y, as defined by some utility surface.

The proposed method works by approximating the conditional probability density function and then using this to obtain the optimal prediction for each case according to Eq. 5. To obtain an approximation of $f_{Y|X}(y|X = \mathbf{x})$ the authors use the method presented in [11,12] that uses ordinal classification to achieve this goal. This utility optimization method has shown significant advantage in a diversity of utility settings and domains.

It is also important to mention that recently, pre-processing solutions have been proposed for the problem of imbalanced regression (e.g. [13]). These pre-processing methods were developed specifically for imbalanced regression tasks which are a sub-problem of utility-based regression where the end-user is more interested in the performance of cases that are scarcely represented in the available data. We must highlight that, although being related, these two problems are different. An extensive review on imbalanced domain learning can be found in [14].

The existing methods for dealing with utility in regression tasks were either developed for specific scenarios, and therefore are not generally applicable, or are based on the minimization of the conditional risk which allows to obtain solutions that are not interpretable. Our proposal allows to address any utility-based regression problem while providing models more interpretable.

4   The MetaUtil Algorithm

In this section we describe our proposal for maximizing utility in regression tasks, the MetaUtil algorithm. Our method is inspired by the MetaCost algorithm by Domingos [5], modifying it to be applicable to regression tasks with non-uniform preferences specified through a utility surface. As MetaCost, our method uses a number m of samples with replacement to obtain different models. In MetaUtil these models are used to obtain m approximations (M_i) of the conditional probability density function, $f_{Y|X}(y|X = \mathbf{x})$, through the same procedure described in [10]. These approximations are averaged to obtain the final estimate of $f_{Y|X}(y|X = \mathbf{x})$ that is used in Eq. 5 to obtain the optimal y value for a given case. These optimal y values are calculated for each original training case and replace the original value in the training set, as it is done in the MetaCost algorithm. In summary, as MetaCost our method uses a sampling procedure to obtain an optimal target variable value for each training case, according to the preference biases of the user. These preference biases in MetaCost are expressed through a cost matrix, and the optimal values calculated using the conditional risk (Eq. 4). In our method the user preferences are described by a utility surface and the optimal values calculated using Eq. 5. In our implementation we also use a parameter ϵ that sets the used granularity for computing all the required approximations. Both MetaCost and our proposed MetaUtil have as outcome a new, modified training set, where the target variable values were changed in accordance with the user preference biases. Our proposal is fully described in Algorithm 1.

An important advantage of MetaUtil (and also MetaCost) lies on the fact that it allows to obtain more interpretable models. This happens because the original data set is changed in a way that is more related with the user preferences, and therefore, the learned model will be obtained in a standard way but will also reflect the user preferences as expressed through the utility surface. Interpretability of the models is a key feature in several real world domains where good predictive performance is not sufficient to convince end-users that a model is reliable (e.g. [15,16]).

Algorithm 1. MetaUtil.

1: **function** METAUTIL($\mathcal{D}$, L, U, m, n, ϵ)
2: //$\mathcal{D}$ - training set
3: //A - regression learning algorithm
4: //U - utility surface
5: //m - number of samples to generate
6: //n - number of examples in each sample
7: //ϵ - granularity parameter
8:
9: **for** $i = 1$ **to** m **do**
10:     $S_i \leftarrow$ sample with replacement of $\mathcal{D}$ with size n
11:     $M_i \leftarrow \{\widetilde{f}_{Y|X}\}_{\mathbf{x} \in \mathcal{D}}$ using S_i and the method in [11,12] with parameter ϵ
12: **end for**
13: **for each** example $\langle \mathbf{x}, y \rangle \in \mathcal{D}$ **do**
14:     $M'(\mathbf{x}) \leftarrow$ average of $M_i(\mathbf{x})$
15:     $y \leftarrow \text{argmax}_{z \in Y} \int M'(\mathbf{x}).U(y, z) \, dy$ approximated with a granularity of ϵ
16: **end for**
17: $M \leftarrow$ model obtained from applying A to the new modified training set
18: **return** M
19: **end function**

---

5   Experimental Evaluation

In this section we describe the experimental evaluation conducted for assessing the effectiveness of the MetaUtil algorithm in maximizing the utility. The main results obtained are presented and discussed. To ensure the easy replication of our work, all code, used data sets and main results are available in https://github.com/paobranco/MetaUtil. All experiments were carried out using the free open source R environment [17].

5.1   Evaluation of MetaUtil Algorithm

The main goal of our experiments is to assess the effectiveness of MetaUtil algorithm in the task of maximizing the utility of predictions. We selected 14 regression data sets from different domains whose main characteristics are described in Table 1. For each of these data sets we have obtained a relevance function through the automatic method [1] we have described before. This method assigns higher relevance to high and low extreme values of the target variable using the quartiles and the inter-quartile range of the target variable sample distribution[2]. Ideally the relevance function should be provided by domain experts. However, given that this information is not available, we used the described automatic method that is suitable for common real world settings where the most important cases are rare and are located at the extremes of the target variable values. In order to test the MetaUtil algorithm in different utility settings, we obtained 3

---

[2] Further details available in [1].

utility surfaces for each data set by using the automatic method proposed by [1] and changing the parameter p described in Sect. 2. As we have mentioned, p allows to assign a different penalization to different types of errors (see examples of different utility surfaces obtained by changing the value of parameter p in Figs. 1 and 2). We used the following values for parameter p: $\{0.2, 0.5, 0.8\}$.

Table 1. Characteristics of the 14 used data sets. (N: Nr of cases; $pred$: Nr of predictors; nom: Nr of nominal predictors; num: Nr numeric predictors; $nRare$: nr. cases with $\phi(Y) > 0.8$; $\%Rare$: $100 \times nRare/N$).

Data Set	N	pred	nom	num	nRare	% Rare
servo	167	4	2	2	34	20.4
a6	198	11	3	8	33	16.7
Abalone	4177	8	1	7	679	16.3
a3	198	11	3	8	32	16.2
a4	198	11	3	8	31	15.7
a1	198	11	3	8	28	14.1
a7	198	11	3	8	27	13.6
boston	506	13	0	13	65	12.8
a2	198	11	3	8	22	11.1
a5	198	11	3	8	21	10.6
fuelCons	1764	37	12	25	164	9.3
bank8FM	4499	8	0	8	288	6.4
Accel	1732	14	3	11	89	5.1
airfoild	1503	5	0	5	62	4.1

We selected two base regression learners: Support Vector Machines (SVM) and Random Forests (RF). For these learners we tested several parameter variants. The algorithms, set of tested parameters and respective used R packages, are described in Table 2.

Table 2. Regression algorithms and their parameter values, and the respective R packages.

Learner	Parameter variants	R package
Support vector machines (SVM)	$cost = \{10, 150\}$	**e1071**[18]
	$gamma = \{0.01, 0.001\}$	
Random forests (RF)	$mtry = \{5, 7\}$	**randomForest**[19]
	$ntree = \{500, 750, 1500\}$	

We applied each of the 10 learning approaches (4 SVM + 6 RF) to each of the 42 problems (14 data sets × 3 utility surface settings). To allow a fair comparison we tested the original regression algorithms (Orig), the MetaUtil algorithm and the strategy for maximizing the utility (UtilOptim) proposed in [10].

The granularity parameter ϵ was set to 0.1. Regarding the required probabilistic classifier, we selected the classification learner most closely related to the regression algorithm being compared against. The motivation for this choice is related with the negative impact in the observed performance when there is a mismatch between the probability estimator and the used classifier [20]. Moreover, we will assume, as done in [5], that the user is able to select the regression scheme that best adapts to the task that is being considered. The same described scheme is used on both tested algorithms for estimating $f_{Y|X}$. In MetaUtil algorithm we set the number of samples with replacement to generate (parameter m) to 20 and the number of examples in each sample (parameter n) to the training set size.

All the described alternatives were evaluated using the NMU measure described in Sect. 2. We selected a normalized measure because it allows to obtain comparable results across different data sets. The NMU values were estimated by 2 repetitions of a 10-fold stratified cross validation process as implemented in R package `performanceEstimation` [21]. In addition to reporting the NMU scores, we also assessed the statistical significance of the observed differences using the non-parametric Friedman F-test together with a post-hoc Nemenyi test with a significance level of 95%.

5.2   Main Results and Discussion

The 10 learning variants were applied to the 42 regression problems (14 data sets using 3 different utility surface settings) and 3 strategies for utility optimization (Orig + UtilOptim + MetaUtil). Thus, we tested 1260 ($10 \times 42 \times 3$) combinations. Tables 3, 4 and 5 show the mean NMU results of the variants of each learner, obtained for each utility setting, i.e., when considering different values for parameter p in the generation of the utility surface.

From the overall analysis of the NMU results we notice that the MetaUtil algorithm shows a competitive performance. This method displays several times the best average performance specially for utility surfaces with higher values of p.

We proceeded with the application of the non-parametric Friedman F-test for assessing the statistical significance of the results. The F-test results allowed the rejection of the null hypothesis that all the tested approaches exhibit the same performance. We then applied the post-hoc Nemenyi test with a significance level of 95% to verify which approaches are statistically different. The critical difference diagrams (CD diagram) [22] with the results aggregated by type of utility surface setting and by learner are displayed in Figs. 3 and 4. In the CD diagrams, lower ranks indicate a better performance and when the lines of two algorithms are connected by a bold horizontal line it means that the their average ranks are not significantly different, i.e. their performance difference is

Table 3. Mean NMU results of the variants of each learner by data set for the value of parameter p set to 0.2 (sd: standard deviation).

	SVM			RF		
	Orig	UtilOptim	MetaUtil	Orig	UtilOptim	MetaUtil
servo	0.4891	**0.5585**	0.4772	**0.5712**	0.5624	0.5666
a6	0.5157	**0.5223**	0.5203	0.5123	**0.5140**	0.5051
Abalone	0.5751	**0.5862**	0.5818	0.5786	0.5849	**0.5855**
a3	0.5039	**0.5108**	0.5085	0.5002	**0.5079**	0.4843
a4	0.5200	**0.5307**	0.5303	0.5269	0.5272	**0.5275**
a1	0.5335	**0.5371**	0.5370	0.5485	**0.5498**	0.5490
a7	0.4983	0.5075	**0.5083**	0.4735	**0.5055**	0.4543
boston	0.5748	**0.5770**	0.5738	0.5784	**0.5806**	0.5784
a2	0.5236	0.5294	**0.5312**	0.5289	0.5276	0.5260
a5	0.5226	0.5284	**0.5288**	0.5269	0.5250	0.5222
fuelCons	0.6138	**0.6183**	0.6164	0.6175	**0.6257**	0.6213
bank8FM	0.5694	0.5699	**0.5720**	0.5703	0.5690	**0.5707**
Accel	0.5582	**0.5606**	0.5593	0.5638	**0.5655**	0.5641
airfoild	0.4566	**0.4853**	**0.4853**	0.4599	**0.4853**	**0.4853**
Mean±sd	0.532±0.042	**0.544±0.036**	0.538±0.039	0.540±0.044	**0.545±0.038**	0.539±0.046

Table 4. Mean NMU results of the variants of each learner by data set for the value of parameter p set to 0.5 (sd: standard deviation).

	SVM			RF		
	Orig	UtilOptim	MetaUtil	Orig	UtilOptim	MetaUtil
servo	0.4875	**0.5601**	0.4778	**0.5718**	0.5655	0.5660
a6	0.5072	**0.5207**	0.5187	**0.5140**	0.5069	0.5116
Abalone	0.5705	**0.5893**	0.5829	0.5764	**0.5884**	0.5883
a3	0.4927	**0.5051**	0.5046	**0.5048**	0.4958	0.4944
a4	0.5140	0.5313	**0.5336**	0.5284	0.5331	**0.5346**
a1	0.5297	0.5394	**0.5425**	0.5479	0.5531	**0.5533**
a7	0.4859	0.4975	**0.4976**	0.4810	**0.4840**	0.4652
boston	0.5743	**0.5775**	0.5748	0.5782	**0.5814**	0.5792
a2	0.5180	0.5289	**0.5309**	0.5269	0.5242	**0.5298**
a5	0.5172	0.5282	**0.5293**	0.5257	0.5261	**0.5269**
fuelCons	0.6135	**0.6194**	0.6170	0.6171	**0.6259**	0.6222
bank8FM	0.5692	0.5702	**0.5723**	0.5703	0.5694	**0.5710**
Accel	0.5580	**0.5611**	0.5601	0.5638	**0.5655**	0.5643
airfoild	0.4508	**0.4633**	**0.4633**	0.4575	**0.4635**	0.4631
Mean±sd	0.528±0.044	**0.542±0.041**	0.536±0.043	0.540±0.043	**0.542±0.045**	0.541±0.046

not significantly different. The SVM results confirm that for this learner there is no statistical significance between the performance of the UtilOptim and MetaUtil algorithms in all tested utility settings. The UtilOptim algorithm achieves a lower rank for the utility surface with the lower value of parameter p, while the MetaUtil has a better rank for the most balanced utility setting.

Table 5. Mean NMU results of the variants of each learner by data set for the value of parameter p set to 0.8 (sd: standard deviation).

	SVM			RF		
	Orig	UtilOptim	MetaUtil	Orig	UtilOptim	MetaUtil
servo	0.4859	**0.5624**	0.4840	**0.5723**	0.5661	0.5664
a6	0.4987	0.5251	**0.5259**	0.5156	0.5128	**0.5239**
Abalone	0.5660	**0.5961**	0.5866	0.5743	**0.5955**	0.5939
a3	0.4814	0.5132	**0.5152**	**0.5093**	0.4960	0.5067
a4	0.5081	0.5412	**0.5481**	0.5299	0.5473	**0.5484**
a1	0.5258	0.5490	**0.5539**	0.5473	0.5589	**0.5612**
a7	0.4735	**0.4950**	0.4926	**0.4886**	0.4728	0.4793
boston	0.5737	**0.5783**	0.5762	0.5780	**0.5819**	0.5804
a2	0.5123	0.5330	**0.5352**	0.5249	0.5293	**0.5373**
a5	0.5119	0.5326	**0.5353**	0.5244	0.5312	**0.5353**
fuelCons	0.6131	**0.6207**	0.6175	0.6167	**0.6260**	0.6233
bank8FM	0.5691	0.5708	**0.5725**	0.5702	0.5699	**0.5713**
Accel	0.5577	**0.5618**	0.5612	0.5637	**0.5656**	0.5646
airfoild	0.4450	**0.4460**	0.4422	**0.4550**	0.4509	0.4509
Mean±sd	0.523±0.047	**0.545±0.044**	0.539±0.046	0.541±0.042	0.543±0.048	**0.546±0.045**

Regarding the RF results, the better performance of UtilOptim algorithm is also confirmed for the lower value of p in the tested utility surface settings. For the remaining values of p, the MetaUtil algorithm has a better performance as it provides lower ranks in the CD diagrams, although not always with statistical significance when compared against UtilOptim algorithm.

When considering the performance of the tested learner variants we can conclude that: (i) using the original learning algorithm is worst with statistical significance under all tested utility settings ; (ii) for the variants of the SVM learner the differences between UtilOptim and MetaUtil algorithms are not statistical significant, although UtilOptim displays a lower rank for the utility settings with a lower value of parameter p and MetaUtil has a lower rank on the remaining utility surfaces; (iii) for the RF learner, UtilOptim is better with statistical significance for the lower value of p, while MetaUtil displays a lower rank on the remaining utility surface settings, although only for the $p = 0.8$ this is statistically significant.

Overall, the results show that MetaUtil is very competitive with the current state of the art in utility optimization (the algorithm UtilOptim). Although our proposal can not be seen as providing clearly better results we should stress that there is a significant difference between the approaches in terms of interpretability. In effect, the MetaUtil algorithm produces models that are biased towards the utility preferences of the user, as they are obtained with a biased training set. This means that the user can check the model to understand why some value was predicted, particularly if an interpretable modelling algorithm is used. This is not true for UtilOptim and this can be a key advantage of MetaUtil for applications were the end user requires interpretable models. In effect, UtilOp-

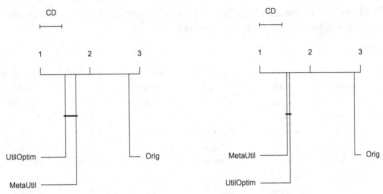

(a) Results for Utility Surface with $p = 0.2$. (b) Results for Utility Surface with $p = 0.5$.

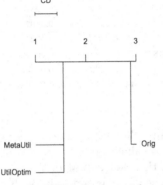

(c) Results for Utility Surface with $p = 0.8$.

Fig. 3. Critical Difference diagram of average NMU results for SVM learner.

tim produces models using the original training set and thus the models are not biased toward the utility preferences. The UtilOptim method works as a post-processing method by changing the predicted value using Eq. 5 to match the preferences of the user. However, the post-processed predictions are not explainable by the learned models and thus the approach is less interpretable.

6   Conclusions

In this paper we propose a new algorithm named MetaUtil for tackling the problem of utility maximization in regression tasks. The proposed method changes the value of the target variable in the training cases to the value that maximizes the expected utility. This new training set can then be used to learn a model using any standard regression algorithm. When compared to competing methods, the MetaUtil algorithm has the advantage of providing more interpretable models, which is a key advantage for several real world applications.

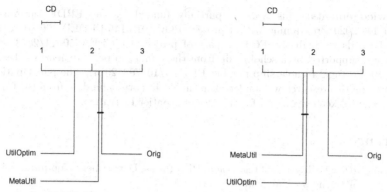

(a) Results for Utility Surface with $p = 0.2$. (b) Results for Utility Surface with $p = 0.5$.

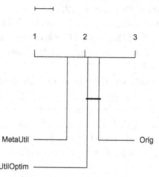

(c) Results for Utility Surface with $p = 0.8$.

Fig. 4. Critical Difference diagram of average NMU results for RF learner.

A large set of experiments was carried out using two learning algorithms, several regression data sets, and a different utility surfaces. The obtained results highlight the advantages of our proposal, when compared to using the original regression algorithm and also against a competing method. The advantages of MetaUtil against the latter are not statistically significant for most of the situations.

The key contributions of his paper are as follows: (i) a new algorithm for tackling the problem of maximizing the utility in regression tasks; (ii) comparison of the proposed approach against standard regression algorithms and a competing method; and (iii) the analysis of the impact of different utility surface settings in the performance of the approaches.

As future work, we plan to explore the performance of the proposed algorithm with other regression tools. We will also study the connections between the data or utility surfaces characteristics, and the performance achieved by the MetaUtil algorithm.

Acknowledgements. This work is partially funded by the ERDF through the COMPETE 2020 Programme within project POCI-01-0145-FEDER-006961, and by National Funds through the FCT as part of project UID/EEA/50014/2013. Paula Branco was supported by a scholarship from the Fundação para a Ciência e Tecnologia (FCT), Portugal (scholarship number PD/BD/105788/2014). The participation of Luis Torgo on this research was undertaken thanks in part to funding from the Canada First Research Excellence Fund for the Ocean Frontier Institute.

References

1. Ribeiro, R.P.: Utility-based regression. PhD thesis, Department Computer Science, Faculty of Sciences - University of Porto (2011)
2. Elkan, C.: The foundations of cost-sensitive learning. In: IJCAI'01: Proceedings of the 17th International Joint Conference of Artificial Intelligence, vol. 1, pp. 973–978. Morgan Kaufmann Publishers (2001)
3. Torgo, L., Ribeiro, R.: Utility-based regression. In: Kok, J.N., Koronacki, J., Lopez de Mantaras, R., Matwin, S., Mladenič, D., Skowron, A. (eds.) PKDD 2007. LNCS (LNAI), vol. 4702, pp. 597–604. Springer, Heidelberg (2007). https://doi.org/10.1007/978-3-540-74976-9_63
4. Ling, C.X., Sheng, V.S.: Cost-sensitive learning. In: Sammut, C., Webb, G.I. (eds.) Encyclopedia of Machine Learning, pp. 231–235. Springer, US, Boston, MA (2011)
5. Domingos, P.: Metacost: a general method for making classifiers cost-sensitive. In: KDD'99: Proceedings of the 5th International Conference on Knowledge Discovery and Data Mining, pp. 155–164. ACM Press (1999)
6. Duda, R.O., Hart, P.E., Stork, D.G.: Pattern Classification. Wiley, New York (2012)
7. Bansal, G., Sinha, A.P., Zhao, H.: Tuning data mining methods for cost-sensitive regression: a study in loan charge-off forecasting. J. Manag. Inf. Syst. **25**(3), 315–336 (2008)
8. Zhao, H., Sinha, A.P., Bansal, G.: An extended tuning method for cost-sensitive regression and forecasting. Decis. Support Syst. **51**(3), 372–383 (2011)
9. Hernández-Orallo, J.: Probabilistic reframing for cost-sensitive regression. ACM Trans. Knowl. Discov. Data **8**(4), 17:1–17:55 (2014)
10. Branco, P., Torgo, L., Ribeiro, R.P., Frank, E., Pfahringer, B., Rau, M.M.: Learning through utility optimization in regression tasks. In: 2017 IEEE International Conference on Data Science and Advanced Analytics, DSAA 2017, Tokyo, Japan, pp. 30–39 (2017). Accessed 19–21 Oct 2017
11. Frank, E., Bouckaert, R.R.: Conditional density estimation with class probability estimators. In: Asian Conference on Machine Learning, pp. 65–81. Springer (2009)
12. Rau, M.M., et al.: Accurate photometric redshift probability density estimation-method comparison and application. Mon. Not. R. Astron. Soc. **452**(4), 3710–3725 (2015)
13. Branco, P., Torgo, L., Ribeiro, R.P.: SMOGN: a pre-processing approach for imbalanced regression. In: First International Workshop on Learning with Imbalanced Domains: Theory and Applications, pp. 36–50 (2017)
14. Branco, P., Torgo, L., Ribeiro, R.P.: A survey of predictive modeling on imbalanced domains. ACM Comput. Surv. (CSUR) **49**(2), 31 (2016)
15. Poursabzi-Sangdeh, F., Goldstein, D.G., Hofman, J.M., Vaughan, J.W., Wallach, H.: Manipulating and measuring model interpretability. arXiv preprint arXiv:1802.07810 (2018)

16. Doshi-Velez, F., Kim, B.: Towards a rigorous science of interpretable machine learning. arXiv preprint arXiv:1702.08608 (2017)
17. R Core Team: R: A Language and Environment for Statistical Computing. R Foundation for Statistical Computing, Vienna, Austria (2018)
18. Dimitriadou, E., Hornik, K., Leisch, F., Meyer, D., Weingessel, A.: e1071: Misc Functions of the Department of Statistics (e1071), TU Wien (2011)
19. Liaw, A., Wiener, M.: Classification and regression by randomforest. R News **2**(3), 18–22 (2002)
20. Domingos, P.: Knowledge acquisition from examples via multiple models. In: Machine Learning - International Workshop Then Conference -, Morgan Kaufmann Publishers, INC., pp. 98–106 (1997)
21. Torgo, L.: An infra-structure for performance estimation and experimental comparison of predictive models in R. In: CoRR arXiv:abs/1412.0436 (2014)
22. Demšar, J.: Statistical comparisons of classifiers over multiple data sets. J. Mach. Learn. Res. **7**, 1–30 (2006)

Predicting Rice Phenotypes with Meta-learning

Oghenejokpeme I. Orhobor[1(✉)] [iD], Nickolai N. Alexandrov[2],
and Ross D. King[1] [iD]

[1] The University of Manchester, Manchester M13 9PL, United Kingdom
 oghenejokpeme.orhobor@manchester.ac.uk
[2] The International Rice Research Institute, Los Baños, Philippines

Abstract. The features in some machine learning datasets can naturally be divided into groups. This is the case with genomic data, where features can be grouped by chromosome. In many applications it is common for these groupings to be ignored, as interactions may exist between features belonging to different groups. However, including a group that does not influence a response introduces noise when fitting a model, leading to suboptimal predictive accuracy. Here we present two general frameworks for the generation and combination of meta-features when feature groupings are present. We evaluated the frameworks on a genomic rice dataset where the regression task is to predict plant phenotype. We conclude that there are use cases for both frameworks.

Keywords: Rice · Bioinformatics · Machine learning · Meta-learning

1 Introduction

Machine learning algorithms are increasingly being adapted for the prediction of plant phenotypes [17]. This task is most commonly regression based as most agronomic phenotypes are quantitative. This observation is true of rice [38], the most agronomically important crop in the world, as a significant proportion of the global population relies on it for their dietary needs [26]. With a growing global population, estimates suggest that we need to double rice yields over the next few decades [34,42]. Therefore, it is crucial that we develop high yielding varieties that are resilient to an increase in biotic and abiotic stresses caused by climate change [39]. The predictive phenotype models built for such plant populations are most commonly used in genomic selection (GS). In GS, these predictive models are used to estimate the likelihood that an individual in a population will express a trait of interest. This likelihood is expressed as a genomic estimated breeding value (GEBV) and is used by plant breeders to select individuals that will serve as parents for the next generation of progeny. Therefore, it is desirable that the models used to estimate GEBVs are as accurate as possible.

© Springer Nature Switzerland AG 2018
L. Soldatova et al. (Eds.): DS 2018, LNAI 11198, pp. 144–158, 2018.
https://doi.org/10.1007/978-3-030-01771-2_10

GS has only been recently adopted in rice [16], and a model which is based on a single learning algorithm is often used for phenotype prediction, most commonly a variant of the best linear unbiased predictor [16,31]. In this context, we propose the use of meta-learning, which seeks to improve overall predictive accuracy by leveraging the predictive power of multiple learning algorithms, and has been shown in other domains to outperform a single learning algorithm if the goal is to optimize predictive accuracy [21]. The process can be broadly split into two main steps, a meta-feature generation step and a meta-feature integration step. In the former, a set of base models are built using a collection of learning algorithms. Each base model is then used to predict meta-features, which are predictions of a phenotype of interest. In the latter, the meta-features generated in the previous step are combined using another learning algorithm to form the final prediction.

A vital consideration we make is that of the nature of the attributes or features present in the input data used in building phenotype prediction models. The input data is often genomic, with features that are representative of the genetic diversity present in a population and are at different loci across an organism's genome [38]. These features are themselves representative of genes which control phenotypes and are located in different chromosomes. Therefore, the features in such genomic data can naturally be grouped by chromosome. In typical predictive experiments, the feature groupings by chromosome in the genomic data are ignored when models are built. The advantage of this approach is that potential interactions between features belonging to different chromosomes are captured. However, this may lead to suboptimal predictive accuracy if the features are in a chromosome with genes that are not associated with a phenotype, which introduces noise in a built model. Therefore, it might be the case that systematically diminishing the effects of features in irrelevant chromosomes might be more optimal. To address this problem, we propose two meta-learning frameworks which seek to improve phenotype prediction accuracy. The first ignores the feature groupings present in the input genomic data, and the other does not.

The remainder of this paper is as follows. In Sect. 2 we present the different considerations in meta-feature generation and integration, and in Sect. 3, we describe the proposed frameworks. In Sect. 4, our experimental setup is given, detailing the learners used in our evaluation. In Sect. 5 we discuss the outcome of evaluating the proposed frameworks, where our results show that there are use cases for both. Lastly, we conclude in Sect. 6.

2   Background

Rather than using a single learning algorithm, we seek to improve the predictive accuracy of models used to predict phenotype by combining the predictive power of a set of base learners utilizing a combining/meta-level learner. For example, assume a rice population with input genomic data (learning set) where one is interested in predicting grain width. Furthermore, assume that the goal is to improve predictive accuracy by combining the predictive power of random forests

[6] (RF) and support vector regression [10] (SVR) using simple linear regression (LR). Therefore, RF and SVR are the base learners while LR is the combining learner.

To amalgamate the predictive power of RF and SVR, they are both independently used to build a model to predict grain width, and the predictions made by these models are considered as grain width meta-features. Meta-features are typically generated by resampling the learning set using v-fold cross-validation [5,32], where each fold serves as a validation set and the remainder as a training set. We adopt this approach in the proposed frameworks. The first advantage v-fold cross-validation offers is in computational expense with regards to time. Given the advances in genotyping and sequencing technologies, the genomic data used in phenotype prediction experiments typically have input features in the order of a million features [1]. Therefore, building a single model takes a substantial amount of time, so other resampling methods like the Monte-Carlo cross-validation [44] may be infeasible. The second advantage is in the reduction of overfitting. As stated earlier, genomic data can have the order of a million input features; therefore there is potential for overfitting as it is often the case that the number of features far outnumber the number of samples ($p>>n$). Using our example, assume 3-fold cross-validation in the meta-feature generation step. In this case, both RF and SVR are used to build three models each on the different training sets and used to predict three meta-feature vectors on the validation sets. This means that we end up with three independent meta-feature matrices with columns corresponding with the number of base learners. Therefore, three sets of combining weights can be learned using LR and applied to the predictions made on unseen data. By doing this, we get combining weights that are not closely fit to one set of examples. A similar approach has been applied to positive effect in super learners [24].

The diversity of the set of base models used in generating the set of meta-features is vital, as it is desirable for the base models to be incorrect in different ways [7]. That is, given a set of base models, it is better for their predictions on some test set to be wrong on different samples so that the amalgamation of their predictions yield improved results. There are two main ways of achieving this. The first is to use a set of different base learners, which has been alluded to in our example, as they would make different assumptions about the nature of the relationships between the features in the input data [12]. For example, RF might make predictions based on nonlinear interactions amongst the features, whereas nearest neighbour techniques [2] which consider the level of relatedness between samples might yield a unique perspective. The second way of achieving model diversity is by varying the input data. That is, the input data can be split into multiple datasets which have different subsets of the features from the original. A base learner can then be used to build models on each of these new datasets, which are then used in the generation of meta-features. This approach is used in the stacked interval partial least squares framework [29], where meta-features are combined from various intervals in spectral data using partial least squares. We generally adopt the first approach and use it in the two proposed frameworks. The second is used only in the framework for which feature groupings are considered. The main difference between what we propose and the work using partial least squares [29] is that we use an ensemble of base learners for each input data subset.

Having generated a set of meta-features the next step is to integrate them, creating the final prediction. Using our example, this entails integrating the meta-feature predictions by RF and SVR. Several integration methods have been proposed. However, most are better suited to classification rather than regression problems [11,41]. In a regression setting, meta-feature integration is done using weights. These weights are coefficients which determine how much each base learner's meta-feature will influence the final prediction. A constant or dynamic weighting approach can be used [28]. Constant weighting in its simplest form involves averaging the meta-feature values for each sample. If the meta-features generated by the base models are incorrect on different samples but are all mostly accurate, averaging the meta-features improves overall accuracy by enhancing the incorrectly predicted samples. A more sophisticated constant weighting approach is to learn the weights using a combining learner, which is LR in our example. Note that on a test set, the learned weights are uniformly applied to every sample. We utilize both of these constant weighting approaches in the proposed procedures. In contrast to constant weighting, dynamic weighting assigns individual weights to each sample in a test set. This is done by learning individual weights for each sample in the test set using only the most closely related samples in the learning set [35]. This approach is computationally expensive in terms of time, and we do not use it in the proposed procedures. However, we conjecture that it may yield interesting results, and will be a subject of future study.

The natural feature groupings present in the genomic data used for phenotype prediction can also be thought of as views in multi-view learning. This assertion is based on the fact that the groups in this context are chromosomes which have genes that may influence a phenotype of interest. Therefore, each group of features represents a different perspective/view in terms of gene-phenotype associations. Several approaches have been proposed in multi-view learning [43], and multiple kernel learning (MKL) [37] is the most closely related to the current discourse. In typical multi-view learning problems, the views are often distinct, with different underlying structures and distributions of the input features. In MKL, learning algorithms that are best suited to each distinct view are used, and their predictions are then combined [9,25]. This approach is similar to what we propose, in that a combining learner is used to integrate the meta-features of different learners. However, our proposal differs in that multiple learners are used within each group or view to form a consensus on their influence on a trait.

3   Proposed Frameworks

In this section, we describe the proposed meta-learning frameworks. The first is for a situation in which the feature groupings present in an input dataset are ignored, and the second is for a situation in which feature groupings are considered, frameworks A and B respectively.

3.1   Framework A

The motivation for this framework is the overall improvement of phenotype prediction accuracy by leveraging the predictive power of multiple learning algorithms. In this case, we assume that although the features in an input dataset can be grouped by chromosome, these groupings are ignored when building a predictive model. Regarding the description of the procedure, we first give a description using an example, followed by a more formal one.

Assume a scenario where there is a learning and test genomic dataset with the goal of predicting grain width. The test set contains samples for which we want to predict their phenotype, and it is not used to build models. The two base learners are RF and SVR, and the combining learner is LR. We also assume v number of folds. For the meta-feature generation step, first split the learning data into v folds. Using each fold as a validation set and the remainder as a training set, build an RF and SVR model for grain width on the training set then predict learning meta-features using the validation set and also predict the test meta-features using the test set. At the end of this, v sets of learning and test meta-feature matrices are generated, all with two columns which correspond to predictions made by RF and SVR.

For the integration step, form a single test meta-feature matrix, $\mathbf{T}_{avg}$, by averaging the v predictions made by each base model (RF and SVR). Using LR, learn combining weights with each of the v learning meta-feature matrices. This produces v sets of weights. Apply each of these weights to $\mathbf{T}_{avg}$, producing v predictions. Finally, average these v predictions to form the final prediction for grain width. More formally:

Assume a learning set, a test set with samples for which we want to predict their phenotype, a set of base learners, a combining learner, and v cross-validation folds.

Step 1.

1. Split the learning set into v folds, aiming for approximately equal number of samples in each fold.
2. For each v fold:
   (a) *validation set* = current fold.
   (b) *training set* = the combination of the other folds.
   (c) build b base models using base learners on the training set.
   (d) predict the validation response using base models, generating a meta-feature matrix $\mathbf{V}_v \in \mathbb{R}^{m \times b}$, where m is the number of samples in the vth fold and b is number of base models.

    (e) predict the test response using base models, generating a meta-feature matrix $\mathbf{T}_v \in \mathbb{R}^{n \times b}$, where n is the number of samples in the test set and b is number of base models.
3. Output:
    (a) a set of validation meta-features $\mathcal{V} = (\mathbf{V}_1, \ldots, \mathbf{V}_v)$.
    (b) a set of test meta-features $\mathcal{T} = (\mathbf{T}_1, \ldots, \mathbf{T}_v)$.

Step 2. Using $\mathcal{V}$ and $\mathcal{T}$ from step 1 and a combining learner ϕ:

1. For each base model ψ, with $\psi_1, \ldots \psi_v$ predictions in $(\mathbf{T}_1, \ldots, \mathbf{T}_v) \in \mathcal{T}$, $\psi_{avg} = 1/v \sum_{i=1}^{v} \psi_i$. Therefore the average predictions for all base models in $\mathcal{T}$ can be represented as $\mathbf{T}_{avg} \in \mathbb{R}^{n \times b}$, where n is the number of samples and b is number of base models.
2. Learn combining weights on each validation meta-feature set in $\mathcal{V}$ using the combining learner ϕ. This produces v weight sets which are applied to $\mathbf{T}_{avg}$, producing $\phi_1, \ldots \phi_v$ predictions. The final prediction is given by $\phi_{avg} = 1/v \sum_{i=1}^{v} \phi_i$.
3. Output ϕ_{avg}.

3.2 Framework B

Like framework A, the motivation for this framework is also to improve overall phenotype predictive accuracy by leveraging the predictive power of multiple learning algorithms. However, in contrast to framework A, feature groupings present in the input genomic data are considered. The rationale for this is that for phenotype prediction, including features which are in regions that have genes that are not associated with a trait might only serve to introduce noise in a built model, leading to suboptimal predictive accuracy. Therefore, systematically diminishing the influence of such features might be more optimal.

For a general genomic dataset, it is assumed that the group to which each feature belongs is known, and all features in the dataset have been separated into their respective groups, c. That is, for a general dataset $\mathbf{D} \in \mathbb{R}^{m \times f}$, where m is the number of samples and f is number of features, $\mathbf{D}$ has been separated into c subsets, $\mathcal{D} = \mathbf{D}^1, \ldots, \mathbf{D}^c$, such that the intersection between the features in any pair of subsets must be empty and the union of the features in all subsets must be equal to the features in $\mathbf{D}$.

The procedure for this framework can be described using the same example in Sect. 3.1. However, we assume that both the learning and test datasets have been split into their c subsets by chromosome. For the meta-feature generation step, first split the learning set into v number of folds, ensuring that the same samples are in all v splits across all c data subsets. Using each fold as a validation set and the remainder as a training set in all c subsets, build an RF and SVR model for grain width on each c training set and then predict the learning meta-features using the corresponding c validation set and also predict the test meta-features using the corresponding c test set. At the end of this, v sets of learning and test meta-feature matrices are generated for the c subsets, all with two columns which correspond to predictions made by RF and SVR. Therefore, there are $v \times c$ meta-feature matrices for the learning and test sets. For the learning meta-feature matrices, merge all c subsets for each v fold. This produces v learning meta-feature sets, where each set has c pairs of RF and SVR meta-features. For the test meta-feature matrices, first form a single test meta-feature matrix for each c subset, $\mathbf{T}_{avg}^c$, by averaging the v predictions made by each base model (RF and SVR) within each c subset. These c averaged test meta-feature matrices are then merged in the same order the learning meta-feature matrices were, forming $\mathbf{T}_{merged}$.

Using LR, learn combining weights with each of the v merged learning meta-feature matrices. This produces v sets of weights. Apply each of these weights to $\mathbf{T}_{merged}$, producing v predictions. Finally, average these v predictions to form the final prediction for grain width. More formally:

Assume a learning and a test set that have been split into their c subsets using the chromosome to which features belong, a set of base learners, a combining learner, and v cross-validation folds.

Step 1.

1. Split all c learning set subsets into v-folds, aiming for approximately equal number of samples in each fold, and ensuring that the same samples are in each fold for each subset.
2. For each v fold and in each c subset:
   (a) *validation set* = current fold.
   (b) *training set* = the combination of the other folds.
   (c) build b base models using base learners on the training set.
   (d) predict the validation response using all trained models, generating a meta-feature matrix $\mathbf{V}_v^c \in \mathbb{R}^{m \times b}$, where m is the number of samples in the vth fold and b is number of base models.
   (e) predict the test response using all trained models, generating a meta-feature matrix $\mathbf{T}_v^c \in \mathbb{R}^{n \times b}$, where n is the number of samples in the test set and b is number of base models.
3. Generating:
   (a) a set of validation meta-features for each c subset, $\mathcal{V}^1, \ldots, \mathcal{V}^c$, where $\mathcal{V}^c = (\mathbf{V}_1^c, \ldots, \mathbf{V}_v^c)$.
   (b) a set of test meta-features for each c subset, $\mathcal{T}^1, \ldots, \mathcal{T}^c$, where $\mathcal{T}^c = (\mathbf{T}_1^c, \ldots, \mathbf{T}_v^c)$.
4. Merge $\mathcal{V}^1, \ldots, \mathcal{V}^c$ in order for all v validation meta-feature sets, creating v merged validation meta-feature sets $\mathcal{V}_{merged} = (\mathbf{V}_1, \ldots, \mathbf{V}_v) \in \mathbb{R}^{m \times p}$, where p is $b \times c$.
5. For each test meta-feature set subset $\mathcal{T}^1, \ldots, \mathcal{T}^c$, average the v predictions of each base learner in $\mathbf{T}_1^c, \ldots, \mathbf{T}_v^c$. This produces the average prediction matrices of all base models for all c subsets, $\mathbf{T}_{avg}^1, \ldots, \mathbf{T}_{avg}^c$. Merge all c average prediction matrices in order to form $\mathbf{T}_{merged} \in \mathbb{R}^{n \times p}$, where p is $b \times c$.
6. Output:
   (a) the set of v merged validation meta-feature matrices $\mathcal{V}_{merged}$.

(b) the merged test meta-feature matrix $\mathbf{T}_{merged}$.

Step 2. Using $\mathcal{V}_{merged}$ and $\mathbf{T}_{merged}$ from step 1 and a combining learner ϕ:

1. Learn combining weights on each validation meta-feature set in $\mathcal{V}_{merged}$ using the combining learner ϕ. This produces v weight sets which are applied to $\mathbf{T}_{merged}$, producing $\phi_1, \ldots \phi_v$ predictions. The final prediction is given by $\phi_{avg} = 1/v \sum_{i=1}^{v} \phi_i$.
2. Output ϕ_{avg}.

4   Experimental Setup

In this section, we discuss the dataset and methods used in our evaluation.

4.1   Dataset

We evaluated the proposed procedures using data from the 3000 rice genomes project [1], downloaded from http://SNP-Seek.irri.org/_download.zul. For the genotype data, we used version 0.4 of the core single nucleotide polymorphism (SNP) subset of 3000 rice genomes, which consists of 3023 samples and 996,009 markers. It is a filtered SNP set with a fraction of missing data at <20%. Using linkage disequilibrium in Plink [33], we pruned this dataset using a window of 50 SNPs, a step size of 5, and with an r^2 value of 0.001. This generated a smaller dataset with 12,286 features which represent the twelve rice chromosomes. The total proportion of missing values in this dataset is approximately 7%. We converted each SNP call for all varieties to numeric values; class 1 homozygotes are represented with 1, class 2 homozygotes as -1, and heterozygotes with 0. Missing values were imputed using column means, as it has been shown that mean imputation is sufficient in cases where less than 20% of the data for each marker is missing [36].

Twelve quantitative traits were considered: culm diameter, culm length, culm number, grain length, grain width, grain weight, days to heading, ligule length, leaf length, leaf width, panicle length, and seedling height. Only 2266 samples in the genotype data are represented in the trait data. Of this 2266 samples in the trait data, some of them have missing values for some traits. For each trait experiment, we excluded samples with unavailable or missing trait data. The raw and processed forms of the data used in our experiments are available in the Mendeley Data Repository at http://dx.doi.org/10.17632/86ygms76pb.1.

4.2   Setup

We used $v = 5$ folds and split the dataset into learning (75%), and testing (25%) sets with random sampling. Predictive accuracy was calculated as the coefficient of determination (R^2). All experiments were performed in R [20] and the code is available at https://github.com/oghenejokpeme/DS2018. For the learners that require parameter tuning, we performed parameter selection using a grid search and cross-validation on the training data. We opted for grid search over random search [3] as the parameters which require tuning and the range of values we explored for these parameters were modest. This can be seen in the provided source code. We considered three sets of learners. Learners that take feature groupings into account, a set of base learners which do not take groupings into account and a set of combining learners.

4.3 Group Learners

In our evaluation, we considered learners which take feature groupings into account. These learners are the group least absolute selection and shrinkage operator [14] (GLASSO), group bridge-penalized regression [19] (GBRGE), and group minimax concave penalty [4] (GMCP). The optimal value for lambda along the regularization path was chosen using five-fold internal cross-validation for GLASSO. For GBRIDGE and GMCP, the Akaike Information Criteria was used as it has been shown to produce slightly better accuracies [30].

4.4 Base Learners

The base learners used are the ridge regression best linear unbiased predictor [13] (RBLUP), random forests (RF), gradient boosted machines [15] (GBM), support vector regression [10] (SVR), k nearest neighbors [2] (KNN), and extreme gradient boosting [8] (XGB). RBLUP is specially designed for genomic predictions and has no parameters that require tuning. For RF the default of 1/3 the total number of variables is considered at each split, five observations are used for each terminal node, and 1000 trees were grown for each forest. For GBM we used a shrinkage parameter of 0.1, interaction depth of 6, 15 minimum number of observations in each node, and 1500 trees were grown. For SVR we used a radial basis kernel, and the hyperparameters were tuned using a grid search. XGB was also tuned with a grid search. Lastly, the optimal number of neighbors, n, used in the KNN models were chosen using cross-validation, where $1 \leq n \leq 30$.

4.5 Combining Learners

The combining learners used are linear regression (LR), gradient descent [23] (GD), kernel regularized least squares [18] (KRLS), ridge regression [40] (RR), and principal component regression [22] (PCR). The regularization parameter for RR was selected using internal cross-validation. A radial basis kernel was used with KRLS, and the bandwidth and regularization parameters were chosen using a grid search. For PCR the number of components used was chosen using internal cross-validation.

5   Results

In this section, we first discuss the performance of the group and base learners. We then discuss the performance of the combining learners on the proposed frameworks and then compare the performance of the base learners to the combining learners.

5.1   Group and Base Learner Performance

Here we discuss the group and base learner performances which serve as a baseline for the performance of the combining learners on the proposed frameworks. For the twelve rice traits considered, a base learner which does not take feature groupings into account outperforms all other learners on eleven of the twelve traits (Table 1). In general SVM and XGB outperform all other learners, even outperforming RBLUP, a learner designed for genomic predictions. We argue that this is the case for two reasons, (1) the traits considered are controlled by features with strong nonlinear interactions which RBLUP does not detect, and (2) SVM and XGB are better able to deal with a large number of irrelevant features. This is significant as recent advances in genotyping and sequencing technologies mean that genomic data is now being generated with the order of a million features, most of which are irrelevant in a built model. Therefore, rather than using traditional methods like RBLUP for phenotype prediction, more sophisticated methods like XGB should also be considered if one wants to use a single learning algorithm. The best performing group learner was GLASSO, which excludes features belonging to groups with low signal by assigning a zero coefficient to all features in such groups. It outperforms all other learners on one trait, seedling height, suggesting that it is indeed the case that some traits might benefit from excluding features from certain chromosomes.

Table 1. Predictive accuracy (R^2) of the group and base learners. The best performing learner for each group is in boldface and the overall performing learner is underlined.

Trait	GLASSO	GBRGE	GMCP	RBLUP	RF	GBM	SVR	KNN	XGB
Culm diameter	**0.164**	-	-	0.163	0.155	0.100	0.179	0.097	**_0.171_**
Culm length	**0.549**	0.512	0.318	0.544	0.533	0.516	**_0.559_**	0.529	0.552
Culm number	0.213	-	-	0.216	0.218	0.191	0.217	0.217	**_0.219_**
Grain length	0.379	**0.387**	0.380	0.370	0.337	0.306	0.383	0.249	**_0.387_**
Grain width	**0.462**	0.458	0.455	0.483	0.446	0.439	0.480	0.379	**_0.489_**
Grain weight	**0.363**	0.325	0.318	0.370	0.353	0.299	**_0.379_**	0.281	**_0.379_**
Days to heading	**0.657**	0.615	0.591	0.674	0.660	0.654	0.680	0.691	**_0.693_**
Ligule length	**0.368**	0.282	0.236	0.374	0.355	0.327	**_0.380_**	0.310	0.370
Leaf length	**0.390**	0.291	0.081	0.400	0.398	0.365	**_0.419_**	0.375	0.397
Leaf width	**0.404**	0.344	0.334	0.399	0.403	0.395	0.413	0.364	**_0.423_**
Panicle length	**0.411**	0.349	0.302	0.412	0.405	0.383	**_0.437_**	0.342	0.428
Seedling height	**_0.225_**	-	-	**0.221**	0.188	0.173	0.207	0.168	0.199

5.2 Combining Learner Performance

In our evaluation of the proposed frameworks, the six base learners outlined in Sect. 4.4 were used to generate meta-features for twelve rice traits. To evaluate the frameworks five learning algorithms were then used as combining learners to integrate the generated meta-features. Comparing frameworks A and B based on the performance of the combining learners showed that for LR, framework A outperforms B on eleven of the twelve traits. For GD, framework A outperforms B in nine of the twelve traits. For KRLS, framework A outperforms B on eight of the twelve traits. For RR, framework A outperforms B on ten traits, they perform equally well on one trait, and framework B outperforms A on one trait. For PCR, framework A outperforms B on nine of the twelve traits, they perform equally well on two traits, and framework B outperforms A on one trait. See Table 2 for the results. These results suggest that on a per learner basis, framework A, in which feature groupings are ignored, is generally the better meta-learning approach.

Table 2. Predictive accuracy (R^2) of the combining learners on frameworks A and B. The best performing framework for each learner is in boldface. The overall best performing learner-framework pair is underlined.

Trait	LR		GD		KRLS		RR		PCR	
	A	B	A	B	A	B	A	B	A	B
Culm diameter	**0.175**	0.119	**0.178**	0.170	**0.178**	0.170	**0.177**	0.172	**0.177**	0.168
Culm length	**0.561**	0.552	**0.561**	-	0.566	**0.569**	**0.564**	0.563	**0.566**	**0.566**
Culm number	**0.236**	0.214	0.232	**0.242**	0.235	**0.239**	**0.233**	0.231	**0.236**	**0.236**
Grain length	**0.391**	0.378	**0.378**	0.348	**0.397**	0.388	**0.398**	0.388	**0.402**	0.383
Grain width	**0.497**	0.472	**0.477**	0.425	**0.499**	0.490	**0.497**	0.488	**0.498**	0.488
Grain weight	**0.379**	0.333	**0.371**	0.338	**0.376**	0.362	**0.382**	0.365	**0.380**	0.356
Heading date	0.692	**0.703**	**0.692**	-	0.698	**0.710**	0.699	**0.708**	0.699	**0.705**
Ligule length	**0.380**	0.374	0.381	**0.383**	0.381	**0.382**	**0.381**	0.375	**0.380**	0.372
Leaf length	**0.411**	0.385	**0.412**	0.398	**0.420**	0.411	**0.416**	0.411	**0.415**	0.409
Leaf width	**0.419**	0.389	**0.416**	0.401	0.419	**0.420**	**0.419**	0.416	**0.419**	0.409
Panicle length	**0.439**	0.394	0.429	**0.443**	**0.431**	0.429	**0.437**	**0.437**	**0.439**	0.438
Seedling height	**0.219**	0.168	**0.218**	0.193	**0.218**	0.214	**0.215**	0.210	**0.217**	0.210

However, evaluating the performance of the frameworks on a per trait basis irrespective of combining learner tells a different story. In this case, framework A and B perform best on six traits each. The results show that no particular learner performs best on any trait-framework pair. This suggests that if the proposed approaches are to be used, combining learners should be chosen based on the framework of choice and the trait one is interested in predicting. One way of making this decision might be to modify the well-known model selection procedure used to select a single model from a set of competing models. However, we acknowledge that this will be computationally expensive given the number of models that are built in both frameworks.

Furthermore, this result demonstrates that in a meta-learning setting, some traits benefit when the feature groupings are ignored in the meta-feature generation and integration steps, while others benefit from having the feature groupings considered. We argue that the latter case occurs for two reasons. Firstly, each group has its own unique set of meta-features, generated by its own set of models. Therefore, noise is not introduced in these models from groups that may not be strongly associated with a phenotype. Secondly, the meta-features for a group represent the degree of association that group has with a phenotype. Therefore, generating meta-features for each feature group in isolation before learning combining weights aids a combining learner in estimating the amount of influence each group has on a phenotype.

For each trait, we compared the best performing combining learner on both frameworks to the best performing base learner. For framework A, we found that the best performing combining learner performs just as well or outperforms the best performing base learner on ten of twelve traits. For framework B, we found that the best performing combining learner performs just as well or outperforms the best performing base learner on nine of twelve traits. See Table 3. These results show that it is not always the case that one of the meta-learning approaches outperforms a single base model. However, the best performing combining learner on at least one of the proposed meta-learning approaches outperforms the best performing single base learner on ten of the twelve traits. Therefore, we conclude that the proposed frameworks generally increase the accuracy by which plant phenotype can be predicted by leveraging the predictive power

Table 3. Predictive accuracy (R^2) of the best performing combining learners on frameworks A and B in comparison to the best performing base learner. The best performing meta-learning or single model approach is in boldface.

Trait	A	B	Base
Culm diameter	**0.178**	0.172	0.171
Culm length	0.566	**0.569**	0.559
Culm number	0.236	**0.242**	0.219
Grain length	**0.402**	0.388	0.387
Grain width	**0.499**	0.490	0.489
Grain weight	**0.382**	0.365	0.379
Heading date	0.699	**0.710**	0.693
Ligule length	0.381	**0.383**	0.380
Leaf length	**0.420**	0.411	0.419
Leaf width	0.419	0.420	**0.423**
Panicle length	0.439	**0.443**	0.437
Seedling height	0.219	0.214	**0.221**

of multiple learning algorithms, in scenarios where the feature groupings present in genomic data are considered and ignored.

6 Conclusion

In this paper, we investigated the prediction of rice phenotypes. We argued that because rice is the most agronomically important crop in the world, the models used by plant breeders for the selection of the parents that will produce progeny with desirable traits should be as accurate as possible. We proposed that meta-learning, which leverages the predictive power of multiple learning algorithms could improve the accuracy by which rice and plant phenotypes, in general, can be predicted. We noted that the genomic datasets often used in predicting phenotype consists of features that can naturally be separated into groups by chromosome and argued that including features from chromosomes which may not influence a trait might lead to suboptimal predictive accuracy, as it introduces noise in a built model. With this in mind, we proposed two meta-learning frameworks, one which does not consider feature groupings (framework A) and another which does (framework B). Our results show that framework A generally outperforms framework B on a per learner level of analysis, but that they perform equally well on a per trait level of analysis. But more importantly, the results show that the best performing meta-learner on at least one of the proposed meta-learning approaches outperforms the best performing single base learner on ten of the twelve traits. Therefore, we conclude that these frameworks, if adopted by plant breeders, has the potential to ensure food security for millions of people.

In future work, we intend to apply the proposed procedures to other agronomically relevant crops like wheat and barley, and possibly on human population data. Furthermore, we intend to extend the proposed procedures by introducing meta-feature pruning, which aids in the selection of the meta-features that will eventually be integrated [27]. There are several methods [7] that can be used to perform meta-feature pruning, and we conjecture that the different techniques will perform differently on the proposed frameworks. As stated in the discussion of considerations we made in developing the proposed frameworks (Sect. 2), we also intend to extend the proposed frameworks by introducing dynamic weighting for the integration of meta-features.

References

1. Alexandrov, N., et al.: SNP-seek database of SNPs derived from 3000 rice genomes. Nucl. Acids Res. **43**(D1), D1023–D1027 (2015)
2. Altman, N.S.: An introduction to kernel and nearest-neighbor nonparametric regression. Am. Stat. **46**(3), 175–185 (1992)
3. Bergstra, J., Bengio, Y.: Random search for hyper-parameter optimization. J. Mach. Learn. Res. **13**(Feb), 281–305 (2012)
4. Breheny, P., Huang, J.: Penalized methods for bi-level variable selection. Stat. Interface **2**(3), 369 (2009)

5. Breiman, L.: Stacked regressions. Mach. Learn. **24**(1), 49–64 (1996)
6. Breiman, L.: Random forests. Mach. Learn. **45**(1), 5–32 (2001)
7. Caruana, R., Niculescu-Mizil, A., Crew, G., Ksikes, A.: Ensemble selection from libraries of models. In: Proceedings of the Twenty-first International Conference on Machine Learning, p. 18. ACM (2004)
8. Chen, T., He, T.: xgboost: extreme gradient boosting. R package version 0.4-2 (2015)
9. Cortes, C., Mohri, M., Rostamizadeh, A.: Learning non-linear combinations of kernels. In: Advances in Neural Information Processing Systems, pp. 396–404 (2009)
10. Cortes, C., Vapnik, V.: Support-vector networks. Mach. Learn. **20**(3), 273–297 (1995)
11. Džeroski, S., Ženko, B.: Stacking with multi-response model trees. In: Roli, F., Kittler, J. (eds.) MCS 2002. LNCS, vol. 2364, pp. 201–211. Springer, Heidelberg (2002). https://doi.org/10.1007/3-540-45428-4_20
12. Džeroski, S., Ženko, B.: Is combining classifiers with stacking better than selecting the best one? Mach. Learn. **54**(3), 255–273 (2004)
13. Endelman, J.B.: Ridge regression and other kernels for genomic selection with r package rrBLUP. Plant Genome **4**(3), 250–255 (2011)
14. Friedman, J., Hastie, T., Tibshirani, R.: A note on the group lasso and a sparse group lasso. arXiv preprint arXiv:1001.0736 (2010)
15. Friedman, J.H.: Greedy function approximation: a gradient boosting machine. Ann. Stat. **29**(5), 1189–1232 (2001)
16. Grenier, C., et al.: Accuracy of genomic selection in a rice synthetic population developed for recurrent selection breeding. PloS ONE **10**(8), e0136594 (2015)
17. Grinberg, N.F., et al.: Implementation of genomic prediction in Lolium perenne (L.) breeding populations. Front. Plant Sci. **7**, 133 (2016)
18. Hainmueller, J., Hazlett, C.: Kernel regularized least squares: Reducing misspecification bias with a flexible and interpretable machine learning approach. Polit. Anal. mpt019 (2013)
19. Huang, J., Ma, S., Xie, H., Zhang, C.H.: A group bridge approach for variable selection. Biometrika **96**(2), 339–355 (2009)
20. Ihaka, R., Gentleman, R.: R: a language for data analysis and graphics. J. Comput. Graph. Stat. **5**(3), 299–314 (1996)
21. Jahrer, M., Töscher, A., Legenstein, R.: Combining predictions for accurate recommender systems. In: Proceedings of the 16th ACM SIGKDD International Conference on Knowledge Discovery and Data Mining, pp. 693–702. ACM (2010)
22. Jolliffe, I.T.: A note on the use of principal components in regression. Appl. Stat. **31**(3) 300–303 (1982)
23. Kivinen, J., Warmuth, M.K.: Exponentiated gradient versus gradient descent for linear predictors. Inf. Comput. **132**(1), 1–63 (1997)
24. Van der Laan, M.J., Polley, E.C., Hubbard, A.E.: Super learner. Stat. Appl. Genet. Mol. Biol. **6**(1), 1544–6115 (2007)
25. Lanckriet, G.R., Cristianini, N., Bartlett, P., Ghaoui, L.E., Jordan, M.I.: Learning the kernel matrix with semidefinite programming. J. Mach. Learn. Res. **5**(Jan), 27–72 (2004)
26. Maclean, J., Hardy, B., Hettel, G.: Rice almanac: source book for one of the most important economic activities on earth. In: IRRI (2013)
27. Mendes-Moreira, J., Soares, C., Jorge, A.M., Sousa, J.F.D.: Ensemble approaches for regression: a survey. ACM Comput. Surv. (CSUR) **45**(1), 10 (2012)
28. Merz, C.J.: Classification and regression by combining models. Ph.D. thesis, University of California Irvine (1998)

29. Ni, W., Brown, S.D., Man, R.: Stacked partial least squares regression analysis for spectral calibration and prediction. J. Chemom. **23**(10), 505–517 (2009)

30. Ogutu, J.O., Piepho, H.P.: Regularized group regression methods for genomic prediction: Bridge, MCP, SCAD, group bridge, group lasso, sparse group lasso, group MCP and group SCAD. In: BMC Proceedings. vol. 8, p. S7. BioMed Central (2014)

31. Onogi, A., et al..: Exploring the areas of applicability of whole-genome prediction methods for asian rice (oryza sativa l.). Theor. Appl. Genet. **128**(1), 41–53 (2015)

32. Parmanto, B., Munro, P.W., Doyle, H.R.: Reducing variance of committee prediction with resampling techniques. Connect. Sci. **8**(3–4), 405–426 (1996)

33. Purcell, S., et al.: Plink: a tool set for whole-genome association and population-based linkage analyses. Am. J. Hum. Genet. **81**(3), 559–575 (2007)

34. Ray, D.K., Mueller, N.D., West, P.C., Foley, J.A.: Yield trends are insufficient to double global crop production by 2050 (2013)

35. Rooney, N., Patterson, D., Anand, S., Tsymbal, A.: Dynamic integration of regression models. In: Roli, F., Kittler, J., Windeatt, T. (eds.) MCS 2004. LNCS, vol. 3077, pp. 164–173. Springer, Heidelberg (2004). https://doi.org/10.1007/978-3-540-25966-4_16

36. Rutkoski, J.E., Poland, J., Jannink, J., Sorrells, M.E.: Imputation of unordered markers and the impact on genomic selection accuracy. G3: Genes Genomes Genet. **3**(3), 427–439 (2013)

37. Sonnenburg, S., Rätsch, G., Schäfer, C., Schölkopf, B.: Large scale multiple kernel learning. J. Mach. Learn. Res. **7**(Jul), 1531–1565 (2006)

38. Spindel, J., et al.: Genomic selection and association mapping in rice (Oryza sativa): effect of trait genetic architecture, training population composition, marker number and statistical model on accuracy of rice genomic selection in elite, tropical rice breeding lines. PLoS Genet. **11**(2), e1004982 (2015)

39. Tai, A.P., Martin, M.V., Heald, C.L.: Threat to future global food security from climate change and ozone air pollution. Nat. Clim. Chang. **4**(9), 817–821 (2014)

40. Tibshirani, R.: Regression shrinkage and selection via the lasso. J. R. Stat. Society. Ser. B (Methodol.) **58**(1) 267–288 (1996)

41. Ting, K.M., Witten, I.H.: Issues in stacked generalization. J. Artif. Intell. Res. (JAIR) **10**, 271–289 (1999)

42. Un, U.N.: World population prospects: the 2015 revision, key findings and advance tables. Working Paper, No. ESA/P/WP. 241. (2015)

43. Xu, C., Tao, D., Xu, C.: A survey on multi-view learning. arXiv preprint arXiv:1304.5634 (2013)

44. Xu, L., Jiang, J.H., Zhou, Y.P., Wu, H.L., Shen, G.L., Yu, R.Q.: MCCV stacked regression for model combination and fast spectral interval selection in multivariate calibration. Chemom. Intell. Lab. Syst. **87**(2), 226–230 (2007)

Reinforcement Learning

Preference-Based Reinforcement Learning
Using Dyad Ranking

Dirk Schäfer[(⊠)] and Eyke Hüllermeier

Department of Computer Science, Paderborn University,
Paderborn, Germany
dirkschaefer.jivas@gmail.com, eyke@upb.de

Abstract. Preference-based reinforcement learning has recently been
introduced as a generalization of conventional reinformcement learning.
Instead of numerical rewards, which are often difficult to specify, the
former assumes weaker feedback in the form of qualitative preferences
between states or trajectories. A specific realization of preference-based
reinforcement learning is approximate policy iteration using label ranking. We propose an extension of this method, in which label ranking is
replaced by so-called *dyad ranking*. The main advantage of this extension
is the ability of dyad ranking to learn from feature descriptions of actions,
which are often available in reinforcement learning. Several simulation
studies are conducted to confirm the usefulness of the approach.

1   Introduction

Reinforcement learning (RL) is an established machine learning methodology for
modeling and optimizing the behavior of an automous agent acting in a dynamic
environment [14]. A key component of RL is a numerical reward function that
is used to provide positive or negative (and possibly delayed) feedback signals
for the agent's actions. This quest for numerical information impedes the use of
RL in situations where precise rewards are difficult to specify.

This observation has been the main motivation for so-called *preference-based
reinforcement learning* (PBRL), which has recently been introduced as a generalization of conventional RL [1,3]. Instead of numerical rewards, it assumes weaker
feedback in the form of qualitative preferences between states or trajectories. A
specific realization of preference-based reinforcement learning is a combination
of *approximate policy iteration* [4] and a preference learning method called *label
ranking* [16]. Roughly speaking, label ranking is used to generalize training information of the form "in state s, taking action a appears to be better than action
a'", so that a ranking of all available actions can be predicted for all states of
the agent's state space.

In this paper, we propose an enhancement of this method, in which label
ranking is replaced by so-called *dyad ranking* [10]. The main advantage of this
extension is the ability of dyad ranking to learn from feature descriptions of

© Springer Nature Switzerland AG 2018
L. Soldatova et al. (Eds.): DS 2018, LNAI 11198, pp. 161–175, 2018.
https://doi.org/10.1007/978-3-030-01771-2_11

actions, i.e., properties of the actions a, which are often available in reinforcement learning. This is not possible in standard label ranking, where choice alternatives (labels) are merely identified by their name but not characterized in terms of attributes. Our speculation is that exploiting feature descriptions will improve learning by helping to generalize, not only over the state space, but also over the action space.

The paper starts with a section about conventional RL and approximate policy iteration, followed by a section about preference-based RL and approximate policy iteration based on label ranking. Our new approach, approximate policy iteration based on dyad ranking, is then introduced in Sect. 4. In order to confirm the usefulness of the approach, several simulation studies are presented in Sect. 5, prior to concluding the paper in Sect. 6.

2   Reinforcement Learning

Conventional reinforcement learning assumes a scenario in which an agent moves through a (finite) *state space* S by repeatedly selecting *actions* from a set $A = \{a_1, \ldots, a_k\}$. A Markovian *state transition* function $\delta : S \times A \longrightarrow \mathbb{P}(S)$, where $\mathbb{P}(S)$ denotes the set of probability distributions over S, randomly takes the agent to a new state, depending on the current state and the chosen action. Occasionally, the agent receives feedback about its actions in the form of a reward signal $r : S \times A \longrightarrow \mathbb{R}$, where $r(s, a)$ is the reward the agent receives for performing action a in state s. The goal of the agent is to choose its actions so as to maximize its expected total reward.

The most common task is to learn a *policy* $\pi : S \longrightarrow A$ that prescribes the agent how to act optimally in each situation (state). More specifically, the goal is often defined as maximizing the expected sum of rewards (given the initial state s), with future rewards being discounted by a factor $\gamma \in [0, 1]$:

$$V^\pi(s) = E\left[\sum_{t=0}^{\infty} \gamma^t r(s_t, \pi(s_t)) \mid s_0 = s\right] \qquad (1)$$

where $(s_0, s_1, s_2, \ldots)$ is a trajectory of π through the state space. With $V^*(s)$ the best possible value that can be achieved for (1), a policy is called optimal if it achieves the best value in each state s. Thus, one possibility to learn an optimal policy is to learn an evaluation of states in the form of a value function [13], or to learn a so-called Q-function which returns the expected reward for a given state-action pair [18]: $Q^\pi(s, a) = r(s, a) + \gamma \cdot V^\pi(\delta(s, a))$.

2.1   Approximate Policy Iteration

Instead of determining optimal actions *indirectly* through learning the value function or the Q-function, one may try to learn a policy *directly* in the form of a mapping from states to actions. A particularly interesting approach in this regard is *approximate policy iteration* (API) with rollouts [4,8]. The key idea of this approach is to use a generative model of the underlying process to perform simulations that in turn allow for approximating the value of an action a in a given state s. To this end, the action is performed, resulting in a state $s_1 = \delta(s, a)$. The value of this state is estimated by performing so-called *rollouts*, i.e., by repeatedly selecting actions following a policy π for at most T steps, and finally accumulating the observed rewards. This is repeated several times, and the average reward over the rollouts is returned as an approximate Q-value $\tilde{Q}^\pi(s, a)$ for taking action a in state s (leading to s_1) and following policy π thereafter.

The rollouts are then used in a policy iteration loop, which iterates through each of the sample states, simulates all actions in a state, and determines the action a^* that promises the highest Q-value. If a^* is significantly better than all alternative actions in this state, a training example (s, a^*) is added to a training set $\mathcal{T}$, suggesting that a^* is the best action to take in state s. Eventually, $\mathcal{T}$ is used for a *policy generalization* step, i.e., to induce a state-action mapping $S \longrightarrow A$ that forms the new policy π'; the underlying problem to be solved is a standard (multi-class) classification problem. This process is repeated several times, until some stopping criterion is met (e.g., if the policy does not improve from one iteration to the next).

3   Preference-Based Reinforcement Learning

The key idea of *preference-based reinforcement learning* (PBRL) is to replace the (quantitative) evaluation of individual actions by the (qualitative) comparison between pairs of actions [3,5]. Comparisons of that kind are in principle enough to make optimal decisions. Besides, they are often more natural and less difficult to acquire, especially in applications where the environment does not provide numerical rewards in a natural way.

The basic piece of information we consider is a pairwise preference of the form $a_i \succ_s a_j$ or, more specifically, $a_i \succ_s^\pi a_j$, suggesting that in state s, taking action a_i (and following policy π afterwards) is better than taking action a_j. Evaluating a trajectory $t = (s_0, s_1, s_2, \dots)$ in terms of its (expected) total reward (1) reduces the comparison of trajectories to the comparison of real numbers; thus, comparability is enforced and a total order on trajectories is induced. More generally, and arguably more in line with the idea of qualitative feedback, one may assume a partial order relation $\sqsupseteq$ on trajectories, which means that trajectories t and t' can also be incomparable. A contextual preference can then be defined as follows:

$$ a_i \succ_s^\pi a_j \quad \Leftrightarrow \quad \mathbf{P}\big(t(a_i) \sqsupseteq t(a_j)\big) > \mathbf{P}\big(t(a_j) \sqsupseteq t(a_i)\big) \;, \tag{2} $$

where $t(a_i)$ denotes the (random) trajectory produced by taking action a_i in state s and following π thereafter, and $\mathbf{P}(t \sqsupseteq t')$ is the probability that trajectory t is preferred to t'. As an in-depth discussion of PBRL is beyond the scope of this paper, we refer the reader to [5] for more technical details; see also [19] for a recent survey of the topic.

3.1 API with Label Ranking

In [5], preference-based reinforcement learning is realized in the form of a preference-based variant of API, namely a variant in which, instead of a classifier $S \longrightarrow A$, a so-called *label ranker* is trained for policy generalization. In the problem of label ranking, the goal is to learn a model that maps instances to rankings over a finite set of predefined choice alternatives [16]. In the context of PBRL, the instance space is given by the state space S, and the set of labels corresponds to the set of actions A. Thus, the goal is to learn a mapping $S \longrightarrow \Pi(A)$, which maps states to total orders (permutations) of the available actions A. In other words, the task is to learn a function that is able to rank all available actions in a state according to their preference (2).

More concretely, a method called *ranking by pairwise comparison* (RPC) is used for training a label ranker [7]. RPC accepts training information in the form of binary (action) preferences $(s, a_k \succ a_j)$, indicating that in state s, action a_k is preferred to action a_j. Information of that kind can be produced thanks to the assumption of a generative model as described in Sect. 2.1. Subsequently, we refer to this approach as API-LR.

4 PBRL Using Dyad Ranking

In comparison to the original, classification-based approach to approximate policy iteration (Sect. 2.1), the ranking-based method outlined in Sect. 3.1 exhibits several advantages, notably the following:

- Pairwise preferences are normally easier to elicit for training than examples for unique optimal actions a^*. In particular, a comparison of only two actions is less difficult than "proving" the optimality of one among a possibly large set of actions.
- The preference-based approach allows for better exploiting the gathered training information. For example, it utilizes pairwise comparisons $a \succ a'$ between two actions even if both of them are suboptimal. As opposed to this, the original approach eventually only uses information about the (presumably) optimal action a^*.

In both approaches, however, actions a_i are treated as distinct elements, with no relation to each other; indeed, neither classification nor label ranking do consider any structure on the set of classes A (apart from the trivial discrete structure). Yet, if classes are actions in the context of RL, A is often equipped with a nontrivial structure, because actions can be described in terms of properties/features and can be more or less similar to each other. For example, if an action is an acceleration in a certain direction, like in the mountain car problem (see Sect. 5 below), then "fast to the right" is obviously more similar to "slowly to the right" than to "fast to the left".

Needless to say, the exploitation of feature-descriptions of actions is a possible way to improve learning in (preference-based) RL, and to generalize, not only over the state space S but also over the action space A. It may allow, for example, to predict the usefulness of actions that have never been tried before. To realize this idea, we make use of so-called *dyad ranking*, a generalization of label ranking that is able to exploit feature-descriptions of labels [11].

4.1   Dyad Ranking

Formally, a dyad is a pair of feature vectors $z = (x, y) \in Z = \mathbb{X} \times \mathbb{Y}$, where the feature vectors are from two (not necessarily different) domains $\mathbb{X}$ and $\mathbb{Y}$. A single training observation ρ_n $(1 \le n \le N)$ takes the form of a dyad ranking

$$\rho_n : z_1 \succ z_2 \succ \cdots \succ z_{M_n}, \quad M_n \ge 2, \tag{3}$$

of length M_n, which can vary between observations in the data set $\mathcal{D} = \{\rho_n\}_{n=1}^N$. The task of a dyad ranking method is to learn a ranking function that accepts as input any set of (new) dyads and produces as output a ranking of these dyads.

An important special case, called *contextual dyad ranking*, is closely related to label ranking [10]. As already mentioned, the label ranking problem is about learning a model that maps instances to rankings over a finite set of predefined choice alternatives. In terms of dyad ranking, this means that all dyads in an observation share the same context x, i.e., they are all of the form $z_j = (x, y_j)$; in this case, (3) can also be written as $\rho_n : (x, y_1) \succ (x, y_2) \succ \cdots \succ (x, y_{M_n})$. Likewise, a prediction problem will typically consist of ranking a subset $\{y_1, y_2, \ldots, y_M\} \subseteq \mathbb{Y}$ in a given context x.

4.2   Bilinear Plackett-Luce Model

The Plackett-Luce (PL) model is a statistical model for rank data. Given a set of alternatives $o_1, \ldots, o_K$, it represents a parameterized probability distribution on the set of all rankings over the alternatives. The model is specified by a parameter vector $v = (v_1, v_2, \ldots v_K) \in \mathbb{R}_+^K$, in which v_i accounts for the "strength" of the option o_i. The probability assigned by the PL model to a ranking is represented by a permutation π, where $\pi(i)$ is the index of the option put on position i, is given by

$$\mathbf{P}(\pi \mid v) = \prod_{i=1}^{K} \frac{v_{\pi(i)}}{v_{\pi(i)} + v_{\pi(i+1)} + \cdots + v_{\pi(K)}}. \tag{4}$$

In dyad ranking, the options o_i to be ranked are dyads $z = (x, y)$. Thus, a model suitable for dyad ranking can be obtained by specifying the PL parameters as a function of the feature vectors x and y [10]:

$$v(z) = v(x, y) = \exp\left(\langle w, \Phi(x, y) \rangle\right), \tag{5}$$

where Φ is a joint feature map [15]. A common choice for such a feature map is the Kronecker product:

$$\Phi(x, y) = x \otimes y = (x_1 \cdot y_1, x_1 \cdot y_2, \ldots, x_r \cdot y_c), \tag{6}$$

Algorithm 1 Approximate Policy Iteration based on Dyad Ranking

---

Require: sample states $\mathcal{S}$, initial (random) policy π_0, max. number of policy iterations p, subroutine **Evaluate Dyad Ranking** for determining dyad rankings for a given state and a set of permissible actions in that state.

1: **function** API-DR$(\mathcal{S}, \pi_0, p)$
2:     $\pi \leftarrow \pi_0,\ i \leftarrow 0$
3:     **repeat**
4:         $\pi' \leftarrow \pi,\ \mathcal{D} \leftarrow \emptyset$
5:         **for all** $s \in \mathcal{S}$ **do**
6:             $\rho_s \leftarrow$**Evaluate Dyad Ranking** $(A(s), \pi)$
7:             $\mathcal{D} \leftarrow \mathcal{D} \cup \{\rho_s\}$
8:         **end for**
9:         $\pi \leftarrow$ **Train Dyad Ranker** $(\mathcal{D}),\ i \leftarrow i + 1$
10:     **until Stopping Criterion** (π, p)
11:     **return** π
12: **end function**

---

which is a vector consisting of all pairwise products of the components of x and y. The Eq. (6) can equivalently be rewritten as a bilinear form $x^\top \mathbf{W} y$ with a matrix $\mathbf{W} = (w_{i,j})$; the entry $w_{i,j}$ can be considered as the weight of the interaction term $x_i y_j$. This choice of the joint-feature map yields the following bilinear version of the PL model, which we call BilinPL:

$$v(z) = v(x, y) = \exp\left(x^\top \mathbf{W} y\right) \tag{7}$$

Given a set of training data in the form of a set of dyad rankings (3), the learning task comes down to estimating the weight matrix $\mathbf{W}$. Thanks to the probabilistic nature of the model, this can be accomplished by leveraging the principle of maximum likelihood; for details of this approach, we refer to [11].

Due to the bilinearity assumption, BilinPL comes with a relatively strong bias. This may or may not turn out as an advantage, depending on whether the assumption holds sufficiently well, but in any case requires a proper feature engineering. As an alternative to the Kroncker product, $v(x, y)$ in (5) can also be represented in terms of a neural network [9]. This approach, called PLNet, allows for *learning* a highly nonlinear joint-feature representation; again, we refer to [11] for details.

4.3   API Using Dyad Ranking

We are now ready to introduce approximate policy iteration based on dyad ranking (API-DR) as a generalization of API-LR. The former is quite similar to the latter, except that a dyad ranker is trained instead of a label ranker. To this end, training data is again produced by executing a number of rollouts on states, starting with a specified action and following the current policy; see Algorithm 1.

In addition to the representation of actions in terms of features, API-DR has another important advantage. Thanks to the use of the (bilinear) PL model, it is not only able to predict a presumably best action in each state, but also informs about the degree of confidence in that prediction. More specifically, it provides a complete probability distribution over all rankings of actions in each state. Information of this kind is useful for various purposes, as will be discussed next.

Algorithm 2 Probabilistic Rollout Procedure

Require: Initial state s_0, initial action a_0, policy π, discount factor γ, number of rollouts K, max. length(horizon) of each trajectory L, generative environment model E

1: **function** ROLLOUT(π, s_0, a_0, K, L)
2:     **for** $k \leftarrow 1$ to K **do**
3:         **while** t $< L$ and ¬**Terminal State**(s_{t-1}) **do**
4:             $(s_t, r_t) \leftarrow$ **Simulate**(E, s_{t-1}, a_{t-1})
5:             $(a_t, p_t) \leftarrow$ **Utilize Policy**(π, s_t)
6:             $\widetilde{Q}_k \leftarrow \widetilde{Q}_k + \gamma^t r_t$
7:             $t \leftarrow t + 1$
8:         **end while**
9:         // Remaining rollouts can be skipped if p_t-values are high
10:     **end for**
11:     $\widetilde{Q} \leftarrow \frac{1}{k} \sum_{i=1}^{k} \widetilde{Q}_i$
12:     **return** $\widetilde{Q}$
13: **end function**

Exploration versus Exploitation. The rollout procedure (Algorithm 2) is invoked by the subroutine *Evaluate Dyad Ranking* (line 6 of Algorithm 1). Here, the PL model is used in its role as a policy, which means that it has to prescribe a single action a^* for each state s. The most obvious approach is to compute, for each action a, the probability

$$P(a \mid W, s) = \frac{\exp(s^\top W a)}{\sum_{i=1}^{K} \exp(s^\top W a)} \tag{8}$$

of being ranked first, and to choose the action maximizing this probability.

Adopting the presumably best action in each state corresponds to pure *exploitation*. It is well known, however, that successful learning requires a proper balance between *exploration* and exploitation. Interestingly, our approach suggests a very natural way of realizing such a balance, simply by replacing the maximization by a "soft-max" operation, i.e., by selecting each action a according to its probability (8).

As an aside, we note that a generalization of the PL model can be used to control the degree of exploration in a more flexible way:

$$P(a \mid W, s) = \frac{\exp(c \cdot s^\top W a)}{\sum_{i=1}^{K} \exp(c \cdot s^\top W a)} \tag{9}$$

for a constant $c \geq 0$; the larger c, the stronger the strategy focuses on the best actions.

Uncertainty Sampling. Another interesting opportunity to exploit probabilistic information is for *active learning* via *uncertainty sampling*. Uncertainty sampling is a general strategy for active learning in which those training examples are requested for which the learner appears to be maximally uncertain [12]. In binary classification, for example, these are typically the instances that are located closest to the (current) decision boundary.

In our case, the distribution (8) informs about the certainty or uncertainty of the learner regarding the best course of action in a given state s (or, alternatively, the uncertainty about the true ranking of all actions in that state). This uncertainty can be quantified, for instance, in terms of the entropy of that distribution, or the margin between the probability of the best and the second-best action. Correspondingly, those states can be selected as sample states $\mathcal{S}$ in Algorithm 1 for which the uncertainty is highest.

5   Experiments

In this section, we illustrate the performance of PBRL-DS by means of several case studies, essentially following and replicating the experimental setup of [5]. Section 5.1 starts with two benchmark problems that are well-known in the field of RL, and which could in principle also be solved using conventional RL methods. In Sect. 5.2, we tackle a problem in which preferences are indeed purely qualitative, and states only partially comparable; this is a typical example of applications in the realm of preference-based RL. Finally, we add another case study, in which we illustrate the use of PBRL for the configuration of image processing pipelines. Here, the motivation comes from the fact that comparing two images (in terms of their quality) is often much easier for a user than evaluating a single image. The data and code used for the experiments can be accessed under the following URL: https://github.com/disc5/dyad-config-rl.

5.1   Standard Benchmarks

Inverted Pendulum. The inverted pendulum (also known as *cart pole*) problem (IP) is to balance a pendulum which is attached on top of a cart. The only way to stabilize the pendulum is by moving the cart, which is placed on a planar ground, to the left or to the right. We adopt the experimental setting from Lagoudakis and Parr [8], in which the position of the cart in space is not taken into account.

In the original formulation, there are three actions possible which are mapped, respectively, onto the forces of $\{-10, 0, 10\}$ Newtons. The state space is continuous and two-dimensional. The first dimension captures the angle θ between the pole and the vertical axis, whereas the second dimension describes the angle velocity $\dot{\theta}$. The transitions of the physical model are determined by the nonlinear dynamics of the system; they depend on the current state $s = (\theta, \dot{\theta})$ and current action value a, respectively:

$$\ddot{\theta} = \frac{g\sin(\theta) - \alpha ml(\dot{\theta})^2 \sin(2\theta)/2 - \alpha\cos(\theta)a}{4l/3 - \alpha ml\cos^2(\theta)},$$

where $\alpha = 1/(m + M)$ and the residual parameters are chosen as in Lagoudakis and Parr (see Table 1).

Table 1. Inverted pendulum model parameters.

Parameter	Symbol	Value	Unit
Gravity	g	9.81	m/s^2
Cart mass	M	8.0	kg
Pendulum mass	m	2.0	kg
Pendulum length	l	0.5	m

Mountain Car. The mountain car problem (MC) consists of driving an underpowered car out of a valley. The agent must learn a policy which takes the momentum of the car into account when driving the car along the valley sides. It can basically power or throttle forwards and backwards. At each time step, the system dynamics depend on a state $s_t = (x_t, \dot{x}_t)$ and an action a_t. It is described by the following equations:

$$x_{t+1} = b_1(x_t + \dot{x}_{t+1})$$
$$\dot{x}_{t+1} = b_2(\dot{x}_t + 0.001a_t - 0.025\cos(3x_t)),$$

where b_1 is a function that restricts the position x to the interval $[-1.2, 0.5]$ and b_2 restricts the velocity to the interval $[-0.07, 0.07]$. In case the agent reaches $x_t = -1.2$, an inelastic collision is simulated by setting the velocity $\dot{x}$ to zero. The gravity depends on the local slope of the mountain, which is simulated with the term $0.025\cos(3x_t)$. As long as the position x is less then 0.5, the agent receives zero reward. If the car hits the right bound ($x = 0.5$), the goal is achieved, the episode ends, and the agent obtains reward 1.

In both problems, the actions are simulated to be noisy, which results in non-deterministic state transitions. Thus, the learner is required to perform multiple rollouts. In particular, we add random noise from the intervals $[-0.2, 0.2]$ and $[-0.01, 0.01]$ to the raw action signals for IP and MC, respectively.

Experiments. Our main evaluation measure is the *success rate* (SR), i.e., the percentage of learned policies that are sufficient. In the case of IP, a policy is considered *sufficient* when being able to balance the pendulum longer than

1000 steps (100 s). For MC, a sufficient policy is one that needs less than 75 steps to reach the goal. More specifically, following [3,4], we plot the cumulative distribution of success rates over a measure of complexity, i.e., the number of actions needed throughout the API procedure for generating a policy that solves a task successfully. The number is obtained by summing up the average numbers of actions performed for each of the K rollouts realized on initial (state, action) pairs[1]. A point (x, y) in these plots can be interpreted as the minimum number of actions x required to reach a success rate of y.

We hypothesize that the incorporation of action features can improve the quality of learned policies, especially in situations where data is scarce. To this end, the quality of policies learned by API-LR[2] and API-DR (implemented in the BilinPL variant) are measured under different conditions. We chose a moderate number of 17 actions on both environments by dividing the original number range into 17 equally sized parts. Thus, the action spaces for IP and MC are given, respectively, as follows:

$$A_{IP} = \{-10, -8.75, -7.5, -6.25, \ldots, 10\}$$
$$A_{MC} = \{-1, -.875, -.75, -.625, \ldots, .875, 1\}$$

Recall that, while API-DR is able to interpret the actions as numbers, and hence to exploit the metric structure of the real numbers, API-LR merely considers all actions as distinct alternatives.

We furthermore defined three conditions referred to as *complete*, *partial* and *duel*. Under the first condition, preferences about the entire action set are available per state. In the *partial* condition, the learner can only learn from three randomly drawn actions per state. In the last condition, only two actions are drawn, leading to only one preference per state. Under all condition the number of sampled states $|\mathcal{S}|$ was set to 50 for the MC task and 100 for the IP task. The results depicted in Fig. 1 clearly confirm our expectations.

5.2   Cancer Clinical Trials Simulation

Preference-based reinforcement learning has been specifically motivated by the example of optimal therapy design in cancer treatment [3]. The concrete scenario is based on a mathematical model of [21] that captures the tumor growth during a treatment, the level of toxicity (inversely related to the wellness of the patient) due to the chemotherapy, the effect of the treatment and the interaction between drug and tumor size. A state is described by the variables tumor size S and toxicity X, while actions correspond to the dosage level $D \in [0, 1]$ of the drug. The model is described by a system of difference equations $S_{t+1} = S_t + \Delta S_t$ and $X_{t+1} = X_t + \Delta X_t$, where $\Delta S_t = (a_1 \cdot \max(X_t, X_0) - b_1 \cdot (D_t - d_1)) \cdot \mathbb{1}_{S_t > 0}$, $\Delta X_t =$

---

[1] Note that the number of actions is not fixed per rollout and rather depends on the quality of the current policy. This includes the case that rollouts can stop prematurely before the maximal trajectory length L is reached.

[2] Throughout all experiments we used the RPC method in conjunction with logistic regression.

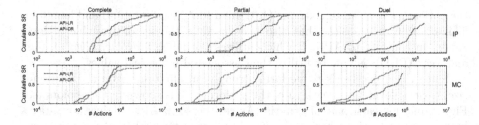

Fig. 1. Performance of the methods for the inverted pendulum (first row) and the mountain car task (second row).

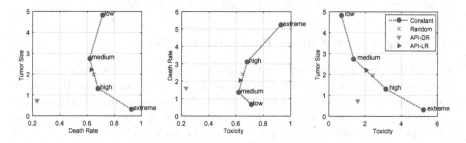

Fig. 2. Results of the cancer clinical trials simulation.

$a_2 \cdot \max(S_t, S_0) + b_2 \cdot (D_t - d_2)$. The probability for a patient to die in the t-th month follows a Bernoulli distribution with parameter $p = 1 - \exp(-\gamma(t))$ using the hazard function $\log \gamma(t) = c_0 + c_1 S_t + c_2 X_t$. Following the recommendation of [21], we fix the parameters of the difference equation as follows: $a_1 = 0.15$, $a_2 = 0.1$, $b_1 = b_2 = 1.2$, $d_1 = d_2 = 0.5$ and $c_0 = -4$, $c_1 = c_2 = 0.5$. The problem is how to choose appropriate dosage levels during a therapy of 6 months.

To circumvent the problem of reward function specification, we propose a preference-based comparison of policies π and π' as follows: π is preferred to π' if a patient survives with policy π and dies under π'. If a patient does not survive under either of the policies, then these are considered to be incomparable. If a patient survives under both policies, we define the preference via Pareto dominance as follows: $\pi \succeq \pi' \Leftrightarrow (C_X \leq C'_X)$ and $(C_S \leq C'_S)$, in which C_X denotes the maximal toxicity level occurred within a 6 month treatment under policy π, and analogously C'_X for π'. C_S and C'_S denote the tumor sizes at the end of the therapy after corresponding to policies π and π', respectively.

We applied API-DR (again in the BilinPL variant) with the feature representation $\boldsymbol{x} = (1, S, X)$ and $\boldsymbol{y} = (1, D, D^2, D^3)$. Moreover, the experimental protocol follows that of [3], in which virtual patients were generated by sampling initial states independently and uniformly from the interval $(0, 2)$. For training, 1000 patients were taken, and the quality of the learned policies was then tested on 200 new patients. In addition to API-LR, we also included a random policy and several constant policies also baselines. While the former selects dosages uniformly at random, these latter always prescribe the same dosage level regardless the patient's health state: extreme (1.0), high (0.7), medium (0.4) and low (0.1). This division of 4 dosages has also been used as the set of available actions. Again, in contrast to API-LR which utilizes the labels *extreme*, *high*, *medium* and *low*, API-DR is able to utilize their associated numerical values.

Since the objective is to perform strongly on all three criteria, i.e., tumor size, toxicity level, and death rate, the performance is shown in three plots (see Fig. 2), one for each pair of criteria. As can be seen, API-DR has advantages in comparison to the other approaches in all aspects: final tumor size, average toxicity level, and the death rate. In comparison with the constant approaches, API-DR is worse than the extreme constant dosage level in terms of final tumor size but superior in terms of death rate and toxicity levels.

5.3   Configuration of Image Processing Pipelines

Our last case study elaborates on the idea of using PBRL for the purpose of algorithm selection and configuration [2], especially in domains where the results produced by an algorithm might be difficult to asses numerically. As an example, we consider the problem of configuring image processing pipelines, with the goal to enhance the quality of an input image. The idea is that, for a human, a comparison between two candidate pictures x, x' is again easier than an absolute quality assessment (here, we mimic such a comparison by applying a similarity measure, defining preference for x in terms of proximity to some reference x^*).

An image processing pipeline is a sequence of possibly parameterized operators, where each operator takes an image as input and produces an image as output. The quality of a pipeline in influenced by the choice of operator types, the number of operators, their order, and of course the parameterization. We consider the choice of an operator with certain parameters as an action, which is taken by a policy learned with API-DR. The approach is outlined in Algorithm 3 and slightly differs from the basic version of Sect. 4.3. Note that, with the judgments on the quality of the pipelines, the function in line 15 extracts pairwise preferences on state/action pairs, and all these preference pairs are added to the training set $\mathcal{T}$. The policy model is trained in a supervised way on these preferences at the end of each round and can then be used for the next round for further improvement.

Experimental Protocol. The policy model in this scenario is PLNet, which is capable of learning non-linear relationships between the preferences of state-operator configuration pairs (x, y). The input consists of a 1-of-K encoding for the pipeline operator positions and another 1-of-K encoding for the operator-parameter combination. Furthermore, PLNet is configured with 3 layers, including one hidden layer with 10 neurons. All weights of the network are initialized randomly between -0.1 and 0.1, and the actual training is performed via stochastic gradient descent using 20 epochs and an initial learning rate of 0.1.

The set of image operators the learner can choose from consists of the logarithmic operator [6], the γ operator, and the brightness operator. Each of those

---

Algorithm 3 Pipeline Policy Training Algorithm

Require: Input $\mathcal{D} = \{(x_n, x_n^*)\}_{n=1}^{N}$, max pipeline length L
1: Initialize random policy model π
2: **repeat**
3:     Sample a number of training examples $\mathcal{S} \subset \mathcal{D}$
4:     $\mathcal{T} = \emptyset$
5:     **for** $n = 1$ **to** $|\mathcal{S}|$ **do**
6:         **for** $l = 1$ **to** L **do**
7:             **if** $l = 1$ **then** $x_n^{(l)} = x_n$
8:             **else** $x_n^{(l)} = x'^{(l-1)}_n$
9:             **end if**
10:             **for** $i = 1$ **to** $|A|$ **do**
11:                 $x'_{n_i} = \text{apply operator}(x_n^{(l)}, a_i)$
12:                 $\hat{x}_{n_i} = \text{rollout}(x'_{n_i}, \pi)$
13:             **end for**
14:             $\rho_n = \text{evaluate pipeline outputs } \{\hat{x}_{n_i}\}_{i=1}^{|A|}$ ▷ human or machine (with x_n^*)
15:             $\mathcal{T}_n = \text{generate pairwise preferences}(\rho_n)$
16:             $\mathcal{T} = \mathcal{T} \cup \mathcal{T}_n$
17:             $x'^{(l)}_n = \text{choose subsequent state of the best performing pipeline } (\rho_n)$
18:         **end for**
19:     **end for**
20:     Train $(\pi, \mathcal{T})$
21:     Evaluate policy $(\pi, \mathcal{D})$
22: **until** No policy improvements
23: **return** π

---

can be parameterized with different values (real numbers). Additionally, three other operators are available, namely an unsharping mask filter, histogram normalization, and a stop operator, which have no parameters. The stop operator enables a policy to control the length of a pipeline; it is usually applied when the outputs are good enough.

The images that are processed with the pipeline stem from the Fashion-MNIST data set [20]. It consists of 60k training and 10k gray scale images, where each image has 28x28 pixels and belongs to one of ten classes. The first hundred images from the original training set are used to create a pipeline training data set that consists of distorted and ground truth image pairs. A distorted image x is generated from a ground truth image by applying the pipeline $Op_1(2.5) \rightarrow Op_2(1.4) \rightarrow Op_1(1.5) \rightarrow Op_1(2.0)$ in reverse order on ground truth images x^*. This essentially serves the purpose to examine whether or not the learner is able to recover the distortion. A test data set is generated in the same way on the first hundred images of the original test data set.

As for the evaluation, we make use of the structured similarity (SSIM) measure [17]. The overall quality of the policy model is measured in terms of the mean average error (MAE) between the produced and the ground truth images. The approach is implemented in Matlab and the results of the experiments can be accessed under the following URL: https://github.com/disc5/dyad-config-rl.

Results. The (averaged) learning curve of the learned policies is shown in Fig. 3. It reflects the reduction in the error with an increasing number of rounds. The learning algorithm first enters an exploration phase, taking advantage of the (Boltzmann) exploration strategy of PBRL-DR as described in Sect. 4.3. The latter is also responsible for the cool-down phase and the convergence of the policy.

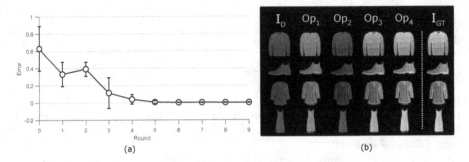

(a)   (b)

Fig. 3. (a) Learning curve of the policy model over a number of rounds. (b) Image processing pipeline with intermediate results. I_D refers to a damaged input and I_{GT} to the ground truth image.

6   Conclusion

We proposed a combination of preference-based reinforcement learning and dyad ranking that is applicable in situations where qualitative instead of quantitative preference information on state-action trajectories is available. This setting extends an existing preference-based variant of approximate policy iteration by incorporating feature descriptions of actions and considering rankings of dyads, i.e., state/action pairs, instead of rankings of actions given states. Thus, it becomes possible to generalize over the state and the action space simultaneously. The advantages of this approach and its ability to improve performance have been demonstrated in several case studies.

Going beyond the approach of approximate policy iteration, our next step is to elaborate on the usefulness of dyad ranking in other approaches to preference-based reinforcement learning.

Acknowledgements. This work was supported by the German Research Foundation (DFG) within the Collaborative Research Center "On-The-Fly Computing" (SFB 901). We are grateful to Javad Rahnama for his help with the case study on image pipeline configuration.

References

1. Akrour, R., Schoenauer, M., Sebag, M.: Preference-based policy learning. In: Proceedings of ECML/PKDD-2011, Athens, Greece (2011)
2. Brazdil, P., Giraud-Carrier, C.G.: Metalearning and algorithm selection: progress, state of the art and introduction to the 2018 special issue. Mach. Learn. **107**(1), 1–14 (2018)
3. Cheng, W., Fürnkranz, J., Hüllermeier, E., Park, S.H.: Preference-based policy iteration: leveraging preference learning for reinforcement learning. In: Proceedings of ECML/PKDD-2011, Athens, Greece (2011)
4. Dimitrakakis, C., Lagoudakis, M.G.: Rollout sampling approximate policy iteration. Mach. Learn. **72**(3), 157–171 (2008)
5. Fürnkranz, J., Hüllermeier, E., Cheng, W., Park, S.H.: Preference-based reinforcement learning: a formal framework and a policy iteration algorithm. Mach. Learn. **89**(1–2), 123–156 (2012)
6. Gonzalez, R.C., Woods, R.E.: Digital Image Processing, 2nd edn. Prentice Hall, Englewood Cliffs (2002)
7. Hüllermeier, E., Fürnkranz, J., Cheng, W., Brinker, K.: Label ranking by learning pairwise preferences. Artif. Intell. **172**, 1897–1917 (2008)
8. Lagoudakis, M., Parr, R.: Reinforcement learning as classification: leveraging modern classifiers. In: Proceedings of ICML, 20th International Conference on Machine Learning, vol. 20, pp. 424–431. AAAI Press (2003)
9. Schäfer, D., Hüllermeier, E.: Plackett-Luce networks for dyad ranking. In: Workshop LWDA, Lernen, Wissen, Daten, Analysen, Potsdam, Germany (2016)
10. Schäfer, D., Hüllermeier, E.: Dyad ranking using a bilinear Plackett-Luce model. In: Appice, A., Rodrigues, P.P., Santos Costa, V., Gama, J., Jorge, A., Soares, C. (eds.) ECML PKDD 2015. LNCS (LNAI), vol. 9285, pp. 227–242. Springer, Cham (2015). https://doi.org/10.1007/978-3-319-23525-7_14
11. Schäfer, D., Hüllermeier, E.: Dyad ranking using Plackett-Luce models based on joint feature representations. Mach. Learn. (2018)
12. Settles, B.: Active learning literature survey. Technical Report 1648, University of Wisconsin-Madison (2008)
13. Sutton, R.S.: Learning to predict by the methods of temporal differences. Mach. Learn. **3**(1), 9–44 (1988)
14. Sutton, R.S., Barto, A.G.: Reinforcement Learning: An Introduction. MIT Press, Cambridge (1998)
15. Tsochantaridis, I., Joachims, T., Hofmann, T., Altun, Y.: Large margin methods for structured and interdependent output variables. J. Mach. Learn. Res. **6**, 1453–1484 (2005)
16. Vembu, S., Gärtner, T.: Label ranking: a survey. In: Fürnkranz, J., Hüllermeier, E., (eds.) Preference Learning. Springer (2010)
17. Wang, Z., Bovik, A.C., Sheikh, H.R., Simoncelli, E.P.: Image quality assessment: from error visibility to structural similarity. IEEE Trans. Image Process. **13**(4), 600–612 (2004)
18. Watkins, C.J., Dayan, P.: Q-learning. Mach. Learn. **8**(3), 272–292 (1992)
19. Wirth, C., Akrour, R., Neumann, G., Fürnkranz, J.: A survey of preference-based reinforcement learning methods. J. Mach. Learn. Res. **18**, 136:1–136:46 (2017)
20. Xiao, H., Rasul, K., Vollgraf, R.: Fashion-MNIST: a novel image dataset for benchmarking machine learning algorithms (2017), arXiv:1708.07747
21. Zhao, Y., Kosorok, M.R., Zeng, D.: Reinforcement learning design for cancer clinical trials. Stat. Med. **28**(15), 1982–1998 (2009)

Streams and Time Series

COBRAS^TS: A New Approach to Semi-supervised Clustering of Time Series

Toon Van Craenendonck^(✉), Wannes Meert, Sebastijan Dumančić, and Hendrik Blockeel

Department of Computer Science, KU Leuven, Leuven, Belgium
{toon.vancraenendonck,wannes.meert,sebastijan.dumancic,
hendrik.blockeel}@kuleuven.be

Abstract. Clustering is ubiquitous in data analysis, including analysis of time series. It is inherently subjective: different users may prefer different clusterings for a particular dataset. Semi-supervised clustering addresses this by allowing the user to provide examples of instances that should (not) be in the same cluster. This paper studies semi-supervised clustering in the context of time series. We show that COBRAS, a state-of-the-art active semi-supervised clustering method, can be adapted to this setting. We refer to this approach as COBRAS^TS. An extensive experimental evaluation supports the following claims: (1) COBRAS^TS far outperforms the current state of the art in semi-supervised clustering for time series, and thus presents a new baseline for the field; (2) COBRAS^TS can identify clusters with separated components; (3) COBRAS^TS can identify clusters that are characterized by small local patterns; (4) actively querying a small amount of semi-supervision can greatly improve clustering quality for time series; (5) the choice of the clustering algorithm matters (contrary to earlier claims in the literature).

1 Introduction

Clustering is ubiquitous in data analysis. There is a large diversity in algorithms, loss functions, similarity measures, etc. This is partly due to the fact that clustering is inherently subjective: in many cases, there is no single correct clustering, and different users may prefer different clusterings, depending on their goals and prior knowledge [17]. Depending on their preference, they should use the right algorithm, similarity measure, loss function, hyperparameter settings, etc. This requires a fair amount of knowledge and expertise on the user's side.

Semi-supervised clustering methods deal with this subjectiveness in a different manner. They allow the user to specify constraints that express their subjective interests [18]. These constraints can then guide the algorithm towards solutions that the user finds interesting. Many such systems obtain these constraints by asking the user to answer queries of the following type: *should these two elements be in the same cluster?* A so-called must-link constraint is obtained if the answer is yes, a cannot-link otherwise. In many situations, answering this

© Springer Nature Switzerland AG 2018
L. Soldatova et al. (Eds.): DS 2018, LNAI 11198, pp. 179–193, 2018.
https://doi.org/10.1007/978-3-030-01771-2_12

type of questions is much easier for the user than selecting the right algorithm, defining the similarity measure, etc. Active semi-supervised clustering methods aim to limit the number of queries that is required to obtain a good clustering by selecting informative pairs to query.

In the context of clustering *time series*, the subjectiveness of clustering is even more prominent. In some contexts, the time scale matters, in other contexts it does not. Similarly, the scale of the amplitude may (not) matter. One may want to cluster time series based on certain types of qualitative behavior (monotonic, periodic, . . .), local patterns that occur in them, etc. Despite this variability, and although there is a plethora of work on time series clustering, semi-supervised clustering of time series has only very recently started receiving attention [7].

In this paper, we show that COBRAS, an existing active semi-supervised clustering system, can be used practically "as-is" for time series clustering. The only adaptation that is needed, is plugging in a suitable similarity measure and a corresponding (unsupervised) clustering approach for time series. Two plug-in methods are considered for this: spectral clustering using dynamic time warping (DTW), and k-Shape [11]. We refer to COBRAS with one of these plugged in as COBRAS^{TS} (COBRAS for Time Series). We perform an extensive experimental evaluation of this approach.

The main contributions of the paper are twofold. First, it contributes a novel approach to semi-supervised clustering of time series, and two freely downloadable, ready-to-use implementations of it. Second, the paper provides extensive evidence for the following claims: (1) COBRAS^{TS} outperforms cDTW^{SS} (the current state of the art) by a large margin; (2) COBRAS^{TS} can identify clusters with separated components; (3) COBRAS^{TS} can identify clusters that are characterized by small local patterns; (4) actively querying a small amount of supervision can greatly improve results in time series clustering; (5) the choice of clustering algorithm matters, it is not negligible compared to the choice of similarity. Except for claim 4, all these claims are novel, and some are at variance with the current literature. Claim 4 has been made before, but with much weaker empirical support.

2   Related Work

Semi-supervised clustering has been studied extensively for clustering attribute-value data, starting with COP-KMeans [18]. Most semi-supervised methods extend unsupervised ones by adapting their clustering procedure [18], their similarity measure [20], or both [2]. Alternatively, constraints can also be used to select and tune an unsupervised clustering algorithm [13].

Traditional methods assume that a set of pairwise queries is given prior to running the clustering algorithm, and in practice, pairs are often queried randomly. Active semi-supervised clustering methods try to query the most informative pairs first, instead of random ones [9]. Typically, this results in better clusterings for an equal number of queries. COBRAS [15] is a recently proposed method that was shown to be effective for clustering attribute-value data. In this paper, we show that it can be used to cluster time series with little modification. We describe COBRAS in more detail in the next section.

In contrast to the wealth of papers in the attribute-value setting, only one method has been proposed specifically for semi-supervised time series clustering with active querying. cDTW^{SS} [7] uses pairwise constraints to tune the warping width parameter w in constrained DTW. We compare COBRAS^{TS} to this method in the experiments.

In contrast to semi-supervised time series clustering, semi-supervised time series *classification* has received significant attention [19]. Note that these two settings are quite different: in semi-supervised classification, the set of classes is known beforehand, and at least one labeled example of each class is provided. In semi-supervised clustering, it is not known in advance how many classes (clusters) there are, and a class may be identified correctly even if none of its instances have been involved in the pairwise constraints.

3   Clustering Time Series with COBRAS

3.1   COBRAS

We describe COBRAS only to the extent necessary to follow the remainder of the paper; for more information, see Van Craenendonck et al. [14,15].

COBRAS is based on two key ideas. The first [14] is that of *super-instances*: sets of instances that are temporarily assumed to belong to the same cluster in the unknown target clustering. In COBRAS, a clustering is a set of clusters, each cluster is a set of super-instances, and each super-instance is a set of instances. Super-instances make it possible to exploit constraints much more efficiently: querying is performed at the level of super-instances, which means that each instance does not have to be considered individually in the querying process. The second key idea in COBRAS [15] is that of the *automatic detection of the right level* at which these super-instances are constructed. For this, it uses an iterative refinement process. COBRAS starts with a single super-instance that contains all the examples, and a single cluster containing that super-instance. In each iteration the largest super-instance is taken out of its cluster, split into smaller super-instances, and the latter are reassigned to (new or existing) clusters. Thus, COBRAS constructs a clustering of super-instances at an increasingly fine-grained level of granularity. The clustering process stops when the query budget is exhausted.

We illustrate this procedure using the example in Fig. 1. Panel A shows a toy dataset that can be clustered according to several criteria. We consider differentiability and monotonicity as relevant properties. Initially, all instances belong to a single super-instance (S_0), which constitutes the only cluster (C_0). The second and third rows of Fig. 1 show two iterations of COBRAS.

In the first step of iteration 1, COBRAS refines S_0 into 4 new super-instances, which are each put in their own cluster (panel B). The refinement procedure uses k-means, and the number of super-instances in which to split is determined based on constraints; for details, see [15]. In the second step of iteration 1, COBRAS

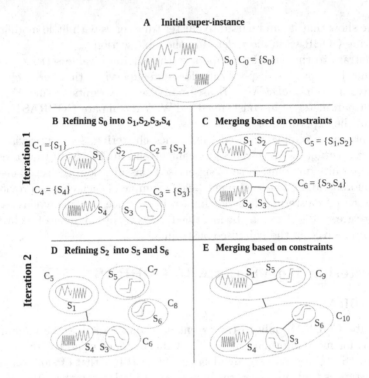

Fig. 1. An illustration of the COBRAS clustering procedure.

determines the relation between new and existing clusters. To determine the relation between two clusters, COBRAS queries the pairwise relation between the medoids of their closest super-instances. In this example, we assume that the user is interested in a clustering based on differentiability. The relation between $C_1 = \{S_1\}$ and $C_2 = \{S_2\}$ is determined by posing the following query: *should* ⋀⋀ and ⌐ *be in the same cluster?* The user answers yes, so C_1 and C_2 are merged into C_5. Similarly, COBRAS determines the other pairwise relations between clusters. It does not need to query all of them, many can be derived through transitivity or entailment [15]. The first iteration ends once all pairwise relations between clusters are known. This is the situation depicted in panel C. Note that COBRAS has not produced a perfect clustering at this point, as S_2 contains both differentiable and non-differentiable instances.

In the second iteration, COBRAS again starts by refining its largest super-instance. In this case, S_2 is refined into S_5 and S_6, as illustrated in panel D. A new cluster is created for each of these super-instances, and the relation between new and existing clusters is determined by querying pairwise constraints. A must-link between S_5 and S_1 results in the creation of $C_9 = \{S_1, S_5\}$. Similarly, a must-link between S_6 and S_3 results in the creation of $C_{10} = \{S_3, S_4, S_6\}$. At this point, the second iteration ends as all pairwise relations between clusters are known. The

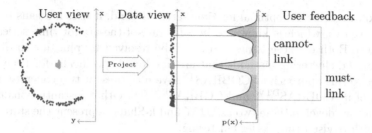

Fig. 2. Clusters may contain separated components when projected on a lower-dimensional subspace.

clustering consists of two clusters, and a data granularity of 5 super-instances was needed.

In general, COBRAS keeps repeating its two steps (refining super-instances and querying their pairwise relations) until the query budget is exhausted.

Separated Components. A noteworthy property of COBRAS is that, by interleaving splitting and merging, it can split off a subcluster from a cluster and reassign it to another cluster. In this way, it can construct clusters that contain *separated components* (different dense regions that are separated by a dense region belonging to another cluster). It may, at first, seem strange to call such a structure a "cluster", as clusters are usually considered to be coherent high-density areas. However, note that a coherent cluster may become incoherent when projected onto a subspace. Figure 2 illustrates this. Two clusters are clearly visible in the XY-space, yet projection on the X-axis yields a trimodal distribution where the outer modes belong to one cluster and the middle mode to another. In semi-supervised clustering, it is realistic that the user evaluates similarity on the basis of more complete information than explicitly present in the data; coherence in the user's mind may therefore not translate to coherence in the data space[1].

The need for handling clusters with multi-modal distributions has been mentioned repeatedly in work on time series anomaly detection [5], on unsupervised time series clustering [11], and on attribute-value semi-supervised constrained clustering [12]. Note, however, a subtle difference between having a multi-modal distribution and containing separated components: the first assumes that the components are separated by a low-density area, whereas the second allows them to be separated by a dense region of instances from another cluster.

3.2 COBRAS^{DTW} and COBRAS$^{k-Shape}$

COBRAS is not suited out-of-the-box for time series clustering, for two reasons. First, it defines the super-instance medoids w.r.t. the *Euclidean* distance, which

---

[1] Note that Fig. 2 is just an illustration; it can be difficult to express the more complete information explicitly as an additional dimension, as is done in the figure.

is well-known to be suboptimal for time series. Second, it uses k-means to refine super-instances, which is known to be sub-state-of-the-art for time series clustering [11]. Both of these issues can easily be resolved by plugging in distance measures and clustering methods that are developed specifically for time series. We refer to this approach as COBRAS^{TS}. We now present two concrete instantiations of it: COBRAS^{DTW} and COBRAS$^{k\text{-Shape}}$. Other instantiations can be made, but we develop these two as DTW and k-Shape represent the state of the art in unsupervised time series clustering.

Algorithm 1 COBRAS^{DTW}

Input: A dataset, the DTW warping window width w, the γ parameter used in converting distances to similarities and access to an oracle answering pairwise queries
Output: A clustering
1: Compute the full pairwise DTW distance matrix
2: Convert each distance d to an affinity a: $a_{i,j} = e^{-\gamma d_{i,j}}$
3: Run COBRAS, substituting k-means for splitting super-instances with spectral clustering on the previously computed affinity matrix

---

COBRAS^{DTW} uses DTW as its distance measure, and spectral clustering to refine super-instances. It is described in Algorithm 1. DTW is commonly accepted to be a competitive distance measure for time series analysis [1], and spectral clustering is well-known to be an effective clustering method [16]. We use the constrained variant of DTW, cDTW, which restricts the amount by which the warping path can deviate from the diagonal in the warping matrix. cDTW offers benefits over DTW in terms of both runtime and solution quality [7,11], if run with an appropriate window width.

COBRAS$^{k\text{-Shape}}$ uses the shape-based distance (SBD, [11]) as its distance measure, and the corresponding k-Shape clustering algorithm [11] to refine super-instances. k-Shape can be seen as a k-means variant developed specifically for time series. It uses SBD instead of the Euclidean distance, and comes with a method of computing cluster centroids that is tailored to time series. k-Shape was shown to be an effective and scalable method for time series clustering in [11]. Instead of the medoid, COBRAS$^{k\text{-Shape}}$ uses the instance that is closest to the SBD centroid as a super-instance representative.

4 Experiments

In our experiments we evaluate COBRAS^{DTW} and COBRAS$^{k\text{-Shape}}$ in terms of clustering quality and runtime, and compare them to state-of-the-art semi-supervised (cDTW^{SS} and COBS) and unsupervised (k-Shape and k-MS) competitors. Our experiments are fully reproducible: we provide code for COBRAS^{TS} in a public repository[2], and a separate repository for our experimental setup[3]. The experiments are performed on the public UCR collection [6].

---

[2] https://bitbucket.org/toon_vc/cobras_ts or using `pip install cobras_ts`.
[3] https://bitbucket.org/toon_vc/cobras_ts_experiments.

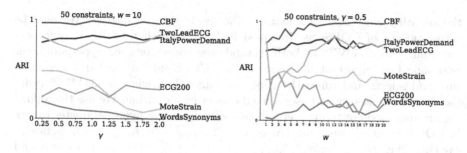

Fig. 3. Sensitivity to γ and w for several datasets.

4.1 Methods

COBRAS^{TS}. COBRAS$^{\text{k-Shape}}$ has no parameters (the number of clusters used in k-Shape to refine super-instances is chosen based on the constraints in COBRAS). We use a publicly available Python implementation[4] to obtain the k-Shape clusterings. COBRAS^{DTW} has two parameters: γ (used in converting distances to affinities) and w (the warping window width). We use a publicly available C implementation to construct the DTW distance matrices [10]. In our experiments, γ is set to 0.5 and w to 10% of the time series length. The value $w = 10\%$ was chosen as Dau et al. [7] report that most datasets do not require w greater than 10%. We note that γ and w could in principle also be tuned for COBRAS^{DTW}. There is, however, no well-defined way of doing this. We cannot use the constraints for this, as they are actively selected during the execution of the algorithm (which of course requires the affinity matrix to already be constructed). We did not do any tuning on these parameters, as this is also hard in a practical clustering scenario, but observed that the chosen parameter values already performed very well in the experiments. We performed a parameter sensitivity analysis, illustrated in Fig. 3, which shows that the influence of these parameters is highly dataset-dependent: for many datasets their values do not matter much, for some they result in large differences.

cDTW^{SS}. cDTW^{SS} uses pairwise constraints to tune the w parameter in cDTW. In principle, the resulting tuned cDTW measure can be used with any clustering algorithm. The authors in [7] use it in combination with TADPole [4], and we do the same here. We use the code that is publicly available on the authors' website[5]. The cutoff distances used in TADPole were obtained from the authors in personal communication.

COBS. COBS [13] uses constraints to select and tune an unsupervised clustering algorithm. It was originally proposed for attribute-value data, but it can trivially be modified to work with time series data as follows. First, the full pairwise distance matrix is generated with cDTW using $w = 10\%$ of the time series length. Next, COBS generates clusterings by varying the hyperparameters of several standard unsupervised clustering methods, and selects

---

[4] https://github.com/Mic92/kshape.
[5] https://sites.google.com/site/dtwclustering/.

the resulting clustering that satisfies the most pairwise queries. We use the active variant of COBS, as described in [13]. Note that COBS is conceptually similar to cDTW^{SS}, as both methods use constraints for hyperparameter selection. The important difference is that COBS uses a fixed distance measure and selects and tunes the clustering algorithm, whereas cDTW^{SS} tunes the similarity measure and uses a fixed clustering algorithm. We use the following unsupervised clustering methods and corresponding hyperparameter ranges in COBS: spectral clustering ($K \in [\max(2, K_{true} - 5), K_{true} + 5]$), hierarchical clustering ($K \in [\max(2, K_{true} - 5), K_{true} + 5]$, with both average and complete linkage), affinity propagation (`damping` $\in [0.5, 1.0]$) and DBSCAN ($\epsilon \in [$`min pairwise dist.`, `max. pairwise dist`$]$, `min_samples` $\in [2, 21]$). For the continuous parameters, clusterings were generated for 20 evenly spaced values in the specified intervals. Additionally, the γ parameter in converting distances to affinities was varied in $[0, 2.0]$ for clustering methods that take affinities as input, which are all of them except DBSCAN, which works with distances. We did not vary the warping window width w for generating clusterings in COBS. This would mean a significant further increase in computation time, both for generating the DTW distance matrices, and for generating clusterings with all methods and parameter settings for each value of w.

k-Shape and k-MS. Besides the three previous semi-supervised methods, we also include k-Shape [11] and k-MultiShape (k-MS) [11] in our experiments as unsupervised baselines. k-MS [11] is similar to k-Shape, but uses multiple centroids, instead of one, to represent each cluster. It was found to be the most accurate method in an extensive experimental study that compares a large number of unsupervised time series clustering methods on the UCR collection [11]. The number of centroids that k-MS uses to represent a cluster is a parameter; following the original paper we set it to 5 for all datasets. The k-MS code was obtained from the authors.

4.2 Data

We perform experiments on the entire UCR time series classification collection [6], which is the largest public collection of time series datasets. It consists of 85 datasets from a wide variety of domains. The UCR datasets come with a predefined training and test set. We use the test sets as our datasets as they are often much bigger than the training sets. This means that whenever we refer to a dataset in the remainder of this text, we refer to the test set of that dataset as defined in [6]. This procedure was also followed by Dau et al. [7].

As is typically done in evaluating semi-supervised clustering methods, the classes are assumed to represent the clusterings of interests. When computing rankings and average ARIs, we ignored results from 21 datasets where cDTW^{SS} either crashed or timed out after 24 h.[6]

---

[6] These datasets are listed at https://bitbucket.org/toon_vc/cobras_ts_experiments.

4.3   Methodology

We use 10-fold cross-validation, as is common in evaluating semi-supervised clustering methods [3,9]. The full dataset is clustered in each run, but the methods can only query pairs of which both instances are in the training set. The result of a run is evaluated by computing the Adjusted Rand Index (ARI) [8] on the instances of the test set. The ARI measures the similarity between the generated clusterings and the ground-truth clustering, as indicated by the class labels. It is 0 for a random clustering, and 1 for a perfect one. The final ARI scores that are reported are the average ARIs over the 10 folds.

We ensure that cDTW^SS and COBS do not query pairs that contain instances from the test set by simply excluding such candidates from the list of constraints that they consider. For COBRAS^TS, we do this by only using training instances to compute the super-instance representatives.

COBRAS^TS and COBS do not require the number of clusters as an input parameter, whereas cDTW^SS, k-Shape and k-MS do. The latter three were given the correct number of clusters, as indicated by the class labels. Note that this is a significant advantage for these algorithms, and that in many practical applications the number of clusters is not known beforehand.

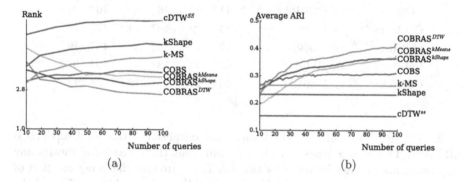

Fig. 4. (a) Average rank over all clustering tasks. Lower is better. (b) Average ARI. Higher is better.

4.4   Results

Clustering quality. Figure 4(a) shows the average ranks of the compared methods over all datasets. Figure 4(b) shows the average ARIs. Both plots clearly show that, on average, COBRAS^TS outperforms all the competitors by a large margin. Only when the number of queries is small (roughly < 15), is it outperformed by COBS and k-MS.

For completeness, we also include vanilla COBRAS (denoted as COBRAS^kMeans) in the comparison in Fig. 4. Given enough queries (roughly > 50), COBRAS^kMeans outperforms all competitors other than COBRAS^TS.

This indicates that the COBRAS approach is essential. As expected, however, COBRAS^{DTW} and COBRAS^{kShape} significantly outperform COBRAS^{kMeans}.

These observations are confirmed by Table 1, which reports the number of times COBRAS^{DTW} wins and loses against the alternatives. The differences with cDTW^{SS} and k-Shape are significant for all the considered numbers of queries (Wilcoxon test, $p < 0.05$). The difference between COBRAS^{DTW} and COBS is significant for 50 and 100 queries, but not for 25. The same holds for COBRAS^{DTW} vs. k-MS. This confirms the observation from Fig. 4(a), which showed that the performance gap between COBRAS^{DTW} and the competitors becomes larger as more queries are answered. The difference between COBRAS^{DTW} and COBRAS$^{k\text{-}Shape}$ is only statistically significant for 100 queries.

Table 1. Wins and losses over the 64 datasets. An asterisk indicates that the difference is significant according to the Wilcoxon test with $p < 0.05$.

	25 queries		50 queries		100 queries	
	Win	Loss	Win	Loss	Win	Loss
COBRAS^{DTW} vs. COBRAS$^{k\text{-}Shape}$	**35**	29	**37**	27	**41***	23
COBRAS^{DTW} vs. COBRAS$^{k\text{-}Means}$	**41***	23	**36**	28	**40***	24
COBRAS^{DTW} vs. k-MS	**35**	29	**40***	24	**47***	14
COBRAS^{DTW} vs. COBS	**37**	27	**42***	22	**45***	19
COBRAS^{DTW} vs. cDTW^{SS}	**62***	2	**53***	11	**55***	9
COBRAS^{DTW} vs. k-Shape	**40***	24	**46***	18	**50***	14

Surprisingly, the unsupervised baselines outperform the semi-supervised cDTW^{SS}. This is inconsistent with the claim that the choice of w dwarfs any improvements by the k-Shape algorithm [7]. To ensure that this is not an effect of the evaluation strategy (10-fold CV using the ARI, compared to no CV and the Rand index (RI) in [7]), we have also computed the RIs for all of the clusterings generated by k-Shape and compared them directly to the values provided by the authors of cDTW^{SS} on their webpage[7]. In this experiment k-Shape attained an average RI of 0.68, cDTW^{SS} 0.67. We note that the claim in [7] was based on a comparison on two datasets. Our experiments clearly indicate that it does not generalize towards all datasets.

Runtime. COBRAS^{DTW}, cDTW^{SS} and COBS require the construction of the pairwise DTW distance matrix. This becomes infeasible for large datasets. For example, computing one distance matrix for the ECG5000 dataset took ca. 30h in our experiments, using an optimized C implementation of DTW.

k-Shape and k-MS are much more scalable [11], as they do not require computing a similarity matrix. COBRAS$^{k\text{-}Shape}$ inherits this scalability, as it uses

---

[7] https://sites.google.com/site/dtwclustering/.

k-Shape to refine super-instances. In our experiments, COBRAS$^{k\text{-Shape}}$ was on average 28 times faster than COBRAS^{DTW}.

5   Case Studies: CBF, TwoLeadECG and MoteStrain

To gain more insight into why COBRAS^{TS} outperforms its competitors, we inspect the clusterings that are generated for three UCR datasets in more detail: CBF, TwoLeadECG and MoteStrain. CBF and TwoLeadECG are examples for which COBRAS^{DTW} and COBRAS$^{k\text{-Shape}}$ significantly outperform their competitors, whereas MoteStrain is one of the few datasets for which they are outperformed by unsupervised k-Shape clustering. These three datasets illustrate different reasons why time series clustering may be difficult: CBF because one of the clusters comprises two separated subclusters; TwoLeadECG, because only limited *subsequences* of the time series are relevant for the clustering at hand, and the remaining parts obfuscate the distance measurements; and MoteStrain because it is noisy.

CBF. The first column of Fig. 5 shows the "true" clusters as they are indicated by the class labels. It is clear that the classes correspond to three distinct patterns (horizontal, upward and downward). The next columns show the clusterings that are produced by each of the competitors. Semi-supervised approaches are given a budget of 50 queries. COBRAS^{DTW} and COBRAS$^{k\text{-Shape}}$ are the only methods that provide a near perfect solution (ARI = 0.96). cDTW^{SS} mixes patterns of different types in each cluster. COBS find pure clusters, but too many: the plot only shows the largest three of 15 clusters for COBS. k-Shape and k-MS mix horizontal and downward patterns in their third cluster. To clarify this mixing of patterns, the figure shows the instances in the third k-Shape and k-MS clusters again, but separated according to their true class.

   Figure 6 illustrates how repeated refinement of super-instances helps COBRAS^{TS} deal with the complexities of clustering CBF. It shows a super-instance in the root, with its subsequent refinements as children. The super-instance in the root, which is itself a result of a previous split, contains horizontal and upward patterns. Clustering it into two new super-instances does not yield a clean separation of these two types: a pure cluster with upward patterns is created, but the other super-instance still mixes horizontal and upward patterns. This is not a problem for COBRAS^{TS}, as it simply refines the latter super-instance again. This time the remaining instances are split into nearly pure ones separating horizontal from upward patterns. Note that the two super-instances containing upward patterns correspond to two distinct subclusters: some upward patterns drop down very close to the end of the time series, whereas the drop in the other subcluster occurs much earlier.

   The clustering process just mentioned illustrates the point made earlier, in Sect. 3.1, about COBRAS's ability to construct clusters with separated components. It is clear that this ability is advantageous in the CBF dataset. Note that being able to deal with *separated* components is key here; k-MS, which is able to find multi-modal clusters, but not clusters with modes that are separated by

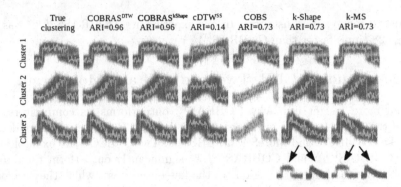

Fig. 5. The first column shows the true clustering of CBF. The remaining columns show the clusterings that are produced by all considered methods. For COBS, only the three largest of 15 clusters are shown. All the cluster instances are plotted, the prototypes are shown in red. For COBRAS^{DTW}, cDTW^{SS} and COBS the prototypes are selected as the medoids w.r.t. DTW distance. For the others the prototypes are the medoids w.r.t. the SBD distance. (Color figure online)

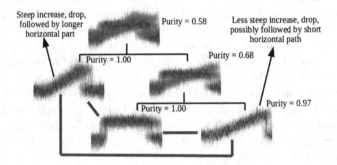

Fig. 6. A super-instance that is generated while clustering CBF, and its refinements. The green line indicates a must-link, and illustrates that these two super-instances will be part of the same multi-modal cluster (that of upward patterns). The red lines indicate cannot-links. The purity of a super-instance is computed as the ratio of the occurrence of its most frequent class, over its total number of instances. (Color figure online)

a mode from another cluster, produces a clustering that is far from perfect for CBF.

Figure 6 also illustrates that COBRAS's super-instance refinement step is similar to top-down hierarchical clustering. Note, however, that COBRAS uses constraints to guide this top-down splitting towards an appropriate level of granularity. Furthermore, this refinement is only one of COBRAS's components; it is interleaved with a bottom-up merging step to combine the super-instances into actual clusters [15].

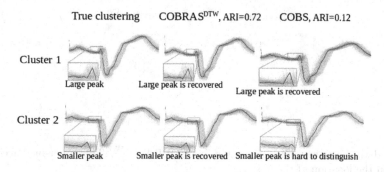

Fig. 7. The first column shows the "true" clustering of TwoLeadECG. The second column shows the clustering produced by COBRAS^{DTW}. The third column shows the clustering produced by COBS, which is the best competitor for this dataset. Prototypes are shown in red, and are the medoids w.r.t. the DTW distance. (Color figure online)

TwoLeadECG. The first column in Fig. 7 shows the "true" clusters for TwoLeadECG. Cluster 1 is defined by a large peak before the drop, and a slight bump in the upward curve after the drop. Instances in cluster 2 typically only show a small peak before the drop, and no bump in the upward curve after the drop. For the remainder of the discussion we focus on the peak as the defining pattern, simply because it is easier to see than the more subtle bump.

The second column in Fig. 7 shows the clustering that is produced by COBRAS^{DTW}; the one produced by COBRAS^{k-Shape} is highly similar. They are the only methods able to recover these characteristic patterns. The last column in Fig. 7 shows the clustering that is produced by COBS, which is the best of the competitors. This clustering has an ARI of 0.12, which is not much better than random. From the zoomed insets in Fig. 7, it is clear that this clustering does not recover the defining patterns: the small peak that is characteristic for cluster 2 is hard to distinguish.

This example illustrates that by using COBRAS^{TS} for semi-supervised clustering, a domain expert can discover more accurate explanatory patterns than with competing methods. None of the alternatives is able to recover the characteristic patterns in this case, potentially leaving the domain expert with an incorrect interpretation of the data. Obtaining these patterns comes with relatively little additional effort, as with a good visualizer answering 50 queries only takes a few minutes. This time would probably be insignificant compared to the time that was needed to collect the 1139 instances in the TwoLeadECG dataset.

MoteStrain. In our third case study we discuss an example for which COBRAS^{TS} does not work well, as this provides insight into its limitations. We consider the MoteStrain dataset, for which the unsupervised methods perform best. k-MS attains an ARI of 0.62, and k-Shape of 0.61. COBRAS^{k-Shape} ranks third with an ARI of 0.51, and COBRAS^{DTW} fourth with an ARI of 0.48. These results are surprising, as the COBRAS algorithms have access to more information than the unsupervised k-Shape and k-MS. Figure 8 gives a reason

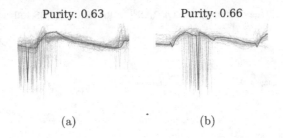

Purity: 0.63   Purity: 0.66

(a)   (b)

Fig. 8. Two super-instances generated by COBRAS^{DTW}. The super-instances are based on the location of the noise.

for this outcome; it shows that COBRAS^{TS} creates super-instances that are based on the location of the noise. The poor performance of the COBRAS^{TS} variants can in this case be explained by their large variance. The process of super-instance refinement is much more flexible than the clustering procedure of k-Shape, which has a stronger bias. For most datasets, COBRAS^{TS}'s weaker bias led to performance improvements in our experiments, but in this case it has a detrimental effect due to the large magnitude of the noise. In practice, the issue could be alleviated here by simply applying a low-pass filter to remove noise prior to clustering.

6   Conclusion

Time series arise in virtually all disciplines. Consequently, there is substantial interest in methods that are able to obtain insights from them. One of the most prominent ways of doing this, is by using clustering. In this paper we have presented COBRAS^{TS}, an novel approach to time series clustering. COBRAS^{TS} is semi-supervised: it uses small amounts of supervision in the form of must-link and cannot-link constraints. This sets it apart from the large majority of existing methods, which are unsupervised. An extensive experimental evaluation shows that COBRAS^{TS} is able to effectively exploit this supervision; it outperforms unsupervised and semi-supervised competitors by a large margin. As our implementation is readily available, COBRAS^{TS} offers a valuable new tool for practitioners that are interested in analyzing time series data.

Besides the contribution of the COBRAS^{TS} approach itself, we have also provided insight into why it works well. A key factor in its success is its ability to handle clusters with separated components.

Acknowledgements. We thank Hoang Anh Dau for help with setting up the cDTW^{SS} experiments. Toon Van Craenendonck is supported by the Agency for Innovation by Science and Technology in Flanders (IWT). This research is supported by Research Fund KU Leuven (GOA/13/010), FWO (G079416N) and FWO-SBO (HYMOP-150033).

References

1. Bagnall, A., Lines, J., Bostrom, A., Large, J., Keogh, E.: The great time series classification bake off: a review and experimental evaluation of recent algorithmic advances. Data Min. Knowl. Discov. **31**(3), 606–660 (2017)
2. Basu, S., Banerjee, A., Mooney, R.J.: Active semi-supervision for pairwise constrained clustering. In: Proceedings of SDM (2004)
3. Basu, S., Bilenko, M., Mooney, R.J.: A probabilistic framework for semi-supervised clustering. In: Proceedings of the Tenth ACM SIGKDD International Conference on Knowledge Discovery and Data Mining, pp. 59–68. ACM (2004)
4. Begum, N., Ulanova, L., Wang, J., Keogh, E.: Accelerating dynamic time warping clustering with a novel admissible pruning strategy. In: Proceedings of SIGKDD (2015)
5. Cao, H., Tan, V.Y.F., Pang, J.Z.F.: A parsimonious mixture of Gaussian trees model for oversampling in imbalanced and multimodal time-series classification. IEEE Trans. Neural Netw. Learn. Syst. **25**(12), 2226–2239 (2014)
6. Chen, Y., et al.: The UCR time series classification archive (2015), http://www.cs.ucr.edu/~eamonn/time_series_data/
7. Dau, H.A., Begum, N., Keogh, E.: Semi-supervision dramatically improves time series clustering under dynamic time warping. In: Proceedings of CIKM (2016)
8. Hubert, L., Arabie, P.: Comparing partitions. J. Classif. (1985)
9. Mallapragada, P.K., Jin, R., Jain, A.K.: Active query selection for semi-supervised clustering. In: Proceedings of ICPR (2008)
10. Meert, W.: DTAIDistance (2018), https://doi.org/10.5281/zenodo.1202379
11. Paparrizos, J., Gravano, L.: Fast and accurate time-series clustering. ACM Trans. Database Syst. **42**(2), 8:1–8:49 (2017)
12. Śmieja, M., Wiercioch, M.: Constrained clustering with a complex cluster structure. Adv. Data Anal. Classif. **11**(3), 493–518 (2017)
13. Van Craenendonck, T., Blockeel, H.: Constraint-based clustering selection. In: Machine Learning. Springer (2017)
14. Van Craenendonck, T., Dumančić, S., Blockeel, H.: COBRA: a fast and simple method for active clustering with pairwise constraints. In: Proceedings of IJCAI (2017)
15. Van Craenendonck, T., Dumančić, S., Van Wolputte, E., Blockeel, H.: COBRAS: fast, iterative, active clustering with pairwise constraints (2018), https://arxiv.org/abs/1803.11060, under submission
16. von Luxburg, U.: A tutorial on spectral clustering. Stat. Comput. **17**(4), 395–416 (2007)
17. von Luxburg, U., Williamson, R.C., Guyon, I.: Clustering: science or art? In: Workshop on Unsupervised Learning and Transfer Learning (2014)
18. Wagstaff, K., Cardie, C., Rogers, S., Schroedl, S.: Constrained K-means clustering with background knowledge. In: Proceedings of ICML (2001)
19. Wei, L., Keogh, E.: Semi-supervised time series classification. In: Proceedings of ACM SIGKDD (2006)
20. Xing, E.P., Ng, A.Y., Jordan, M.I., Russell, S.: Distance metric learning, with application to clustering with side-information. In: NIPS 2003 (2002)

Exploiting the Web for Semantic Change Detection

Pierpaolo Basile[1]([✉]) and Barbara McGillivray[2,3]

[1] Department of Computer Science, University of Bari Aldo Moro, Bari, Italy
pierpaolo.basile@uniba.it
[2] Modern and Medieval Languages, University of Cambridge, Cambridge, UK
bmcgillivray@turing.ac.uk
[3] The Alan Turing Institute, London, UK

Abstract. Detecting significant linguistic shifts in the meaning and usage of words has gained more attention over the last few years. Linguistic shifts are especially prevalent on the Internet, where words' meaning can change rapidly. In this work, we describe the construction of a large diachronic corpus that relies on the UK Web Archive and we propose a preliminary analysis of semantic change detection exploiting a particular technique called Temporal Random Indexing. Results of the evaluation are promising and give us important insights for further investigations.

Keywords: Semantic change detection
Diachronic analysis of language · Time series

1 Introduction

Languages can be studied from two different and complementary viewpoints: the diachronic perspective considers the evolution of a language over time, while the synchronic perspective describes the language rules at a specific point of time, without taking its history into account [8]. During the last decade, the surge in available data spanning different epochs has inspired a new analysis of cultural, social, and linguistic phenomena from a temporal perspective. Language is dynamic and evolves, it varies to reflect the shift in topics we talk about, which in turn follow cultural changes. So far, the automatic analysis of language has largely been based on datasets that represented a snapshot of a given domain or time period (*synchronic approach*). However, since the rise of big data, which has made large corpora of data spanning several periods of time available, large-scale diachronic analysis of language has emerged as a new approach to study linguistic and cultural trends over time by analysing these new sources of information. One of the largest sources of information is the Web, which has been exploited to build corpora used in linguistics or in Natural Language Processing (NLP) tasks. Generally, these corpora are built using a synchronic approach without taking into account temporal information.

L. Soldatova et al. (Eds.): DS 2018, LNAI 11198, pp. 194–208, 2018.
https://doi.org/10.1007/978-3-030-01771-2_13

In this paper, we propose to analyze the Web using a *diachronic approach* by relying on the UK Web Archive project [15]. The goal of this project is to analyse the change in language over time as reflected in the textual content of UK websites. We focus on one specific kind of language change, namely semantic change, aiming to develop a computational system that is able to detect which words have changed meaning over the period of time covered by the corpus of UK websites.

Semantic change is a very common phenomenon in language. Over time, words can acquire new meanings or lose existing ones. For example, the original meaning of the verb *tweet*, according to the Oxford English Dictionary (OED), is transitive, defined as follows:

```
Of a bird: to communicate (something) with a brief
high-pitched sound or call, or a series of such sounds.
```

According to the OED, this meaning was first recorded in writing in 1851. On the other hand, the OED assigns the first written usage of the related intransitive meaning to 1856:

```
Of a bird: to make a brief high-pitched sound or call, or
a series of such sounds. Also in extended use.
```

The OED also lists two additional senses, which are much more recent. The transitive one is defined as follows:

```
To post (a message, image, link, etc.) on the social
networking service Twitter. Also: to post a message
to (a particular person, organization, etc.).
```

This meaning was first recorded in 2006. The intransitive one is defined as:

```
To make a posting on Twitter. Also: to use Twitter
regularly or habitually.
```

and was first recorded in 2007.

Semantic change detection systems allow for large-scale analyses that identify cultural and social trends. For example, when the contexts of the word *sleep* are compared between 1960s and 1990s, it has been shown through distributional semantics models that this word acquired more negative connotations linked to sleep disorders [12]. Moreover, such systems have a range of applications in NLP. For example, they can improve sentiment analysis tools because they can identify positive or negative content expressed via newly emerged meanings, such as the positive slang sense of *sick* meaning "awesome".

The use of the Web as a source of data for diachronic semantic analysis poses an important challenge that we aim to tackle in this paper: the massive size of the dataset requires efficient computational approaches which are able to scale up to process terabytes of data. In this scenario, Distributional Semantic Models (DSMs) represent a promising solution. DSMs are able to represent words as points in a geometric space, generally called WordSpace [22,23] by simply analysing how words are used in a corpus. However, a WordSpace represents a snapshot of a specific corpus and it does not take into account temporal information. For this reason, we rely on a particular method, called Temporal Random Indexing (TRI), that enables the analysis of the time evolution of the meaning of a word [4,16]. TRI is able to efficiently build WordSpaces taking into account temporal information. We exploit this methodology in order to build geometrical spaces of word meanings that span over several periods of time. The TRI framework provides all the necessary tools to build WordSpaces over different time periods and perform such temporal linguistic analysis. The system has been tested on several domains, such as a collection of Italian books, English scientific papers [3], the Italian version of the Google N-gram dataset [2], and Twitter [16].

The paper is structured as follows: Sect. 2 provides details about our methodology, while Sect. 3 describes the dataset that we have developed and the results of a preliminary evaluation. Related work is provided in Sects. 4, and 5 reports final remarks and future work.

2   Method

This section provides details about the methodology adopted during our research work. In particular, we build a diachronic corpus using data coming from the Web. Relying on this corpus, we build a semantic distributional model that takes into account temporal information. The last step is to build time series in order to track how the meaning of a word change over time. These time series are created by exploiting information extracted from the distributional semantic models. In the following sub-sections we provide details about each of the aforementioned steps.

2.1   Corpus Creation

The first step is to create a diachronic corpus starting from data coming from the web. The web collection under consideration is the JISC UK Web Domain Dataset (1996–2013) [15] which collects resources from the Internet Archive (IA) that were hosted on domains ending in .uk, and those that are required in order to render .uk pages.

The JISC dataset is composed of two parts: (1) the first part contains resources from 1996 to 2010 for a total size of 32TB; (2) the second one contains resources from 2011–2013 for a total size of 30TB. The JISC dataset cannot be made generally available, but can be used to generate secondary datasets. For that reason we provide the corpus in the form of co-occurrence matrices extracted from it. The dataset contains resources crawled by the IA Web Group for different archiving partners, the Web Wide crawls and other miscellaneous crawls run by IA, as well as data donations from Alexa and other companies or institutions. So it is impossible to know all the crawling configuration used by the different partners. However the dataset contains not only HTML pages and textual resources but also video, images and other types of files.

The first step of the corpus creation consists in filtering the JISC dataset in order to extract only the textual resources. For this purpose, we extract the text from textual resources (e.g. TXT files) and parse HTML pages in order to extract their textual content. We adopt the jsoup library[1] for parsing HTML pages.

The original dataset stores data in the ARC and WARC formats, which are standard formats used by the Internet Archive project for storing data crawled from the web as sequences of content blocks. The WARC format is an enhancement of ARC for supporting metadata, detect duplicate events and more. We process ARC and WARC archives in order to extract the textual content and store data in the WET format. WET is a standard format for storing plain text extracted from in ARC/WARC archives. We transform the original dataset in the standard WET format which contains only textual resources. The output of this process provides about 5TB of WET archives.

The second step consists in tokenizing the WET archives in order to produce a tokenized version of the textual content. We exploit the StandardAnalyzer.[2] provided by the Apache Lucene API[3] This analyzer provides also a standard list of English stop words. The size of the tokenized corpus is approximately 3TB.

In the third step, we create co-occurrence matrices, which store co-occurrences information for each word token. In order to track temporal information, we build a co-occurrence matrix for each year from 1996 to 2013. Each matrix is stored in a compressed text format, one row per token. Each row reports the target token and the list of tokens co-occurring with it. An example for the word *linux* is reported in Fig. 1, which shows that the token *swapping* co-occurs 4 times with *linux*, the word *google* 173 times, and so on. We extract co-occurrences taking into account a window of five words to the left and to the right of the target word. For the construction of co-occurrence matrices, we exploit only words that occur at least 4,500 times in the dataset. We do not apply any text processing step such us lemmatization or stemming for two reasons: (1) the idea is to build a language independent tool and (2) in this first evaluation we want to reduce the number of parameters and focus our attention on the change point detection strategy. Finally, we obtain a vocabulary of about one million words and the total size of compressed matrices is about 818GB.

```
linux   swapping   4   google   173   xp   454   manufacturer
237   job   64   install   255   security   137   cgi   47
operating   705   host   69   performance   44   sharing
56...
```

Fig. 1. Co-occurrence matrix

---

[1] https://jsoup.org/.

[2] https://lucene.apache.org/core/7_3_1/core/index.html.

[3] https://lucene.apache.org/core/.

The whole process is described in Fig. 2: WARC/ARC archives are converted into WET files in order to extract the text and they are tokenized; the tokenized text is exploited by the *Matrix Builder* for building the co-occurrence matrices; matrices are the input for *TRI* that performs Temporal Random Indexing and provides a WordSpace for each time period; finally WordSpaces are used to build time series. The last part of the chart sketches the process used to detect semantic changepoints (see Sect. 2.2) and the evaluation step described in Sect. 3.

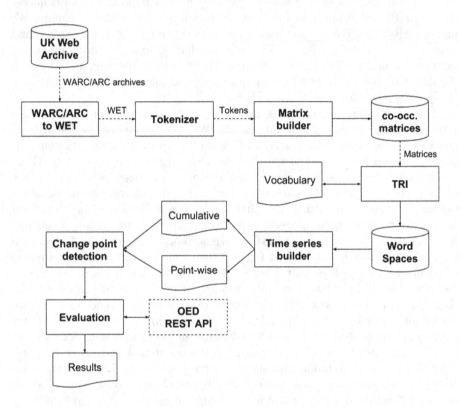

Fig. 2. Flowchart of the whole semantic change detection process.

2.2 Semantic Change Detection

Our method for semantic change detection relies on a previous model based on Temporal Random Indexing (TRI) [3,4]. In particular, we further develop the TRI approach in three directions: (1) we improve the system in order to manage very large datasets, such as the JISC UK Web Domain Dataset; (2) we introduce a new way to weight terms in order to reduce the impact of very frequent terms; (3) we introduce new methods for detecting semantic shift from time series analysis techniques.

The idea behind TRI is to build different WordSpaces for each time period under investigation. The peculiarity of TRI is that word vectors over different time periods are directly comparable because they are built using the same random vectors. TRI works as follows:

1. Given a corpus C of documents and a vocabulary V of terms[4] extracted from C, the method assigns a random vector r_i to each term $t_i \in V$. A random vector is a vector that has values in $\{-1, 0, 1\}$ and is sparse with few non-zero elements randomly distributed along its dimensions. The sets of random vectors assigned to all terms in V are near-orthogonal;
2. The corpus C is split into different time periods T_k using temporal information, for example the year of publication;
3. For each period T_k, a WordSpace WS_k is built. All the terms of V occurring in T_k are represented by a semantic vector. The semantic vector sv_i^k for the i-th term in T_k is built as the sum of all the random vectors of the terms co-occurring with t_i in T_k. When computing the sum, we apply some weighting to the random vector. In our case, to reduce the impact of very frequent terms, we use the following weight: $\sqrt{\frac{th \times C_k}{\#t_i^k}}$, where C_k is the total number of occurrences in T_k and $\#t_i^k$ is the occurrences of the term t_i in T_k. The parameter th is set to 0.001.

This way, the semantic vectors across all time periods are comparable since they are the sum of the same random vectors.

In order to track the words' meaning change over time, for each term t_i we build a time series $\Gamma(t_i)$. A time series is a sequence of values, one value for each time period, and it indicates the semantic shift of that term in the given period. We adopt several strategies for building the time series. The first strategy is based on term log-frequency; each value in the series is defined as:
$$\Gamma_k(t_i) = log(\frac{\#t_i^k}{C_k}).$$
In order to exploit the ability of our methods in computing vectors similarity over time periods, we define two strategies for building the time series:

point-wise: $\Gamma_k(t_i)$ is defined as the cosine similarity between the semantic vector of t_i in the time period k, sv_i^k, and the semantic vector of t_i in the previous time period, sv_i^{k-1}. This way, we aim to capture semantic change between two time periods;

cumulative: we build a cumulative vector $sv_i^{C_{k-1}} = \sum_{j=0}^{k-1} sv_i^j$ and compute the cosine similarity of this cumulative vector and the vector sv_i^k. The idea behind the cumulative approach is that the semantics of a word at point $k-1$ depends on the semantics of the word in all the previous time periods. The cumulative vector is the semantic composition of all the previous word vectors, the composition is performed through the vector sum [20].

Given a time series, we need a method for finding significant changepoints in the series, which we interpret as indications that semantic change has taken place. We adopt three strategies:

---

[4] V contains the terms that we want to analyse, typically, the most n frequent terms.

1. the *Mean shift model* [26], proposed in [17], defines a mean shift of a general time series Γ pivoted at time period j as:

$$K(\Gamma) = \frac{1}{l-j} \sum_{k=j+1}^{l} \Gamma_k - \frac{1}{j} \sum_{k=1}^{j} \Gamma_k \qquad (1)$$

In order to determine if a mean shift is relevant at time j, we adopt a bootstrapping [10] approach, under the null hypothesis that there is no change in the mean. In particular, a confidence level is computed by constructing B bootstrap samples by permuting $\Gamma(t_i)$. Finally, we estimate changepoints by considering the time points with a confidence value above a predefined threshold;

2. the *valley model*, in which any point j that has a value lower than the previous point $j-1$ in the time series is considered a changepoint. The idea is that if we observe a decrease in the similarity between the semantic vector of a word at a given point in time and the semantic vector of the same word in the previous time point, then this indicates that the word's semantics is changing;

3. the *variance model*, in which the difference between the value in the time series at a point j and the value at the point $j-1$ is compared with the variance of the time series; when the difference is higher than one, two or four times the variance, the point is considered a changepoint.

2.3   System Output and Neighborhood Analysis

The system's output consists of lists of candidate words which are predicted to have undergone semantic change, together with the year in which this change is predicted to have happened. In addition, for each candidate, we can extract its corpus neighbours, defined as the top n words whose semantic vectors have the highest cosine similarity with the vector of the candidate word.

To take an example, our system considered *blackberry* as a candidate for semantic change, with three changepoints, in the years 1998, 2007, and 2009. The original sense of *blackberry* refers to the *"edible berry-like fruit of the bramble, Rubus fruticosus"*, the *"The trailing plant Rubus fruticosus"*, and *"Any of various other dark-coloured edible berries"*, according to the OED. However, a more recent sense emerged in 1999, defined in the OED as *"A proprietary name for: a type of pager or smartphone capable of sending and receiving email messages"*.

If we look at the top 20 neighbours of *blackberry* in 1999 extracted by TRI from the UK Web Archive JISC dataset 1996–2013 corpus, we see that the majority of them are words related to original sense (highlighted in bold face the list below), either as collocates (like *pie*) or as distributionally similar nouns to *blackberry* (like *strawberry*):

cherry, **berries**, **strawberry**, **blossom**, **pie**, **blueberry**, **blackcurrant**, brierley, **pudding**, beacon, **red**, **raspberry**, hill, lion, mill, green, **chestnut**, brick, **ripe**, **scent**

On the other hand, the top 20 corpus neighbours of *blackberry* in 2003 include some words from the domain of mobile phones, highlighted in bold face below[5]:

> blueberry, plum, **phones**, **cellphones**, **handsets**, loganberry, ripe, strawberry, **devices**, orange, **phone**, currant, gooseberry, **gprs**, wings, blackcurrant, damson, **bluetooth**, berries, blackberries

By 2004, the majority of corpus neighbours of *blackberry* involve words related to mobile phones, indicating that this has become the predominant sense of the word in the corpus, as shown by the following list of top 20 neighbours in 2004:

> handspring, **handsets**, **tmobile**, **justphones**, nec, **handset**, **payg**, **tarriffs**, **lg**, **cellphones**, **pickamobile**, **phonesnokia**, **prepay**, **sim**, **tariffs**, **phones**, **phoneid**, **findaphone**, **mobilechooser**, **unlock**

The system's output lists contain several thousand candidates, a set which is too large to assess by hand. Therefore, we devised a novel automatic evaluation framework, outlined in the next section.

3   Evaluation

There is no general framework for evaluating the accuracy of semantic change detection systems. Previous work has evaluated semantic change systems either indirectly via their performance on related tasks (e.g. [11]), or via a small-scale qualitative analysis (e.g. [13]). In order to measure how well our system achieves the intended aim to identify words that have changed their meaning over the time covered by the UK Web Archive JISC dataset 1996–2013, we developed a novel evaluation framework. We evaluated our semantic change system and a baseline system against a dictionary-based gold standard. In the baseline system, we used a time series consisting of the frequency counts of each word form in the corpus. The evaluation of this baseline was aimed to detect any contribution given by the cosine similarity scores and TRI in our system.

We used the data from the Oxford English dictionary (OED) API as gold standard. The OED contains a diachronic record of the semantics of the words in the English lexicon. Each entry corresponds to a lemma and part-of-speech pair, and contains the list of its senses, each with a definition, the year when each sense was first recorded in writing, a corresponding quotation, following optionally more dated quotations which illustrate the use of the word with that sense at different points in time.

---

[5] We did not highlight *orange* in the list because in this context it could refer to the fruit, in which case it would be related to the fruit sense of *blackberry*, or to the mobile phone company, in which case it would be related to the cellphone sense of *blackberry*.

We performed the evaluation of each system in two steps. First, we calculated the accuracy of the semantic changepoint detection component, with the aim to measure how well the system detected semantic change candidates at the correct point in time. For each semantic change candidate outputted by each system, we checked that it appeared in the OED with a first usage dated from 1995 or later[6]. If this was not the case, we excluded the candidate word and the changepoint year from the analysis, as we were not able to assess whether the word changed meaning in the time span under consideration. We also only considered words that had a frequency of at least 100 in the corpus. We compared the changepoint year of semantic change according to our system with the year when the sense was first recorded according to the OED. The candidate and its changepoint were considered correct if the changepoint year was no earlier than the year when it was first recorded according to the OED. For example, the OED records the first usage of the verb *follow* with the transitive meaning of *"To track the activities or postings of (a person, group, etc.) by subscribing to their account on a social media website or application."*, and dates it from 2007. Our system suggested *follow* as a candidate for semantic change, with a changepoint in 2009. According to our evaluation approach, this counted as a correct candidate.

The results of the first evaluation step are summarized in Table 1. Semantic change detection is a very difficult task, especially when measured against a highly-curated resource like the OED, which relies on an evidence basis that is much broader in scope compared to the UK Web Archive. Therefore, it is not surprising that the precision scores are low. Of the several tens of thousands candidates outputted by our system or the baseline, only less than 400 were correct, in all configurations of the parameters. The precision scores range between 0.003 and 0.005. Given that the number of words in the gold standard is 462, the recall scores range between 0.104 and 0.849, with the highest score being associated to the point-wise and cumulative time series and the valley model for changepoint detection. It is important to note that methods reporting the highest recall (cumulative/valley and point-wise/valley) provide a high number of candidates (about 77,515) but these represent only the 7.7% of the whole dictionary exploited by our system (about one million). Overall, we can say that the valley model for changepoint detection yields the highest recall scores and outperforms the mean shift model and the variance model, and that the system with cumulative and pointwise time series outperforms the system with frequency-based time series (baseline). We are not able to provide a comparison with methods based on word embeddings due to the difficult to scale-up these approaches on our large corpus. We plan to perform this comparison as future work.

For the second evaluation step, we focussed on the candidates that were considered correct according to the method explained above. For those candidates,

---

[6] As the earliest texts in the corpus date from 1996, we allowed for a one-year buffer between this date and the date of first usage according to the OED, under the assumption that a sense first recorded in the OED in 1995 could be recorded with sufficient evidence in our corpus at least one year later.

Table 1. Summary of evaluation metrics of our systems and the baseline against the gold standard (OED). The first column details the time series construction type; the second column details the changepoint detection approach. The variance approach is followed by a numeric parameter: 'Variance 1' means that the changepoint is identified when the difference between the value in the time series at a point j and the value at the point $j - 1$ is higher than the variance of the time series; 'Variance 2' means that the changepoint is identified when the difference between the value in the time series at a point j and the value at the point $j - 1$ is higher than twice the variance of the time series.

System	Changepoint	# correct	Candidates	Precision	Recall	F1-score
Baseline	Mean shift	76	14,176	0.005	0.165	0.010
Baseline	Valley	378	77,493	0.005	0.818	0.010
Baseline	Variance 1	0	145	0	0	0
Baseline	Variance 2	0	52	0	0	0
Cumulative	Mean shift	48	15,266	0.003	0.104	0.006
Cumulative	**Valley**	392	77,515	0.005	**0.848**	0.010
Cumulative	Variance 1	165	47,389	0.003	0.357	0.007
Cumulative	Variance 2	56	14,452	0.004	0.121	0.008
Point-wise	Mean shift	74	23,855	0.003	0.161	0.006
Point-wise	**Valley**	392	77,515	0.005	**0.848**	0.010
Point-wise	Variance 1	382	76,061	0.005	0.827	0.010
Point-wise	Variance 2	340	69,492	0.005	0.736	0.010

we measured the accuracy of the output from the point of view of their semantics. In other words, we checked that the new meanings of the correct candidate words identified by the system corresponded to the new meanings as recorded in the gold standard. For each semantic change candidate word (and corresponding changepoint year) which was considered correct according to the approach illustrated above, we assessed how closely the new meaning of the candidate matched the senses in the OED first recorded after 1995. We measured this by collecting two sets of words. For the first set, we approximated the semantics of the new meaning as detected by the system with the 100 closest corpus neighbours to the candidate word, measuring proximity between words with the cosine distance. For the second set, we approximated the semantics of the OED senses with a bag-of-words approach. We pre-processed all words appearing in the definition and quotation text of each OED sense by stemming and lower-casing them. Then, we compared the two sets by calculating their Jaccard index, defined as the ratio between the number of elements in the intersection between the two sets, and the number of elements in the union of the two sets. Finally, we extracted the rank of each sense by according to the Jaccard index with the corpus neighbours (in decreasing order), and reported the rank of the correct candidate as an evaluation measure. For this evaluation, we focussed on the best-performing models according to the recall measure, as precision scores were low in all cases. These models involved collecting the time series with the pointwise and cumulative methods, and calculating the changepoint with the valley method and led to the highest recall score of 0.848.

Let us take the example of *mobile*, which the system predicted changed its meaning in 2000. *Mobile* has two post-1995 senses in the OED, the first is first recorded in 1998, and the second is first recorded in 1999. Their definitions from the OED are, respectively:

1. A person's mobile phone number; cf. mobile phone number *n.*
2. As a mass noun. Mobile phone technology, networks, etc., esp. considered as a means to access the Internet; the Internet as accessed from mobile phones, tablet computers, and other portable wireless devices. Frequently with on, over, via, etc.

The top 20 corpus neighbours for *mobile* in 2000 include the words *phones, phone, connected, devices*, which are shared with the OED definition and quotation of the second sense.

Table 2 shows the results of the neighbourhood-based evaluation on the best performing models according to recall scores. Although the Jaccard indices between the corpus neighbours of the candidates and the bag-of-words from the OED definition and quotation texts are usually very low, with an average of only 0.008, when we matched the semantics of the candidates (as measured by their top 100 corpus neighbours) with the OED senses first recorded after 1995, we found that the OED senses corresponding to the model's candidates (i.e. those OED senses whose first usage was no later than the candidates' changepoints) tended to be ranked first. This indicates that the models are accurate not only at spotting the correct changepoint for a word, but also its new semantic features.

Table 2. Results of the neighbourhood-based evaluation on the two models with highest recall scores. The third column shows the average rank of the matching OED senses of the candidates. The fourth column shows the average number of OED senses included in the ranking. The fifth column shows the average rank of the matching OED senses excluding the cases in which there is only one OED sense for the candidates. The last column shows the average number of OED senses included in the ranking, excluding the cases in which there is only one OED sense for the candidates. The ranking is based on the Jaccard index between the corpus neighbours of the candidate and the bag-of-words of the OED definition and quotation text.

System	Changepoint	Av. rank	Av. OED senses	Av. rank (>1 sense)	# OED senses (>1)
Cumulative	Valley	1.206	1.336	1.811	2.324
Point-wise	Valley	1.206	1.332	1.799	2.290

In conclusion, analyzing the results we can notice that both cumulative and point-wise methods are able to overcome the baseline even though generally the precision is low due to the task difficulty. Evaluating semantic shift detection approaches is an open challenge, and researchers often rely on self-created test sets, or even simply manually inspecting the results. Moreover, our approach is able to correctly identify the semantics of the change according to the definition in the dictionary. We believe that this is the first work that tries to systematically analyze the semantic aspect of the changepoint.

4 Related Work

Over the past decade, semantic change detection has been an emerging research area within NLP, and a variety of different approaches have been developed. Recent surveys on the current state of the art in this field have also been produced [18,24].

A significant portion of the research in this area has focused on detecting semantic change in diachronic corpora spanning over several centuries [9,11,12, 14,27,28]. One of the most commonly used corpora is the multilingual Google Books N-gram Corpus [19], which covers the last five centuries and contains the N-grams from texts of over 6% of books ever published. Other researchers have used the 1800–1999 portion of this dataset, which consists of $8.5 * 10^{11}$ tokens [13].

A smaller set of previous studies focus on the more difficult task of detecting semantic change over a shorter time period, and use corpora which cover relatively short spans. Examples include a corpus consisting of articles from the New York Times published between 1990 and 2016 [29], a corpus based on the issues of the Chinese newspaper "People's Daily" from 1946 to 2004 [25], the British National Corpus (100 million words, 1990s) [7], and data from the French newspaper "Le Monde" between 1997 and 2007 [6].

Concerning the methods employed, previous work includes a range of methods, from neural models to Bayesian learning [11] to various algorithms for dynamic topic modeling [5]. A significant part of the literature employ methods based on word embeddings [9,13]. Very recently, dynamic embeddings have been shown as an improvement over using classical static embeddings for the task of semantic change detection [1,21]. In this method, embedding vectors are inferred for each time period and a joint model is trained over all intervals, while simultaneously allowing word and context vectors to drift.

All previous works based on word embeddings have in common the fact that they build a different semantic space for each period taken into consideration; this approach does not guarantee that each dimension bears the same semantics in different spaces [16], especially when embedding techniques are employed. In order to overcome this limitation, Jurgens and Stevens [16] introduced Temporal Random Indexing technique as a means to discover semantic changes associated to different events in a blog stream. Our methodology relies on the technique introduced by Jurgens and Stevens, but with a different aim. While Jurgens and

Stevens exploit TRI for the specific task of event detection, we setup a framework for semantic change detection relying on previous studies where TRI was applied on collection of Italian books, English scientific papers [3] and the Italian version of the Google N-gram dataset [2]. Moreover, it is important to stress that word embeddings techniques are based on word/context prediction that requires a learning step. On the other hand, TRI is based on counting words in context that is less computationally expensive and allows to scale up the method on a large Web collection.

5   Conclusions

In this work, we proposed several methods based on Temporal Random Indexing (TRI) for detecting semantic changepoints in the Web. We built a diachronic corpus exploiting the JISC UK Web Archive Dataset (1996–2013) which collects resources from the Internet Archive (IA) that were hosted on domains ending in .uk. We extracted about 5TB of textual data and we performed a preliminary evaluation using the Oxford English Dictionary (OED) as the gold standard. Results show that methods based on TRI are able to overcome baselines based on word occurrences, however, we obtain low precision due to a large number of detected changepoints. Moreover, for the first time, we propose a systematical approach for evaluating the semantics of detected changepoints by using both the neighborhood and the word meaning definition extracted from the OED. The precision of our model is low, which can be explained by several factors. First, the evaluation was based on an external resource, the OED, which relies on different data sources compared to web pages. This means that a semantic change recorded by our system is likely not to be necessarily reflected in the OED. Second, the task of semantic change detection is very hard, and our contribution is the first one to provide an evaluation based on a dictionary, so low precision values are not surprising. On the other hand, recall reaches a maximum value of 84%, which we consider an encouraging result. Overall, the results we report show that our approach is not only able to detect the correct time period, but also it is able to capture the correct semantics associated with the changepoint. As future work we plan to investigate other time series approaches for reducing the number of detected changepoints with the aim of increasing the precision.

Acknowledgments. This research was undertaken with the support of the Alan Turing Institute (EPSRC Grant Number EP/N510129/1). The access to the Oxford English Dictionary API was provided by Oxford University Press via a licence for non-commercial research.

References

1. Bamler, R., Mandt, S.: Dynamic word embeddings. In: Precup, D., Teh, Y.W. (eds.) Proceedings of the 34th International Conference on Machine Learning. Proceedings of Machine Learning Research, vol. 70, pp. 380–389. PMLR, International Convention Centre, Sydney, Australia (06–11 Aug 2017), http://proceedings.mlr.press/v70/bamler17a.html

2. Basile, P., Caputo, A., Luisi, R., Semeraro, G.: Diachronic analysis of the italian language exploiting google ngram (2016)

3. Basile, P., Caputo, A., Semeraro, G.: Analysing word meaning over time by exploiting temporal random indexing. In: Basili, R., Lenci, A., Magnini, B. (eds.) First Italian Conference on Computational Linguistics CLiC-it 2014. Pisa University Press (2014)

4. Basile, P., Caputo, A., Semeraro, G.: Temporal random indexing: a system for analysing word meaning over time. Ital. J. Comput. Linguist. **1**(1), 55–68 (2015)

5. Blei, D.M., Lafferty, J.D.: Dynamic topic models. In: ICML, pp. 113–120 (2006)

6. Boussidan, A., Ploux, S.: Using topic salience and connotational drifts to detect candidates to semantic change. In: Proceeding IWCS '11 Proceedings of the Ninth International Conference on Computational Semantics, pp. 315–319 (2011)

7. Cook, P., Stevenson, S.: Automatically identifying changes in the semantic orientation of words. In: Proceedings of the Seventh conference on International Language Resources and Evaluation, Valletta, Malta (2010)

8. De Saussure, F.: Course in General Linguistics. Open Court, La Salle, Illinois (1983)

9. Dubossarsky, H., Weinshall, D., Grossman, E.: Outta control: laws of semantic change and inherent biases in word representation models. In: Proceedings of the 2017 Conference on Empirical Methods in Natural Language Processing, pp. 1136–1145 (2017)

10. Efron, B., Tibshirani, R.J.: An Introduction to the Bootstrap. Chapman and Hall/CRC, Boca Raton (1994)

11. Frermann, L., Lapata, M.: A bayesian model of diachronic meaning change. Trans. Assoc. Comput. Linguist. **4**, 31–45 (2016)

12. Gulordava, K., Baroni, M.: A distributional similarity approach to the detection of semantic change in the Google Books Ngram corpus. In: Proceedings of the EMNLP 2011 Geometrical Models for Natural Language Semantics (GEMS 2011) Workshop, pp. 67–71 (2011). http://clic.cimec.unitn.it/marco/publications/gems-11/gulordava-baroni-gems-2011.pdf

13. Hamilton, W.L., Leskovec, J., Jurafsky, D.: Diachronic word embeddings reveal statistical laws of semantic change. arXiv preprint arXiv:1605.09096 (2016)

14. Jatowt, A., Duh, K.: A framework for analyzing semantic change of words across time. In: Proceedings of the ACM/IEEE Joint Conference on Digital Libraries, pp. 229–238 (2014). https://doi.org/10.1109/JCDL.2014.6970173

15. JISC, the Internet Archive: Jisc uk web domain dataset (1996–2013) (2013). https://doi.org/10.5259/ukwa.ds.2/1

16. Jurgens, D., Stevens, K.: event detection in blogs using temporal random indexing. In: Proceedings of the Workshop on Events in Emerging Text Types, pp. 9–16. Association for Computational Linguistics (2009)

17. Kulkarni, V., Al-Rfou, R., Perozzi, B., Skiena, S.: Statistically significant detection of linguistic change. In: Proceedings of the 24th International Conference on World Wide Web, pp. 625–635. ACM (2015)

18. Kutuzov, A., Øvrelid, L., Szymanski, T., Velldal, E.: Diachronic word embeddings and semantic shifts: a survey (2018). arXiv:1806.03537

19. Lin, Y., Michel, J.B., Aiden, E.L., Orwant, J., Brockman, W., Petrov, S.: Syntactic annotations for the google books ngram corpus. In: Proceedings of the 50th Annual Meeting of the Association for Computational Linguistics, Jeju, Republic of Korea, 8–14 July 2012, pp. 169–174. Association for Computational Linguistics (2012)

20. Mitchell, J., Lapata, M.: Composition in distributional models of semantics. Cogn. Sci. **34**(8), 1388–1429 (2010)

21. Rudolph, M., Blei, D.: Dynamic embeddings for language evolution. In: Proceedings of the 2018 World Wide Web Conference on World Wide Web (2018)

22. Sahlgren, M.: The word-space model: using distributional analysis to represent syntagmatic and paradigmatic relations between words in high-dimensional vector spaces (2006)

23. Schiitze, H.: Word space. Adv. Neural Inf. Process. Syst. **5**, 895–902 (1993)

24. Tang, X.: A State-of-the-Art of Semantic Change Computation. arXiv preprint arXiv:1801.09872 (Cl), 2–37 (2018)

25. Tang, X., Qu, W., Chen, X.: Semantic change computation: a successive approach. World Wide Web-Internet Web Inf. Syst. **19**(3), 375–415 (2016). https://doi.org/10.1007/s11280-014-0316-y

26. Taylor, W.A.: Change-Point Analysis: A Powerful New Tool for Detecting Changes. Taylor Enterprises, Inc. (2000)

27. Wijaya, D.T., Yeniterzi, R.: Understanding semantic change of words over centuries. In: Proceedings of the 2011 International Workshop on DETecting and Exploiting Cultural diversiTy on the Social Web - DETECT '11, p. 35 (2011). https://doi.org/10.1145/2064448.2064475, http://dl.acm.org/citation.cfm?doid=2064448.2064475

28. Xu, Y., Kemp, C.: A computational evaluation of two laws of semantic change. Proc. CogSci **2015**, 1–6 (2015)

29. Yao, Z., Sun, Y., Ding, W., Rao, N., Xiong, H.: Dynamic Word Embeddings for Evolving Semantic Discovery. Technical report (2017). https://doi.org/10.1145/3159652.3159703, arXiv:1703.00607

Online Gradient Boosting for Incremental Recommender Systems

João Vinagre[1,3]($\boxtimes$) (iD), Alípio Mário Jorge[1,3] (iD), and João Gama[2,3] (iD)

[1] FCUP - University of Porto, Porto, Portugal
[2] FEP - University of Porto, Porto, Portugal
[3] LIAAD - INESC TEC, Porto, Portugal
jnsilva@inesctec.pt, amjorge@fc.up.pt, jgama@fep.up.pt

Abstract. Ensemble models have been proven successful for batch recommendation algorithms, however they have not been well studied in streaming applications. Such applications typically use incremental learning, to which standard ensemble techniques are not trivially applicable. In this paper, we study the application of three variants of online gradient boosting to top-N recommendation tasks with implicit data, in a streaming data environment. Weak models are built using a simple incremental matrix factorization algorithm for implicit feedback. Our results show a significant improvement of up to 40% over the baseline standalone model. We also show that the overhead of running multiple weak models is easily manageable in stream-based applications.

Keywords: Recommender systems · Boosting · Online learning
Data streams

1 Introduction

The increasing amount and rate of data that is generated in modern online systems is overwhelming. The demand for techniques and algorithms that allow the timely extraction of knowledge is clearly increasing. However some of the most popular data analysis techniques rely on batch data processing, and are not suitable for continuous flows of data. Ensemble learning is a popular technique to improve the accuracy of machine learning algorithms. Although valuable contributions have been made in online ensemble learning for classification and regression problems, there is few work available in the literature that studies online ensembles for recommender systems. In this paper, we propose online boosting for incremental recommendation algorithms.

We focus on a top-N recommendation task with implicit feedback streams. The data stream consists of a continuous flow of user-item pairs (u, i) that indicate a positive preference of user u for item i. The top-N recommendation task consists of presenting a ranked list of arbitrary size N to any known user u with the items that are more likely to match her preferences.

© Springer Nature Switzerland AG 2018
L. Soldatova et al. (Eds.): DS 2018, LNAI 11198, pp. 209–223, 2018.
https://doi.org/10.1007/978-3-030-01771-2_14

Algorithms that deal with this problem setting must fulfill the following two requirements:

- **Online learning**: ability to maintain models with fast incremental updates;
- **Implicit data processing**: ability to learn from implicit feedback;

In [21] we propose ISGD, a fast incremental matrix factorization algorithm that we see especially well adjusted for online ensemble learning. In addition to fulfilling the above two requirements, it is highly competitive in terms of accuracy and at least one order of magnitude faster than its alternatives, making it naturally suitable for online ensemble learning. ISGD relies on stochastic gradient descent to learn a model from a stream of (u, i) pairs. This model is able to predict a numeric score for any (u, i) pair using a regression approach. These scores are then used to produce a personalised ranking of items for each user. In an ensemble, scores can be aggregated across its members. The item ranking is produced according to this aggregated score.

We evaluate if online boosting approaches designed for regression problems are able to improve the accuracy of recommender systems in top-N recommendation tasks. To our best knowledge this is the first work to use online boosting in recommendation problems.

1.1   Boosting in Machine Learning

Boosting is a convenient ensemble method to improve the predictive ability of machine learning algorithms. It is designed as a stagewise additive model where each base learner – or weak learner – tries to correct for the deficiencies of the previous one. Aggregating the contributions of all weak learners, we obtain a *strong* learner.

There are fundamentally two approaches to boosting. The first is proposed in [8] in the Adaboost algorithm. This algorithm works by re-weighting data points according to their classification error. If an example is misclassified by a weak learner, its weight is increased, otherwise, it is decreased. The following weak learner will then put relatively more effort on misclassified examples, and less effort on the correctly classified ones. Adaboost is proposed for binary classification. To use it with multi-class classification or regression, it is necessary to recast the original problem as binary classification.

The second approach is proposed in [9] with Stochastic Gradient Boosting. It directly tackles regression using an optimization framework. The first weak learner tries to learn the original values of the target variable, and every subsequent weak learner targets the residuals of the previous one. Predictions are obtained by summing the predictions of all weak models. This approach is naturally suitable for regression, but it can also be used for classification, e.g. using logistic regression.

1.2   Related Work

Bagging [4], Boosting [8] and Stacking [24] are three well-known ensemble methods used with recommendation algorithms. In the field of recommender systems all three technique have been studied in the past. Boosting is experimented by [7,14,18,19], bagging is studied also by [14,19], and stacking by [20]. In all of these contributions, ensemble methods work with batch learning algorithms only. In this paper, we aim to build a boosting model online, over a data stream, so we need stream-based methods.

Stream-based ensemble learning has been widely studied in classification and regression. The Random Forest algorithm [5] is a widely known ensemble model that has been successfully used in data stream mining [10]. Several online algorithms have been proposed for bagging [2,3,17] and boosting [1,3,6,13,16,17]. Two up-to-date comprehensive surveys on ensemble methods for classification and regression over data streams are available in [12,15].

We have recently proposed online bagging for recommendation problems in [22]. To our best knowledge, this is the only available work in the literature on online ensemble learning for recommendation. In this paper, we use a similar approach to study online boosting for top-N recommendation.

2   Online Boosting

An online version of Adaboost is proposed by Oza and Russel in [17]. However, this is primarily designed for binary classification. To tackle our problem (see Sect. 1), an approach for regression is more suitable. In [1,13], online gradient boosting algorithms for regression are proposed – respectively Algorithms 1 and 2.

Algorithm 1: OBoostH - Hu's Online Gradient Boosting

Data: stream $D = \{(\boldsymbol{x}_1, y_1), (\boldsymbol{x}_2, y_2), \ldots\}$
input : no. weak models M, learn rate γ

for $(\boldsymbol{x}, y) \in D$ **do**
    $\hat{y} \leftarrow 0$
    $\tilde{y} \leftarrow y$
    $\delta \leftarrow 0$
    for $m \leftarrow 1$ **to** M **do**
        $\hat{y} \leftarrow \hat{y} - \gamma \hat{y}_m$
        $\tilde{y} \leftarrow L(\hat{y}, \tilde{y})$
        Pass $(\boldsymbol{x}, \delta + \tilde{y})$ to weak model m
        $\delta \leftarrow \delta + \tilde{y} - \hat{y}$

In both algorithms, we maintain the pseudo-residuals $\tilde{y}$ through the iterations over the M weak models. In the first iteration the corresponding model learns

---

Algorithm 2: OBoostB - Beygelzimer's Online Gradient Boosting

---

Data: stream $D = \{(\boldsymbol{x}_1, y_1), (\boldsymbol{x}_2, y_2), \ldots\}$
input : no. weak models M, learn rate γ

for $m \leftarrow 1$ to M do
    $\sigma_m \leftarrow 0$
for $(\boldsymbol{x}, y) \in D$ do
    $\hat{y} \leftarrow 0$
    $\tilde{y} \leftarrow y$
    for $m \leftarrow 1$ to M do
        $\hat{y} \leftarrow (1 - \sigma_m)\hat{y} + \gamma\hat{y}_m$
        Get loss $\tilde{y} \leftarrow L(\hat{y}, \tilde{y})$
        Pass $(\boldsymbol{x}, \tilde{y})$ to weak model m with lear rate γ
        Update σ_m with online gradient descent

---

the target $\tilde{y} = y$, very much like a standalone model would do. In subsequent iterations, the pseudo-residual is set to the outcome of a loss function, so the corresponding model learns the residual of the previous one.

There are two main differences between both algorithms. Algorithm 2 *optionally* uses of a set of M dynamically updated shrinkage factors σ_m that force the partial values of $\hat{y}$ to follow the gradient of the loss function. Algorithm 1 keeps track of the overall residual δ of the ensemble, and adds it to the partial residuals $\tilde{y}$ at the learning step.

2.1 Online Boosting for Recommendation

Both [1, 13] show that online boosting is able to outperform standalone models in standard regression problems. However, it is impossible to extrapolate those results to top-N recommendation problems for three main reasons.

First, recommendation is fundamentally different from regression. At most, in our problem setting, we can look at the recommendation model as a multiplicity of regression problems – one for each user, or one for each item, depending on the viewpoint – that are jointly learned from a stream of user-item relations.

Second, the top-N recommendation problem involves ranking items for each user. The accuracy of the model is not measured directly by the success in minimizing the objective function, but rather by evaluating the actual recommendation lists, which naturally adds a degree of separation.

Finally, gradient boosting models for regression are typically built using *weak* models. The term *weak* is used because these models are only required to be better than random guessing. In our problem, we are using an algorithm that is able to achieve good results on its own, as shown in [21], obviously much better than random guessing. Since it has been shown in the past that boosting applied to strong learners is frequently non-productive or even counter-productive [23], we need to assess whether ISGD has room for improvement in a boosting framework.

Our base algorithm is ISGD [21], a simple online matrix factorization method for implicit positive-only data. It is designed for streams of user-item pairs (u, i)

that indicate a positive interaction between user u and item i. Examples of positive interactions are users buying items in an online store, streaming music tracks from an online music streaming service, or simply visiting web pages. This is a much more widely available form of user feedback, than for example, ratings data, which is only available from systems in which users are allowed to rate items (e.g. in a 5 star scale). Matrix factorization for implicit data works by decomposing a user-item matrix R in two latent factor matrices $P^{u \times z}$ and $Q^{i \times z}$ that span u known users and i known items respectively, in z common latent features, such that:

$$r_{ui} \approx \hat{r}_{ui} = p_u q_i^T \tag{1}$$

We assume the value $r_{ui} = 1$ for each positive interaction – i.e. each user-item pair (u, i) occurring in the data – and $r_{ui} = 0$ otherwise. The decomposition is obtained by minimizing the squared error between 1 and $\hat{r}_{ui}$ (2) for all known examples in a data stream D. Note that $r_{ui} = 1$ iff $(u, i) \in D$.

$$\min_{P., Q.} \sum_{(u,i) \in D} (1 - p_u q_i^T)^2 + \lambda(||p_u||^2 + ||q_i||^2) \tag{2}$$

Algorithm 3: ISGD

Data: stream $D = \{(u, i)_1, (u, i)_2, \ldots\}$
input : latent features z, iterations $iter$, regularization λ, learn rate η
output: factor matrices P and Q

for $(u, i) \in D$ **do**
    if $u \notin \mathrm{Rows}(P)$ **then**
        $p_u \leftarrow \mathbf{Vector}(\mathrm{size} : z)$
        $p_u \sim \mathcal{N}(0, 0.1)$
    if $i \notin \mathrm{Rows}(Q)$ **then**
        $q_i \leftarrow \mathbf{Vector}(\mathrm{size} : z)$
        $q_i \sim \mathcal{N}(0, 0.1)$
    for $n \leftarrow 1$ **to** $iter$ **do**
        $\epsilon_{ui} \leftarrow 1 - p_u q_i^T$
        $p_u \leftarrow p_u + \eta(\epsilon_{ui} q_i - \lambda p_u)$
        $q_i \leftarrow q_i + \eta(\epsilon_{ui} p_u - \lambda q_i)$

In (2), the regularization term $\lambda(||p_u||^2 + ||q_i||^2)$ penalizes overly complex models that tend to overfit. The hyperparameter λ controls the amount of regularization.

ISGD – Algorithm 3 – uses Stochastic Gradient Descent to solve (2). The algorithm continuously updates factor matrices P and Q, correcting the model to fit to the incoming user-item pairs. If (u, i) occurs in the stream, then the model prediction $\hat{r}_{ui} = p_u q_i^T$ should be close to 1. Top-N recommendations to any user u is obtained by a ranking function $f = |1 - \hat{r}_{ui}|$ for all items i in ascending order, and taking the top N items.

Applying Algorithms 1 and 2 to ISGD, we obtain Algorithms 5 and 4, respectively. In both algorithms, we calculate the subgradients based on the square loss. Note that for ISGD to work within a boosting framework, we need to allow it to receive arbitrary target values $\tilde{y}$ for training, instead of a fixed value 1.

Algorithm 4: OBoostH with ISGD

Data: stream $D = \{(u,i)_1, (u,i)_2, \ldots\}$
input : latent features z, iterations $iter$, regularization λ, learn rate η, no. nodes M, boosting learn rate γ
output: set of factor matrices $P^{1..M}$ and $Q^{1..M}$

for $(u,i) \in D$ do
    $\hat{y} \leftarrow 0, \tilde{y} \leftarrow 1, \delta \leftarrow 0$
    for $m \leftarrow 1$ to M do
        if $u \notin \text{Rows}(P^m)$ then
            $p_u^m \leftarrow \text{Vector}(\text{size} : z)$
            $p_u^m \sim \mathcal{N}(0, 0.1)$
        if $i \notin \text{Rows}(Q^m)$ then
            $q_i^m \leftarrow \text{Vector}(\text{size} : z)$
            $q_i^m \sim \mathcal{N}(0, 0.1)$
        $\hat{y} \leftarrow \hat{y} + \gamma p_u^m (q_i^m)^T$
        for $n \leftarrow 1$ to $iter$ do
            $\epsilon_{ui} \leftarrow \delta + \tilde{y} - p_u^m (q_i^m)^T$
            $p_u^m \leftarrow p_u^m + \eta(\epsilon_{ui} q_i^m - \lambda p_u^m)$
            $q_i^m \leftarrow q_i^m + \eta(\epsilon_{ui} p_u^m - \lambda q_i^m)$
        $\delta \leftarrow \delta + \tilde{y} - \hat{y}$
        $\tilde{y} \leftarrow \tilde{y} - \hat{y}$

To score a new (u,i) pair in the stream, we simply aggregate the contributions from the weak models, scaled by the boosting learn rate γ:

$$\hat{r}_{ui} = \gamma \sum_{m=1}^{M} p_u^m (q_i^m)^T \tag{3}$$

3   Evaluation

In our experimental work we wish to assess if online boosting is able to improve over the baseline ISGD. We also wish to compare between several alternatives:

- OBoostH - Algorithm 4;
- OBoostB1 - Algorithm 5 with adaptive shrinkage factor;
- OBoostB2 - Algorithm 5 **without** shrinkage – i.e. $\sigma = 0$.

Algorithm 5: OBoostB with ISGD

Data: stream $D = \{(u,i)_1, (u,i)_2, \ldots\}$
input : latent features z, iterations $iter$, regularization λ, learn rate η, no.
        nodes M, boosting learn rate γ
output: set of factor matrices $P^{1..M}$ and $Q^{1..M}$

for $m \leftarrow 1$ **to** M **do**
    $\sigma_m \leftarrow 0$
for $(u,i) \in D$ **do**
    $\hat{y} \leftarrow 0, \tilde{y} \leftarrow 1$
    for $m \leftarrow 1$ **to** M **do**
        if $u \notin \text{Rows}(P^m)$ **then**
            $p_u^m \leftarrow \text{Vector}(size : z)$
            $p_u^m \sim \mathcal{N}(0, 0.1)$
        if $i \notin \text{Rows}(Q^m)$ **then**
            $q_i^m \leftarrow \text{Vector}(size : z)$
            $q_i^m \sim \mathcal{N}(0, 0.1)$
        $\hat{y} \leftarrow (1 - \sigma_m \gamma)\hat{y} + \gamma p_u^m (q_i^m)^T$
        for $n \leftarrow 1$ **to** $iter$ **do**
            $\epsilon_{ui} \leftarrow \tilde{y} - p_u^m (q_i^m)^T$
            $p_u^m \leftarrow p_u^m + \eta(\epsilon_{ui} q_i^m - \lambda p_u^m)$
            $q_i^m \leftarrow q_i^m + \eta(\epsilon_{ui} p_u^m - \lambda q_i^m)$
        $\tilde{y} \leftarrow \tilde{y} - \hat{y}$
        $\sigma_m \leftarrow \sigma_m + \frac{\tilde{y} \cdot \hat{y}}{\sqrt{|D|}}$

3.1 Datasets

To simulate a streaming environment we need datasets that maintain the natural order of the data points, as they were generated. We use 4 implicit preference datasets with naturally ordered events, described in Table 1. ML1M is based on the Movielens-1M movie rating dataset[1]. To obtain the YHM-6KU, we sample 6000 users randomly from the Yahoo! Music dataset[2]. LFM-50U is a subset consisting of a random sample of 50 users taken from the Last.fm[3] dataset[4]. PLC-STR [5] consists of the music streaming history taken from Palco Principal[6], a portuguese social network for non-mainstream artists and fans.

All of the 4 datasets consist of a chronologically ordered sequence of positive user-item interactions. However, ML1M and YHM-50U are obtained from ratings datasets. To use them as positive-only data, we retain the user-item pairs for which the rating is in the top 20% of the rating scale. This means retaining only the rating 5 in ML1M and rating of 80 or more in the YHM-6KU dataset.

---

[1] http://www.grouplens.org/data [Jan 2013].
[2] https://webscope.sandbox.yahoo.com/catalog.php?datatype=r [Jan 2013].
[3] http://last.fm/.
[4] http://ocelma.net/MusicRecommendationDataset [Jan 2013].
[5] https://rdm.inesctec.pt/dataset/cs-2017-003, file: `playlisted_tracks.tsv`.
[6] http://www.palcoprincipal.com/.

Table 1. Dataset description

Dataset	Events	Users	Items	Sparsity
PLC-STR	588 851	7 580	30 092	99.74%
LFM-50U	1 121 520	50	159 208	85.91%
YHM-6KU	476 886	6 000	127 448	99.94%
ML1M	226 310	6 014	3 232	98.84%

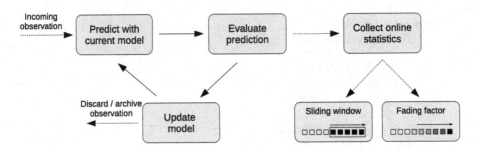

Fig. 1. Prequential evaluation

3.2   Prequential Evaluation

We run a set of experiments using prequential evaluation [11] as described in [21]. For each incoming observation in the stream, we make a prediction with our current model and score that prediction matching it to the actual observation. Then we update the model with the observation and advance to the following example. Statistics on the scores can be maintained, for example, using a sliding window or a fading factor. The prequential process is depicted in Fig. 1.

In our recommendation environment, each observation in the dataset consists of a simple user-item pair (u, i) that indicates a positive interaction between user u and item i. The following steps are performed in the prequential evaluation process:

1. If u is a known user, use the current model to recommend a list of items to u, otherwise go to step 3;
2. Score the recommended list given the observed item i;
3. Update the model with (u, i);
4. Proceed to the next observation

We use the Hit Ratio at cutoffs $N \in \{1, 10, 20\}$ – denoted as HR@N. This is obtained by recommending a list of the N best items found by the algorithm to user u at step 1 of the prequential process. Then in step 2, we score the list with 1 if the item i is within the list, and 0 otherwise. The hit ratio is obtained by averaging the scores. This can be done at the end of the experiment, or online, using a moving average.

Table 2. Average hit rate of baseline and boosted ISGD with $M \in \{2, 4, 8, 16, 32\}$. The first line of each dataset shows the results for the baseline ISGD (without boosting).

Dataset	M	HR@1			HR@10			HR@20		
		OBoostH	OBoostB1	OBoostB2	OBoostH	OBoostB1	OBoostB2	OBoostH	OBoostB1	OBoostB2
PLC-STR	ISGD		0.133			0.278			0.304	
	2	0.134	0.125	0.139	0.302	0.307	0.260	0.331	0.340	0.324
	4	0.097	0.085	0.117	0.286	0.249	0.319	0.336	0.294	0.357
	8	0.101	0.100	0.122	0.292	0.317	0.335	0.340	0.366	0.377
	16	0.099	0.093	0.119	0.285	0.299	0.336	0.336	0.352	0.381
	32	0.093	0.098	0.094	0.278	0.306	0.248	0.330	0.360	0.281
LFM-50U	ISGD		0.032			0.048			0.050	
	2	0.035	0.035	0.035	0.053	0.054	0.052	0.055	0.057	0.054
	4	0.035	0.033	0.036	0.053	0.061	0.058	0.055	0.064	0.061
	8	0.035	0.032	0.038	0.053	0.062	0.064	0.055	0.066	0.067
	16	0.035	0.033	0.038	0.053	0.064	0.065	0.055	0.067	0.068
	32	0.035	0.035	0.033	0.053	0.065	0.060	0.055	0.068	0.063
YHM-6KU	ISGD		0.032			0.089			0.112	
	2	0.035	0.035	0.035	0.099	0.102	0.096	0.124	0.127	0.120
	4	0.017	0.024	0.036	0.056	0.099	0.110	0.074	0.132	0.137
	8	0.018	0.020	0.035	0.056	0.099	0.120	0.073	0.132	0.151
	16	0.018	0.019	0.034	0.056	0.091	0.124	0.073	0.121	0.156
	32	0.017	0.011	0.031	0.056	0.039	0.112	0.074	0.047	0.141
ML1M	ISGD		0.006			0.033			0.050	
	2	0.006	0.007	0.007	0.037	0.038	0.035	0.057	0.059	0.054
	4	0.006	0.005	0.006	0.040	0.036	0.041	0.063	0.059	0.065
	8	0.006	0.005	0.006	0.040	0.040	0.043	0.064	0.065	0.070
	16	0.006	0.005	0.006	0.040	0.040	0.044	0.064	0.066	0.071
	32	0.006	0.006	0.005	0.040	0.042	0.034	0.064	0.069	0.053

Hyperparameters for the base algorithm were obtained using a grid search over the first 10% of the data in each dataset using prequential evaluation. The optimal hyperparameters found for the four datasets are presented in Table 4.

3.3 Overall Results

Table 2 presents all results from our experiments. Values in Table 2 are obtained by averaging hit rate obtained at all prequential evaluation steps. Results that are significantly better than standalone ISGD are highlighted in italic. To assess the significance of the differences we use the signed McNemar test over a sliding window with size $n = 10000$. Given that this is an online learning process, the significance of the differences between algorithms can vary during the learning process.

We also present a graphical overview of the relative improvement in HR@20 of the three variants of boosting with respect to the standalone version of ISGD in Fig. 2 (a) through (d).

By observing the plots, the most obvious observation is that in the majority of cases boosting substantially improves the accuracy of the top-N recommendation task, with improvements up to 40% over the baseline with the YHM-6KU

Table 3. Average update and recommendation time of baseline and boosted ISGD with $M \in \{2, 4, 8, 16, 32\}$. The first line of each dataset shows the times for the baseline ISGD (without boosting).

Dataset	M	Update time (ms)			Rec. time (ms)		
		OBoostH	OBoostB1	OBoostB2	OBoostH	OBoostB1	OBoostB2
PLC-STR	ISGD	0.3			19		
	2	0.8	0.6	0.5	40	35	32
	4	1.5	1.0	0.8	75	55	51
	8	2.8	2.2	1.9	147	108	97
	16	5.7	5.1	4.7	302	274	236
	32	10.5	9.7	9.1	593	531	489
LFM-50U	ISGD	2.7			84		
	2	7.5	6.4	6.1	231	208	200
	4	16.1	13.7	12.0	428	386	360
	8	28.9	25.6	23.1	808	726	689
	16	54.3	58.8	52.3	1.5 s	1.5 s	1.4 s
	32	104.1	110.6	104.4	2.9 s	3.1 s	2.8 s
YHM-6KU	ISGD	2.7			85		
	2	13.8	12.1	11.3	220	200	186
	4	29.2	26.6	24.5	413	379	361
	8	52.5	56.4	52.0	758	740	729
	16	100.4	107.7	103.2	1.4 s	1.5 s	1.5 s
	32	194.0	208.3	203.2	2.8 s	3.0 s	2.9 s
ML1M	ISGD	0.2			5		
	2	0.5	0.3	0.3	9	8	8
	4	1.0	0.5	0.4	18	13	11
	8	1.8	1.1	0.9	35	27	25
	16	3.3	2.1	2.0	76	54	50
	32	6.4	4.2	3.8	149	108	87

Table 4. Hyperparameter settings for ISGD

Dataset	z	$iter$	η	λ
PLC-STR	200	6	0.35	0.5
LFM-50U	160	4	0.5	0.4
YHM-6KU	200	9	0.25	0.45
ML1M	160	8	0.1	0.4

dataset. This is obtained using OBoostB2, with fixed parameters $\sigma_m = 0$. Shrinkage seems to help in same cases and hurt in others. For example, comparing Fig. 2 (a) and (c), we see that shrinkage helps in the first case – with PLC-STR –, but

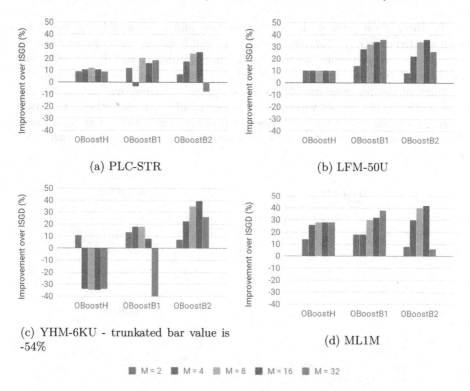

(a) PLC-STR

(b) LFM-50U

(c) YHM-6KU - trunkated bar value is -54%

(d) ML1M

■ M = 2   ■ M = 4   ■ M = 8   ■ M = 16   ■ M = 32

Fig. 2. Improvements obtained in HR@20 with respect to standalone ISGD

heavily hurts in the second, with YHM-6KU. OboostB1 and OBoostB2 yield the best results. OBoostH has the most inconsistent outcome: it has the capability of improving more than 10% over the baseline in PLC-STR and ML1M – Fig. 2 (a) and (d) –, however it is counter-productive with YHM-6KU – Fig. 2 (c).

Increasing the number of weak models is beneficial up to a certain point. In several cases, we see a degradation when increasing the number of base models from 16 to 32. However, this phenomenon is not consistent across datasets and algorithms. OBoostH does not benefit much from a large number of models. Increasing from $M = 8$ to $M = 16$ or $M = 32$ barely has noticeable impact.

Another interesting observation is that ensembles with only 2 base models are able to outperform the standalone version by up to 10%. By continuously doubling the number of weak models, the gain in accuracy, if any, is not proportional to the additional effort of maintaining a twice as many models. This behavior is similar to online bagging [22], but much less consistent.

The lack of consistency in results may be caused by a number of factors, including local minima, noise and even overfitting to local phenomena, such as concept drifts. Another possible cause is the fact that ISGD is already a strong learner.

In Table 3 we present the average update and recommendation times for each algorithm in each dataset. One downside of boosting with respect to other ensemble alternatives such as bagging, is that training cannot be trivially parallelized, given its stage-wise nature, where every weak model learns from the outcome of the previous one. However, although the time overhead grows linearly with the number of weak models – expectedly –, update times are easily manageable, since ISGD already has fast updates. The overhead at recommendation is much more relevant and can be problematic in applications with strict requirements.

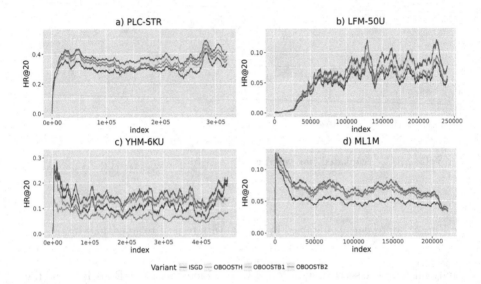

Fig. 3. Prequential outcome of HR@20 with $M = 16$ over a sliding with size $n = 10000$.

3.4   Evolving Results

To have a better insight into the learning process, we also depict it using a moving average of the collected hit ration over a sliding window in Fig. 3. This visualization is useful because overall average results hide the evolution of the process, that may considerably vary over time. In this case, the evolution of the learning process of alternative methods is steadily consistent with the overall results shown in Table 2.

3.5   Statistical Significance

Although differences in Figs. 2 and 3 are clearly visible, there is no guarantee about their statistical significance. To assess this, we use the signed McNemar test over a sliding window as described in [21].

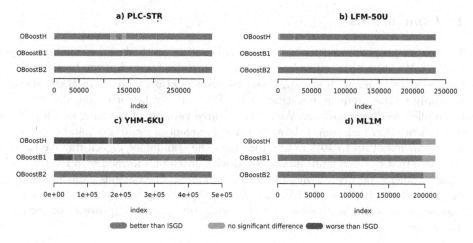

Fig. 4. Statistical significance with McNemar for HR@20 with $M = 16$ over a sliding with size $n = 10000$.

Given two alternative algorithms A and B, at any data point j, we take the HR@N in the current window and formulate the null hypothesis that there is no significant difference between the two sequences, i.e. that the two algorithm have equivalent performance. The test works by keeping count of two quantities: the number of instances n_{10} for which the prediction of A is correct and the prediction of B is wrong, and the number of instances n_{01} for which the opposite occurs. These quantities are used to calculate the statistic:

$$Mcn = \frac{(n_{10} - n_{01})^2}{n_{10} + n_{01}} \tag{4}$$

Mcn follows a χ^2 distribution with one degree of freedom. For a significance level $\alpha = 0.01$, the critical value $Mcn = 6.635$ is used. If $Mcn > 6.635$ the null hypothesis – that there is no significant difference between the two alternatives – is rejected. In our setting, we perform pairwise tests between ISGD and all variants of online boosting.

As a result, we obtain a sequence of values for the McNemar statistic, that allow us to test for the significance of differences between ISGD and each of the three boosting methods. This is illustrated in Fig. 4. In the green regions, the corresponding algorithm is significantly better than ISGD, the gray regions correspond to non-significant differences, and the regions where ISGD is significantly better are plotted in red. This visualization complements the one in Fig. 3. In this case – boosting with $M = 16$ and with the HR@20 metric –, the only cases where boosting does not consistently improve ISGD are using OBoostH and partially OBoostB1 with the YHM-6KU dataset.

4   Conclusions

This paper proposes and evaluates online boosting methods for recommender systems. We show that online boosting algorithms for regression can be successfully used together with stream-based recommendation models. In this paper we apply online gradient boosting to ISGD, a matrix factorization algorithm for implicit feedback streams. We evaluate three variants of boosting in a top-N recommendation task over 4 datasets using prequential evaluation, and observe improvements of up to 40% in accuracy over the standalone algorithm. We further note that optimal gains are achieved with ensembles formed by a relatively small number – between 2 and 16 – of base models. The obvious gains in accuracy, together with the relatively small cost of learning multiple models show that online boosting is a viable and promising approach to improve recommendations over implicit feedback streams.

Acknowledgments. This work is financed by the European Regional Development Fund (ERDF), through the Incentive System to Research and Technological development, within the Portugal2020 Competitiveness and Internationalization Operational Program – COMPETE 2020 – within project PushNews (POCI-01- 0247-FEDER-0024257). The work is also financed by the ERDF through COMPETE 2020 within project POCI-01-0145-FEDER-006961, and by national funds through the Portuguese Foundation for Science and Technology (FCT) as part of project UID/EEA/50014/2013.

References

1. Beygelzimer, A., Hazan, E., Kale, S., Luo, H.: Online gradient boosting. In: Advances in Neural Information Processing Systems 28: Annual Conference on Neural Information Processing Systems 2015, December 7–12, 2015, Montreal, Quebec, Canada, pp. 2458–2466 (2015)
2. Bifet, A., Holmes, G., Pfahringer, B.: Leveraging bagging for evolving data streams. In: Balcázar, J.L., Bonchi, F., Gionis, A., Sebag, M. (eds.) ECML PKDD 2010. LNCS (LNAI), vol. 6321, pp. 135–150. Springer, Heidelberg (2010). https://doi.org/10.1007/978-3-642-15880-3_15
3. Bifet, A., Holmes, G., Pfahringer, B., Kirkby, R., Gavaldà, R.: New ensemble methods for evolving data streams. In: Proceedings of the 15th ACM SIGKDD International Conference on Knowledge Discovery and Data Mining, Paris, France, June 28 - July 1, 2009, pp. 139–148. ACM (2009)
4. Breiman, L.: Bagging predictors. Mach. Learn. **24**(2), 123–140 (1996)
5. Breiman, L.: Random forests. Mach. Learn. **45**(1), 5–32 (2001)
6. Chen, S., Lin, H., Lu, C.: An online boosting algorithm with theoretical justifications. In: Proceedings of the 29th International Conference on Machine Learning, ICML 2012, Edinburgh, Scotland, UK, June 26 - July 1, 2012. icml.cc / Omnipress (2012)
7. Chowdhury, N., Cai, X., Luo, C.: BoostMF: boosted matrix factorisation for collaborative ranking. In: Appice, A., Rodrigues, P.P., Santos Costa, V., Gama, J., Jorge, A., Soares, C. (eds.) ECML PKDD 2015. LNCS (LNAI), vol. 9285, pp. 3–18. Springer, Cham (2015). https://doi.org/10.1007/978-3-319-23525-7_1

8. Freund, Y., Schapire, R.E.: Experiments with a new boosting algorithm. In: Proceedings of the 13th Intl. Conference on Machine Learning ICML '96, pp. 148–156. Morgan Kaufmann (1996)

9. Friedman, J.H.: Stochastic gradient boosting. Comput. Stat. Data Anal. **38**(4), 367–378 (2002)

10. Gama, J., Medas, P., Rocha, R.: Forest trees for on-line data. In: Proceedings of the 2004 ACM Symposium on Applied Computing (SAC), Nicosia, Cyprus, March 14–17, 2004, pp. 632–636. ACM (2004)

11. Gama, J., Sebastião, R., Rodrigues, P.P.: On evaluating stream learning algorithms. Mach. Learn. **90**(3), 317–346 (2013)

12. Gomes, H.M., Barddal, J.P., Enembreck, F., Bifet, A.: A survey on ensemble learning for data stream classification. ACM Comput. Surv. **50**(2), 23:1–23:36 (2017)

13. Hu, H., Sun, W., Venkatraman, A., Hebert, M., Bagnell, J.A.: Gradient boosting on stochastic data streams. In: Proceedings of the 20th International Conference on Artificial Intelligence and Statistics, AISTATS 2017, 20–22 April 2017, Fort Lauderdale, FL, USA. Proceedings of Machine Learning Research, vol. 54, pp. 595–603. PMLR (2017)

14. Jahrer, M., Töscher, A., Legenstein, R.A.: Combining predictions for accurate recommender systems. In: Proceedings of the 16th ACM SIGKDD International Conference on Knowledge Discovery and Data Mining, KDD 2010, pp. 693–702. ACM (2010)

15. Krawczyk, B., Minku, L.L., Gama, J., Stefanowski, J., Wozniak, M.: Ensemble learning for data stream analysis: a survey. Inf. Fusion **37**, 132–156 (2017)

16. Lee, H.K.H., Clyde, M.A.: Lossless online bayesian bagging. J. Mach. Learn. Res. **5**, 143–151 (2004)

17. Oza, N.C., Russell, S.J.: Experimental comparisons of online and batch versions of bagging and boosting. In: Proceedings of the 7th ACM SIGKDD International Conference on Knowledge Discovery and Data Mining, KDD 2001, pp. 359–364. ACM (2001)

18. Schclar, A., Tsikinovsky, A., Rokach, L., Meisels, A., Antwarg, L.: Ensemble methods for improving the performance of neighborhood-based collaborative filtering. In: Proceedings of the 2009 ACM Conference on Recommender Systems, RecSys 2009, pp. 261–264. ACM (2009)

19. Segrera, S., Moreno, M.N.: An experimental comparative study of web mining methods for recommender systems. In: Proceedings of the 6th WSEAS Intl. Conf. on Distance Learning and Web Engineering, pp. 56–61. WSEAS (2006)

20. Sill, J., Takács, G., Mackey, L.W., Lin, D.: Feature-weighted linear stacking. CoRR (2009). arXiv:0911.0460

21. Vinagre, J., Jorge, A.M., Gama, J.: Fast incremental matrix factorization for recommendation with positive-only feedback. In: Dimitrova, V., Kuflik, T., Chin, D., Ricci, F., Dolog, P., Houben, G.-J. (eds.) UMAP 2014. LNCS, vol. 8538, pp. 459–470. Springer, Cham (2014). https://doi.org/10.1007/978-3-319-08786-3_41

22. Vinagre, J., Jorge, A.M., Gama, J.: Online bagging for recommender systems. Expert Syst. **35**(4) (2018). https://doi.org/10.1111/exsy.12303

23. Wickramaratna, J., Holden, S.B., Buxton, B.F.: Performance degradation in boosting. In: Kittler, J., Roli, F. (eds.) MCS 2001. LNCS, vol. 2096, pp. 11–21. Springer, Heidelberg (2001). https://doi.org/10.1007/3-540-48219-9_2

24. Wolpert, D.H.: Stacked generalization. Neural Netw. **5**(2), 241–259 (1992)

Selection of Relevant and Non-Redundant Multivariate Ordinal Patterns for Time Series Classification

Arvind Kumar Shekar[1,2(✉)], Marcus Pappik[2], Patricia Iglesias Sánchez[1],
and Emmanuel Müller[2]

[1] Robert Bosch GmbH, Stuttgart, Germany
{arvindkumar.shekar,patricia.iglesiassanchez}@de.bosch.com
[2] Hasso Plattner Institute, Potsdam, Germany
marcus.pappik@student.hpi.uni-potsdam.de,emmanuel.mueller@hpi.de

Abstract. Transformation of multivariate time series into feature spaces are common for data mining tasks like classification. Ordinality is one important property in time series that provides a qualitative representation of the underlying dynamic regime. In a multivariate time series, ordinalities from multiple dimensions combine together to be discriminative for the classification problem. However, existing works on ordinality do not address the multivariate nature of the time series. For multivariate ordinal patterns, there is a computational challenge with an explosion of pattern combinations, while not all patterns are relevant and provide novel information for the classification. In this work, we propose a technique for the extraction and selection of relevant and non-redundant multivariate ordinal patterns from the high-dimensional combinatorial search space. Our proposed approach **Ord**inal feature **ex**traction (*ordex*), simultaneously extracts and scores the relevance and redundancy of ordinal patterns without training a classifier. As a filter-based approach, *ordex* aims to select a set of relevant patterns with complementary information. Hence, using our scoring function based on the principles of Chebyshev's inequality, we maximize the relevance of the patterns and minimize the correlation between them. Our experiments on real world datasets show that ordinality in time series contains valuable information for classification in several applications.

1 Introduction

Time series classification is predominant in several application domains such as health, astrophysics and economics [3,4,21]. In particular, for automotive applications, the time series data is transmitted from the vehicle to a remote location. In such cases, the transmission costs are large for lengthy and high-dimensional time series signals. Feature-based approaches handle this problem by transforming the lengthy time series into compact feature sets. The transformation of time series can be done based on several properties, e.g., frequency and amplitude properties of the time series are captured using a Fast Fourier Transform (FFT).

© Springer Nature Switzerland AG 2018
L. Soldatova et al. (Eds.): DS 2018, LNAI 11198, pp. 224–240, 2018.
https://doi.org/10.1007/978-3-030-01771-2_15

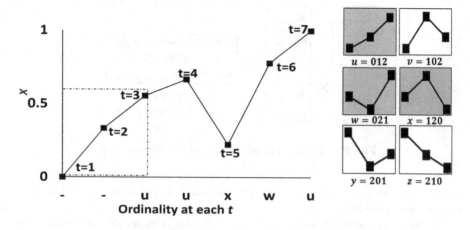

Fig. 1. Example of univariate ordinality and the all ordinalities of $d = 3$

Several time series applications need to capture the structural changes instead of the exact values at each instant of time [16,21]. A transformation based on the ordinality of the time series effectively captures these structural changes in a dynamic system [1,4,16]. Let us consider a simple univariate time series X (c.f. Fig. 1) of length $l = 7$, where $X[t]$ denotes the value of X at time t. To evaluate the ordinality at each time step t, a collection of $d - 1$ (where $d \geq 2$), preceding values in the time series are used [1]. For $d = 3$, the ordinality at $t = 3$[1] is $X(t) > X(t - 1) > X(t - 2)$, which is represented as 012. As shown in Fig. 1, for a fixed d, there are at most $d!$ unique ordinalities that exist in a time series and we denote each of them with an unique symbol. Hence, the ordinalities of X at $t = 3, ..., 7$ are denoted as (u, u, x, w, u). Given $d!$ ordinalities, an ordinal pattern is a subset of ordinalities, e.g., $\{u, x\}$ is an univariate ordinal pattern. Thus, there are at most $2^{d!}$ patterns present in a univariate time series.

In a multivariate time series classification task, there can be co-occurrence of patterns between multiple dimensions that are more relevant for the class prediction than individual patterns. For example (c.f. Fig. 2), in automotive applications, an increasing pattern (u) of engine torque and declining (z) temperature combined together indicates a specific component failure. However, the increasing torque in combination with other ordinalities (e.g., v_{temp}) is not relevant for classification. In such cases, for m dimensions, the number of possible multivariate pattern combinations scales up to $2^{(d! \cdot m)}$. Following the traditional feature-based approach [3] of transforming all pattern combinations into numeric features and performing feature selection to identify the relevant patterns is computationally inefficient.

Thus, the first challenge is to efficiently extract these multivariate patterns and estimate their relevance simultaneously. However, none of the existing works

---

[1] As $t = 1$ and 2 have less than $d - 1$ preceding values.

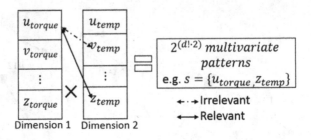

Fig. 2. Example of multivariate pattern combination

on ordinal patterns [1,4,16] consider the influence of ordinalities in multivariate time series datasets.

Additionally, multiple patterns can have similar information (redundant) for the class prediction. For example, for a declining engine torque pattern, the engine speed also exhibits a declining pattern. This implies that both patterns provide redundant information for classification. In such cases, it is necessary to ensure that the extracted patterns have complementary information to each other. Thus, the second challenge lies in estimating the novelty of the features extracted using ordinal patterns. Nevertheless, existing feature-based transformation techniques [3,12,13,18] do not focus on considering both challenges: relevance w.r.t. classes and redundancy of the extracted features. In this work, we introduce **Ord**inal feature **ex**traction (*ordex*), a feature-based approach for multivariate time series classification using the property of ordinality in the time series.

After conversion of the raw multivariate time series dataset into its ordinal representation, we define a method to extract multivariate ordinal patterns. To estimate the relevance of these patterns, *ordex* introduces a measure. This measure estimates the recurrence of an extracted pattern in a given class and its uniqueness w.r.t. other classes. The relevance estimation is followed by the redundancy calculation. Given a set of relevant patterns, *ordex* scores the non-redundancy of each pattern based on its correlation with other relevant patterns. Finally, both scores are combined such that the unified score exemplifies relevance and non-redundancy. Experiments on real world and synthetic datasets show that our approach is beneficial for several application domains.

2   Related Work

We group the time series classification techniques into two categories, i.e., feature-based [3,12,13,18] and sequence-based [19,21]. *Ordex* is a feature-based approach for multivariate time series classification.

Feature-based: The work of [13,17,18] aims to extract features based on time series properties such as mean, variance, kurtosis, Lyapunov exponent and skewness of the time series. In addition, FFT based approaches [12] capture the recurring patterns in a time series which can be useful for classification. Recent works also apply time warping distance (DTW) [8] and symbolic aggregate approximation (SAX) [11] for transformation of time series. All aforementioned works perform feature extraction without considering the relevance of the extracted features. For a high-dimensional time series, this often leads to extraction of features that are redundant and not relevant for classification. For this problem, the recent work HCTSA [3] applies feature selection. However, the process of feature generation and selection is computationally expensive. *Ordex* is efficient by simultaneously generating and evaluating the features for its relevance and redundancy without additional post-processing such as feature selection.

Exploiting ordinality as a property for feature extraction in time series is yet unexplored. Ordinality was introduced as a complexity measure to compare time series [1] and later extended for change detection [16] and variability assessment in ECG signals [4]. All aforementioned works on ordinality focus on univariate time series. On contrary, *ordex* introduces the novel concept of multivariate ordinal patterns and a relevance measure to estimate its relevance for the classification task.

Sequence-based: Shapelet technique classifies new time series based on the distance between the subsequence of a time series (shapelet) and the new time series [21]. The work was extended in MCMR [19] to extract non-redundant shapelets for univariate time series classification. Recurrent neural network frameworks such as Long Short-Term Memory (LSTM) are used for multivariate sequence classification tasks [5]. However, the large training times of LSTM's is a drawback. In contrast, *ordex* efficiently generates features based on relevant multivariate ordinal pattern combination by also evaluating the redundancy.

3   Problem Overview

A multivariate ordinal pattern s is a set of ordinalities from multiple dimensions, e.g., in Fig. 2, $s = \{u_{torque}, z_{temp}\}$. In a multivariate time series dataset, a large number of pattern combinations exist and several of them are irrelevant for classification and redundant to each other. We denote $error : s \mapsto \mathbb{R}$ as the error function of the classifier trained using an ordial pattern s. The classification error using a relevant pattern s_1 is lower in comparison to that of an irrelevant pattern s_2, i.e., $error(s_1) < error(s_2)$. On the other hand, using redundant patterns for classification does not improve the prediction accuracy. That is, for a set of patterns S, where $s_i \in S$ has redundant information to other elements in S, $error(S) \approx error(S \setminus s_i)$. Irrelevant and redundant features lead to large feature space and lower prediction quality [15]. Hence, the contributions of this work are two-fold:

(1) Including and defining the multivariate nature of ordinal patterns for time series classification.

(2) A novel score for evaluating the relevance and redundancy of ordinal patterns without training a classifier.

From a pool of large number of ordinal patterns, we aim to select a set of o patterns $S = \{s_1, \cdots, s_o\}$ that are relevant for classification and are non-redundant w.r.t. other elements in the set. Hence, we maximize the sum of the individual relevancies and minimize the correlation between the ordinal patterns. This requires a scoring function that can efficiently estimate the ability of a multivariate pattern to discriminate between different classes, i.e.,

$$rel : s \in S \mapsto \mathbb{R}.$$

Secondly, a redundancy scoring function to ensure that the elements in S have complementary information to contribute for the classifier, i.e.,

$$red : (s \in S, S \setminus s) \mapsto \mathbb{R}.$$

Notations: As we aim to extract and evaluate ordinal patterns from multivariate time series, we begin with the conversion of raw time series into its ordinal domain. In the work of [1], ordinality of degree $d \geq 2 \mid d \in \mathbb{N}$ at each instant of time $t \mid (d-1) < t \leq l$ for a univariate times series $X = (x_1, \cdots, x_l)$ of length l is defined as,

$$\mathbb{O}_d(X, t) = \big(rank(X[t]), rank(X[t-1]), ..., rank(X[t-(d-1)])\big), \quad (1)$$

where $rank(X[t])$ is the position of $X[t]$ after sorting the values[2] of $(X[t], ..., X[t-(d-1)])$. Thus, the ordinal representation of a univariate time series X is a new series $ord_d(X) = (\mathbb{O}_d(X, t), \cdots, \mathbb{O}_d(X, l))$, where the ordinality $\mathbb{O}_d(X, t)$ at each instant of time t is assigned as a symbol. The resulting series can have a maximum of $d!$ distinct symbols and a length of $l' = l - (d-1)$. For example, in Fig. 1, $ord_3(X) = (u, u, x, w, u)$ and $l' = 7 - (3-1) = 5$.

A m-dimensional time series sample $T^j = \langle X_1, \cdots, X_m \rangle$ is a m-tuple of univariate time series. Finally, a multivariate time series dataset $D = \{T^1, \cdots, T^n\}$ consists of n such multivariate time series samples. As a supervised approach, each sample $T^j \in D$ is assigned a class from a set of possible classes $C = \{c_1, \cdots, c_k\}$. The ith dimension in the jth sample of a dataset is denoted as T_i^j. The ordinal representation of a multivariate time series dataset D is a collection of the ordinal representations of all univariate time series, i.e., $ord_d(D) = \{\langle ord_d(T_1^j), ..., ord_d(T_m^j) \rangle \mid j = 1, \cdots, n\}$. For the ease of notation, we use a fixed length l for all time series, but this is not a formal requirement.

4   Ordinal Feature Extraction (*Ordex*)

Ordex is a heuristic approximation algorithm that includes evaluation of relevance and redundancy of ordinal patterns. A m-dimensional time series dataset D is converted to its ordinal representation of defined degree d, i.e., $ord_d(D)$ (c.f. Sect. 3). From the ordinal search space, *ordex* aims to extract multivariate ordinal patterns. Hence, we begin with the introduction of multivariate ordinal

---

[2] In Fig. 1, $\mathbb{O}_{d=3}(X, t=4) = X(t) > X(t-1) > X(t-(3-1)) = 012$.

patterns. This section is followed by our relevance and non-redundancy scoring function for ordinal patterns. Finally, we elaborate on the algorithmic component of our approach.

4.1   Extraction of Multivariate Ordinal Patterns

As shown in Fig. 2, a multivariate ordinal pattern is a subset of ordinalities from multiple dimensions. We introduce multivariate ordinal pattern set with our formal definition.

Definition 1. *Multivariate Ordinal Pattern set*
Let $\mathcal{I} = \{1, \cdots, m\}$ be the set of dimensions and $\Omega_i = \bigcup_{1 \leq j \leq n} ord_d(T_i^j) \mid i \in \mathcal{I}$ is a set of ordinalities in the ith dimension of all samples in D. Given the search space $\Omega = \{\Omega_i \mid \forall i \in \mathcal{I}\}$ and a subset of $m' \mid m' \leq m$ dimensions, i.e., $\mathcal{I}' \subseteq \mathcal{I} \mid |\mathcal{I}'| = m'$, we define a multivariate ordinal pattern set as,

$$s = \{\Pi_i \subseteq \Omega_i \mid \forall i \in \mathcal{I}'\}.$$

Example 1. Assume a time series dataset $D = \{T^1, T^2\}$ with three dimensions, (i.e., $\mathcal{I} = \{1, 2, 3\}$) and two samples (i.e., $n = 2$) of length $l = 8$.

Using Fig. 3, we show one possible multivariate ordinal pattern extracted from D in Example 1 by applying Definition 1. As the first step, the time series data is converted into its ordinal representation of $d = 3$ by assigning its ordinality at each instant of time (c.f. Eq. 1). For a set of ordinalities Ω_i in the ith dimension of all time series samples, e.g., $\Omega_1 = \bigcup_{1 \leq j \leq 2} ord_3(T_1^j)$, a multivariate ordinal pattern of size $m' = 2$ is a subset of ordinalities from m' dimensions. In our example in Fig. 3, we select a random subset of dimensions $\mathcal{I}' = \{1, 3\}$. From each selected dimension, a subset of ordinalities are drawn to form a multivariate ordinal pattern set, i.e., $s = \{\Pi_1 \subseteq \Omega_1, \Pi_3 \subseteq \Omega_3\}$. In Fig. 3 we show one possible multivariate ordinal pattern set s, where ordinalities u and w are drawn from Ω_1. Similarly, ordinalities y and x are drawn from Ω_3.

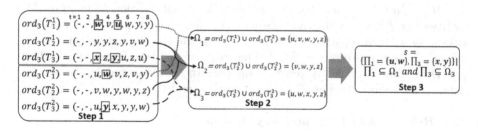

Fig. 3. Illustration of multivariate ordinal pattern set

As discussed in the Sect. 3, evaluating every possible pattern set is computationally inefficient. In this work, we handle this challenge by using the Monte-Carlo approach [9], where a random multivariate pattern set is extracted for each iteration.

In order to score the relevance of the extracted multivariate ordinal pattern set for classification, we transform the multivariate symbolic representation of ordinalities into a numeric feature. As our approach uses the ordinal representation of time series and not the actual values, it is not possible to perform transformation based on standard operations such as mean or median. Following the literature of probabilistic sequential mining [20], we perform transformation based on the occurrences of a pattern set. For the extracted s, we compute its probability in each time series sample $j \mid j = 1, ..., n$ based on our definition below.

Definition 2. *Transformation function*
Let $T = (T[1], \cdots, T[l])$ be a m-dimensional time series sample of length l, i.e., $T[t] \in \mathbb{R}^m$, and $\mathcal{I}'$ is a set of dimensions from which a multivariate ordinal pattern set s is extracted. The pattern s occurs in T at time t iff $ord(T_i[t]) \in \Pi_i, \forall i \in \mathcal{I}'$. The transformation function assigns the probability of s in a time series sample, i.e., $P : (s, T) \mapsto \mathbb{R}$ and we define the transformation function as,

$$P(s, T) = \frac{|\{t \mid s \ occurs \ in \ T \ at \ time \ t\}|}{l - (d - 1)}.$$

Hence, for a time series dataset with n-samples, the defined transformation function generates a n-dimensional numeric feature vector $f = (P(s, T^1), \cdots, P(s, T^n))$.

Example 2. Assume we apply our transformation function (c.f. Definition 2) to transform the multivariate ordinal pattern set s in Fig. 3 into a numeric feature.

The Definition 2 transforms a multivariate pattern into a numeric feature by evaluating the co-occurrence of ordinalities from multiple dimensions. In Fig. 3, s occurs at $t = 3, 5$ in T^1, i.e., $ord_3(T_1^1[3]) = w \in \Pi_1, ord_3(T_3^1[3]) = x \in \Pi_3$ and $ord_3(T_1^1[5]) = u \in \Pi_1, ord_3(T_3^1[5]) = y \in \Pi_3$. Thus, the occurrence of s in T^1 is $P(s, T^1) = \frac{2}{6} = 0.33$. The pattern s occurs in T^2 once at $t = 4$, i.e., $ord_3(T_1^2[4]) = w \in \Pi_1, ord_3(T_3^2[4]) = y \in \Pi_3$. On applying the transformation function on T^2, we have $P(s, T^2) = \frac{1}{6} = 0.16$ and the generated feature vector is $f = (0.33, 0.16)$. Hence, for a set of o patterns $S = \{s_1, \cdots, s_o\}$, the transformation generates a numeric feature space of size $\mathbb{R}^{n \times o}$. Thus, the defined transformation function (c.f. Definition 2) efficiently converts the pattern set into numeric features for datasets with large number of dimensions and samples.

4.2   Relevance and Redundancy Scoring

The transformed feature is based on the pattern set s drawn by a Monte-Carlo iteration and its relevance for classification is necessary to be evaluated. With our defined transformation function, a naïve solution is to convert all patterns into numeric features and perform feature selection. As such an approach is computationally expensive, it is necessary to evaluate the relevance of an ordinal pattern set right after the transformation. By estimating the misclassification rate of a classifier trained for each transformed feature, it is possible to evaluate the feature relevance. However, we aim to efficiently score the relevance and redundancy of a transformed feature without training a classifier. Hence, we estimate the misclassification rate of a feature f by applying principles of Chebyshev's inequality [7]. *Ordex* is applicable for more than two classes. For ease of understanding, we assume a binary classification task with classes $C = \{c_a, c_b\}$ and a feature f generated using the multivariate ordinal pattern set s. From the theory of Chebychev inequality [7], the misclassification using feature f is represented by the variance $Var[f \mid c \in C]$ and expected value $E[f \mid c \in C]$ as,

$$error(f) = \frac{Var[f|c_a] + Var[f|c_b]}{2 \cdot (|E[f|c_a] - E[f|c_b]|)^2}. \tag{2}$$

The Eq. 2 has statistical properties similar to a two-sample t-test. Its detailed proof is provided in the supplementary[3] material and we explain the intuition behind the equation with an example.

Example 3. Assume two multivariate ordinal patterns s_1 and s_2, where s_1 is relevant and s_2 is irrelevant for the classification.

Each ordinal pattern set in Example 3 is transformed into numeric features f_1 and f_2 respectively (c.f. Definition 2). As a relevant pattern, s_1 has a higher discriminative power, i.e., it occurs in every time series of one class (e.g., c_a) with a high probability and never occurs for the other class. Therefore, the distributions of the transformed feature f_1 for each class, exhibits a minimal variance, i.e., $Var[f_1 \mid c_a]$ and $Var[f_1 \mid c_b]$. On contrary, an irrelevant multivariate ordinal pattern set s_2, without any discriminative power to classify, occurs in different time series randomly. Hence, the distribution of the transformed numeric feature $f_2 \mid c_a$ and $f_2 \mid c_b$ has random peaks and lows. This leads to a larger variance in the respective distributions $Var[f_2 \mid c_a]$ and $Var[f_2 \mid c_b]$. This means, the classification error is high when the sum of the variances are large.

In real world applications, due to factors such as noise, it is possible that s_1 (which has high occurrence for class c_a) occurs in a few samples of class c_b, i.e., $Var[f_1 \mid c_b]$ is not exactly equal to zero. Hence, in addition to the variance, the distance between the expected values of the distributions is estimated, i.e., $|E[f|c_b] - E[f|c_a]|$. As we aim to extract the most distinguishing pattern set between two classes, the expected value of a relevant feature for each class will have a larger difference, i.e., $E[f_1 \mid c_a] >> E[f_1 \mid c_b]$. This large difference in the expected values helps the classification boundaries to be well-separated. This means, the classification error is large if the difference between the expected values are small.

Definition 3. *Relevance scoring*
For a classification task with $C = \{c_1, \cdots, c_k\}$ classes, the ability of a transformed feature f to distinguish any pair of classes $c_a \in C$ and $c_b \in C$ is,

---

[3] https://hpi.de//mueller/ordex.html.

$$dis_{c_a,c_b}(f) = 1 - error(f)$$

and we define its relevance as the lowest value of all pairwise dis scores, i.e.,
$rel(f) = min\{dis_{c_a,c_b}(f) \mid c_a \neq c_b\}.$

Assume a classification task with classes c_a, c_b, c_c for which the $dis_{c_a,c_b}(f)$, $dis_{c_a,c_c}(f)$ and $dis_{c_b,c_c}(f)$ are computed. The three values denote the accuracy of each class. The relevance of f is defined as the minimum of the three dis scores in Definition 3. Intuitively, it means that feature relevance is the lowest accuracy of all pairwise scores. Hence, maximizing $rel(f)$ implies maximizing the lowest accuracy of all pairs of classes.

As explained in Sect. 1, there are large number of multivariate ordinal patterns in a time series dataset. However, multiple pattern combinations can be redundant to each other, i.e., they do not provide novel information for classification. Such redundant ordinal patterns lead to low accuracy and large feature sets. The relevance estimation does not include the effect of redundancy. This means, two redundant patterns are scored the same based on their relevance scores.

A transformed feature f represents the probability of a particular pattern in each multivariate time series sample and two features are redundant if their occurrence distribution is discriminative for the same class. Assume two redundant ordinal patterns s_1 and s_2, such that their numeric transformations are f_1 and f_2 respectively (c.f. Definition 2).

Table 1. Illustrative example of ordinal pattern redundancy

j	1	2	3	4	5	6	7	8	9
f_1	**0.8**	**0.88**	**0.95**	0.1	0.5	0.3	0.4	0.35	0.19
f_2	**0.2**	**0.12**	**0.05**	0.9	0.5	0.7	0.6	0.65	0.81
class	c_a	c_a	c_a	c_b	c_b	c_b	c_c	c_c	c_c

Feature f_1 signifies that the pattern s_1 occurs with a higher probability for class c_a, i.e., its values can be used to differentiate class c_a from $\{c_b, c_c\}$ (c.f. Table 1). On contrary, feature f_2 signifies that the pattern s_2 occurs with a low probability for class c_a and its values can also classify c_a from other classes. Hence, f_1 and f_2 are redundant to each other as they are discriminative for the same class and they exhibit a monotonic relationship. To quantify the redundancy between two features, we measure the monotonicity between them. In this work, we instantiate the redundancy function with Spearmans-ρ as a measure of monotonicity [6], i.e., $red(f_i, f_j) = |\rho(f_i, f_j)|$, as it does not assume the underlying distribution of the variable. By defining the redundancy between features as an absolute value, our redundancy measure ranges between $[0, 1]$. However, other measures of monotonicity are also applicable.

For a set of o transformed features $F = \{f_1, \cdots, f_o\}$, the redundancy of $f \in F$ against all elements in the set, i.e., $F \setminus f$, is the maximal imposed redundancy of f on the other features in the set. Hence, we compute the pairwise redundancy of f against all features in $F \setminus f$ and use its maximum. Multiple possibilities exist for combining the relevance and redundancy scores. For example, in the work of [15], the relevance of a feature is penalized for its magnitude of redundancy by computing the harmonic mean between them. Other options include subtracting the magnitude of feature redundancy from its relevance score [2]. From experimental evaluation[3], we understand that both penalization techniques work well for $ordex$. Hence, we choose the latter, i.e., $score(f, F) = rel(f) - red(f, F \setminus f)$, for its simplicity. The unified $score$ represents the relevance of f for classification and its redundancy w.r.t. other elements in F. Finally, the unified $score$ for a set of features is the sum of all individual feature's $score$.

$$score(F) = \sum_{f \in F} score(f, F \setminus f) \tag{3}$$

5 Algorithm

From a given dataset, Algorithm 1 aims to select o relevant and non-redundant patterns by transforming them into numeric features. As mentioned in Sect. 1, it is computationally not feasible to evaluate every ordinal pattern combination. To address this computational challenge, we perform I Monte-Carlo iterations. Each Monte-Carlo iteration extracts a random ordinal pattern set s which is converted into its numeric representation using Definition 2 (c.f. Line 4). For the first o Monte-Carlo iterations, the algorithm draws o random pattern sets which are not scored for relevance or redundancy (c.f. Line 5). Thereon, each newly extracted pattern replaces the worst performing pattern from the set of selected patterns (c.f. Lines 8–13). The scoring of F in each iteration is performed using Eq. 3.

For high-dimensional time series, this random pattern selection leads to the inclusion of several irrelevant (for class prediction) dimensions. Hence, in Line 3, we regulate the selection process by setting the maximum number of selected dimensions to m', i.e., $|\mathcal{I}'| \leq m'$ (c.f. Definition 1). The selection of s is a random process, this leads to the selection of different pattern sets in every execution. To avoid this and make the random process stable [9], the overall occurrence probability of s is approximately $\alpha \in [0, 1]$. Assuming independence between dimensions, each $\Pi_i \in s$ is selected with an occurrence probability of $\alpha^{\frac{1}{|\mathcal{I}'|}}$. The influence of m' and α on the stability and prediction accuracy will be evaluated in the experimental section. The theoretical time complexity of Algorithm1 is presented in the supplementary material (see footnote 3).

Algorithm 1 Ordinal feature extraction

Input: D, o, m', α, I
1: Initialize $F = \emptyset$
2: **for** I **do**
3:     $Draw \ s = \{\Pi_i \subseteq \Omega_i \mid \forall i \in \mathcal{I}'\} \mid probability(\Pi_i) = \alpha^{\frac{1}{|\mathcal{I}'|}}$ (c.f. Definition 1)
4:     Transform s to numeric f (c.f. Definition 2)
5:     **if** $|F| < o$ **then** $F = \{F\} \cup \{f\}$
6:     **else**
7:         $max\_score = score(F)$ and $F\_best = F$
8:         **for** $f' \in F$ **do**
9:             **if** $score(\{F \setminus f'\} \cup f) > max\_score$ **then**
10:                 $F\_best = \{F \setminus f'\} \cup \{f\}$ and $max\_score = score(F\_best)$
11:             **end if**
12:         **end for**
13:         $F = F\_best$
14:     **end if**
15: **end for**
16: **return** F

---

6 Experiments

In this section we evaluate the efficiency and quality of $ordex$[4] on multivariate synthetic, real world datasets from the UCI repository [10] and a dataset from our automotive domain. Following the previous works [3,13,17,21], we use accuracy on the test dataset as a quality measure. As a non-deterministic approach, we execute $ordex$ five times on each dataset and plot the mean test data accuracy and run times in the experimental section below. For both synthetic and real world experiments, we use K-NN (with $K = 5$) classifier for the training and testing of the transformed features. Other $ordex$ parameters used in our experiments are provided in the supplementary (See footnote 3) material.

For generation[5] of multivariate synthetic time series datasets, we made adaptations to the well-known cylinder-bell-funnel time series generator [14]. Using the data generator, we generate separate training and test datasets. As real world datasets we use the character trajectory (3 dimensions and 20 classes), activity recognition (6 dimensions and 7 classes), indoor user movement (4 dimensions and 2 classes), occupancy detection (5 dimensions and 2 classes) and EMG Lower Limb data (5 dimensions and 2 classes) from the UCI repository [10]. The EMG data was recorded with three different experimental settings, called 'pie', 'mar' and 'sen', which we treat as three different data sets. For confidentiality reasons we do not publicly provide or discuss the Bosch dataset (25 dimensions and 2 classes) we used in this work.

---

[4] https://figshare.com/s/4023c66f7a87b59628b4.
[5] https://github.com/KDD-OpenSource/data-generation.

As a feature-based approach, we compare *ordex* with the various competitor techniques of the same paradigm. As competitors that extract features from the time series without evaluating its relevance to the target classes, we test Nanopoulos [13], DTW [8], SAX [11], Wang [17] and Fast Fourier Transforms. As a competitor that evaluates the feature relevance after extraction, we consider HCTSA [3] approach. As a multivariate neural network based approach, we test LSTM as a competitor.

6.1 Scalability Experiments

We evaluate the scalability of *ordex* w.r.t. increasing dimensionality and a fixed number of time series samples. Figure 4(a) shows the breakdown analysis of time elapsed for each phase in *ordex*, i.e., conversion of training data into ordinal representation, selection of relevant ordinal pattern sets and transformation of relevant ordinal pattern sets into numeric features on test dataset.

Our experiments in Fig. 4(a) show that the run time of *ordex* scales linearly w.r.t. increasing number of dimensions. After selection of the relevant pattern sets from the training dataset, the time taken for transformation of the relevant patterns into numeric features on a test dataset is negligible. This is desirable as new samples will be transformed into static features efficiently. Scalability of *ordex* w.r.t. increasing time series length (l) and samples (n) show similar behavior and the results are available in the supplementary[3] material.

6.2 Robustness

In this section we analyze the robustness of our approach against increasing number of irrelevant dimensions. For synthetic datasets with different dimensionality ($m = 40, 70, 130, 160$), of which only five are relevant for classification, we aim to identify the influence of *ordex* on prediction accuracy.

For datasets with a large number of irrelevant features, Fig. 4(b) shows that the random selection process has a higher probability of selecting irrelevant ordinal patterns in the early iterations of the selection phase. This demands several iterations (I) to reach the best accuracy. For example, a dataset with 130 dimensions required 60 iterations to reach the best accuracy and a dataset with 40 dimensions required only 20 iterations to reach the same accuracy.

6.3 Parameter Analysis

Ordex has two major parameters, m' and α. The parameter m' decides the maximum number of dimensions to include for the extraction of pattern set s. Large values of m' include several irrelevant dimensions and setting m' to very small values restrict the search space of pattern combinations to evaluate. Thus, both cases requires a higher number of iterations to identify the best combination. From experimental analysis (c.f. Fig. 4(c)), we observe $1 < m' \leq 5$ to be a reasonable range to set for an optimal trade-off between quality and runtime. All real world experiments in the forthcoming section will use m' values within this range.

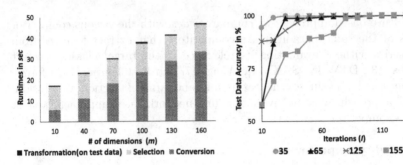

(a) Scalability w.r.t. increasing m, where $l = 600$, $n = 200$

(b) Robustness of *ordex* with varying number of irrelevant dimensions and fixed number (5) of relevant dimensions

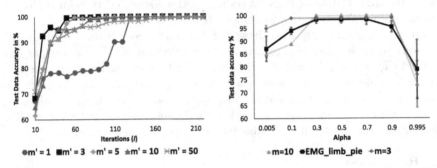

(c) Effect of m' on prediction quality, where $\alpha = 0.3$

(d) Effect of α on prediction quality, where $m' = 3$

Fig. 4. Synthetic data experiments, where $d = 5$

The α parameter decides the number of ordinalities to include from each dimension. Setting α to a large value leads to the inclusion of several irrelevant ordinalities for classification. Hence, large α values lead to inconsistent results (higher standard deviation) and lower test data prediction quality. Setting α to lower values does not largely affect the average prediction quality. However, their standard deviation over five test runs was high. Our experiments on synthetic data in Fig. 4(d) shows $0.3 \leq \alpha \leq 0.9$ range to be a reasonable α value for datasets with different dimensionality. In addition, using EMG Lower Limb Pie dataset, the Fig. 4(d) shows that this range of alpha value is practically applicable for real world data.

6.4   Real World Datasets

Table 2 compares the prediction accuracy of various approaches against *ordex*. Overall, we observe that considering relevance and redundancy during feature extraction improves the prediction quality. In addition, by including the multi-variate nature of ordinalities, *ordex* shows better prediction accuracy w.r.t. the competitor approaches on several datasets. In the character dataset, *ordex* was the second best amidst competitor approaches falling behind DTW. However, DTW approach [8] has higher run times for dataset with large number of samples, e.g., the DTW approach took more than a day for computations on our Bosch dataset with 5722 time series samples.

Table 2. Test data accuracy in % with *ordex* $d = 5$ and $m' = 3$. SAX word size and alphabet size is 3. LSTM of maximum epochs 100 and mini-batch size 10. Experiments that had run times more than one day are denoted as **

Dataset	*ordex*	Nanopoulos	DTW	SAX	Wang	FFT	HCTSA	LSTM
EMG limb sen	**93.33 ± 3.1**	33.3		83.3	33.3	92.3	66.7	50	50 ± 0
EMG limb pie	**96.67 ± 6.6**	16.7		33.3	50	66.6	33.3	50	66.6 ± 0
EMG limb mar	**95 ± 3.6**	83.3		66.7	95	66.6	66.7	92	63.5 ± 0
Character	75.37 ± 1.7	27.1		**88.3**	8.2	70	17.6	25.4	11.98 ± 5.1
Activity recognition	**100± 0**	44.5		100	2.8	91	100	17.4	100 ± 0
User Movement	**57.98 ± 1.9**	45.2		46.8	52.4	45.2	42.9	50.8	47.6 ± 0.8
Occupancy	**94.1 ± 1.9**	63.6		**94.1**	78.4	78.4	70.6	75.4	84.7 ± 8.13
Bosch	**97.08 ± 1.5**	37.7		**	60.2	95.3	59.2	**	56.6 ± 3.4

Table 3. Runtime in *sec*, experiments that had run times more than one day are denoted as **

Dataset	*ordex*	Nanopoulos	DTW	SAX	Wang	FFT	HCTSA	LSTM
EMG limb sen	130.3	8.3	840.3	298.1	1512	8.6	9498	372
EMG limb pie	130	7.9	830.5	266.9	1087	7.9	4088	450.6
EMG limb mar	100.3	7	619.4	278.9	1232	7.1	11999	272
Character	105.3	23	852	458.3	5020	22.06	5511	263
Activity recognition	210.3	5.6	19.9	166.3	1235	5.2	797.2	1808
User Movement	155	2.1	46.8	111.61	428.4	2.8	180.15	174
Occupancy	126.7	1.2	15.6	113.4	49.38	1.1	399.4	125
Bosch	2775.7	344.3	**	4920.4	6876	265.2	**	7335

Table 3 compares the run times (testing and training) of the various approaches against *ordex*. As discussed in Sect. 3, *ordex* evaluates a combinatorial search space. Considering the complexity of the challenge, *ordex* performs reasonable w.r.t. run times in Table 3. By performing the feature extraction and evaluation simultaneously, *ordex* has lower run times in comparison to HCTSA that performs feature selection after extraction of a high-dimensional feature space from the time series. As shown in Sect. 6.1, the major execution time of *ordex* is dominated by the conversion and selection process. Considering the improvement in the prediction quality with negligible time for transforming the relevant and non-redundant ordinalities into numeric features (c.f. Fig. 4(a)), *ordex* is a better choice than the competitor approaches.

6.5   Redundancy Evaluation

The ground truth of feature redundancy is unknown for real world datasets. Using redundant features does not provide novel information for classification, i.e., redundant features do not improve the classification accuracy. Thus, following the work of [15], we evaluate redundancy based on the classifier accuracy in Fig. 5. For a set of o best features extracted using $ordex$, the top scored features of $ordex$ are relevant and non-redundant. Hence, the initial features have increasing prediction quality in Fig. 5. For example, EMG Limb Pie dataset requires 6 features, after which the features are relevant but have redundant information and the classifier accuracy does not improve.

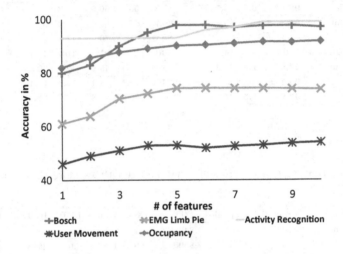

Fig. 5. Accuracy of top 10 features of $ordex$

7   Conclusion and Future Works

In this work we proposed a feature-based time series classification approach $ordex$, that is purely based on the ordinality of the raw time series. The results of various state-of-the-art feature-based algorithms on the synthetic and real world datasets show that our method is suitable for multivariate time series datasets. By scoring relevance and non-redundancy, $ordex$ achieves better prediction quality with fewer features.

As $ordex$ operates on ordinal domain, two signals can have the same ordinality at different amplitudes. Hence, as future work we aim to extend $ordex$ to include the effect of the signal amplitude in addition to its ordinality.

References

1. Bandt, C., Pompe, B.: Permutation entropy: a natural complexity measure for time series. Phys. Rev. Lett. **88**(17), 174102 (2002)
2. Ding, C., Peng, H.: Minimum redundancy feature selection from microarray gene expression data. J. Bioinf. Comput. Biol. **3**(02), 185–205 (2005)
3. Fulcher, B.D., Jones, N.S.: Highly comparative feature-based time-series classification. IEEE Trans. Knowl. Data Eng. **26**(12), 3026–3037 (2014)
4. Graff, G., et al.: Ordinal pattern statistics for the assessment of heart rate variability. Eur. Phys. J. Spec. Top. **222**(2), 525–534 (2013)
5. Hochreiter, S., Schmidhuber, J.: Long short-term memory. Neural Comput. **9**(8), 1735–1780 (1997)
6. Hollander, M., Wolfe, D.A., Chicken, E.: Nonparametric Statistical Methods. Wiley, New York (2013)
7. Karlin, S., Studden, W.J.: Tchebycheff Systems: With Applications in Analysis and Statistics. Interscience, New York (1966)
8. Kate, R.J.: Using dynamic time warping distances as features for improved time series classification. Data Min. Knowl. Discov. **30**(2), 283–312 (2016)
9. Keller, F., Müller, E., Bohm, K.: Hics: high contrast subspaces for density-based outlier ranking. In: 2012 IEEE 28th International Conference on Data Engineering, pp. 1037–1048. IEEE (2012)
10. Lichman, M.: UCI Machine Learning Repository (2013). http://archive.ics.uci.edu/ml
11. Lin, J., Khade, R., Li, Y.: Rotation-invariant similarity in time series using bag-of-patterns representation. J. Intell. Inf. Syst. **39**(2), 287–315 (2012)
12. Mörchen, F.: Time series feature extraction for data mining using DWT and DFT (2003)
13. Nanopoulos, A., Alcock, R., Manolopoulos, Y.: Feature-based classification of time-series data. Int. J. Comput. Res. **10**(3), 49–61 (2001)
14. Saito, N.: Local feature extraction and its applications using a library of bases. Topics in Analysis and Its Applications: Selected Theses, pp. 269–451 (2000)
15. Shekar, A.K., Bocklisch, T., Sánchez, P.I., Straehle, C.N., Müller, E.: Including multi-feature interactions and redundancy for feature ranking in mixed datasets. In: Ceci, M., Hollmén, J., Todorovski, L., Vens, C., Džeroski, S. (eds.) ECML PKDD 2017. LNCS (LNAI), vol. 10534, pp. 239–255. Springer, Cham (2017). https://doi.org/10.1007/978-3-319-71249-9_15
16. Sinn, M., Ghodsi, A., Keller, K.: Detecting change-points in time series by maximum mean discrepancy of ordinal pattern distributions. In: UAI 2012 Proceedings of the Twenty-Eighth Conference on Uncertainty in Artificial Intelligence (2012)
17. Wang, X., Smith, K., Hyndman, R.: Characteristic-based clustering for time series data. Data Min. Knowl. Discov. **13**(3), 335–364 (2006)
18. Wang, X., Wirth, A., Wang, L.: Structure-based statistical features and multivariate time series clustering. In: Seventh IEEE International Conference on Data Mining, 2007, ICDM 2007, pp. 351–360. IEEE (2007)
19. Wei, Y., Jiao, L., Wang, S., Chen, Y., Liu, D.: Time series classification with max-correlation and min-redundancy shapelets transformation. In: 2015 International Conference on Identification, Information, and Knowledge in the Internet of Things (IIKI), pp. 7–12. IEEE (2015)

20. Xi, X., Keogh, E., Wei, L., Mafra-Neto, A.: Finding motifs in a database of shapes. In: Proceedings of the 2007 SIAM International Conference on Data Mining, pp. 249–260. SIAM (2007)

21. Ye, L., Keogh, E.: Time series shapelets: a new primitive for data mining. In: Proceedings of the 15th ACM SIGKDD International Conference on Knowledge Discovery and Data Mining, pp. 947–956. ACM (2009)

Self Hyper-Parameter Tuning for Data Streams

Bruno Veloso[1,2], João Gama[1,3], and Benedita Malheiro[4,5(✉)]

[1] LIAAD - INESC TEC, Porto, Portugal
[2] UPT - University Portucalense, Porto, Portugal
[3] FEP - University of Porto, Porto, Portugal
[4] ISEP - Polytechnic of Porto, Porto, Portugal
mbm@isep.ipp.pt
[5] CRAS - INESC TEC, Porto, Portugal

Abstract. The widespread usage of smart devices and sensors together with the ubiquity of the Internet access is behind the exponential growth of data streams. Nowadays, there are hundreds of machine learning algorithms able to process high-speed data streams. However, these algorithms rely on human expertise to perform complex processing tasks like hyper-parameter tuning. This paper addresses the problem of data variability modelling in data streams. Specifically, we propose and evaluate a new parameter tuning algorithm called Self Parameter Tuning (SPT). SPT consists of an online adaptation of the Nelder & Mead optimisation algorithm for hyper-parameter tuning. The method explores a dynamic size sample method to evaluate the current solution, and uses the Nelder & Mead operators to update the current set of parameters. The main contribution is the adaptation of the Nelder-Mead algorithm to automatically tune regression hyper-parameters for data streams. Additionally, whenever concept drifts occur in the data stream, it re-initiates the search for new hyper-parameters. The proposed method has been evaluated on regression scenario. Experiments with well known time-evolving data streams show that the proposed SPT hyper-parameter optimisation outperforms the results of previous expert hyper-parameter tuning efforts.

Keywords: Parameter tuning · Hyper-parameters · Optimisation
Nelder-Mead · Regression

1 Introduction

Due to the increasing popularity of data stream sources, *e.g.*, crowdsourcing platforms, social networks and smart sensors and devices, data stream or on-line processing has become indispensable. The main goal of data stream processing is to timely extract meaningful knowledge, *e.g.*, to build and update the models of the entities generating the data, from an incoming sequence of events. However, existing data stream modelling algorithms are still not fully automated,

© Springer Nature Switzerland AG 2018
L. Soldatova et al. (Eds.): DS 2018, LNAI 11198, pp. 241–255, 2018.
https://doi.org/10.1007/978-3-030-01771-2_16

e.g., model hyper-parameter tuning still relies in batch or off-line processing techniques. This work addresses this issue by proposing a novel method to tune dynamically the model hyper-parameters according to the incoming events and, thus, contributing to the broad topic of "Data streams, evolving data and models".

In the literature, the hyper-parameter optimisation problem has been addressed using grid-search [12], random-search [1] and gradient descent [19] algorithms. However, these approaches have been applied to off-line rather than on-line processing scenarios since they require train and validation stages. To overcome this limitation, we argue that an on-line processing scenario such as hyper-parameter optimisation for data stream requires some automation level.

Our Self Parameter Tuning (SPT) proposal consists of the use of a direct-search algorithm to find optimal solutions on a search space. Specifically, we apply the Nelder-Mead algorithm [21] to dynamic size data stream samples, continuously searching for the optimal hyper-parameters.

The main contribution of this paper is the proposal of an algorithm that optimises regression hyper-parameters for data streams based on the Nelder-Mead algorithm. It not only processes successfully regression problems, but is, to the best of our knowledge, the single one which effectively works with data streams and reacts to the data variability. Consequently, SPT contributes to the full automation of stream modelling algorithms.

This paper contains five sections. In Sect. 2 we describe the related automatic machine learning work. In Sect. 3 we describe the proposed solution for the identified problem. Section 4 describes the experiments and discusses the results obtained. Finally, Sect. 5 presents the conclusions and suggests future developments.

2   Related Work

The topic of auto machine learning is relatively new and few contributions are found in the literature. We identified contributions addressing auto-ML tools [1, 8,27], model selection algorithms [5,6], hyper-parameter optimisation algorithms [9,16,23] and Nelder-Mead optimisation solutions [7,14,24].

In 2012, Berdstra and Bengio developed a python library for hyper-parameter optimisation named Hyperopt [1]. Internally, it adopts a Bayesian optimizer and uses cross-validation evaluation to orient the search. While it can be used together with scikit-learn to optimise models, it does not work with data streams. In 2013, Thornton *et al.* presented a framework for classification problems called Auto-Weka [27]. It allows hyper-parameter tuning, using a Bayesian optimizer and a cross-validation evaluation mechanism. Feurer *et al.* proposed, in 2015, an extension to SkLearn called Auto-SkLearn which takes into account the performance of similar data sets, making it more data efficient [8]. More recently [17] presented Auto Weka 2.0 for regression algorithms.

In 1995, Kohavi and John described a method to automatically select a hyper-parameter [16]. This method relies on the minimization of the estimated error and applies a grid search algorithm to find local minima. The problem of this solution is that is has an exponential complexity. Escalante *et al.* used, in 2009, a particle swarm optimisation algorithm to select the best model [6]. The algorithm is simple and the optimisation surface contains multiple optimal solutions. One year later, the same authors proposed to build ensemble classification models [5], using this particle swarm model selection algorithm. Nichol and Schulman proposed in 2018 a scalable meta learning algorithm to initialise the parameters of future tasks [23]. The algorithm uses, repeatedly, Stochastic Gradient Descent (SGD) on the training task to tune the parameters.

Koenigstein *et al.* adopted, in 2011, the Nelder-Mead direct search method to optimise more than twenty hyper-parameters of an incremental algorithm with multiple bias [14]. Fernandes *et al.* proposed a batch method for estimating the parameters and the initialisation of a PARAFAC tensor decomposition based link predictor [7]. In 2016, Kar *et al.* applied an exponentially decay centrifugal force to all vertices of the Nelder-Mead algorithm [13]. Although this approach produces improved results, it needs more iterations to converge when compared with the standard Nelder-Mead algorithm. Pfaffe *et al.* addressed the problem of the on-line selection and tuning of algorithms [24]. The authors suggested the adoption of an e-greedy strategy, a well known reinforcement learning technique, to select the best algorithm and Nelder-Mead optimisation to tune the parameters of the chosen algorithm. Both stages are iterative, requiring, in the cases presented, 100 iterations to ensure convergence. This can be a serious drawback for the on-line processing of many data streams.

SPT differs from the above proposals because it automatically adjusts the hyper-parameters of the stream modelling algorithms according to the incoming events. Although SPT relies on Nelder-Mead to dynamically tune the stream-based modelling algorithms under analysis, rather than being iterative, it executes whenever significant data changes in the data stream are perceived.

3   Self Parameter Tuning Method

This paper presents the SPT algorithm which was designed to optimise a set of hyper-parameters in vast search spaces. To make our proposal robust and easier

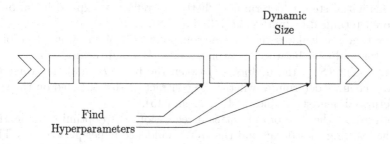

Fig. 1. Application of the proposed algorithm to the data stream.

to use, we adopt a direct-search algorithm, using heuristics to avoid algorithms which rely on hyper-parameters. Specifically, we adapt the Nelder-Mead method [21] to work with data streams.

Figure 1 represents the application of the proposed algorithm. In particular, to find a solution for n hyper-parameters, it requires $n + 1$ input models, *e.g.*, to optimise two hyper-parameters, the algorithm needs three alternative input models.

The Nelder-Mead algorithm processes dynamically each data stream sample, using a previously saved copy of the models, until the input models converge. Each model represents a vertex of the Nelder-Mead algorithm and is computed in parallel to reduce the time response. The initial model vertexes are randomly selected and the Nelder-Mead operators are applied at dynamic intervals. The following subsections describe the implemented Nelder-Mead algorithm, including the dynamic sample size selection.

3.1   Nelder-Mead Optimization Algorithm

This algorithm is a simplex search algorithm for multidimensional unconstrained optimization without derivatives. The vertexes of the simplex, which define a convex hull shape, are iteratively updated in order to sequentially discard the vertex associated with the largest cost function value.

The Nelder-Mead algorithm relies on four simple operations: *reflection, shrinkage, contraction* and *expansion*. Figure 2 illustrates the four corresponding Nelder-Mead operators R, S, C and E. Each black bullet represents a model containing a set of hyper-parameters. The vertexes (models under optimisation) are ordered and named according to the root mean square error (RMSE) value: best (B), good (G), which is the closest to the best vertex, and worst (W). M is a mid vertex (auxiliary model). Algorithms 1 and 2 describe the application of the four operators.

Algorithm 1 presents the reflection and extension of a vertex and Algorithm 2 presents the contraction and shrinkage of a vertex. For each Nelder-Mead operation, it is necessary to compute an additional set of vertexes (midpoint M, reflection R, expansion E, contraction C and shrinkage S) and verify if the calculated vertexes belong to the search space. First, Algorithm 1 computes the midpoint (M) of the best face of the shape as well as the reflection point (R). After this initial step, it determines whether to reflect or expand based on the set of predetermined heuristics (lines 3, 4 and 8).

Algorithm 2 calculates the contraction point (C) of the worst face of the shape – the midpoint between the worst vertex (W) and the midpoint M – and shrinkage point (S) – the midpoint between the best (B) and the worst (W) vertexes. Then, it determines whether to contract or shrink based on the set of predetermined heuristics (lines 3, 4, 8, 12 and 15).

The goal, in the case of data stream regression, is to optimise the learning rate, the learning rate decay and the split confidence hyper-parameters. These hyper-parameters are constrained to values between 0 and 1. The violation of this constraint results in the adoption of the nearest lower or upper bound.

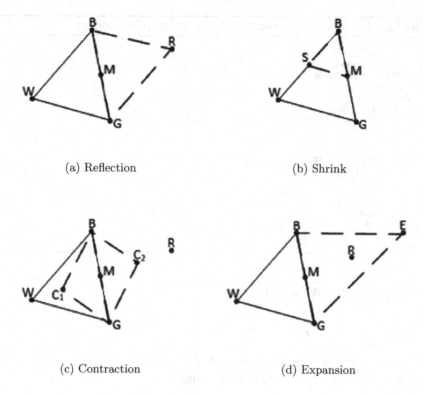

(a) Reflection

(b) Shrink

(c) Contraction

(d) Expansion

Fig. 2. Nelder-Mead Operators.

Algorithm 1 Nelder-Mead - reflect (a) or expand operators (d).

1: $M = (B + G)/2$
2: $R = 2M - W$
3: **if** $f(R) < f(G)$ **then**
4:     **if** $f(B) < f(R)$ **then**
5:         $W = R$
6:     **else**
7:         $E = 2R - M$
8:         **if** $f(E) < f(B)$ **then**
9:             $W = E$
10:         **else**
11:             $W = R$
12:         **end if**
13:     **end if**
14: **end if**

3.2   Dynamic Sample Size

The dynamic sample size, which is based on the RMSE metric, attempts to identify significant changes in the streamed data. Whenever such a change is detected, the Nelder-Mead compares the performance of the $n + 1$ models under analysis to choose the most promising model. The sample size S_{size} is given by Equation 1 where σ represents the standard deviation of the RMSE and M the

---

Algorithm 2 Nelder-Mead - contract (c) or shrink (b) operators.

1: $M = (B + G)/2$
2: $R = 2M - W$
3: **if** $f(R) \geq f(G)$ **then**
4:     **if** $f(R) < f(W)$ **then**
5:         $W = R$
6:     **else**
7:         $C = (W + M)/2$
8:         **if** $f(C) < f(W)$ **then**
9:             $W = C$
10:         **else**
11:             $S = (B + W)/2$
12:             **if** $f(S) < f(W)$ **then**
13:                 $W = S$
14:             **end if**
15:             **if** $f(M) < f(G)$ **then**
16:                 $G = M$
17:             **end if**
18:         **end if**
19:     **end if**
20: **end if**

---

desired error margin. We use $M = 95\%$.

$$S_{size} = \frac{4\sigma^2}{M^2} \tag{1}$$

However, to avoid using small samples, that imply error estimations with large variance, we defined a lower bound of 30 samples.

3.3   Stream-Based Implementation

The adaptation of the Nelder-Mead algorithm to on-line scenarios relies extensively on parallel processing. The main thread launches the $n + 1$ model threads and starts a continuous event processing loop. This loop dispatches the incoming events to the model threads and, whenever it reaches the sample size interval, assesses the running models and calculates the new sample size. The model assessment involves the ordering of the $n + 1$ models by RMSE value and the application of the Nelder-Mead algorithm to substitute the worst model. The Nelder-Mead parallel implementation creates a dedicated thread per Nelder-Mead operator, totalling seven threads. Each Nelder-Mead operator thread generates a new model and calculates the incremental RMSE using the instances of the last sample size interval. The worst model is substituted by the Nelder-Mead operator thread model with lowest RMSE.

4   Experimental Evaluation

The following subsections describe the experiments performed, including the data sets, the evaluation metrics and protocol, the tests and the results. The experiments were performed with an Intel Xeon CPU E5-2680 2.40 GHz Central

Processing Unit (CPU), 32 GiB DDR3 Random Access Memory (RAM) and 1 TiB of hard drive platform running the Ubuntu 16.04.

The open-source Massive Online Analysis (MOA) framework [2] was selected for the experiments due to the variety of implemented stream-based algorithms as well as the respective evaluation metrics. Moreover, it allows easy benchmarking with the pre-existing implementations.

The adaptive model rules regression algorithm was chosen due to the inherent expressiveness of decision rules and to the fact that each rule uses the Page-Hinkley test to detect and react to changes in data stream [4]. Since the parameters with higher impact on the algorithm output are the split confidence, learning rate and learning rate decay, the algorithm will attempt to tune the three.

The SPT approach was compared against a default hyper-parameter initialisation – hereafter called baseline. The baseline hyper-parameter initialisation was 0.1 for the split confidence, 1.0 for the learning rate and 0.1 for the learning rate decay, which are the MOA framework default hyper-parameters. The default values were used for baseline since we were unable to find previous results regarding the tuning of adaptive model rules hyper-parameters for data streams. The alternative would be to perform off-line hyper-parameter tuning with the instances received so far, e.g., using grid search.

4.1   Data Sets

The evaluation was performed using only real and public data sets with a minimum number of 100000 instances: (*i*) the YearPredictionMSD [29] data set, holding 515344 instances; (*ii*) the Twitter [18] data set with 583251 instances; and (*iii*) the SGEMM GPU kernel performance [28] data set, containing 240600 instances.

4.2   Evaluation Protocol

The evaluation protocol defines the data ordering, partitions and distribution. To assess the proposed method we applied two different protocols: holdout evaluation [15] and the predictive sequential (prequential) evaluation [11]. First, we use the holdout evaluation protocol to find an optimal solution for the hyper-parameters and assess the reproducibility of the algorithm with multiple experiments. Then, we apply the prequential evaluation to the data as a stream. We determine the incremental RMSE adopted by Takács *et al.* (2009) [26], which is calculated incrementally after each new instance.

Figure 3 presents holdout data partition. The data is ordered temporally and, then, partitioned in two halves: 50% to "Train" and the remaining 50% to "Test". First, the holdout algorithm finds an optimal solution for the selected hyper-parameters using the train data. Then, it builds a model using the train data and the identified optimal hyper-parameters. Finally, the holdout algorithm updates and evaluates the created model using the test data.

In the case of the prequential protocol, the entire data is used for training and testing. The algorithm scans the available data, and for each example the

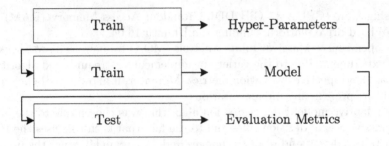

Fig. 3. Holdout – data splitting and processing.

current decision model makes a prediction. After, it receives the true label, and updates the decision model.

In the case of the prequential protocol, the entire data is used for training and testing as represented in Fig. 4. First the data is ordered temporally, then it is used to build incrementally the model and, finally, is evaluated with a sliding window of 1000 instances, as proposed by [10].

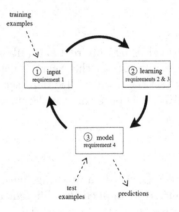

Fig. 4. Prequential – data splitting and processing [2].

The holdout and prequential tests were repeated 30 times to compute the average and standard deviation of the evaluation metrics. This number of repetitions is a compromise between the time required to process each data set and the number of runs required to compute the statistical values with confidence.

4.3 Significance Tests

To detect the statistical differences between the proposed and the baseline approaches we applied three different significance tests: (*i*) the Wilcoxon Test [30] to verify if the mean ranks of two samples differ; (*ii*) the McNemar Test

[20] to assess if a statistically significant change occurs on a dichotomous trait at two time points on the same population; and (*iii*) the critical distance (CD) measure proposed by [3] for a graphical interpretation of the statistical results. We define a 5% of significance level p for all tests. The goal of the Wilcoxon and McNemar tests is to reject the null-hypothesis, *i.e.*, that both approaches have the same performance. We run 30 trials for each experiment. For a significance of 5%, the critical value of McNemar Test (MT_{crit}) is 3.84 and the critical value of the Wilcoxon Test (WT_{crit}) is 137. In the case of the McNemar, two samples are statistically different if $MT_{stat} > MT_{crit}$, whereas, in the case of Wilcoxon, two samples are statistically different if the $|WT_{stat}| > WT_{crit}$.

4.4   Regression Experiments

First, we added our hyper-parameter optimisation algorithm to the MOA framework by defining a new regression task which uses the regression algorithm of Duarte *et al.* (2016) [4]. When we launch the task, it initialises four identical regression models with randomly selected values for the three hyper-parameters and applies our algorithm.

Figure 5 illustrates the convergence of the optimisation of the three hyper-parameters with the three regression data sets. While the four models rapidly converge with the Twitter data set, with the YearPredictionMSD and SGEMM GPU data set, only three of the four models converge. This means that, as the number of degrees of freedom of the objective function increases, the algorithm requires more instances to converge to a solution.

The holdout evaluation, after verifying the convergence of the models, assesses the performance of the regression algorithm with both the baseline (B) and the SPT hyper-parameters. This step was repeated thirty times with randomly shuffled data variations to compute the average and standard deviation of the prediction RMSE (see Table 1). The RMSE has a negligible decrease of 0.3% with the Twitter data set, drops 4.3% with the YearPredictionMSD data set and shows a reduction of 34.4% with the SGEMM GPU data set.

Table 2 presents the statistical results of the Wilcoxon and McNemar tests. The results shows that the null hypothesis is rejected for all data sets, meaning that the results with SPT and baseline are statistically different.

Figure 6 displays the critical distance between the proposed and baseline approaches, showing that they are statistically different. The critical distance was calculated using the Nemenyi test [22].

Figure 7 compares prequential evaluation results of the baseline and SPT approaches. The charts, which display $log(RMSE_B/RMSE_{SPT})$, indicate that the hyper-parameters found by SPT were not the best for all stream instances.

Based on these conclusions, we decided to make our algorithm responsive to concept drift and receive feedback from the Page-Hinkley test [25] which detects data changes. Whenever a drift occurs, our optimisation algorithm re-initiates the search for new hyper-parameters, *i.e.*, takes into account the variability of the data. Finally, we applied again the prequential evaluation protocol to assess

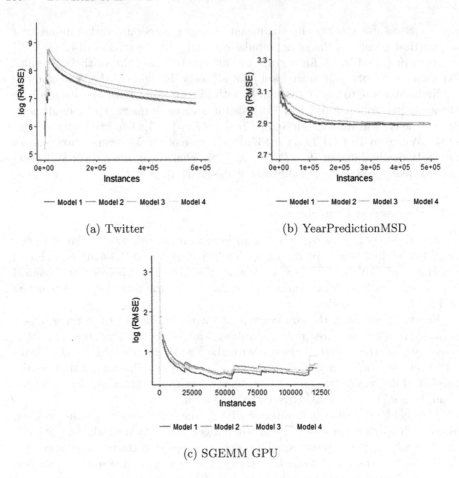

(a) Twitter

(b) YearPredictionMSD

(c) SGEMM GPU

Fig. 5. Regression – Model convergence

Table 1. Regression results

Dataset	Approach	RMSE (μ)	CV (%)
YearPredictionMSD	B	0.178×10^2	0.459
	SPT	0.178×10^2	0.344
Twitter	B	1.646×10^2	0.013
	SPT	1.576×10^2	0.014
SGEMM GPU	B	0.090×10^2	0.837
	SPT	0.059×10^2	0.937

the response of our algorithm to data changes. Figure 8 shows that it reacts to concept drifts, improving our initial results.

Table 2. Regression statistical results

Dataset	Test		Value	p-value
YearPredictionMSD	McNemar	MT_{stat}	24.3	8.244×10^{-7}
	Wilcoxon	WT_{stat}	463	5.588×10^{-9}
Twitter	McNemar	MT_{stat}	24.3	8.244×10^{-7}
	Wilcoxon	WT_{stat}	435	3.79×10^{-6}
SGEMM GPU	McNemar	MT_{stat}	28.0	1.19×10^{-7}
	Wilcoxon	WT_{stat}	465	1.86×10^{-6}

Fig. 6. Regression – Critical Distance

Computationally, the on-line tuning of three parameters requires four threads and, temporally, during model assessment, plus seven threads, against the single thread of the baseline algorithm. The duration of the assessment phase depends on the number of instances of the dynamic sample size interval.

5 Conclusions

This paper describes the SPT – Self Parameter tuning – approach, a hyper-parameter optimisation algorithm for data streams. SPT explores the adoption of a simplex search mechanism combined with dynamic data samples and concept drift detection to tune and find good parameter configuration that minimise the objective function.

The main contribution of this paper is an extension of the Nelder-Mead optimisation algorithm which is, to the best of our knowledge, the single one which effectively works with data streams and reacts to the data variability. The SPT algorithm is, in terms of existing hyper-parameter optimisation algorithms, less computationally expensive than Bayesian optimisers, stochastic gradients or even grid search algorithms.

We applied SPT to regression data sets and concluded that the selection of the hyper-parameters has a substantial impact in terms of accuracy. The performance of our algorithm with regression problems was affected by the data variability and, consequently, we enriched it with a concept drift detection functionality.

Our algorithm is able to operate over data streams, adjusting hyper-parameters based on the variability of the data, and does not require an iterative approach to converge to an acceptable minimum. We test our approach

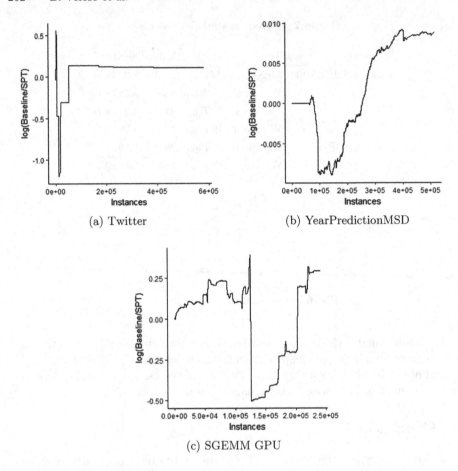

(a) Twitter

(b) YearPredictionMSD

(c) SGEMM GPU

Fig. 7. Regression – Prequential evaluation without concept drift detection

extensively on regression problems against baseline methods that do not perform automatic adjustments of hyper-parameters, and found that our approach consistently and significantly outperforms them.

Future work will includes three key points: (*i*) apply the algorithm to classification and recommendation algorithms; (*ii*) enrich the algorithm with the ability to select not only hyper-parameters but also models; and (*iii*) make a thorough comparison with other optimisation algorithms.

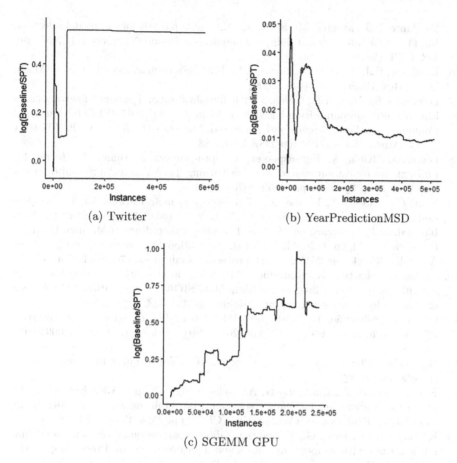

(a) Twitter   (b) YearPredictionMSD

(c) SGEMM GPU

Fig. 8. Regression – Prequential evaluation with concept drift detection

Acknowledgements. This work is partially funded by the ERDF through the COMPETE 2020 Programme within project POCI-01-0145-FEDER-006961, and by National Funds through the FCT as part of project UID/EEA/50014/2013.

References

1. Bergstra, J., Bengio, Y.: Random search for hyper-parameter optimization. J. Mach. Learn. Res. **13**(1), 281–305 (2012). http://dl.acm.org/citation.cfm?id=2503308.2188395
2. Bifet, A., Holmes, G., Kirkby, R., Pfahringer, B.: MOA: massive online analysis. J. Mach. Learn. Res. **11**(May), 1601–1604 (2010)
3. Demšar, J.: Statistical comparisons of classifiers over multiple data sets. J. Mach. Learn. Res. **7**, 1–30 (2006). http://dl.acm.org/citation.cfm?id=1248547.1248548
4. Duarte, J., Gama, J., Bifet, A.: Adaptive model rules from high-speed data streams. ACM Trans. Knowl. Discov. Data **10**(3), 30:1–30:22 (2016). http://doi.acm.org/10.1145/2829955

5. Escalante, H.J., Montes, M., Sucar, E.: Ensemble particle swarm model selection. In: The 2010 International Joint Conference on Neural Networks (IJCNN), pp. 1–8. IEEE (2010)

6. Escalante, H.J., Montes, M., Sucar, L.E.: Particle swarm model selection. J. Mach. Learn. Res. **10**(Feb), 405–440 (2009)

7. Fernandes, S., Tork, H.F., Gama, J.: The initialization and parameter setting problem in tensor decomposition-based link prediction. In: 2017 IEEE International Conference on Data Science and Advanced Analytics (DSAA), pp. 99–108 (Oct 2017). https://doi.org/10.1109/DSAA.2017.83

8. Feurer, M., Klein, A., Eggensperger, K., Springenberg, J., Blum, M., Hutter, F.: Efficient and robust automated machine learning. In: Advances in Neural Information Processing Systems, pp. 2962–2970 (2015)

9. Finn, C., Abbeel, P., Levine, S.: Model-agnostic meta-learning for fast adaptation of deep networks. In: Precup, D., Teh, Y.W. (eds.) Proceedings of the 34th International Conference on Machine Learning. Proceedings of Machine Learning Research, vol. 70, pp. 1126–1135. PMLR, International Convention Centre, Sydney, Australia (06–11 Aug 2017). http://proceedings.mlr.press/v70/finn17a.html

10. Gama, J., Sebastião, R., Rodrigues, P.P.: Issues in evaluation of stream learning algorithms. In: Proceedings of the 15th ACM SIGKDD International Conference on Knowledge Discovery and Data Mining, pp. 329–338. ACM (2009)

11. Gama, J.: Sebastião, R., Rodrigues, P.P.: On evaluating stream learning algorithms. Mach. Learn. **90**(3), 317–346 (2013). https://doi.org/10.1007/s10994-012-5320-9

12. Hsu, C.W., Chang, C.C., Lin, C.J., et al.: A practical guide to support vector classification (2003)

13. Kar, R., Konar, A., Chakraborty, A., Ralescu, A.L., Nagar, A.K.: Extending the nelder-mead algorithm for feature selection from brain networks. In: 2016 IEEE Congress on Evolutionary Computation (CEC), pp. 4528–4534. IEEE (2016)

14. Koenigstein, N., Dror, G., Koren, Y.: Yahoo! music recommendations: modeling music ratings with temporal dynamics and item taxonomy. In: Proceedings of the Fifth ACM Conference on Recommender Systems, pp. 165–172. ACM (2011)

15. Kohavi, R.: A study of cross-validation and bootstrap for accuracy estimation and model selection. In: Proceedings of the 14th International Joint Conference on Artificial Intelligence - Volume 2, pp. 1137–1143. IJCAI 1995. Morgan Kaufmann Publishers Inc., San Francisco, CA, USA (1995). http://dl.acm.org/citation.cfm?id=1643031.1643047

16. Kohavi, R., John, G.H.: Automatic parameter selection by minimizing estimated error. In: Machine Learning Proceedings 1995, pp. 304–312. Elsevier (1995)

17. Kotthoff, L., Thornton, C., Hoos, H.H., Hutter, F., Leyton-Brown, K.: Auto-weka 2.0: Automatic model selection and hyperparameter optimization in weka. J. Mach. Learn. Res. **18**(1), 826–830 (2017). http://dl.acm.org/citation.cfm?id=3122009.3122034

18. Laboratoire d'Informatique de Grenoble: Twitter data set, http://ama.liglab.fr/resourcestools/datasets/buzz-prediction-in-social-media/, Accessed on March 2018

19. Maclaurin, D., Duvenaud, D., Adams, R.P.: Gradient-based hyperparameter optimization through reversible learning. In: Proceedings of the 32nd International Conference on International Conference on Machine Learning - Volume 37, pp. 2113–2122. ICML 2015, JMLR.org (2015), http://dl.acm.org/citation.cfm?id=3045118.3045343

20. McNemar, Q.: Note on the sampling error of the difference between correlated proportions or percentages. Psychometrika **12**(2), 153–157 (1947). https://doi.org/10.1007/BF02295996
21. Nelder, J.A., Mead, R.: A simplex method for function minimization. Comput. J. **7**(4), 308–313 (1965). https://doi.org/10.1093/comjnl/7.4.308
22. Nemenyi, P.: Distribution-free multiple comparisons. In: Biometrics. vol. 18, p. 263. INTERNATIONAL BIOMETRIC SOC 1441 I ST, NW, SUITE 700, WASHINGTON, DC 20005–2210 (1962)
23. Nichol, A., Schulman, J.: Reptile: a Scalable Metalearning Algorithm. ArXiv e-prints (2018)
24. Pfaffe, P., Tillmann, M., Walter, S., Tichy, W.F.: Online-autotuning in the presence of algorithmic choice. In: 2017 IEEE International Parallel and Distributed Processing Symposium Workshops (IPDPSW), pp. 1379–1388. IEEE (2017)
25. Sebastião, R., Fernandes, J.M.: Supporting the page-hinkley test with empirical mode decomposition for change detection. In: Kryszkiewicz, M., Appice, A., Ślkezak, D., Rybinski, H., Skowron, A., Raś, Z.W. (eds.) ISMIS 2017. LNCS (LNAI), vol. 10352, pp. 492–498. Springer, Cham (2017). https://doi.org/10.1007/978-3-319-60438-1_48
26. Takács, G., Pilászy, I., Németh, B., Tikk, D.: Scalable collaborative filtering approaches for large recommender systems. J. Mach. Learn. Res. **10**, 623–656 (2009). http://dl.acm.org/citation.cfm?id=1577069.1577091
27. Thornton, C., Hutter, F., Hoos, H.H., Leyton-Brown, K.: Auto-weka: Combined selection and hyperparameter optimization of classification algorithms. In: Proceedings of the 19th ACM SIGKDD International Conference on Knowledge Discovery and Data Mining. pp. 847–855. KDD 2013. ACM, New York, NY, USA (2013). http://doi.acm.org/10.1145/2487575.2487629
28. University of California: SGEMM GPU kernel performance data set, https://archive.ics.uci.edu/ml/datasets/SGEMM+GPU+kernel+performance/, Accessed on March 2018
29. University of California: YearPredictionMSD data set, https://archive.ics.uci.edu/ml/datasets/yearpredictionmsd, Accessed on March 2018
30. Wilcoxon, F.: Individual comparisons by ranking methods. Biom. Bull. **1**(6), 80–83 (1945). http://www.jstor.org/stable/3001968

Subgroup and Subgraph Discovery

Compositional Subgroup Discovery on Attributed Social Interaction Networks

Martin Atzmueller[✉]

Department of Cognitive Science and Artificial Intelligence, Tilburg University,
Warandelaan 2, 5037 AB Tilburg, The Netherlands
m.atzmuller@uvt.nl

Abstract. While standard methods for detecting subgroups on plain social networks focus on the network structure, attributed social networks allow compositional analysis, i. e., by exploiting attributive information. Accordingly, this paper applies a compositional perspective for identifying compositional subgroup patterns. In contrast to typical approaches for community detection and graph clustering it focuses on the dyadic structure of social interaction networks. For that, we adapt principles of subgroup discovery – a general data mining technique for the identification of local patterns – to the dyadic network setting. We focus on social interaction networks, where we specifically consider properties of those social interactions, i. e., duration and frequency. In particular, we present novel quality functions for estimating the interestingness of a subgroup and discuss their properties. Furthermore, we demonstrate the efficacy of the approach using two real-world datasets on face-to-face interactions.

1 Introduction

The identification of interesting subgroups (often also called communities) is a prominent research direction in data mining and (social) network analysis, e. g., [2,3,17,21,49]. Typically, a structural perspective is taken, such that specific subgraphs — in a graph representation of the network — induced by a set of edges and/or nodes are investigated. Attributed networks, where nodes and/or edges are labeled with additional information, allow for further dimensions for detecting patterns that describe a specific subset of nodes of the graph representation of a (social) network. However, there are different foci relating to the specific problem and data at hand. The method of subgroup discovery, for example, a powerful and versatile method for exploratory data mining, focuses on detecting subgroups described by specific patterns that are interesting with respect to some target concept and quality function. In contrast, community detection, as a (social) network analysis method, aims at detecting subgroups of individuals, i. e., nodes of a network, that are densely (and often cohesively) connected by a set of links. Thus, the former stresses the compositional notion of a pattern describing a subgroup, i. e., based on attributes/properties of nodes

© Springer Nature Switzerland AG 2018
L. Soldatova et al. (Eds.): DS 2018, LNAI 11198, pp. 259–275, 2018.
https://doi.org/10.1007/978-3-030-01771-2_17

and/or edges, while the latter focuses on structural properties of a pattern, such that specific subgraphs are investigated that induce a specific pattern.

Problem. We formalize the problem of detecting compositional patterns of actor-dyads, i. e., edges connecting two nodes (corresponding to the actors) in a graph representation of an attributed network. We aim to detect the subgroup patterns that are most interesting according to a given interestingness measure. For estimating the interestingness, we utilize a quality function which considers the dyadic structure of the set of dyads induced by the compositional pattern. In particular, we focus on social interaction networks, where we specifically consider properties of social interactions, e. g., duration and frequency. Then, the quality measure should consider those patterns as especially interesting which deviate from the expected "overall" behavior given by a null-model, i. e., modeling dyadic interactions due to pure chance. Then, those models should also incorporate the properties of social interaction networks mentioned above.

Objectives. We tackle the problem of detecting compositional patterns capturing subgroups of nodes that show an interesting behavior according to their dyadic structure as estimated by a quality measure. We present novel approaches utilizing subgroup discovery and exceptional model mining techniques [3,7,18]. Further, we discuss estimation methods for ranking interesting patterns, and we propose two novel quality functions, that are statistically well-founded. This provides for a comprehensive and easily interpretable approach for this problem.

Approach & Methods. For our compositional subgroup discovery approach, we adapt principles of subgroup discovery – a general data analysis technique for exploratory data mining – to the dyadic network setting. In particular, we present two novel quality functions for estimating the interestingness of a subgroup and its specific dyadic interactions and discuss their properties. Furthermore, we demonstrate the efficacy of the approach using two real-world datasets.

Contributions. Our contribution is summarized as follows:

1. We formalize the problem of compositional subgroup discovery and present an approach for detecting compositional subgroup patterns capturing interesting subgroups of dyads, as estimated by a quality function.
2. Based on subgroup discovery and exceptional model mining techniques, we propose a flexible modeling and analysis approach, and present two novel interestingness measures for compositional analysis, i. e., quality functions for subgroup discovery. These enable estimating the quality of subgroup patterns in order to generate a ranking. The proposed quality functions are statistically well-founded, and provide a statistical significance value directly, also easing interpretation by domain specialists.
3. We demonstrate the efficacy of our proposed approach and the presented quality measures using two real-world datasets capturing social face-to-face interaction networks.

Structure. The rest of the paper is structured as follows: Sect. 2 discusses related work. After that, Sect. 4 outlines the proposed approach. Next, Sect. 5

presents results of an exploratory analysis utilizing two real-world social interaction network datasets of face-to-face interactions. Finally, Sect. 6 concludes with a discussion and interesting directions for future work.

2   Related Work

Below, we summarize related work on subgroup discovery, social interaction networks, and community detection, and put our proposed approach into context.

2.1   Subgroup Discovery and Exceptional Model Mining

Subgroup discovery is an exploratory data mining method for detecting interesting subgroups, e. g., [3,29,50]. It aims at identifying descriptions of subsets of a dataset that show an interesting behavior with respect to certain interestingness criteria, formalized by a quality function, e. g., [50]. Here, the concept of exceptional model mining has recently been introduced [18,34]. It can be considered as a variant of subgroup discovery enabling more complex target properties. Applications include mining characteristic patterns [8], mining subgroups of subgraphs [45], or descriptive community mining, e. g., [7]. In contrast to the approaches mentioned above, we adapt subgroup discovery for dyadic analysis on social interaction networks, and propose novel interestingness measures as quality functions on networks for that purpose.

2.2   Mining Social Interaction Networks

A general view on mining social interaction networks is given in [2], captured during certain events, e. g., during conferences. Here, patterns on face-to-face contact networks as well as evidence networks [40]) and their underlying mechanisms, e. g., concerning homophily [11,39,41] are analyzed, however only concerning specific hypotheses or single attributes [46]. Furthermore, [6,38] describe the dynamics of communities and roles at conferences, while [28] focuses on their evolution. This is also the focus of, e. g., [4,37] where exceptional communities/subgroups with respect to sequential transitions are detected. In contrast, this paper targets the detection of interesting patterns describing such dyadic-oriented subgroups in attributed networks, modeling social interactions.

Attributed (or labeled) graphs as richer graph representations enable approaches that specifically exploit the descriptive information of the labels assigned to nodes and/or edges of the graph, in order to detect densely connected groups or clusters, e. g., [16]. In [7], for example, the COMODO algorithm is presented. It applies subgroup discovery techniques for description-oriented community detection. Using additional descriptive features of the nodes contained in the network, the task is to identify communities as sets of densely connected nodes together with a *description*, i. e., a logical formula on the values of the nodes' descriptive features. Here, in contrast, we do not focus on the graph structure, like approaches for community detection, e. g., [7,24,44] or exceptional model mining approaches, e. g., [10,12,15,26] on attributed graphs. Instead, we apply a dyadic perspective on interactions focusing on such parameters such as interaction *frequency* and *duration*. We propose two novel quality functions in such dyadic interaction contexts, i. e., for reliably identifying interesting subsets of dyads using subgroup discovery. To the best of the author's knowledge, no subgroup discovery approach tackling this problem has been proposed so far.

3   Background: Subgroup Discovery

Subgroup discovery [3,50] is a powerful method, e. g., for (data) exploration and descriptive induction, i. e., to obtain an overview of the relations between a so-called target concept and a set of explaining features. These features are represented by attribute/value assignments, i. e., they correspond to binary features such as items known from association rule mining [1]. In its simplest case, the target concept is often represented by a binary variable. However, more complex target concepts can also be modeled, leading to exceptional model mining which targets specifically complex target models. In this work, for subgroup discovery we adopt the general scope proposed in [3,29–31,36,43,50,51], such that subgroup discovery also subsumes exceptional model mining as a special case, enabling more complex target concepts than just, e. g., a single dependent variable. Then, subgroups are ranked using a quality function, e. g., [3,22,29,35,50].

In the context of attributed networks, we formalize the necessary notions in the following. Formally, an *edge – attribute database* $DB = (E, A, F)$ is given by a set of edges E and a set of attributes A. For each attribute $a \in A$, a range $dom(a)$ of values is defined. An attribute/value assignment $a = v$, where $a \in A, v \in dom(a)$, is called a *feature*. We define the feature space V to be the (universal) set of all features. For each edge $e \in E$ there is a mapping $F : E \to 2^V$ describing the set of features that are assigned to an edge. Intuitively, such features can be given by attribute–value paris, (binary) labels such as items in the context of association rule mining, etc.

Basic elements used in subgroup discovery are patterns and subgroups. Intuitively, a *pattern* describes a *subgroup*, i. e., the subgroup consists of the edges (and the respective nodes) that are covered by the respective pattern, i. e., those having the respective set of features. It is easy to see, that a pattern describes a fixed set of edges (inducing a subgroup of nodes), while a subgroup can also be described by different patterns, if there are different options for covering the subgroup' edges. A (subgroup) *pattern* P is defined as a conjunction

$$P = s_1 \wedge s_2 \wedge \cdots \wedge s_n,$$

of (extended) features

$$s_i \subseteq V,$$

which are then called selection expressions, where each s_i selects a subset of the range $dom(a)$ of an attribute $a \in A$. A selection expression s is thus a Boolean function $E \to \{0, 1\}$ that is true if the value of the corresponding attribute is contained in the respective subset of V for the respective edge $e \in E$. The set of all selection expressions is denoted by S.

A *subgroup (extension)*

$$E_P := ext(P) := \{e \in E | P(e) = true\}$$

is the set of all edges which are covered by the pattern P. Using the set of edges, it is straightforward to extract the subset of covered nodes.

The interestingness of a pattern is determined by a quality function

$$q: 2^S \to \mathbb{R}.$$

It maps every pattern in the search space to a real number that reflects the interestingness of a pattern (or the extension of the pattern, respectively). Many quality functions for a single target feature, e.g., in the binary or numerical case, trade-off the size $n = |ext(P)|$ of a subgroup and the deviation $t_P - t_0$, where t_P is the average value of a given target feature in the subgroup identified by the pattern P and t_0 the average value of the target feature in the general population. Thus, standard quality functions are of the form

$$q_a(P) = n^a \cdot (t_P - t_0), \ a \in [0; 1].$$

For binary target concepts, this includes, for example, a *simplified binomial* function $q_a^{0.5}$ for $a = 0.5$, or the *gain* quality function q_a^0 with $a = 0$. However, as we will see below, such simple formalizations (as utilized by standard subgroup discovery approaches) do not cover the specific properties in dyadic network analysis - that is why provide specific adaptations for that case below.

While a quality function provides a *ranking* of the discovered subgroup patterns, often also a statistical assessment of the patterns is useful in data exploration. Quality functions that directly apply a statistical test, for example, the Chi-square quality function, e. g., [3] provide a p-value for simple interpretation.

For network data, there exist several quality measures for comparing a network structure to a null-model. For a given subgroup we can, for example, adapt common community quality measures, e. g., [7] for subgroup discovery. Also, the quadratic assignment procedure [32] (QAP) is a standard approach applying a graph correlation measure: For comparing two graphs G_1 and G_2, it estimates the correlation of the respective adjacency matrices M_1 and M_2 and tests that graph level statistic against a QAP null hypothesis [32]. QAP compares the observed graph correlation of (G_1, G_2) to the distribution of the respective resulting correlation scores obtained on repeated random row and column permutations of the adjacency matrix of G_2. However, this relates to the whole graph and not to specific subgroups of dyads, i. e., a subset of edges.

As we will see below, we can apply similar mechanisms for comparing a sub-network induced by a given subgroup pattern with a set of randomized sub-networks given the same distributional characteristics with respect to the total set of edges. However, in contrast to simple permutation operations, we have to take special care with respect to the social interaction properties, as we discuss below in detail, in order to compare the observed number of edges covered by a subgroup pattern with the expected number given a null-model.

Using a given subgroup discovery algorithm, the result of top-k subgroup discovery is the set of the k patterns $P_1, \ldots, P_k$, where $P_i \in 2^S$, with the highest interestingness according to the applied quality function. A subgroup discovery task can now be specified by the 5-tuple: (DB, c, S, q, k), where c indicates the target concept; the search space 2^S is defined by the set of basic patterns S.

4 Method

We first provide an overview on the proposed approach for the analysis of social interaction networks. Next, we present two novel quality functions for that task.

4.1 Compositional Network Analysis Using Subgroup Discovery

We focus on the analysis of *social interaction networks* [2,42], i. e., user-related social networks capturing social relations inherent in social interactions, social activities and other social phenomena which act as proxies for social user-relatedness. According to Wassermann and Faust [49, p. 37 ff.] social interaction networks focus on *interaction* relations between *people* as the corresponding actors. Then, a dyad, i. e., a link between two actors, models such a dyadic interaction. In a graph representation of the network, the dyad is then represented by an edge between two nodes (corresponding to the respective actors). Given attributed networks, also describing attributes, i. e., properties of nodes and/or edges can be used to characterize subgroups in order to *characterize* or *explain* a certain (observed) behavior, e. g., [21,33,49]. Here, we focus on *compositional network analysis* using subgroup discovery, where subgroups are induced by (a set of) describing attributes. Subgroup discovery enables hypotheses generation by directly exploring a given attribute space in order to identify interesting (compositional) subgroups according to some interestingness measure. As an exploratory method, we can e. g., focus on the top-k subgroups. Such patterns are then *local models* describing "interesting subsets" in terms of their attributes.

In the following, we focus on attributed networks, i. e., edge-attributed graphs with respect to actor attributes, enabling compositional dyadic analysis [49]. The interestingness can be flexibly defined using a quality measure. For social interaction networks, we distinguish between the following two properties:

1. Interaction duration: In social interaction networks, the duration of an interaction can be captured by a weight assigned to a specific link connecting the interacting actors. Then, simple networks that just capture those interactions can be represented by weighted graphs. In the unweighted case, we can just assign a default weight w for an edge e, e. g., $w(e) = 1.0$.
2. Interaction frequency: The frequency of interactions is typically indicated by multiple links between the two interacting actors, represented by a set of edges connecting the respective nodes in a multigraph. In addition, the duration of the interaction can also be captured as described above.

In the scope of this work, we focus on a numeric target feature t_P corresponding to the observed number of edges normalized by the expectation, for pattern P; for the interaction duration, we consider the weighted variant, i. e., taking the edge weights into account. Then, we rank subgroups utilizing the (normalized) mean of that target feature t_P. It is important to note, that we use the number of all possible contacts (edges) for computing the mean of t_P, i. e., including edges with a zero weight. Therefore, we take into account all possible edges between all nodes (actors), as discussed below, for simple graphs (for interation duration), as well as for multigraphs where we also consider interaction frequency.

4.2   Quality Measures

For ranking a set of subgroup patterns, we propose two quality measures. Essentially, we distinguish two cases: First, simple compositional networks represented as simple attributed graphs, which can also be weighted, and second attributed multigraphs. We propose two quality functions for estimating *dyadic means* of a pattern P, corresponding to the numeric target feature t_P discussed above. This is combined with randomization approaches for estimating the significance of the respective values. Altogether, this results in statistically well-founded quality functions, yielding intuitively interpretable values.

Simple Attributed Graphs. In the case of a simple network (without multiple links) we can simply add up the number of (weighted) edges E_P captured by a pattern P, and normalize by the number of all possible edges n_E in the node subset induced by P, i. e., all contributing nodes that are connected by any edge e contained in E_P. That means, for example, that if we consider the mean duration of contacts in a social interaction network as the target t_P, where the duration is indicated by the weight of a (contact) edge between two nodes (i. e., the involved actors), then we normalize by the number of all possible contacts that can occur in that set of nodes. Thus, intuitively, we take contacts of length zero into account for completeness. Thus, for a pattern P, we estimate its quality $q_S(P)$ as follows:

$$q_S(P) = Z\left(\frac{1}{n_E} \cdot \sum_{e \in E_P} w(e)\right), \tag{1}$$

with $n_E = \frac{n_{E_P}(n_{E_P}-1)}{2}$, where n_{E_P} is the number of nodes covered by a pattern P. Z is a function that estimates the statistical significance of the obtained value (i. e., t_P) given a randomized model, which we discuss below in more detail.

Attributed Multigraphs. For more complex attributed networks containing multi-links between actors, we model these as attributed multigraphs. Then, we can additionally take the interaction frequency into account, as discussed above. The individual set of interactions is modeled using a set of links between the different nodes representing the respective actors of the network. Thus, for normalizing the mean of target t_P, we also need to take into account the multiplicity of edges between the individual nodes. Then, with $n_E = \frac{n_{E_P}(n_{E_P}-1)}{2}$ indicating the total number of (single) edges between the individual nodes captured by pattern P, $m_i, i = 1 \ldots n_E$ models the number of multi-edges for an individual edge i connecting two nodes. With that, extending Eq. 1 for a pattern P in the multigraph case, we estimate its quality $q_M(P)$ as follows:

$$q_M(P) = Z(\frac{1}{n_E + m_E} \cdot \sum_{e \in E_P} w(e)), \qquad (2)$$

with $m_E = \sum_{i=1}^{n_E}(m_i - 1)$. It is easy to see that Eq. 2 simplifies to Eq. 1 for a simple attributed network, as a special case.

Randomization-Based Significance Estimation. As summarized above in Sect. 3, standard quality functions for subgroup discovery compare the mean of a certain target concept with the mean estimated in the whole dataset. In the dyadic analysis that we tackle in this paper, however, we also need to take edge formation of dyadic structures into account, such that, e. g., simply calculating the mean of the observed edges normalized by all edges for the whole dataset is not sufficient. In addition, since we use subgroup discovery for identifying a dyadic subgraph (i. e., a set of edges) induced by a pattern, we also aim to confirm the *impact* by checking the statistical significance compared to a null-model. For that, we propose a sampling based procedure: We draw r samples without replacement with the same size of the respective subgroup in terms of the number of edges, i. e., we randomly select r subsets of edges of the whole graph. For the two cases discussed above, i. e., for the simple attributed graph and the multigraph representation, we distinguish two cases:

1. Simple graph network representation: In the simple case, we just take into account the

$$N = \frac{n(n-1)}{2}$$

possible edges between all nodes of the simple graph. Thus, in a sampling vector $R = (r_1, r_2, \ldots, r_N)$, we fill the $r_i, i = 1 \ldots N$ positions with the weights of the corresponding edges of the graph, for which that a non-existing edge in the given graph is assigned a weight of zero.

2. Multigraph network representation: In the multigraph case we also consider the number of all possible edges between all the nodes, however, we also need to take the multi-edges into account, as follows:

$$N = \frac{n(n-1)}{2} + \sum_{i=1}^{n}(m_i - 1),$$

where $m_i, i = 1, \ldots, n$, are the respective multi-edge counts for an individual edge i. As above, we assign the sampling vector R accordingly, where we set the weight entries of non-existing edges to zero.

For selecting the random subsets, we apply sampling without replacement. This is essentially equivalent to a shuffling based procedure, e.g., [19,23]. Then, we determine the mean of the target feature t_R (e.g., mean duration) in those induced r subsets of edges. In that way, we build a distribution of "false discoveries" [19] using the r samples. Using the mean t_P in the original subgroup and the set of r sample means, we can construct a z-score which directly leads to statistical assessment for computing a p-Value. This is modeled using the function $Z(t_P), Z : \mathbb{R} \to \mathbb{R}$ which is then used for estimating the statistical significance of the target t_P of pattern P. In order to ensure that the r samples are approximately normally distributed, we can apply a normality test, for example, the Shapiro-Wilk-test [48]. If normality is rejected, a possible alternative is to compute the empirical p-value of a subgroup [23]. However, in practice often the distribution of the sampled means is approximately normally distributed, so that a p-value can be directly computed from the obtained z-score.

Table 1. Statistics/properties of the real-world datasets: Number of participants $|V|$, unique contacts $|U|$, total contacts $|C|$ average degree, diameter d, density, count of F2F contacts (C), cf. [27] for details.

Network	$|V|$	$|U|$	$|C|$	⌀Degree	d	Density	$|C|$
LWA 2010	77	1004	5154	26.08	3	0.34	5154
HT 2011	69	550	1902	15.94	4	0.23	1902

5 Results

Below, we describe the utilized two real-world datasets on social face-to-face interaction networks and experimental results of applying the presented approach.

5.1 Datasets

We applied social interaction networks captured at two scientific conferences, i.e., at the LWA 2010 conference in Kassel, Germany, and the Hypertext (HT) 2011 conference in Eindhoven, The Netherlands. Using the CONFERATOR system [5], we invited conference participants[1] to wear active RFID proximity tags.[2] When the tags are worn on the chest, tag-to-tag proximity is a proxy for a (close-range) face-to-face (F2F) contact, since the range of the signals is approximately 1.5 m if not blocked by the human body, cf. [14] for details. We record a F2F contact when the length of a contact is at least 20 s. A contact ends when the proximity tags do not detect each other for more than 60 s. This results in time-resolved networks of F2F contacts. Table 1 provides summary statistics of the collected datasets; see [27] for a detailed description.

---

[1] Study participants also gave their informed consent for the use of their data (including their profile) in scientific studies.

[2] http://www.sociopatterns.org.

In addition to the F2F contacts of the participants, we obtained further (socio-demographic) information from their Conferator online profile. In particular, we utilize information on the participants' (1) *gender*, (2) *country* of origin, (3) (university) *affiliation*, (4) academic status – *position* – i. e., professor, postdoc, PhD, student, (5) and their main conference *track* of interest. Note that not all attributes are available for both conferences; e. g., country is not available for the LWA 2010 conference since almost all participants were from Germany; here, we refer to the (university) affiliation instead. In contrast, the country information is very relevant for HT 2011. For those attributes given above, we created features on the edges of the attributed (multi-)graphs in such a way, so that an edge was labeled with "<feature>=EQ" if the respective nodes shared the same value of the feature, e. g., *gender=female* for both nodes. Otherwise, the edge was labeled with "<feature>=NEQ". That means that, for example, the subgroup described by the pattern *gender=EQ* contains the nodes, for which the dyadic actors always agree on their attribute *gender*.

Table 2. Top-20 most exceptional subgroups according to the aggregated duration of face-to-face interactions at LWA 2010 (simple attributed network): The table shows the respective patterns, the covered number of dyads, the mean contact length in seconds and the significance compared to the null-model (Quality (Z)).

Description	Size	∅CLength	Quality (Z)
track=EQ	456	182.05	19.01
affiliation=NEQ	959	245.39	18.91
position=NEQ	885	227.44	17.93
affiliation=NEQ, position=NEQ	868	220.01	17.36
affiliation=NEQ, track=EQ	428	158.18	16.22
position=NEQ, track=EQ	392	145.7	15.71
gender=NEQ	705	182.5	15.43
affiliation=NEQ, position=NEQ, track=EQ	381	139.92	15.2
gender=NEQ, track=EQ	312	123.84	14.01
affiliation=NEQ, gender=NEQ	669	160.01	13.2
gender=NEQ, position=NEQ	627	152.02	12.89
affiliation=NEQ, gender=NEQ, position=NEQ	614	145	12.1
gender=EQ	299	257.69	11.91
gender=EQ, track=EQ	144	189.02	11.75
affiliation=NEQ, gender=NEQ, track=EQ	289	102.15	11.35
affiliation=NEQ, gender=EQ, track=EQ	139	179.23	11.25
affiliation=NEQ, gender=EQ, position=NEQ, track=EQ	120	179.59	11.13
gender=EQ, position=NEQ, track=EQ	123	180.46	11.06
affiliation=NEQ, gender=EQ	290	252.35	11.01
affiliation=EQ, track=EQ	28	298.74	11

Table 3. Top-20 most exceptional according to the non-aggregated duration of face-to-face interactions at LWA 2010 (attributed multigraph): The table shows the respective subgroup patterns, the covered number of dyads, the mean contact length in seconds and the significance compared to the null-model (Quality (Z)).

Description	Size	Length	Quality (Z)
affiliation=EQ, gender=EQ, position=EQ, track=EQ	30	239	793.96
affiliation=EQ, gender=EQ, position=NEQ, track=NEQ	7	71.29	491.59
affiliation=EQ, gender=EQ, position=EQ, track=NEQ	39	164.02	476.73
affiliation=EQ, gender=EQ, track=EQ	39	160.73	475.71
affiliation=EQ, gender=EQ, position=EQ	69	184.37	412.34
affiliation=EQ, gender=EQ, track=NEQ	46	127.68	341.41
affiliation=EQ, gender=NEQ, position=NEQ, track=NEQ	34	105.83	337.98
affiliation=EQ, gender=EQ, position=NEQ, track=EQ	9	44.63	274.97
affiliation=EQ, position=NEQ, track=NEQ	41	91.99	263.29
affiliation=EQ, gender=EQ	85	128.89	257.45
affiliation=EQ, position=EQ, track=NEQ	78	119.78	249.23
affiliation=EQ, gender=NEQ, position=EQ, track=NEQ	39	77.24	226.94
affiliation=EQ, gender=EQ, position=NEQ	16	44.93	203.45
affiliation=EQ, gender=NEQ, track=NEQ	73	86.25	182.48
affiliation=EQ, track=NEQ	119	103.35	171.08
affiliation=EQ, gender=NEQ, position=NEQ, track=EQ	98	92.89	170.31
gender=EQ, position=EQ, track=EQ	142	107.1	165.17
affiliation=NEQ, gender=EQ, position=EQ, track=NEQ	87	83.01	162.58
affiliation=EQ, gender=NEQ, position=EQ, track=EQ	228	135.41	161.12
affiliation=EQ, position=EQ, track=EQ	258	137.37	156.49

5.2   Experimental Results and Discussion

For compositional analysis, we applied subgroup discovery on the attributes described in Sect. 5.1. We utilized the VIKAMINE [9] data mining platform for subgroup discovery[3], utilizing the SD-Map* algorithm [8], where we supplied our novel quality functions for determining the top-20 subgroups.

For the target concept, we investigated the *mean length of contacts* – corresponding to the *duration* of a social interaction in the respective subgroup. We applied both simple attributed networks, and multigraph representations: For the former, social interactions between respective actors were aggregated, such that the corresponding weight is given by the sum of all interactions between those actors. For the multigraph case, we considered the face-to-face interations with their respective durations individually. Tables 2, 3, 4 and 5 show the results.

Overall, we notice several common patterns in those tables, both for LWA 2010 and HT 2011: We observe the relatively strong influence of homophilic features such as *gender*, *track*, *country*, and *affiliation* in the detected patterns,

---

[3] http://www.vikamine.org.

Table 4. Top-20 most exceptional subgroups according to the aggregated duration of face-to-face interactions at HT 2010 (simple attributed network): The table shows the respective patterns, the covered number of dyads, the mean contact length in seconds and the significance compared to the null-model (Quality (Z)).

Description	Size	Length	Quality (Z)
gender=EQ	357	114.76	15.76
gender=EQ, track=EQ	114	83.87	15.32
country=EQ, gender=EQ, track=EQ	35	111.75	14.21
country=EQ, track=EQ	42	89.74	13.89
track=EQ	185	70.4	13.73
country=EQ, gender=EQ, position=NEQ, track=EQ	18	140.52	12.98
country=EQ, gender=EQ	55	70.06	12.75
country=NEQ	470	87.76	12.61
country=EQ	80	56.51	12.59
position=NEQ	365	76.89	11.87
gender=EQ, position=EQ, track=EQ	46	68.43	11.8
country=EQ, position=NEQ, track=EQ	23	99.62	11.62
position=EQ	185	60.15	11.45
position=EQ, track=EQ	60	53.32	11.44
country=EQ, gender=EQ, position=NEQ	30	82.03	11.29
country=NEQ, gender=EQ	302	82.91	11.19
gender=EQ, position=EQ	136	61.91	10.81
gender=EQ, position=NEQ	221	71.43	10.52
gender=EQ, position=NEQ, track=EQ	68	58.42	10.13
track=NEQ	365	70.22	10.03
country=EQ, position=NEQ	50	45.89	9.86

confirming preliminary work that we presented in [11] only analyzing the individual features and their contribution to establishing social interactions. Using compositional subgroup discovery we can analyze those patterns at a more fine-grained level, also taking more complex patterns, i. e., combinations of different features into account. Thus, our results indicate more detailed findings both concerning the individual durations, the influence of repeating interactions, and the impact of complex patterns given by a combination of several features.

Furthermore, we also observe that the compositional multigraph analysis, i. e., focusing on dyadic interactions in the multigraph case focuses on much more specific patterns with many more contributing features, in contrast to more general patterns in the case of the simple attributed network. That is, for the multigraph case smaller subgroups (indicated by the size of the set of involved actors/nodes) are detected that are more specific regarding their descriptions, i. e., considering the length of the describing features. Then, these can provide more detailed insights into, e. g., homophilic processes. We can assess different specializations of competing properties, see e. g., lines #1 and #3 in Table 3.

Table 5. Top-20 most exceptional subgroups according to the non-aggregated duration of face-to-face interactions at HT 2011 (attributed multigraph): The table shows the respective subgroup patterns, the covered number of dyads, the mean contact length in seconds and the significance compared to the null-model (Quality (Z)).

Description	Size	Length	Quality (Z)
country=EQ, gender=NEQ, position=EQ, track=EQ	13	159.57	353.49
country=EQ, gender=NEQ, position=EQ, track=NEQ	32	126.3	173.93
country=EQ, gender=NEQ, position=EQ	45	102.51	120.37
country=EQ, gender=NEQ, position=NEQ, track=EQ	15	45.74	92.91
country=EQ, gender=EQ, position=EQ, track=NEQ	17	42.27	83.02
country=EQ, gender=NEQ, track=EQ	28	49.86	74.91
country=EQ, gender=EQ, position=EQ, track=EQ	113	85.67	65.45
country=EQ, position=EQ, track=EQ	126	85.04	62.09
country=EQ, position=EQ, track=NEQ	49	52.29	61.21
country=EQ, gender=EQ, position=EQ	130	59.27	45.2
country=NEQ, gender=NEQ, position=EQ, track=EQ	32	29.08	42.28
country=EQ, gender=EQ, position=NEQ, track=NEQ	38	31.69	41.84
gender=NEQ, position=EQ, track=EQ	45	30.63	38.17
country=EQ, gender=NEQ, track=NEQ	78	41.06	38.02
country=EQ, gender=EQ, position=NEQ, track=EQ	255	72.55	36.41
country=EQ, position=EQ	175	52.37	35.98
country=NEQ, gender=EQ, position=EQ, track=EQ	166	41.72	32.72
gender=EQ, position=EQ, track=EQ	279	52.69	32.33
country=EQ, gender=EQ, track=EQ	368	66.86	30.3
country=EQ, position=NEQ, track=EQ	270	60.25	30.29
position=EQ, track=EQ	324	43.21	27.79

Also, the "specialization transition" between two patterns provides interesting insights, e. g., considering the patterns *affiliation=EQ, gender=EQ* (line #10) and *affiliation=EQ, gender=EQ, track=EQ* (line #4) shown in Table 3 which indicates the strong homophilic influence of the track feature. A similar pattern also emerges for HT 2011, regarding *country=EQ, gender=NEQ, position=EQ*; here both *track=NEQ* and *track=EQ* improve on the mean contact duration; the latter is considerably stronger, also in line with our expectations, e. g., cf. [11].

6   Conclusions

In this paper, we formalized the problem of detecting compositional patterns in attributed networks, i. e., capturing dyadic subgroups that show an interesting behavior as estimated by a quality measure. We presented a novel approach adapting techniques of subgroup discovery and exceptional model mining [3,7,18]. Furthermore, we discussed estimation methods for ranking interesting patterns, and presented two novel quality measures for that purpose. Finally, we demonstrated the efficacy of the approach using two real-world datasets.

Our results indicate interesting findings according to common principles observed in social interaction networks, e. g., the influence of homophilic features on the interactions. Furthermore, the applied quality functions allow to focus on specific properties of interest according to the applied modeling method, e. g., whether a simple attributed network or a multigraph representation is applied. Furthermore, the proposed quality functions are statistically well-founded, and provide a statistical significance value directly, also easing their interpretation.

For future work, we aim to extend the concepts developed in this work towards multiplex networks, also taking into account temporal network dynamics. For that, we aim to consider methods for analyzing sequential patterns [4] as well as approaches for modeling and analyzing multiplex network approaches, e. g., [25,47]. Finally, methods for testing specific hypothesis and Bayesian estimation techniques, e. g., [4,13,20] are further interesting directions to consider.

Acknowledgements. This work has been partially supported by the German Research Foundation (DFG) project "MODUS" under grant AT 88/4-1.

References

1. Agrawal, R., Srikant, R.: Fast algorithms for mining association rules. In: Proceedings of VLDB, pp. 487–499. Morgan Kaufmann (1994)
2. Atzmueller, M.: Data mining on social interaction networks. JDMDH **1** (2014)
3. Atzmueller, M.: Subgroup discovery. WIREs DMKD **5**(1), 35–49 (2015)
4. Atzmueller, M.: Detecting community patterns capturing exceptional link trails. In: Proceedings of IEEE/ACM ASONAM. IEEE Press, Boston, MA, USA (2016)
5. Atzmueller, M., et al.: Enhancing social interactions at conferences. it - Inf. Technol. **53**(3), 101–107 (2011)
6. Atzmueller, M., Doerfel, S., Hotho, A., Mitzlaff, F., Stumme, G.: Face-to-face contacts at a conference: dynamics of communities and roles. In: Atzmueller, M., Chin, A., Helic, D., Hotho, A. (eds.) MSM/MUSE -2011. LNCS (LNAI), vol. 7472, pp. 21–39. Springer, Heidelberg (2012). https://doi.org/10.1007/978-3-642-33684-3_2
7. Atzmueller, M., Doerfel, S., Mitzlaff, F.: Description-oriented community detection using exhaustive subgroup discovery. Inf. Sci. **329**(C), 965–984 (2016)
8. Atzmueller, M., Lemmerich, F.: Fast subgroup discovery for continuous target concepts. In: Rauch, J., Raś, Z.W., Berka, P., Elomaa, T. (eds.) ISMIS 2009. LNCS (LNAI), vol. 5722, pp. 35–44. Springer, Heidelberg (2009). https://doi.org/10.1007/978-3-642-04125-9_7
9. Atzmueller, M., Lemmerich, F.: VIKAMINE - open-source subgroup discovery, pattern mining, and analytics. In: Flach, P.A., De Bie, T., Cristianini, N. (eds.) ECML PKDD 2012. LNCS (LNAI), vol. 7524, pp. 842–845. Springer, Heidelberg (2012). https://doi.org/10.1007/978-3-642-33486-3_60
10. Atzmueller, M., Lemmerich, F.: Exploratory pattern mining on social media using geo-references and social tagging information. IJWS **2**(1/2), 80–112 (2013)
11. Atzmueller, M., Lemmerich, F.: Homophily at academic conferences. In: Proceedings of WWW 2018 (Companion). IW3C2/ACM (2018)

12. Atzmueller, M., Mollenhauer, D., Schmidt, A.: Big data analytics using local excep-
    tionality detection. In: Enterprise Big Data Engineering, Analytics, and Manage-
    ment. IGI Global, Hershey, PA, USA (2016)
13. Atzmueller, M., Schmidt, A., Kloepper, B., Arnu, D.: HypGraphs: an approach
    for analysis and assessment of graph-based and sequential hypotheses. In: Appice,
    A., Ceci, M., Loglisci, C., Masciari, E., Raś, Z.W. (eds.) NFMCP 2016. LNCS
    (LNAI), vol. 10312, pp. 231–247. Springer, Cham (2017). https://doi.org/10.1007/
    978-3-319-61461-8_15
14. Barrat, A., Cattuto, C., Colizza, V., Pinton, J.F., den Broeck, W.V., Vespignani,
    A.: High resolution dynamical mapping of social interactions with active RFID.
    PLoS ONE 5(7) (2010)
15. Bendimerad, A., Cazabet, R., Plantevit, M., Robardet, C.: Contextual subgraph
    discovery with mobility models. In: International Workshop on Complex Networks
    and their Applications, pp. 477–489. Springer (2017)
16. Bothorel, C., Cruz, J.D., Magnani, M., Micenkova, B.: Clustering attributed
    graphs: models measures and methods. Netw. Sci. 3(03), 408–444 (2015)
17. Burt, R.S.: Cohesion versus structural equivalence as a basis for network subgroups.
    Sociol. Methods Res. 7(2), 189–212 (1978)
18. Duivesteijn, W., Feelders, A.J., Knobbe, A.: Exceptional model mining. Data Min.
    Knowl. Discov. 30(1), 47–98 (2016). Jan
19. Duivesteijn, W., Knobbe, A.: Exploiting false discoveries - statistical validation of
    patterns and quality measures in subgroup discovery. In: Proceedings of ICDM,
    pp. 151–160. IEEE (2011)
20. Espín-Noboa, L., Lemmerich, F., Strohmaier, M., Singer, P.: JANUS: a hypothesis-
    driven bayesian approach for understanding edge formation in attributed multi-
    graphs. Appl. Netw. Sci. 2(1), 16 (2017)
21. Frank, O.: Composition and structure of social networks. Mathématiques et Sci.
    Hum. Math. Soc. Sci. 137 (1997)
22. Geng, L., Hamilton, H.J.: Interestingness measures for data mining: a survey. ACM
    Comput. Surv. 38(3) (2006)
23. Gionis, A., Mannila, H., Mielikäinen, T., Tsaparas, P.: Assessing data mining
    results via swap randomization. ACM Trans. Knowl. Discov. Data (TKDD) 1(3),
    14 (2007)
24. Günnemann, S., Färber, I., Boden, B., Seidl, T.: GAMer: a synthesis of subspace
    clustering and dense subgraph mining. In: KAIS. Springer (2013)
25. Kanawati, R.: Multiplex network mining: a brief survey. IEEE Intell. Inform. Bull.
    16(1), 24–27 (2015)
26. Kaytoue, M., Plantevit, M., Zimmermann, A., Bendimerad, A., Robardet, C.:
    Exceptional contextual subgraph mining. Mach. Learn. 106(8), 1171–1211 (2017)
27. Kibanov, M., et al.: Is web content a good proxy for real-life interaction? A case
    study considering online and offline interactions of computer scientists. In: Pro-
    ceedings of ASONAM. IEEE Press, Boston, MA, USA (2015)
28. Kibanov, M., Atzmueller, M., Scholz, C., Stumme, G.: Temporal evolution of con-
    tacts and communities in networks of face-to-face human interactions. Sci. China
    Inf. Sci. 57(3), 1–17 (2014). March
29. Klösgen, W.: Explora: a multipattern and multistrategy discovery assistant. In:
    Advances in Knowledge Discovery and Data Mining, pp. 249–271. AAAI (1996)
30. Klösgen, W.: Applications and research problems of subgroup mining. In: Raś,
    Z.W., Skowron, A. (eds.) ISMIS 1999. LNCS, vol. 1609, pp. 1–15. Springer, Hei-
    delberg (1999). https://doi.org/10.1007/BFb0095086

31. Klösgen, W.: Handbook of Data Mining and Knowledge Discovery, Chap. 16.3: Subgroup Discovery. Oxford University Press, New York (2002)

32. Krackhardt, D.: QAP partialling as a test of spuriousness. Soc. Netw. **9**, 171–186 (1987)

33. Lau, D.C., Murnighan, J.K.: Demographic diversity and faultlines: the compositional dynamics of organizational groups. Acad. Manag. Rev. **23**(2), 325–340 (1998)

34. Leman, D., Feelders, A., Knobbe, A.: Exceptional model mining. In: Daelemans, W., Goethals, B., Morik, K. (eds.) ECML PKDD 2008. LNCS (LNAI), vol. 5212, pp. 1–16. Springer, Heidelberg (2008). https://doi.org/10.1007/978-3-540-87481-2_1

35. Lemmerich, F., Atzmueller, M., Puppe, F.: Fast exhaustive subgroup discovery with numerical target concepts. Data Min. Knowl. Discov. **30**, 711–762 (2016). https://doi.org/10.1007/s10618-015-0436-8

36. Lemmerich, F., Becker, M., Atzmueller, M.: Generic pattern trees for exhaustive exceptional model mining. In: Flach, P.A., De Bie, T., Cristianini, N. (eds.) ECML PKDD 2012. LNCS (LNAI), vol. 7524, pp. 277–292. Springer, Heidelberg (2012). https://doi.org/10.1007/978-3-642-33486-3_18

37. Lemmerich, F., Becker, M., Singer, P., Helic, D., Hotho, A., Strohmaier, M.: Mining subgroups with exceptional transition behavior. In: Proceedings of ACM SIGKDD, pp. 965–974. ACM (2016)

38. Macek, B.E., Scholz, C., Atzmueller, M., Stumme, G.: Anatomy of a conference. In: Proceedings of ACM Hypertext, pp. 245–254. ACM (2012)

39. McPherson, M., Smith-Lovin, L., Cook, J.M.: Birds of a feather: homophily in social networks. Annu. Rev. Sociol. **27**(1), 415–444 (2001)

40. Mitzlaff, F., Atzmueller, M., Benz, D., Hotho, A., Stumme, G.: Community assessment using evidence networks. In: Atzmueller, M., Hotho, A., Strohmaier, M., Chin, A. (eds.) MSM/MUSE -2010. LNCS (LNAI), vol. 6904, pp. 79–98. Springer, Heidelberg (2011). https://doi.org/10.1007/978-3-642-23599-3_5

41. Mitzlaff, F., Atzmueller, M., Hotho, A., Stumme, G.: The social distributional hypothesis. J. Soc. Netw. Anal. Min. **4**(216), 1–14 (2014)

42. Mitzlaff, F., Atzmueller, M., Stumme, G., Hotho, A.: Semantics of user interaction in social media. In: Complex Networks IV, SCI, vol. 476. Springer (2013)

43. Morik, K.: Detecting interesting instances. In: Hand, D.J., Adams, N.M., Bolton, R.J. (eds.) Pattern Detection and Discovery. LNCS (LNAI), vol. 2447, pp. 13–23. Springer, Heidelberg (2002). https://doi.org/10.1007/3-540-45728-3_2

44. Moser, F., Colak, R., Rafiey, A., Ester, M.: Mining cohesive patterns from graphs with feature vectors. In: SDM, vol. 9, pp. 593–604. SIAM (2009)

45. Neely, R., Cleghern, Z., Talbert, D.A.: Using subgroup discovery metrics to mine interesting subgraphs. In: Proceedings of FLAIRS, pp. 444–447. AAAI (2015)

46. Robins, G., Pattison, P., Kalish, Y., Lusher, D.: An introduction to exponential random graph (p*) models for social networks. Soc. Netw. **29**(2) (2007)

47. Scholz, C., Atzmueller, M., Barrat, A., Cattuto, C., Stumme, G.: New insights and methods for predicting face-to-face contacts. In: Proceedings of ICWSM. AAAI (2013)

48. Shapiro, S.S., Wilk, M.B.: An analysis of variance test for normality (complete samples). Biometrika **52**(3/4), 591–611 (1965)

49. Wasserman, S., Faust, K.: Social Network Analysis: Methods and Applications. Structural Analysis in the Social Sciences, vol. 8, 1st edn. Cambridge university press, Cambridge (1994)

50. Wrobel, S.: An algorithm for multi-relational discovery of subgroups. In: Komorowski, J., Zytkow, J. (eds.) PKDD 1997. LNCS, vol. 1263, pp. 78–87. Springer, Heidelberg (1997). https://doi.org/10.1007/3-540-63223-9_108
51. Wrobel, S., Morik, K., Joachims, T.: Maschinelles Lernen und Data Mining. Handbuch der Künstlichen Intelligenz **3**, 517–597 (2000)

Exceptional Attributed Subgraph Mining to Understand the Olfactory Percept

Maëlle Moranges[1,2], Marc Plantevit[1,3(✉)], Arnaud Fournel[1,4],
Moustafa Bensafi[1,4], and Céline Robardet[1,2]

[1] Université de Lyon, Lyon, France
marc.plantevit@liris.cnrs.fr
[2] INSA Lyon, LIRIS, CNRS UMR5205, Villeurbanne, France
[3] Université Lyon 1, LIRIS, CNRS UMR5205, Villeurbanne, France
[4] CNRS, CRNL, UMR5292, INSERM U1028, Bron, France

Abstract. Human olfactory perception is a complex phenomenon whose neural mechanisms are still largely unknown and novel methods are needed to better understand it. Methodological issues that prevent such understanding are: (1) to be comparable, individual cerebral images have to be transformed in order to fit a template brain, leading to a spatial imprecision that has to be taken into account in the analysis; (2) we have to deal with inter-individual variability of the hemodynamic signal from fMRI images which render comparisons of individual raw data difficult. The aim of the present paper was to overcome these issues. To this end, we developed a methodology based on discovering exceptional attributed subgraphs which enabled extracting invariants from fMRI data of a sample of individuals breathing different odorant molecules.Four attributed graph models were proposed that differ in how they report the hemodynamic activity measured in each voxel by associating varied attributes to the vertices of the graph. An extensive empirical study is presented that compares the ability of each modeling to uncover some brain areas that are of interest for the neuroscientists.

1 Introduction

Olfaction is a chemical sense whose functions is to detect the presence of odorous substances present in the environment in order to modulate appetitive, defensive and social behaviors [6,26]. Olfactory deficits are a common symptom of neurode-generative or psychiatric disorders and clinical research proposed that olfaction could have great potential as an early biomarker of disease [2,25] for example using neuroimaging to investigate the breakdown of structural connectivity profile of the primary olfactory networks. On a fundamental level, whereas olfaction has received much attention over the last decades, human olfactory perception is a complex phenomenon whose mechanisms are still largely unknown.

Neuroscientific investigations revealed that perception of odors results from the interaction between volatile molecules (described by multiple physicochemical descriptors) and olfactory receptors located in the nasal cavity. Once the

© Springer Nature Switzerland AG 2018
L. Soldatova et al. (Eds.): DS 2018, LNAI 11198, pp. 276–291, 2018.
https://doi.org/10.1007/978-3-030-01771-2_18

interaction is done, a neural signal is then transmitted to central areas of the brain to generate a percept called "odor" that is often accompanied by a strong hedonic or emotional tone (either pleasant or unpleasant). Understanding the link between odor (hedonic) perception and its underlying brain activity is an important challenge in the field. Although past brain imaging studies revealed that the brain activation in response to smells is distributed and can represent different attributes of odor perception (from perception of irritation to intensity or hedonic valence) [10,13,16], there is clear need to develop new brain imaging analysis techniques in order to (i) take into account the large variability across individuals in terms odor perception and brain activation, and (ii) refine the network and understand for instance how different sub-parts of a given area are involved in the processing of pleasant and unpleasant odors. This will be the main aims of the present paper.

The most popular method to acquire brain imaging data in humans is called functional Magnetic Resonance Imaging or fMRI. An important issue when performing inter-individual analysis on fMRI data is that each individual image is transformed in order to fit and map onto a template (so that comparison across participants can be made on a unique model of the brain) [11]. Therefore, voxel (i.e., 3d pixel) mapping from the individual to the template may be imprecise, and looking for voxels that have a strong hemodynamic response for all individuals can be unsuccessful. One solution to circumvent this problem is thus to take into account this imprecision by looking for areas whose voxels – although imprecise – have specific hemodynamic response to some odors for a large proportions of individuals compare to the rest of the brain. To achieve this goal, we propose to model fMRI images as an attributed graph where the vertices are the voxels (brain unit), the edges encode the adjacency relationship, and vertex attributes stand for the hemodynamic response to an odor. We propose to analyze such a graph with CENERGETICS [4] which makes possible to identify brain areas with exceptional hemodynamic response in some experimental settings.

Commonly, in order to demonstrate a functional activity of a given voxel, neuroscientists make use of general linear model coupled with massive univariate statistics [12] whereby the mean activity a voxel is compared in a "test condition" (e.g. when participants are asked to breathe odors) and a "control" condition (e.g. when participants are asked to breathe non-odorized air). The statistical comparison is usually performed using a Student t-test. However, this type of comparison presents some weaknesses when trying to take into account inter individual variability. Figure 1 illustrates this issue. The distributions of the t-values associated to the hemodynamic responses that come from the fMRI of two individuals smelling the same odor (EUG, the Eugenol molecule that smells like cloves) are represented. If we consider that a voxel is activated when t is greater than 1.96, that corresponds to an error of type 1 of 5%, in one case there are almost 5% of the voxels that are activated whereas in the other case this number is lower than 10^{-2}%. Thus, if we look for the invariants between individuals for the same odor, there is a good chance that they do not exist.

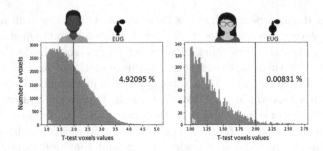

Fig. 1. Distribution of the t-test values associated to the voxels of two individuals smelling EUG odor.

We propose different ways to evaluate the level of activation of a voxel, based on normalized average values, ranks, t-test or pairwise comparisons. In the first approach, the level of activation of a voxel by an odor is evaluated by the average value of the hemodynamic response of this odor. The second proposal captures the average, for all individuals, of the rank of the odor-related response among all the responses of an individual when perceiving different odors. In the third model, the odor activates the voxel if its hemodynamic response is statistically significant according to the Student's t-test. In the fourth model, the attributes are pairwise comparisons of odor responses (e.g. odor1 > odor2) and their values are the number of individuals for which this comparison holds.

Exceptional subgraphs on fMRI data are presented in Sect. 2, as well as the attributed graphs built to model the brain activation during olfactory perception. Such patterns can be mined using CENERGETICS, an algorithm designed to discover connected subgraphs with over-represented and/or under-represented attributes, that was developed to analyze urban data. CENERGETICS is applied on these graphs and the obtained results are compared through an extensive empirical study in Sect. 3. Related work is reviewed in Sect. 4 and concluding remarks are given in Sect. 5.

2 Mining Exceptional Subgraphs for Olfactory Percept Analysis

We propose to use pattern mining techniques to identify relationships between odor perception and brain areas. The data we analyze come from a neuroscience experiment measuring hemodynamic responses when perceiving different odors using fMRI. p individuals participated in the study and each of them inhaled q different odors t times. During each olfactory trial, a brain volume is acquired. Each image reflects the hemodynamic response in each of 3 millimeters cubed unit of the brain, hereafter called voxel. The hemodynamic response function is then modeled as regressors that render hemodynamic activity [23]. Let us denote by $X_k(v, i, j)$ the level of activity measured in the voxel v for individual $i = 1 \ldots p$ while smelling an odor $j = 1 \ldots q$, at time $k = 1 \ldots t$, and $A_k(v, i)$ the activity measured while individual i is breathing air. The specific activity value of an odor perception is obtained by the average difference of X and A:

$$M(v,i,j) = \frac{\sum_k (X_k(v,i,j) - A_k(v,i))}{t}$$

However, the measure M is not bounded, taking positive and negative values, with its intensity depending on the sensitivity of the participant. To be able to compare different individuals, it is usual in neuroscience to normalize the value M that is tranformed into a statistical Student t-value:

$$M'(v,i,j) = \frac{M(v,i,j)}{\frac{\sigma(X-A)}{\sqrt{t}}}$$

with $\sigma(X - A)$ the standard deviation of the measures $X - A$ over k.

In an inter-individual analysis, we are interested in voxel areas (1) that are activated by a given odor for most of the individuals and (2) whose activation level is much higher than that observed in other areas of the brain. To identify such patterns, we propose to mine exceptional subgraphs in an attributed graph which models the brain activation when the subject is stimulated by an odor.

2.1 Mining Activated Areas in the Brain

Brain activity can be modeled as a vertex attributed graph whose vertices V represent voxels and edges E connect adjacent voxels. A set of attribute value pairs P is associated to each vertex and describes the activity of the corresponding voxel. The attributes are denoted $\mathcal{A}$ and take their value x in $\mathbb{R}$:

$$P : V \rightarrow \{(a,x) \mid a \in \mathcal{A}, x \in \mathbb{R}\}$$

Our objective is to identify brain areas whose attribute value pairs distinguish them from the rest of the brain. To this end, we propose to discover connected subgraphs associated to exceptional attribute value pairs. An attribute value pair is considered as exceptional for a subgraph if it has a much higher value in its vertices than in the remaining of the graph. Hence, an exceptional attributed subgraph is defined as a pair (S, K) with $S \subseteq V$, a subset of vertices that induces the subgraph $G[S]$, and $K \subseteq \mathcal{A}$ a subset of attributes whose values are exceptional for this subgraph. This is evaluated by the weighted relative accuracy defined as:

$$\text{WRAcc}(S,K) = \frac{sum(S,\mathcal{A})}{sum(V,\mathcal{A})} \times \left(\frac{sum(S,K)}{sum(S,\mathcal{A})} - \frac{sum(V,K)}{sum(V,\mathcal{A})} \right)$$

with $sum(S,K) = \sum_{v \in S} \sum_{(a,x) \in P(v),\, a \in K} x$.

Definition 1 (Exceptional attributed subgraph). *Given an attributed graph $G = (V, E, \mathcal{A}, P)$ and two thresholds $minV$ and δ, an Exceptional attributed subgraph (S, K) is such that (1) $|S| \geq minV$, (2) $G[S]$ is connected, (3) $WRAcc(S, K) \geq \delta$ and (4) $\forall v \in S, \forall p \in K, WRAcc(v, p) > 0$.*

Condition (1) ensures that patterns involve enough vertices to be of interest. Condition (2) preserves the notion of areas and avoid discontinuity. Condition (3) assesses the exceptionality of the attributes, while condition (4) enforces the subgraph to be cohesive. Such patterns can be mined using the algorithm presented in [4], originally designed to exhibit the predominant activities and their associated urban areas in graphs that model urban areas. The algorithm, named CENERGETICS, mines exceptional subgraph in attributed graphs.

Exceptional attributed subgraph definition can be extended to also catch attributes whose values are exceptionally lower for the subgraph than for the rest of the graph. CENERGETICS enables the possibility of discovering subgraphs with both exceptionally over- or under-represented attributes. In this section, we consider over-represented attribute values. The case of under-represented attributes is discussed in the empirical study.

Several attributed graphs can be constructed based on fMRI data. They differ by the attribute value pairs associated to the vertices, that is to say, by the attribute and by the value associated to them.

2.2   Attributed Graphs that Model Olfactory Perception

The attribute value pairs P of the graph reflects the strength of the hemodynamic response of the corresponding voxel when perceiving the odors. The attributes A can be the odor names, but also another characteristics such as chemical properties or the feelings felt during the perception (for instance their pleasant or unpleasant character). We denote by π the characteristic used to describe the odor (i.e. an injective function from odors to a set of labels). The attributes of A can also be pairs of odors to characterize voxels with pairwise inequalities.

The value of each attribute results of the aggregation of the measurements M obtained for different individuals. Since these measurements may contains errors, which may come from the material used, but also from the brain activity of the participant during the experiment (stress, thoughts), we consider below different ways of aggregating the data. These different approaches attempt to overcome this problem and will be experimentally compared.

Mean of the values: A voxel activation can be characterized by the mean of the values:

$$(\pi[j], x) \in P(v) \text{ with } x = \frac{\sum_i M'(v, i, j)}{p}$$

To limit the effect of high inter-individual variability it can be preferable not to consider the measure M' as an interval scale, but to downgrade the type of measurement scale and only consider the ranks.

Average rank: The voxel activation is evaluated by the average rank of the odor in the individual perceptions:

$$(\pi[j], x) \in P(v) \text{ with } x = \frac{\sum_i rank(v, i, j)}{p}$$

with $rank(v, i, j) = |\{\ell = 1 \ldots q \mid M'(v, i, \ell) \leq M'(v, i, j)\}|$.

t-test based approach: We can also downgrade the measure to consider it as a nominal variable. The discretization can be obtained thanks to a t-test assesses whether a voxel is activated or not. For a voxel v, an individual i and an odor j, if $M'(v, i, j)$ is greater than the critical value with $df = t - 1$ of the student distribution (given the confidence level $\alpha = 0.05$), then the hemodynamic response is considered to be different from the one observed while breathing air and the voxel is esteemed activated:

$$(\pi[j], x) \in P(v) \text{ with } x = |\{i = 1 \ldots p \,|\, (T_{0.05} < M'(v, i, j))\}|$$

Approach based on pairwise inequality: We propose another setting based on the pairwise comparison of the hemodynamic responses. The vertex attributes are pairs of odors (o_1, o_2) and their value is the number of individuals who have a higher value while smelling o_1 than when smelling o_2. Thereby, $q \times (q - 1)$ attributes (the number of pairs of odors) are associated to each vertex v of the graph and their values are:

$$((\pi[o_1], \pi[o_2]), x) \in P(v) \text{ with } x = |\{i = 1 \ldots p \,|\, M'(v, i, o_1) > M'(v, i, o_2)\}|$$

3   Empirical Study

In this section, we report our experimental results. These experiments aim to compare the different ways to build an attributed graph that are described in the previous section. Especially, we want to identify which ones are the most promising to identify exceptional subgraphs and how the related patterns make sense. To this end, we study the main characteristics of the discovered patterns for each modeling. In our experiments, 14 individuals smelled 6 odorants[1] 10 times ($p = 14$, $q = 6$ and $k = 10$.).

As mentioned in the previous section, exceptional attributed subgraph definition can be extended to make possible the discovery of subgraphs whose attribute values are lower than what observed on the rest of the graphs. For mean and rank modelings, we just adapted the WRAcc measure to catch under-represented attributes [4]. In the pairwise inequality based modeling, we consider all pairs of attributes, so it is not necessary to consider under-represented attribute values. For the t-test based method, the under-representedness is captured by the lower tail of the t-test distribution ($\alpha = 0.05$). In the following, the over-represented attributes are called positive and the under-represented ones are named negative.

Experiments were carried out on an Intel Core i7-4770 3.40 GHz machine with 8 GB RAM. Applied on the whole brain (902629 voxels), CENERGETICS takes at most 7 min to discover the exceptional subgraphs. However, in the following, we focus on the piriform cortex, an area made of 662 voxels known as the first olfactory area that receives information of the olfactory bulb. Here, we aim to understand how odors and their perceptual properties (hedonic) are processed in this area compared to the other brain areas.

---

[1] The odorant names are: 3Hex, ACE, DEC, EUG, HEP, MAN.

For this study, the parameters of CENERGETICS are such that the computed exceptional subgraphs contain at least one vertex and have a WRAcc value greater than 0.004. Extractions take at most 201 ms regardless of the considered models. The number of patterns are as follows: 555 for Mean, 238 for Rank, 118 for t-test and 803 for pairwise modelings.

This empirical study aims to answer the following questions: (a) Are the collections of exceptional subgraphs obtained with the 4 modelings different? (distributions) (b) Do the different modeling capture the same phenomena? (same attributes, similar areas) (c) What about considering odor characteristics? (hedonic values) (d) Do the discovered subgraphs make sense? What kind of insights can they provide to neuroscientists?

To this end, we first study the main characteristics of the exceptional attributed subgraphs obtained by each modeling. We then provide a detailed crossed-analysis of the top 3 patterns of the four collections. Finally, we consider other attributes related to the odorants to discuss the potential of each modeling. We also provide neuroscientists' feedback on these patterns. Additional results are provided as supplementary material[2].

3.1   Comparison of Patterns Obtained from the Different Models

Figure 2 reports the distribution of the patterns according to their WRAcc value. The distributions of the four approaches are similar. Nevertheless, t-test and mean based modelings retrieve patterns with the highest WRAcc values. The pairwise inequality based modeling provides patterns with lower WRAcc values. This can be due to the total number of attributes that is larger for the pairwise model (30) than for the other approaches (12).

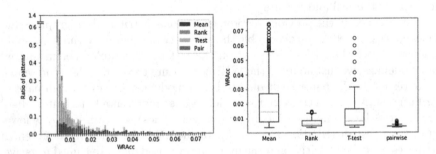

Fig. 2. Distributions of patterns with respect to their WRAcc value for each approach (left). Statistical characteristics of these distributions (right).

Similarly, we show the distribution of patterns according to the number of vertices they contain in Fig. 3. Patterns discovered by the t-test based approach contain less vertices than the ones retrieved by the other modelings. One possible

---

[2] goo.gl/ppJFEX.

consequence of this observation is a greater risk to provide false positive patterns to the end-user. Other methods have a smoother distribution, methods rank and pairwise giving the biggest patterns.

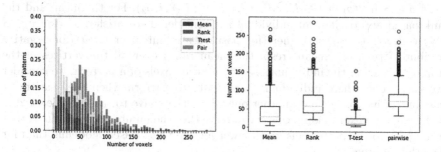

Fig. 3. Distribution of patterns with respect to their number of voxels for each approach (left). Statistical characteristics of these distributions (right).

We also study the distribution of patterns with respect to their number of attributes. As the maximum number of attributes is different for the four modelings, we normalize the observed number of attributes by dividing the observed number of attributes by the maximal possible number of attributes in a pattern – 15 for the pairwise based approach and 6 for the others (opposite attributes cannot appear in a same pattern). Results are given in Fig. 4. The patterns discovered with t-test based approach have, in general, a lower number of attributes than the patterns obtained with other approaches. When normalized, the distribution of the patterns discovered with the pairwise inequality based approach is greater than the others.

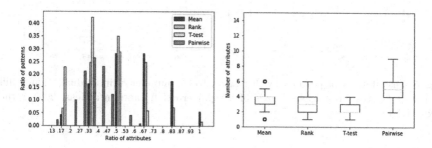

Fig. 4. Distribution of patterns with respect to their number of attributes for each approach (left). Statistical characteristics of these distributions (right).

Figure 5 reports the distributions of patterns according to the number of the individuals that participate to the patterns. In reality, individuals do not directly participate to a pattern, but it is the hemodynamic response measured by their fMRI on the voxels of the pattern (S, K) that indirectly associates an individual to it. For the t-test modeling, an individual i satisfies a voxel $v \in S$ if $\forall j \in K$, $T_{0.05} < M'(v, i, j)$ (for positive attributes) or $T_{0.05} > -M'(v, i, j)$ (for negative attributes). For the pairwise modeling, an individual i satisfies a voxel $v \in S$ if $\forall (o_1, o_2) \in K$, $M'(v, i, o_1) > M'(v, i, o_2)$. For the mean and the rank based approaches, an individual i is considered to satisfies a voxel $v \in S$ if $\forall j \in K$, $M'(v, i, j)$ is higher (for positive attributes) or lower (for negative attributes) than the mean (resp. the mean rank) over all the vertices of the graph. As none of the individuals satisfy all the voxels of a pattern, we consider two cases: one where individuals participate to pattern when they satisfy at least one of its voxels, and another one where they have to satisfy at least 20% of the pattern voxels. Doing this, we observe that the number of individuals that participate to the patterns is much lower for the t-test based modeling than for the other approaches.

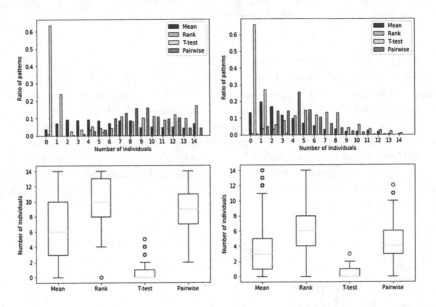

Fig. 5. Distribution of patterns with respect to the number of individuals that fulfill all the attributes on at least 1 voxel of the pattern (left) and on at least 20% of the voxels (right).

The study of the previous distributions leads to some partial conclusions. Patterns discovered by the t-test based approach generally contain less vertices and attributes and are supported by fewer individuals than the other approaches. On the contrary, patterns discovered thanks to modelings that take into account the ranks of the hemodynamic responses (i.e., the rank and the pairwise inequality modelings) involve more vertices, attributes and are supported by more individuals. To discover some inter-individual invariant, t-test based method seems to be less suited than methods that takes into account ranks. This is what it is further investigated in the following qualitative study.

3.2 Qualitative Comparison of the Top-3 Patterns

The top 3 patterns according to the WRAcc measure discovered by the 4 approaches are reported in Table 1. Notice that these patterns are obtained after a post-processing that ensures diversified results by constraining the overlap between the top 3 patterns to be lower than 30% [4]. The notation "$o_1 < o_2$" is used to say that the hemodymamic response of the odor o_1 is lower that the hemodymamic response of the odor o_2. The numbers of individuals reported in Table 1 are the numbers of individuals who participate to at least one voxel of the patterns. Figure 7 shows, for each pattern, the distribution of the percentage of its voxels that are satisfied by the individuals. To evaluate how these patterns overlap each others, we compute the Jaccard similarity of their set of vertices. Figure 6 reports these values when they are greater than 0.30. We can observe that: (a) The best 3 patterns respectively found for the mean, rank and pairwise approaches match each others. They have a strong Jaccard index (between 0.42 and 0.67). Both the mean and the rank based approaches give us exactly the same pieces of information in term attributes. The pairwise based approach is in agreement with these patterns but it provides additional insights. For the first pattern, for instance, three other odorants (3Hex, HEP and MAN) have also a hemodynamic response greater than the ACE's one. (b) The best pattern discovered by the t-test based approach does not match any other patterns (Jaccard index lower than 0.3). Furthermore, only 4 individuals support at least one vertice and only 2 support at least 10% of the vertices. The second one has a small overlap with the third patterns of both mean and rank approaches (0.35) but concerns other odorants. The third t-test based pattern overlaps with the first patterns of the other approaches. Even though it is not in contradiction with the pairwise approach, it concerns different odors compared to the other two methods. (c) The second pattern obtained thanks to the pairwise based modeling overlaps with the third pattern of both mean and rank approaches. It provides additional information compared to these patterns as $DEC < ACE$ and $3Hex < ACE$ relationships are also present in the pattern.

To conclude, the pairwise based approach gives more pieces of information than the other ones and patterns are better supported by individuals.

3.3 Patterns Based on Other Odor Characteristics

The discovery of exceptional attributed subgraphs in which the attributes are the odorants leads to the identification of areas of interest for the neuroscientists. However, the odorant properties are not taken into account in the analysis and thus their interpretation requires much effort. Neuroscientists aim to find links between brain areas and some odorant attributes, especially their hedonic perception during the fMRI measurement. During the experiment, the subjects must express a hedonic judgment regarding the breathed smell and say

Table 1. Top 3 patterns for each modeling (see supplementary material for their visualization).

		Mean	Rank	t-test	Pairwise
1	Positive attributes	DEC, EUG	DEC, EUG	HEP	$ACE < 3Hex, DEC, EUG, HEP, MAN$
	Negative attributes	ACE	ACE	-	-
	Number of voxels	138	125	104	117
	Number of individuals	12	12	4	12
2	positive attributes	ACE	ACE	$3Hex$	$3Hex, DEC, HEP, EUG < ACE$
	Negative attributes	EUG, HEP	EUG, HEP	-	-
	Number of voxels	134	122	154	151
	Number of individuals	7	10	4	12
3	Positive attributes	ACE	ACE	MAN	$DEC < 3Hex, MAN, ACE$
					$3Hex < ACE$
	Negative attributes	DEC	DEC	-	-
	Number of voxels	178	173	123	119
	Number of individuals	12	14	5	14

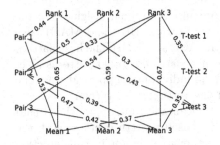

Fig. 6. Graph of the Jaccard similarities between top 3 patterns.

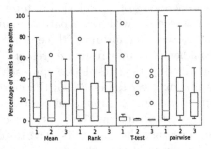

Fig. 7. Distribution of the percentage of voxels that are satisfied by each individual for the top 3 patterns.

whether it is pleasant, unpleasant or neutral. In Fig. 8, the distribution of pleasant/neutral/unpleasant odorants in the patterns discovered by the four methods is reported. There is no pattern capturing only odorants that are all perceived as pleasant (or unpleasant) by a large proportion of individuals. We then consider hedonicity as an attribute and perfom new extractions with CENERGETICS considering the different modelings.

We enforce syntactic constraints to focus on patterns that are of interest for the neuroscientists. For the mean, rank and t-test based approaches, we search patterns verifying one of these conditions: (a) the hemodynamic response of odorant perceived as neutral is higher than those perceived as pleasant and unpleasant; (b) the hemodynamic response of odorant perceived as neutral is lower than those perceived as pleasant and unpleasant; (c) the hemodynamic response

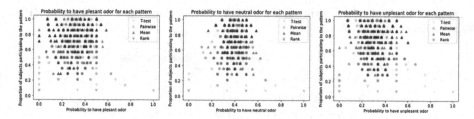

Fig. 8. Percentage of pleasantness of the dominant odors for each number of individual participating to the pattern.

of odorant perceived as pleasant is higher than those perceived as unpleasant; and (d) the hemodynamic response of odorant perceived as unpleasant is higher than those perceived as pleasant. For the pairwise approach, we look for patterns whose attributes describe order between pleasant and unpleasant[3].

CENERGETICS takes less than 32 ms to extract the patterns for each approach. The top 5 patterns w.r.t. WRAcc measure for each method are reported in Table 2 and their brain visualization is given in Fig. 9. The patterns discovered by the t-test based approach are too small to be analyzed (only 1 to 5 voxels) and visualized. Patterns discovered by the different methods overlap. Those that have a Jaccard similarity greater than 0.3 (see supplementary material for more details) capture similar information (e.g. syntactic constraints). Some of these patterns highlight some areas in which polarized hedonic values have a different distribution than neutral hedonic value. This confirms neuroscientists' priors.

Indeed, the fact that the most emotional odors (pleasant and unpleasant) (blue, red and green patterns in Fig. 9) are more represented in the posterior part of the piriform cortex whereas responses to neutral odors (cyan patterns in Fig. 9: $R4$ and $P3$) are more localized anteriorly within the piriform cortex is consistent with previous findings in Neuroscience [13, 16] showing that the posterior part of the piriform cortex represent salient perceptual experience of smells. Note that this posterior area of the piriform cortex is at the neighborhood of another area known to be involved in emotional processing, namely the amygdala.

The mean, rank and pairwise based modelings find similar information, which improves the neuroscientists' confidence in these findings. Furthermore, the pairwise based modeling conveys more information to the neuroscientists than the two others. This approach is promising and could be used on other odor attributes to potentially formulating new hypotheses on the olfactory percept in neuroscience.

4   Related Work

Scientists have always seen Exploratory Data Analysis (EDA) as an important research area since its introduction [27]. Among the various EDA techniques

---

[3] I.e., $neutral > pleasant, unpleasant$, or $neutral < pleasant, unpleasant$, or $pleasant > neutral > unpleasant$, or $unpleasant > neutral > pleasant$.

Table 2. Top 5 patterns with the hedonic attributes : unpleasant (U), pleasant (P) and neutral (N).

		Mean	Rank	t-test	Pairwise
1	Positive attributes	U	U	P	$N < P, U$
	Negative attributes	P	P	U	-
	Number of voxels	134	143	5	43
2	Positive attributes	P	P	N	$P < N < U$
	Negative attributes	U	U	P, U	-
	Number of voxels	137	150	4	44
3	Positive attributes	P, U	U	N	$P, U < N$
	Negative attributes	N	P	P, U	-
	Number of voxels	58	66	1	61
4	Positive attributes	P	N	P	$U < N < P$
	Negative attributes	U	P, U	U	-
	Number of voxels	64	28	1	41
5	Positive attributes	P, U	P, U	P	$N < P, U$
	Negative attributes	N	N	U	-
	Number of voxels	25	24	1	62

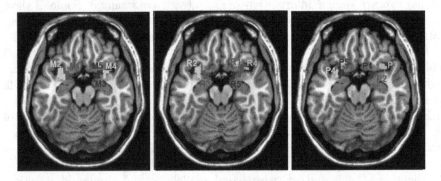

Fig. 9. Upper view of the top 5 patterns satisfying the syntactic constraints for each approach. Patterns of mean based approach (left), patterns of rank based approach (center) and patterns of pairwise inequality based approach (right). The color red represents the unpleasant pattern, green the pleasant pattern, cyan the pattern where the neutral odors are higher than the pleasant and the unpleasant odors and the blue patterns where the neutral odors are lower than the pleasant and the unpleasant odors. (Color figure online)

that aim to maximize insight into datasets and uncover underlying structures, Subgroup Discovery (SD) [18,28] is a generic data mining task concerned with finding regions in the data that stand out with respect to a given target. Many other data mining tasks have similar goals as SD, e.g., emerging patterns [7], contrast sets [3]. However, among these different tasks, SD is known as the most generic one, especially SD is agnostic of the data and the pattern domain. For instance, subgroups can be defined with conjunction of conditions on symbolic [19] or numeric attributes [1, 15] as well as sequences [14]. Furthermore, the single target can be discrete or numeric [22]. Exceptional Model Mining (EMM) [21], while sharing the same exploration space (i.e., the description space), extends SD by offering the possibility to handle complex targets, e.g., several discrete attributes[9, 20], two numeric targets [8] and preferences [24].

Our method is rooted in the SD/EMM framework. Nevertheless, the problem we tackle cannot be directly addressed with an instance of SD/EMM. Indeed a target space is provided instead of explicit targets. Dynamic EMM/SD (i.e., EMM/SD with a non-fixed model) has been recently investigated for different aims. Bosc et al. [5] propose a method to handle multi-label data where the number of labels per objects is much lower than the total number of labels which prevent the use of usual EMM model. Other dynamic EMM approaches aim to discover exceptional attributed sub-graphs [4, 17]. Notice that in these works, exceptional subgraphs must also fulfill a structural constraint (e.g., connectivity). In this paper, we demonstrate that exceptional attributed sub-graph discovery is promising to provide new insights in neuroscience.

5   Conclusion

In this paper, we introduced a new way to analyze fMRI data in order to better understand olfactory perception at the cerebral level. To this end, we modelled fMRI data as an attributed graph whose vertices depict voxels and attributes encode the hemodynamic response related to a series of odorant molecules. We defined four different ways to analyze hemodynamic activity within the studied voxels. Exceptional attributed graphs were then discovered using CENERGET-ICS algorithm. A thorough empirical study compared the different models. The strength of such an approach lies in its ability to be more robust to spatial imprecision and inter-individual variability than classical fMRI analysis approaches: here, the pairwise inequality attribute based modeling seems to be the most promising approach. It makes possible the discovery of areas of interest supported by many individuals while conveying much more semantics than other models. This paper is the first attempt to apply attributed graph analysis to understand the olfactory perception and their neural underpinnings. It opens up several avenues for further research including definition of new pattern domains to fully take into account the brain specificity as well as prior knowledge.

References

1. Atzmüller, M., Puppe, F.: SD-map - a fast algorithm for exhaustive subgroup discovery. In: ECMLPKDD, pp. 6–17 (2006)
2. Baba, T., et al.: Severe olfactory dysfunction is a prodromal symptom of dementia associated with Parkinson's disease: a 3 year longitudinal study. Brain **135**(1), 161–169 (2012)

3. Bay, S.D., Pazzani, M.J.: Detecting group differences: mining contrast sets. Data Min. Knowl. Discov. **5**(3), 213–246 (2001)
4. Bendimerad, A.A., Plantevit, M., Robardet, C.: Mining exceptional closed patterns in attributed graphs. Knowl. Inf. Syst. **56**(1), 1–25 (2018)
5. Bosc, G., et al.: Local subgroup discovery for eliciting and understanding new structure-odor relationships. In: Calders, T., Ceci, M., Malerba, D. (eds.) DS 2016. LNCS (LNAI), vol. 9956, pp. 19–34. Springer, Cham (2016). https://doi.org/10.1007/978-3-319-46307-0_2
6. Croy, I., Bojanowski, V., Hummel, T.: Men without a sense of smell exhibit a strongly reduced number of sexual relationships, women exhibit reduced partnership security-a reanalysis of previously published data. Biol. Psychol. **92**(2), 292–294 (2013)
7. Dong, G., Li, J.: Efficient mining of emerging patterns: discovering trends and differences. In: KDD, pp. 43–52. ACM (1999)
8. Downar, L., Duivesteijn, W.: Exceptionally monotone models—the rank correlation model class for exceptional model mining. Knowl. Inf. Syst. **51**(2), 369–394 (2017). May
9. Duivesteijn, W., Feelders, A.J., Knobbe, A.: Exceptional model mining. Data Min. Knowl. Discov. **30**(1), 47–98 (2016)
10. Fournel, A., Ferdenzi, C., Sezille, C., Rouby, C., Bensafi, M.: Multidimensional representation of odors in the human olfactory cortex. Hum. Brain Mapp. **37**, 2161–2172 (2016)
11. Friston, K., Ashburner, J., Frith, C.D., Poline, J.B., Heather, J.D., Frackowiak, R.S., et al.: Spatial registration and normalization of images. Hum. Brain Mapp. **3**(3), 165–189 (1995)
12. Friston, K.J., Holmes, A.P., Worsley, K.J., Poline, J.P., Frith, C.D., Frackowiak, R.S.: Statistical parametric maps in functional imaging: a general linear approach. Hum. Brain Mapp. **2**(4), 189–210 (1994)
13. Gottfried, J.A., Winston, J.S., Dolan, R.J.: Dissociable codes of odor quality and odorant structure in human piriform cortex. Neuron **49**(3), 467–479 (2006)
14. Grosskreutz, H., Lang, B., Trabold, D.: A relevance criterion for sequential patterns. In: ECMLPKDD, pp. 369–384 (2013)
15. Grosskreutz, H., Rüping, S.: On subgroup discovery in numerical domains. Data Min. Knowl. Discov. **19**(2), 210–226 (2009)
16. Howard, J.D., Plailly, J., Grueschow, M., Haynes, J.D., Gottfried, J.A.: Odor quality coding and categorization in human posterior piriform cortex. Nat. Neurosci. **12**(7), 932 (2009)
17. Kaytoue, M., Plantevit, M., Zimmermann, A., Bendimerad, A., Robardet, C.: Exceptional contextual subgraph mining. Mach. Learn. (2017)
18. Klösgen, W.: Explora: a multipattern and multistrategy discovery assistant. In: Advances in Knowledge Discovery and Data Mining, pp. 249–271. AAAI (1996)
19. Lavrač, N., Kavšek, B., Flach, P., Todorovski, L.: Subgroup discovery with CN2-SD. J. Mach. Learn. Res. **5**(Feb), 153–188 (2004)
20. van Leeuwen, M., Knobbe, A.J.: Diverse subgroup set discovery. Data Min. Knowl. Discov. **25**(2), 208–242 (2012)
21. Leman, D., Feelders, A., Knobbe, A.: Exceptional model mining. In: Daelemans, W., Goethals, B., Morik, K. (eds.) ECML PKDD 2008. LNCS (LNAI), vol. 5212, pp. 1–16. Springer, Heidelberg (2008). https://doi.org/10.1007/978-3-540-87481-2_1
22. Lemmerich, F., Atzmueller, M., Puppe, F.: Fast exhaustive subgroup discovery with numerical target concepts. Data Min. Knowl. Discov. **30**(3), 711–762 (2016)

23. Muthukumaraswamy, S.D., Edden, R.A., Jones, D.K., Swettenham, J.B., Singh, K.D.: Resting gaba concentration predicts peak gamma frequency and fMRI amplitude in response to visual stimulation in humans. Proc. Natl. Acad. Sci. **106**(20), 8356–8361 (2009)

24. Rebelo de Sá, C., Duivesteijn, W., Soares, C., Knobbe, A.: Exceptional preferences mining. In: Calders, T., Ceci, M., Malerba, D. (eds.) DS 2016. LNCS (LNAI), vol. 9956, pp. 3–18. Springer, Cham (2016). https://doi.org/10.1007/978-3-319-46307-0_1

25. Schofield, P.W., Ebrahimi, H., Jones, A.L., Bateman, G.A., Murray, S.R.: An olfactory 'stress test'may detect preclinical alzheimer's disease. BMC Neurol. **12**(1), 24 (2012)

26. Stevenson, R.J.: An initial evaluation of the functions of human olfaction. Chem. Senses **35**(1), 3–20 (2009)

27. Tukey, J.W.: Exploratory Data Analysis, vol. 2. Reading, Mass (1977)

28. Wrobel, S.: An algorithm for multi-relational discovery of subgroups. In: Komorowski, J., Zytkow, J. (eds.) PKDD 1997. LNCS, vol. 1263, pp. 78–87. Springer, Heidelberg (1997). https://doi.org/10.1007/3-540-63223-9_108

Extending Redescription Mining
to Multiple Views

Matej Mihelčić[1]([⊠]), Sašo Džeroski[2], and Tomislav Šmuc[1]

[1] Ruđer Bošković Institute, Bijenička cesta 54, 10000 Zagreb, Croatia
{matej.mihelcic,tomislav.smuc}@irb.hr
[2] Jožef Stefan Institute, Jamova cesta 39, 1000 Ljubljana, Slovenia
Saso.Dzeroski@ijs.si

Abstract. Redescription mining is a data mining task that discovers re-descriptions of different subsets of entities from available data. Locating such re-descriptions is important in many scientific disciplines because it allows detecting different types of associations including synergy of different attributes of interest. There exist a number of redescription mining algorithms, however they are all restricted to use of one or maximally two disjoint sets of attributes (views) to re-describe different subsets of entities. The main reasons for this limitation are computational complexity and potentially large increase in number of produced patterns, in multi-view setting, during redescription mining. In this work we present an algorithm that allows mining redescriptions from multiple views using the CLUS-RM algorithm. Presented algorithm efficiently solves aforementioned problems. Its computational complexity, with respect to attribute operations, increases linearly with the increase of number of views and we present techniques to handle large number of produced redescriptions during redescription mining step.

1 Introduction

Redescription mining [18] discovers multiple descriptions (redescriptions) of different subsets of entities from the data. It is a descriptive, unsupervised task whose main result-redescriptions are presented as tuples of logical formulas (containing conjunction, negation and disjunction operators). Analyses using redescription mining increase understanding of underlying data and allow detecting interesting associations between different subsets of attributes. Many interesting problems (e.g in biology, medicine, pharmacy, economy) involve multiple disjoint sets of attributes. Each set provides information about different aspect of observed entities or provides information derived from a different source.

Redescription mining [18] is related to clustering [3,6,12,22,24], since it discovers groups of entities sharing different properties. It is also related to multi-view clustering [2], multi-layer clustering [10], since it aims to use different sets of disjoint attributes to redescribe entities and to conceptual clustering [5,14] since it aims to describe discovered clusters using a predefined query language. The main advantage of this methodology is its ability to find bi-directional

L. Soldatova et al. (Eds.): DS 2018, LNAI 11198, pp. 292–307, 2018.
https://doi.org/10.1007/978-3-030-01771-2_19

(equivalence-like) relations between attributes which is a stricter version of association than found by association rule mining [1, 11, 26].

Due to prohibitive increase in computational complexity and a large number of patterns produced during mining, current approaches for redescription mining [7, 15, 17, 18, 25, 27] work with maximally two disjoint sets of attributes (views). However, there are applications where more than two views are available and where studying redescriptions created using all views provides insights not available using other available techniques.

In this work, we describe an algorithm for multi-view redescription mining that utilizes the CLUS-RM algorithm [15] and the generalized redescription set construction procedure [16] to find redescriptions from arbitrary many views. This approach finds multi-view redescriptions and its computational complexity grows linearly with the increase of number of views. This is done by finding redescriptions on all pairs of available views and extending these redescriptions (adding missing formulas to their tuple) by using them as targets to guide the search on other views. Thus, two main ideas make multi-view redescription mining feasible in practice: (a) finding redescriptions on pairs of views and (b) utilizing CLUS-RMs multi-target regression (classification) capabilities to use incomplete redescriptions as targets to obtain redescriptions containing information from all available views.

Required notation, generalized definitions and relevant related work are presented in Sect. 2. The proposed algorithm for multi-view redescription mining is presented in Sect. 3. Section 4 presents computational complexity analysis of the proposed algorithm whereas Sect. 5 describes used data. Section 6 presents experiments, parameter setup and obtained results. Finally, Sect. 7 presents conclusions and future work directions.

2   Notation and Related Work

In this section we present notation and definitions used throughout this manuscript and provide relevant related work in redescription mining.

2.1   Notation and Definitions

Dataset in multi-view redescription mining consists of a set of entities E, a set of views $\{W_1, W_2, \ldots, W_n\}$, $n \in \mathbb{N}$, and a corresponding set of variables (attributes) $\{V_1, V_2, \ldots, V_n\}$. Redescription is a tuple of queries (logical formulas) $R = (q_1, q_2, \ldots, q_n)$, where query q_i uses variables from set V_i to describe entities. Queries contain conjunction, negation and disjunction logical operators.

Redescription $R_{ex} = (q_{1_{ex}}, q_{2_{ex}}, q_{3_{ex}}, q_{4_{ex}})$ was obtained on our use case dataset describing world countries by using trading view (view 1), population view (view 2), energy view (view 3) and country development and wealth view (view 4). It is defined as:

$q_{1_{ex}} :\ 1.05 \leq \mathrm{E}/\mathrm{I}_{90} \leq 1.88 \ \wedge\ 0.96 \leq \mathrm{E}/\mathrm{I}_{14} \leq 1.63$
$q_{2_{ex}} :\ -0.2 \leq \mathrm{POP\_GROWTH} \leq 0.5 \ \wedge\ 55.83 \leq \mathrm{InternetUsers} \leq$

$$82.35 \wedge 4.53 \leq \text{AdFertRate} \leq 9.52$$
$q_{3_{ex}}$: $142.0 \leq \text{ConvCrudeOilProd} \leq 5397.0 \wedge 11257.0 \leq \text{ElectricityTotNet-}$
CapPPH ≤ 48934.0
$q_{4_{ex}}$: $29850.0 \leq \text{GNIPerCapita} \leq 49830.0 \wedge 1330.42 \leq \text{RenewableInternal-}$
Freshwater ≤ 6524.32

R_{ex} contains four queries (one per view). Query structure is equivalent as in two-view case (see [15]). It re-describes 6 countries: Spain, Japan, Italy, Germany, France and Austria.

In the continuation, we generalize all redescription evaluation measures defined in [15]. The support set of a query q_i ($supp(q_i)$) is a set of all entities satisfying its condition. A redescription $R = (q_1, q_2, \ldots, q_n)$ describes some entity if this entity is contained in a support set of all redescription queries. Redescription support set is a set of entities described by this redescription, $supp(R) = \cap_{i=1}^{n} supp(q_i)$.

Redescription accuracy is measured by Jaccard index, which is defined in multi-view case as: $J(R) = \frac{|\cap_{i=1}^{n} supp(q_i)|}{|\cup_{i=1}^{n} supp(q_i)|}$.

The p-value (p_{val}), computed from the binomial distribution is defined as: $p_{val}(R) = \sum_{k=|supp(R)|}^{|E|} \binom{|E|}{k} (\prod_{i=1}^{n} p_i)^k \cdot (1 - \prod_{i=1}^{n} p_i)^{|E|-k}$. $|E|$ equals the number of entities in the dataset and $p_1, p_2, \ldots, p_n$ correspond to marginal probabilities of obtaining the query $q_1, q_2, \ldots, q_n$. This p-value represents a probability of obtaining a set of size equal to or larger than $supp(R)$ by combining n randomly selected queries with marginal probabilities $p_1, p_2 \ldots, p_n$.

$attr(R)$ denotes a set of attributes used in redescription queries and the attribute Jaccard index of two redescriptions is: $attJ(R_1, R_2) = \frac{|attr(R_1) \cap attr(R_2)|}{|attr(R_1) \cup attr(R_2)|}$. The average redescription attribute Jaccard index R_i is defined as: $AvgAJ(R_i) = \frac{2 \cdot \sum_{j \neq i} attJ(R_i, R_j)}{n \cdot (n-1)}$. The entity Jaccard index of two redescriptions is defined as $elemJ(R_1, R_2) = \frac{|supp(R_1) \cap supp(R_2)|}{|supp(R_1) \cup supp(R_2)|}$ and the average entity Jaccard index of a redescription is: $AvgEJ(R_i) = \frac{2 \cdot \sum_{j \neq i} elemJ(R_i, R_j)}{n \cdot (n-1)}$. These measures provide information about the redundancy of a redescription with respect to entities and attributes.

2.2 Related Work

Ramakrishnan et al. [18] introduced the field of redescription mining and presented a first redescription mining algorithm, based on alternating CART trees. The developed algorithm, called CARTwheels, used two views containing Boolean attributes to create redescriptions. Similarly Greedy and MID (based on frequent closed itemset mining) approach by Gallo et al. [9] use two views containing Boolean attributes.

Approaches by Zaki [25] (based on lattice of closed itemsets) and Parida [17] (based on relaxation lattice) use one view containing Boolean attributes to create redescriptions. Greedy approach by Galbrun and Miettinen [7], Split trees and Layered trees algorithms developed by Zinchenko [27] and the CLUS-RM algorithm (based on Predictive Clustering trees) developed by Mihelčić et al. [15]

use two views containing Boolean, categorical or numerical attributes to create redescriptions. Siren [8] is a fully interactive redescription mining environment which allows mining redescriptions and contains several visualizations of individual redescriptions. It works on two views containing Boolean, categorical or numerical attributes. As can be seen, all current redescription mining approaches have a serious limitation of being able to work with maximally two views.

3   Algorithm for Multi-view Redescription Mining

In this section we present the algorithm for multi-view redescription mining based on CLUS-RM algorithm [15] that uses generalized redescription set construction procedure [16] to create a redescription set that is returned to the user. To shorten the pseudocode of the proposed algorithm, we use notation CLUS_RM(W_i,W_j,C) to denote the execution of two-view CLUS-RM algorithm as presented in [15] with given view input parameters W_i and W_j and the redescription constraint parameters C (which include minimal Jaccard index, maximal p-value, minimal and maximal redescription support size). The algorithm returns a set of redescriptions R. GRC(R,W,n) denotes the execution of generalized redescription set construction procedure [16] which takes a redescription set R, a user-defined importance weight matrix W and an integer n denoting the required size of output redescription set. The procedure returns one or more optimized redescription sets of size equal to or smaller than n.

The proposed algorithm for multi-view redescription mining (see Algorithm 1) is run for a predefined number of random initializations of input dataset (lines 3–4 in Algorithm 1). The main idea is to use the CLUS-RM algorithm to produce incomplete redescriptions on each pair of available views (lines 5–10 in Algorithm 1) . These redescriptions are incomplete because they do not contain queries from all available views. Produced, incomplete redescriptions are used as targets to produce matching rules on the rest of the views using multi-target regression (classification) capabilities of Predictive Clustering Trees [13] (lines 11–16 in Algorithm 1). Newly produced rules are used to complete potentially incomplete redescriptions (line 17 in Algorithm 1). To avoid blow up in number of produced (especially incomplete) redescriptions, we add two additional parameters to the algorithm, MaxExpansionSize and WorkSetSize. The MaxExpansionSize parameter denotes the absolutely maximum allowed redescription set size between algorithm iterations whereas WorkSetSize denotes the maximum allowed number of candidates to be used for redescription set optimization. These parameters are used to reduce the size of redescription set (lines 18–21 in Algorithm 1). If redescription set size remains to large after reduction of incomplete redescriptions, the size is reduced using generalized redescription set construction procedure (lines 22–24 in Algorithm 1). All redescriptions from a set of redescription sets S are retained and the rest is discarded. Redescription query size is reduced using query size minimization procedure introduced in [15] (line 25 in Algorithm 1). Finally, generalized redescription set construction procedure is used to create a set of reduced redescription sets (line 26 in Algorithm 1) which is returned to the user.

The function `constructTargets` from line 14 in Algorithm 1 works similarly as the target construction procedure defined in [15]. Each incomplete redescription in a set $\mathcal{R}_{all}$ constitutes one target variable in a newly constructed multi-target regression (classification) task. Every entity re-described by some redescription $R_k \in \mathcal{R}_{all}$ has a value 1.0 for the k-th target variable. If an entity is not re-describe by redescription R_k it has a value 0.0 for this variable. Predictive clustering trees model is trained on a dataset containing attributes of a k-th view and aforementioned target variables to construct rules which are used to complete redescriptions.

Algorithm 1. An algorithm for multi-view redescription mining

---

Input: Available views $MW = \{W_1, \ldots, W_n\}$, Constraints $\mathcal{C}$, Settings set
Output: A set of reduced redescription sets $\mathcal{R}$

```
 1: procedure MW-RM
 2:     R_all ← ∅
 3:     for (nrand = 0; nrand<set.NRandomRestarts; nrand++) do
 4:         MW' = {W'_1,...,W'_n} ←initializeViews()
 5:         for (i=0; i<|MW| − 1; i++) do
 6:             for (j=i+1; j<|MW|; j++) do
 7:                 RunInd← 0
 8:                 while (RunInd<set.maxIter) do
 9:                     R_{i,j} ← CLUS_RM(W'_i,W'_j,C)
10:                     R_all ← R_all ∪ R_{i,j}
11:                     for (k=0; k<|MW|; k++) do
12:                         if (k==i || k==j) then
13:                             continue
14:                         DW_k ←constructTargets(W_k,R_all)
15:                         P_k ←PCT(DW_k)
16:                         extractRulesFromPCT(P_k,r_k)
17:                         R_all ←completeRedescriptions(R_all,r_k,set.Op,k)
18:                         if (|R_all| > set.MaxExpansionSize) then
19:                             removeIncomplete(R_all)
20:                     if (|R_all| > set.WorkSetSize) then
21:                         removeIncomplete(R_all)
22:                     if (|R_all| > set.WorkSetSize) then
23:                         S ← GRC(R_all,W,n)
24:                         R_all ← ∪_{R∈S}R
25:         minimizeQueries(R_all)
26:         S ←GRC(R_all,W,n)
27:     return S
```

The pseudocode of a procedure `completeRedescriptions` is presented in Algorithm 2. The procedure iterates over all incomplete redescriptions (line 2 in Algorithm 2) and attempts to complete them with some rule from rule set r (line 3 in Algorithm 2). If adding new query to some existing incomplete redescription R satisfies accuracy restraints, a new redescription is created (lines 4–5 in

Algorithm 2). If the refinement procedure is used (thoroughly explained in [16]) redescription is added to the set of redescriptions only if there is no redescription with equal support and maximal accuracy (lines 8–9 in Algorithm 2) (since in this case it makes no sense to refine the newly created redescription because key attribute from maximally accurate redescription would be transferred to the new redescription).

Algorithm 2. Complete redescriptions

Input: Redescription set $\mathcal{R}$, Rule set r, Operator set op, Settings set, View W_i
Output: A set of redescriptions $\mathcal{R}$

1: **procedure** COMPLETEREDESCRIPTIONS
2:    **for** $(R \in \mathcal{R}, R.q_i = \emptyset)$ **do**
3:       **for** $(r_j \in r)$ **do**
4:          **if** $(\text{JS}(supp(R) \cap supp(r_j), \cup_{q_i \in R} supp(q_i) \cup supp(r_j)) \geq set.minJS)$ **then**
5:             $R_{new} \leftarrow R.insertQuery(r_j, W_i)$
6:             **if** $(set.UseRefinement)$ **then**
7:                **for** $(R \in \mathcal{R})$ **do**
8:                   **if** $(supp(R) = supp(R_{new}) \wedge JS(R) = 1.0)$ **then**
9:                      break
10:                   **if** $(supp(R_{new}) \subseteq supp(R))$ **then**
11:                      $R_{new} \leftarrow (R_{new}.q_1 \wedge R.q_1, \ldots, R_{new}.q_n \wedge R.q_n)$
12:                   **if** $(supp(R_{new}) \in [set.minSupp, set.maxSupp] \wedge$
                         $JS(R_{new}) \geq set.minJS)$ **then**
13:                      $\mathcal{R} \leftarrow \mathcal{R} \cup R_{new}$
14:             **else**
15:                **if** $(supp(R_{new}) \in [set.minSupp, set.maxSupp] \wedge$
                      $JS(R_{new}) \geq set.minJS)$ **then**
16:                   $\mathcal{R} \leftarrow \mathcal{R} \cup R_{new}$
17:          **if** $(set.useNeg \wedge JS(supp(R) \cap supp(\neg r_j), \cup_{q_i \in R} supp(q_i) \cup supp(\neg r_j)) \geq$
               $set.minJS)$ **then**
18:             $R_{new} \leftarrow R.insertQuery(\neg r_j, W_i)$
19:    **if** $set.useDisj$ **then**
20:       **for** $(R \in \mathcal{R})$ **do**
21:          $indMax \leftarrow -1, maxJS \leftarrow 0, indMax1 \leftarrow -1, maxJS1 \leftarrow 0$
22:          **for** $(r_j \in r)$ **do**
23:             $disjJS \leftarrow (\text{JS}(\cap_{q_k \in R, k \neq i} supp(q_k) \setminus supp(R), supp(r_j))$
24:             $disjJS1 \leftarrow (\text{JS}(\cap_{q_k \in R, k \neq i} supp(q_k) \setminus supp(R), supp(\neg r_j))$
25:             **if** $(disjJS > maxJS)$ **then**
26:                maxJS←disjJS, indMax← j
27:             **if** $(disjJS1 > maxJS1)$ **then**
28:                maxJS1←disjJS1, indMax1← j
29:          **if** $(maxJS > 0)$ **then**
30:             $R.q_{W_i} \leftarrow R.q_{W_i} \vee r_{indMax}$
31:          **if** $(maxJS1 > 0)$ **then**
32:             $R.q_{W_i} \leftarrow R.q_{W_i} \vee \neg r_{indMax1}$
33:    **return** $\mathcal{R}$

Otherwise, the newly created redescription is refined with redescriptions whose support set is equal or superset of its support set (lines 10–11 in Algorithm 2). If newly created redescription has required accuracy and support set size characteristics it is added to the set of all redescriptions (lines 12–13 in Algorithm 2). Since it is possible to extend the same redescription with multiple different queries (potentially resulting in redescriptions of different support), this step and equivalent step when negated query is used (lines 17–18 in Algorithm 2) can cause significant growth of number of produced (potentially incomplete) redescriptions. Using disjunction operator is described in lines 19–32 of Algorithm 2. The main idea is to compute how much entities (described by all queries except that corresponding to the selected view) enter redescription support set if a selected rule is combined with the existing query of a redescription (for a selected view) using disjunction operator. Finally, extended set of redescriptions is returned in line 33 of Algorithm 2.

The high level overview of the algorithm for multi-view redescription mining can be seen in Fig. 1.

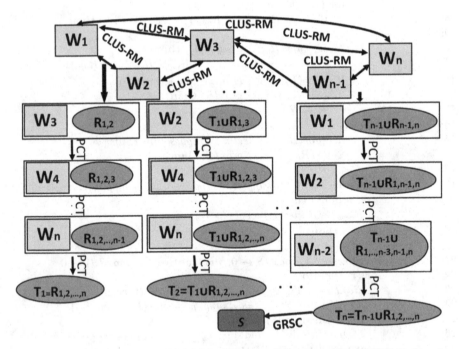

Fig. 1. The algorithm uses CLUS-RM algorithm [15] to create two-view redescriptions on all pairs of views. Views are combined as denoted by numbers (W_1, W_2) first, (W_1, W_3) second, (W_{n-1}, W_n) last. Produced redescriptions are used as targets used to construct PCT model [13] and corresponding rules on other views. The final redescription set T_n is used to create a set of redescription sets S using generalized redescription set construction procedure (GRSC) [16].

4   Computational Complexity

In this section we analyse the computational complexity of the proposed algorithm for multi-view redescription mining.

The computational complexity of a Predictive Clustering tree model is $O(z \cdot m \cdot |E|^2)$ (see [19]), where z denotes the number of nodes in the tree, m the number of attributes and $|E|$ the number of entities contained in the data. As previously demonstrated, the average time complexity of a CLUS-RM algorithm is $O(z \cdot (|V_1| + |V_2|) \cdot |E|^2 + z^2 \cdot |E|)$, where $|V_i|$ denotes the number of attributes in the i-th view. The worst time complexity (given inadequate hashing function) equals $O(z \cdot |E|^2 \cdot (|V_1| + |V_2| + z))$ (for more detailed description see [15]). We can see that the time complexity of CLUS-RM dominates the complexity of PCT. It can also be seen that in between CLUS-RM executions, maximal size of redescription set is constrained and can be considered a constant. Because of this, computational complexity of generalized redescription set construction procedure equals $O(|E|)$. Thus, the average time complexity of the algorithm for multi-view redescription mining is $O(\sum_{i=1}^{n-1} \sum_{j=i+1}^{n}(z \cdot (|V_i| + |V_j|) \cdot |E|^2 + z^2 \cdot |E|))$ which equals $O((n-1) \cdot z \cdot (\sum_{i=1}^{n} |V_i|) \cdot |E|^2 + \frac{n \cdot (n-1)}{2} \cdot z^2 \cdot |E|)$, where n denotes the number of views. Since $n << min|V_i|, |E|, i \leq n$ and $n^2 << |E|$ in most real applications it can be considered a constant. Thus, the average time complexity of the algorithm is $O(z \cdot (\sum_{i=1}^{n} |V_i|) \cdot |E|^2 + z^2 \cdot |E|)$. Similarly, the worst time complexity of the algorithm equals: $O(z \cdot |E|^2 \cdot (\sum_{i=1}^{n} |V_i| + z))$.

5   Data Description

We use two different use case dataset to evaluate the proposed methodology for multi-view redescription mining: (a) Country dataset and (b) River water quality dataset.

(a) The Country dataset contains 141 entities (world countries) which are described with 4 different views. All data reflect values for countries in year 2012. The first view describes country trade using 309 numerical attributes (percentage of export or import a commodity takes in total country export/import and the ratio export to import of these values). The data was obtained from the UNCTAD database [21]. The second view contains a set of 21 numerical attributes describing population of these countries (percentage of rural population, mortality, fertility, migration, education etc.). Part of these data was obtained from the World bank database [23] and a part of the data was obtained from the UN database [20]. The third view contains a set of 47 numerical attributes describing energy production and consumption of these countries (bitumen, oil, gas, coal etc.). The fourth view contains a set of 33 numerical attributes describing different aspects of country development and wealth (agricultural value, workers remittance, GDP growth, GNI, GNIPPP, inflation, improved water source, CO_2 emission etc.). The data contained in the third and fourth view was obtained from the UN database [20]. All views contain missing values.

(b) The River water quality dataset [4] contains 3 views describing 1061 water samples taken from Slovenian rivers by the Hydrometeorological Institute of Slovenia. The first view contains physical and chemical measurements (16 numerical attributes) such as biological oxygen demand, chlorine concentration, CO_2 concentration etc. The second view contains the frequency of occurrence of 7 different plant species whereas the third view contains the frequency of occurrence of 7 different animal species. Frequencies are coded as: 0-not present, 1-incidental occurrence, 3-frequently occurring and 5-abundantly occurring.

6   Experiments and Results

We evaluate the proposed multi-view redescription mining algorithm on two previously described use case datasets: Country dataset and River water quality dataset.

On the Country dataset we create three different redescription sets. The first set is constructed using trade and population view, the second set is constructed using trade, population and energy view whereas the fourth dataset is created using all available views. The goal of this experiment was to see what additional information can be gained by adding additional views on top of trade and population view (which were already studied in previous work [10,15]).

On the River water quality dataset we also create three different redescription sets. The first set is constructed using view containing physical and chemical measurements of water and occurrence frequencies of selected plant species, the second set is created using view containing physical and chemical measurement of water and occurrence frequencies of selected animal species. The third redescription set is created using all available views. In this experiment we aim to determine if there is any connection between water quality and occurrence of different animal and plant species and if there are some specific water environments that benefit or diminish joined animal and plant abundance. The main parameters used to construct all redescription sets are provided in Table 1.

Table 1. Algorithm parameters used to create redescription set on a Country dataset (C) and a River water quality dataset (W).

$\mathcal{D}$	min J	max p	support	$|q_i|$	Rand. rest.	Iter.	Lang.	Trees	$|\mathcal{R}_{red}|$	$|\mathcal{R}_{max}|$	$|\mathcal{R}_{work}|$
C	0.6	0.01	[5, 100]	8	15	10	all	1	≤200	10000	3000
W	0.3	0.01	[5, 800]	8	10	10	Conj, disj	1	≤200	10000	3000

The Conjunctive refinement procedure [16] with minAddJS parameter was used to increase the number of accurate redescriptions and overall accuracy. minAddJS=0.3 was used for the Country dataset and minAddJS=0.2 for the River water quality dataset (minAddJS≤ minJ used to increase diversity and accuracy).

We analyse the structure of each obtained redescription set and compare the characteristics of obtained redescriptions. We also compute the number of produced incomplete redescriptions and present the running time of the algorithm for each constructed redescription set.

Table 2. Basic statistics for all redescription sets produced on the Country (C) and River water quality (W) datasets.

$\mathcal{R}$	$|\mathcal{R}|$	E. Coverage	A. Coverage	$|\mathcal{R}_{inc}|$	Exec. time (minutes)
$\mathcal{R}_{C_{2w}}$	200	0.96	0.66	0	14.73
$\mathcal{R}_{C_{3w}}$	200	1.0	0.59	13834	97.82
$\mathcal{R}_{C_{4w}}$	72	0.99	0.37	209624	807.65
$\mathcal{R}_{W_{w12}}$	200	0.99	0.92	0	13.73
$\mathcal{R}_{W_{w13}}$	200	1.0	0.92	0	12.35
$\mathcal{R}_{W_{w123}}$	104	0.96	0.97	3035	62.63

The execution time of our algorithm (see Table 2) increases as the number of views increases. It also depends on the number of attributes in each view and the predefined size of maximum ($|\mathcal{R}_{max}|$) and work ($|\mathcal{R}_{work}|$) redescription set. Using smaller maximum and working set size will reduce the execution time of the algorithm but will also potentially degrade the quality of the resulting redescription set. The $|\mathcal{R}_{inc}|$ in Table 2 denotes the total number of incomplete redescriptions produced during redescription mining with a predefined number of views. Majority of these redescriptions are discarded in some algorithm iteration. As can be seen, this number rises significantly as the number of views increases.

The distributions of Jaccard index, normalized support size, log p-value and redescription average entity and attribute Jaccard index for the Country dataset are presented in Fig. 2.

It can be seen from Fig. 2 that redescriptions contained in $\mathcal{R}_{C_{3w}}$ and $\mathcal{R}_{C_{4w}}$ have generally smaller p-values then redescriptions in $\mathcal{R}_{C_{2w}}$. This is one of the advantages of using multiple views. Using knowledge contained in multiple views (since it needs to describe equal subset of entities) significantly increases probability that the obtained redescriptions are significant and that the obtained knowledge is meaningful (this also follows from the definition of theoretical p-value). Redescriptions obtained using all views on the River quality dataset have significantly higher p-values when using all views than only two views. The reason for this is that all obtained redescriptions have very large support (larger than 25%), at the same time they mostly have relatively low accuracy (smaller than 0.5). Due to smaller accuracy, at least one redescription query (but usually all) has larger support set size than the corresponding redescription (whose support set size is already very large), thus marginal probabilities of redescription queries are high. This results in high theoretical redescription p-values.

We have found many interesting redescriptions on the Country dataset. Highly developed countries detected in our earlier work [10,15] have been again

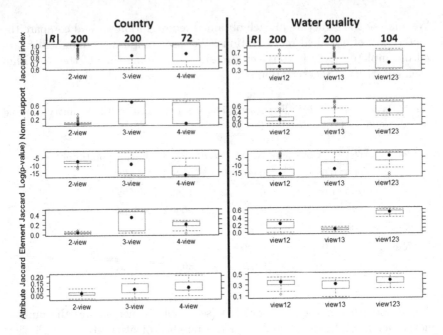

Fig. 2. Comparative boxplots showing various redescription quality measures for redescriptions contained in sets $\mathcal{R}_{C_{2w}}$, $\mathcal{R}_{C_{3w}}$ and $\mathcal{R}_{C_{4w}}$ constructed on a Country dataset and redescription sets $\mathcal{R}_{W_{w12}}, \mathcal{R}_{W_{w13}}$ and $\mathcal{R}_{W_{w123}}$ constructed on the River quality dataset.

discovered and redescribed using trade, population, energy and wealth and progress indicators. Attributes and their values in all views indicate high country development. For example, high export to import ratio of specialised machinery, high percentage of population older than 65 (>17%), high percentage of Internet users (>80%), moderate to high net solar capacity, low to moderate charcoal final consumption, high value value added in services such as wholesale and retail trade, education, healthcare, imputed bank service charges etc.Additional indicator of development and wealth is relatively high percentage of agricultural land.

We present one redescription example from each redescription set obtained on the Country dataset ($R_1 \in \mathcal{R}_{C_{2w}}$, $R_2 \in \mathcal{R}_{C_{3w}}$, $R_3 \in \mathcal{R}_{C_{4w}}$) in Table 3. The corresponding redescription support set of these redescriptions are depicted in Fig. 3.

Redescriptions presented in Table 3, with the exception of Malaysia (which is highly developed country in Asia), mostly described highly developed European countries. All countries described by R_1 have noticeable import of coffee, tea, cocoa and spices, and part of these countries has high export of manufactured

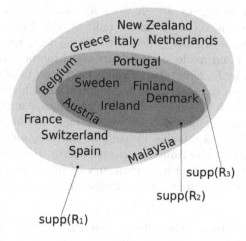

Fig. 3. Redescription support sets of R_1, R_2 and R_3.

Table 3. Example redescriptions obtained on the Country dataset.

R_1 : (q_{1_1}, q_{2_1})

q_{1_1} : $(0.0 \leq \mathtt{I}_{34} \leq 2.0 \;\wedge\; 0.35 \leq \mathtt{E/I}_{83} \leq 4.31 \wedge 0.0 \leq \mathtt{E}_{45} \leq 0.0 \;\wedge$
$0.38 \leq \mathtt{E/I}_6 \leq 3.1 \;\wedge 0.0 \leq \mathtt{E}_{93} \leq 2.0 \;\wedge\; 1.0 \leq \mathtt{I}_{37} \leq 1.0 \;\wedge$
$0.22 \leq \mathtt{E/I}_{15} \leq 1.56) \;\vee\; (0.0 \leq \mathtt{E}_{93} \leq 2.0 \;\wedge\; 0.38 \leq \mathtt{E/I}_6 \leq 3.1 \;\wedge$
$21.0 \leq \mathtt{E}_{23} \leq 70.0 \;\wedge 1.0 \leq \mathtt{I}_{37} \leq 1.0 \wedge\; 0.0 \leq \mathtt{I}_{34} \leq 2.0 \;\wedge$
$0.35 \leq \mathtt{E/I}_{83} \leq 3.1 \;\wedge\; 0.0 \leq \mathtt{E}_{45} \leq 0.0)$

q_{2_1} : $12.46 \leq \mathtt{POP\_MILION\_CITY} \leq 40.42 \wedge\; 2.9 \leq \mathtt{MORT} \leq 8.5 \wedge$
$64.28 \leq \mathtt{POP\_15\_64} \leq 68.19$

quality: $J(R_1) = 1.0$, $p(R_1) = 9.2 \cdot 10^{-12}$, $|supp(R_1)| = 15$

R_2 : $(q_{1_2}, q_{2_2}, q_{2_3})$

q_{1_2} : $0.73 \leq \mathtt{E/I}_{80} \leq 1.27 \wedge\; 0.72 \leq \mathtt{E/I}_{83} \leq 2.79$

q_{2_2} : $-140001.0 \leq \mathtt{NetMigration} \leq 272626.0 \;\wedge\; 2.9 \leq \mathtt{MORT} \leq 4.0 \;\wedge$
$14.8 \leq \mathtt{POP\_MILION\_CITY} \leq 24.95$

q_{3_2} : $1249.0 \leq \mathtt{MotorGasolineFinEngCons} \leq 2910.0 \;\wedge$
$2438.0 \leq \mathtt{HardCoalTotEngSupp} \leq 4602.0$

quality: $J(R_2) = 0.83$, $p(R_2) = 2.2 \cdot 10^{-16}$, $|supp(R_2)| = 5$

R_3 : $(q_{1_3}, q_{2_3}, q_{3_3}, q_{4_3})$

q_{1_3} : $13.0 \leq \mathtt{E}_1 \leq 36.0 \;\wedge\; 0.94 \leq \mathtt{E/I}_7 \leq 3.2 \;\wedge\; 0.72 \leq \mathtt{E/I}_{83} \leq 2.79$

q_{2_3} : $-140001.0 \leq \mathtt{NetMigration} \leq 272626.0 \;\wedge\; 2.9 \leq \mathtt{MORT} \leq 4.2 \;\wedge$
$14.8 \leq \mathtt{POP\_MILION\_CITY} \leq 40.42$

q_{3_3} : $(121.0 \leq \mathtt{FuelOilFinConsm} \leq 380.0 \;\wedge\; 6295.0 \leq \mathtt{Electricity}$
$\mathtt{TotNetCapPPCF} \leq 9564.0) \;\vee\; (121.0 \leq \mathtt{FuelOilFinConsm} \leq 546.0$
$\wedge\; 158.0 \leq \mathtt{BitumenFC} \leq 386.0)$

q_{4_3} : $1.93 \cdot 10^{11} \leq \mathtt{GNIAtlas} \leq 5.59 \cdot 10^{11} \;\wedge\; 31.63 \leq \mathtt{RevenueExcluding}$
$\mathtt{Grants} \leq 43.95$

quality: $J(R_3) = 1.0$, $p(R_3) = 0$, $|supp(R_3)| = 7$

goods. Countries described by R_1, R_2 and R_3 have high percentage of population living in large cities (larger than 1 million), relatively small mortality of children under 5 years. Countries described by R_2 have relatively low energy supply by hard coal and moderate consumption of motor gasoline. Part of the countries described by R_3 have relatively high electricity total net capacity, very high GNIAtlas (which indicates high development) and high value of Revenue excluding grants (which is characteristic to many European countries).

On the River water quality dataset, we obtained several redescriptions that describe water quality parameters in samples that contain high frequency of some animal or plant species from our selected subset. However, we were unable to locate samples that contain high frequency of both some animal and plant species with predefined accuracy. Redescriptions created using all views on this

Table 4. Example redescriptions obtained on the River water quality dataset.

R_{w1} :　$(q_{1_{w1}}, q_{2_{w1}})$
$q_{1_{w1}}$:　$0.43 \leq \text{NO}_3 \leq 18.81 \ \wedge \ 0.24 \leq \text{KMnO}_4 \leq \ 19.1 \wedge \ 0.24 \leq \text{SiO}_2 \leq 6.48 \ \wedge$
　　　　$0.36 \leq \text{Cl} \leq 5.95$
$q_{2_{w1}}$:　$0.0 \leq \text{Gongrosira\_incrustans} \leq 3.0 \wedge \ 3.0 \leq \text{Nitzschia\_palea} \leq 5.0$
quality: $J(R_{w1}) = 0.41$, $p(R_{w1}) = 1.7 \cdot 10^{-12}$, $|supp(R_{w1})| = 241$

R_{w2} :　$(q_{1_{w2}}, q_{2_{w2}})$
$q_{1_{w2}}$:　$0.18 \leq \text{KMnO}_4 \leq 0.28 \ \wedge \ 0.25 \leq \text{Cl}_4 \leq \ 0.39 \wedge \ 0.06 \leq \text{Nh}_4 \leq 0.11 \ \wedge$
　　　　$1.69 \leq \text{NO}_3 \leq 2.95 \ \wedge \ 4.2 \leq \text{hardness} \leq 5.8 \ \wedge \ 4.1 \leq \text{O}_2\text{sat} \leq 4.9$
$q_{2_{w2}}$:　$0.0 \leq \text{Erpobdella\_octoculata} \leq 0.0 \wedge \ 0.0 \leq \text{Diptera\_simulium} \leq 1.0 \ \wedge$
　　　　$1.0 \leq \text{Trichoptera\_rhyacophila} \leq 3.0 \ \wedge \ 0.0 \leq \text{Trichoptera\_}$
　　　　$\text{hydropsyche} \leq 0.0 \ \wedge \ 3.0 \leq \text{Baetis\_rhodani} \leq 5.0 \ \wedge \ 3.0 \leq \text{Gammarus\_}$
　　　　$\text{fosarum} \leq 5.0$
quality: $J(R_{w2}) = 0.44$, $p(R_{w2}) = 1.1 \cdot 10^{-16}$, $|supp(R_{w2})| = 12$

R_{w3} :　$(q_{1_{w3}}, q_{2_{w3}}, q_{3_{w3}})$
$q_{1_{w3}}$:　$0.06 \leq \text{Cl} \leq 0.37 \ \wedge \ 0.0 \leq \text{PO}_4 \leq \ 0.61$
$q_{2_{w3}}$:　$0.0 \leq \text{Stigeoclonium\_tenue} \leq 0.0 \wedge \ 0.0 \leq \text{Nitzchia\_palea} \leq 1.0$
$q_{3_{w3}}$:　$0.0 \leq \text{Erpobdella\_octoculata} \leq 1.0 \wedge \ 0.0 \leq \text{Oligochaeta\_tubifex} \leq 0.0$
quality: $J(R_{w3}) = 0.42$, $p(R_{w3}) = 0.0$, $|supp(R_{w3})| = 362$

R_{w4} :　$(q_{1_{w4}}, q_{2_{w4}}, q_{3_{w4}})$
$q_{1_{w4}}$:　$(0.01 \leq \text{NH}_4 \leq 0.26 \ \wedge \ 3.4 \leq \text{O}_2\text{sat} \leq \ 10.58 \ \wedge \ 3.19 \leq \text{O}_2 \leq 8.49 \ \wedge$
　　　　$0.97 \leq \text{conduct} \leq 4.74) \ \vee \ (0.01 \leq \text{NH}_4 \leq 0.37 \ \wedge \ 0.0 \leq \text{CO}_2 \leq 0.82 \ \wedge$
　　　　$1.85 \leq \text{conduct} \leq 5.64)$
$q_{2_{w4}}$:　$0.0 \leq \text{Stigeoclonium\_tenue} \leq 1.0 \wedge \ 0.0 \leq \text{Nitzchia\_palea} \leq 3.0$
$q_{3_{w4}}$:　$(0.0 \leq \text{Baetis\_rhodani} \leq 1.0 \ \wedge \ 0.0 \leq \text{Erpobdella\_octoculata} \leq 1.0 \ \wedge$
　　　　$0.0 \leq \text{Oligochaeta\_tubifex} \leq 0.0) \vee (0.0 \leq \text{Oligochaeta\_tubifex} \leq 3.0 \ \wedge$
　　　　$0.0 \leq \text{Diptera\_simulium} \leq 0.0) \ \vee \ (1.0 \leq \text{Diptera\_simulium} \leq 5.0 \ \wedge$
　　　　$1.0 \leq \text{Gammarus\_fossarum} \leq 5.0)$
quality: $J(R_{w4}) = 0.76$, $p(R_{w4}) = 5.7 \cdot 10^{-8}$, $|supp(R_{w4})| = 783$

dataset mostly describe samples for which some plant and some animal species do not occur or have very low (incidental) occurrence frequency.

Redescription R_{w1} discovered using view containing physical and chemical measurements of water samples and occurrence frequency of plant species in these samples (presented in Table 4) describes water samples containing frequent and abundant occurrence of *Nitzchia palea*. This species is found in samples containing wide range of NO_3, $KMnO_4$, SiO_2 and Cl concentration, with slightly elevated concentration of Cl. Since Nitzchia palea is used as indicator of moderately polluted to polluted water [4], higher presence of Chlorine may be expected as means to reduce pollution.

We have located 12 samples with frequent and abundant occurrence of animal species *Baetis rhodani* and *Gammarus fosarum* (described by redescription R_{w2} in Table 4). The discovered samples contain elevated levels of NO_3 and have higher hardness. Redescription R_{w3} describes 362 water samples with no or incidental occurrence of plant species *Nitzchia palea* and animal species *Erpobdella octoculata*. These samples do not contain plant species *Stigeoclonium tenue* and animal species *Oligochaeta tubifex*. Redescriptions such as R_{w4} constructed using all three views are more complicated and contain mixed occurrence of plant, animal species and more complex query describing water characteristics.

7   Conclusion and Future Work

We have presented an algorithm for multi-view redescription mining and demonstrated its use on two use case datasets presenting knowledge (via discovered multi-view redescriptions) not obtainable using previously developed tools in the field. Presented technique allows simultaneously exploring trade, demographic, ecological and energy production/consumption information for different world countries and detecting, understanding and relating characteristics of water samples to habitation of both animal and plant species. Performed experiments confirmed that the increase in number of views significantly increases the number of incomplete produced redescriptions, however it is harder to obtain complete redescriptions. This is to be expected since redescriptions created on larger number of views have to satisfy larger set of constraints. Though, such redescriptions have higher significance then redescriptions with equal accuracy and support set size constructed on smaller number of views because information from multiple views reinforce the confidence in obtained redescription.

Future work should involve: (a) reduction of PCT calls and number of tests in the algorithm, (b) utilizing multi-threaded execution to improve execution time.

Acknowledgement. The authors acknowledge the European Commission's support through the projects MAESTRA (Gr. no. 612944) and HBP SGA2 (Gr. no. 785907), support of the Croatian Science Foundation (Pr. no. 9623: Machine Learning Algorithms for Insightful Analysis of Complex Data Structures) and partial support by the European Regional Development Fund under the grant KK.01.1.1.01.0009 (DAT-ACROSS).

References

1. Agrawal, R., Mannila, H., Srikant, R., Toivonen, H., Verkamo, A.I.: Fast discovery of association rules. In: Advances in Knowledge Discovery and Data Mining, pp. 307–328. American Association for Artificial Intelligence (1996)
2. Bickel, S., Scheffer, T.: Multi-view clustering. In: Proceedings of the Fourth IEEE International Conference on Data Mining, pp. 19–26. ICDM 2004, IEEE Computer Society, Washington, DC, USA (2004)
3. Cox, D.R.: Note on grouping. J. Am. Stat. Assoc. **52**(280), 543–547 (1957)
4. Džeroski, S., Demšar, D., Grbović, J.: Predicting chemical parameters of river water quality from bioindicator data. Appl. Intell. **13**(1), 7–17 (2000)
5. Fisher, D.H.: Knowledge acquisition via incremental conceptual clustering. Mach. Learn. **2**(2), 139–172 (1987)
6. Fisher, W.D.: On grouping for maximum homogeneity. J. Am. Stat. Assoc. **53**(284) (1958)
7. Galbrun, E., Miettinen, P.: From black and white to full color: extending redescription mining outside the Boolean world. Stat. Anal. Data Min. **5**(4), 284–303 (2012)
8. Galbrun, E., Miettinen, P.: Siren: an interactive tool for mining and visualizing geospatial redescriptions. In: Proceedings of the 18th ACM SIGKDD International Conference on Knowledge Discovery and Data Mining, pp. 1544–1547. KDD 2012, ACM, New York, NY, USA (2012)
9. Gallo, A., Miettinen, P., Mannila, H.: Finding subgroups having several descriptions: algorithms for redescription mining. In: Proceedings of the SIAM International Conference on Data Mining (SDM), pp. 334–345. SIAM (2008)
10. Gamberger, D., Mihelčić, M., Lavrač, N.: Multilayer clustering: a discovery experiment on country level trading data. In: Džeroski, S., Panov, P., Kocev, D., Todorovski, L. (eds.) DS 2014. LNCS (LNAI), vol. 8777, pp. 87–98. Springer, Cham (2014). https://doi.org/10.1007/978-3-319-11812-3_8
11. Hipp, J., Güntzer, U., Nakhaeizadeh, G.: Algorithms for association rule mining - a general survey and comparison. SIGKDD Explor. Newsl. 58–64 (2000)
12. Jain, A.K., Murty, M.N., Flynn, P.J.: Data clustering: a review. ACM Comput. Surv. **31**(3), 264–323 (1999)
13. Kocev, D., Vens, C., Struyf, J., Džeroski, S.: Tree ensembles for predicting structured outputs. Pattern Recognit. **46**(3), 817–833 (2013)
14. Michalski, R.S.: Knowledge acquisition through conceptual clustering: a theoretical framework and an algorithm for partitioning data into conjunctive concepts. J. Policy Anal. Inf. Syst. **4**(3), 219–244 (1980)
15. Mihelcic, M., Dzeroski, S., Lavrac, N., Smuc, T.: Redescription mining with multitarget predictive clustering trees. In: New Frontiers in Mining Complex Patterns - 4th International Workshop, NFMCP, pp. 125–143. Porto, Portugal (2015)
16. Mihelčić, M., Džeroski, S., Lavrač, N., Šmuc, T.: A framework for redescription set construction. Expert. Syst. Appl. **68**, 196–215 (2017)
17. Parida, L., Ramakrishnan, N.: Redescription mining: structure theory and algorithms. In: AAAI, pp. 837–844. AAAI Press/The MIT Press (2005)
18. Ramakrishnan, N., Kumar, D., Mishra, B., Potts, M., Helm, R.F.: Turning cartwheels: an alternating algorithm for mining redescriptions. In: Proceedings of the 10th ACM SIGKDD International Conference on Knowledge Discovery and Data Mining, pp. 266–275. KDD 2004, ACM, New York, NY, USA (2004)
19. Stojanova, D., Ceci, M., Appice, A., Džeroski, S.: Network regression with predictive clustering trees. Data Min. Knowl. Discov. **25**(2), 378–413 (2012)

20. UN: Un database (2018), http://data.un.org/Explorer.aspx
21. UNCTAD: Unctad database (2014), http://unctadstat.unctad.org/
22. Ward, J.H.: Hierarchical grouping to optimize an objective function. J. Am. Stat. Assoc. **58**(301), 236–244 (1963)
23. WorldBank: World bank database (2014), http://data.worldbank.org/
24. Xu, D., Tian, Y.: A comprehensive survey of clustering algorithms. Ann. Data Sci. **2**(2), 165–193 (2015)
25. Zaki, M.J., Ramakrishnan, N.: Reasoning about sets using redescription mining. In: Proceedings of the 11th ACM SIGKDD International Conference on Knowledge Discovery in Data Mining, pp. 364–373. KDD 2005, ACM, New York, USA (2005)
26. Zhang, M., He, C.: Survey on association rules mining algorithms. In: Advancing Computing, Communication, Control and Management, pp. 111–118. Lecture Notes in Electrical Engineering, Springer, Berlin Heidelberg (2010)
27. Zinchenko, T.: Redescription Mining Over non-Binary Data Sets Using Decision Trees. Master's thesis, Universität des Saarlandes Saarbrücken, Germany (2014)

Text Mining

Author Tree-Structured Hierarchical Dirichlet Process

Md Hijbul Alam[1(✉)], Jaakko Peltonen[1,2(✉)], Jyrki Nummenmaa[1],
and Kalervo Järvelin[1]

[1] University of Tampere, Tampere, Finland
[2] Aalto University, Espoo, Finland
{hijbul.alam,jaakko.peltonen,jyrki.nummenmaa,
kalervo.jarvelin}@uta.fi

Abstract. Three key aspects of online discussion venues are the multitude of participants, the underlying trends of content, and the structure of the venue. However, most models are unable to take into account all three of these. In hierarchically organized message forums, authors may participate differently at multiple levels of sections, with different interests and contributions across the hierarchy. Well-designed probabilistic models of online discussion are applicable to many tasks such as prediction of future content or authorship attribution. However, traditional models such as Hierarchical Dirichlet Processes (HDPs) do not fully take into account authors, and are further unable to fully take into account deep hierarchical venues where documents can arise at all tree nodes. We introduce the Author Tree-structured Hierarchical Dirichlet Process (ATHDP), allowing Dirichlet process based topic modeling of both text content and authors over a given tree structure of arbitrary size and height. Experiments on six hierarchical discussion data sets demonstrate better performance of ATHDP compared to traditional HDP based alternatives in terms of perplexity and authorship attribution accuracy.

Keywords: Hierarchical Dirichlet Processes · Topic Modeling
Message Forum

1 Introduction

Online forums (message boards) are popular social media platforms for information exchange and knowledge sharing, where users ask questions or start discussions by creating a thread, and other users post answers or comments. While some forums are specialized, general-interest forums cover a broad range of interests such as politics, health, beauty, cooking, product reviews, and so on. To help users navigate and participate, forums such as "Suomi24" (www.suomi24. fi) are organized into hierarchical sections. Hierarchical organization also occurs

MHA and JP had equal contributions. The work was supported by Academy of Finland decisions 295694 and 313748.

© Springer Nature Switzerland AG 2018
L. Soldatova et al. (Eds.): DS 2018, LNAI 11198, pp. 311–327, 2018.
https://doi.org/10.1007/978-3-030-01771-2_20

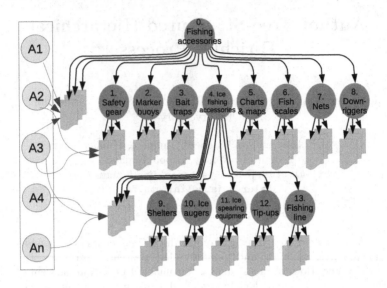

Fig. 1. Hierarchical document organization in a branch of the Amazon product hierarchy.

in online reviews for instance in retailer websites such as Amazon.com, where reviews follow the hierarchy of the products; we use Amazon reviews as a case study and point out dedicated review sites such as Yelp also feature hierarchical organization.

Three crucial aspects of online discussion are the huge diversity of interests being discussed, the huge pool of participants that contribute to the discussions, and the huge but still often structured diversity of online discussion areas where the discussion happens. The key question is how to take all three aspects into account in probabilistic modeling and machine learning of online discussion. In particular, the organization of online discussion areas and the identities of participants are at least partly observed data which can be taken into account for modeling the third aspect, diversity of the underlying topics of discussion.

The three aspects have different characteristics. The interests are expressed in a latent way through the observed text content, the authors are typically observed through author usernames but the pool of authors is unordered, whereas the venue is often both observed and structured: in particular, online discussion often occurs in venues having a prominent *hierarchical* organization for user-generated text content. Hierarchical structure of online forums is designed to cover a subset of prototypical user interests. However, user interests need not match the structure. For example, for an issue touching on multiple interests (say social security and mental health) there might be no dedicated section, and such issues might instead be discussed in multiple sections that each cover one of the interests. Discussion content is typically not regulated to strictly follow their section, hence users may start threads or write replies that deviate from the section theme, and threads commented by multiple users follow a mixture of their interests. Successful modeling of all three aspects of online discussions is important for studies of human online discussion behavior, for tracking trends of ideas and consumer interests, for recommender systems of discussion content or external content like targeted advertising, and for intelligent interfaces to browse and participate in discussions.

The content of the discussions is text data, and probabilistic modeling of text data is often done by generative topic models such as Latent Dirichlet Allocation [5] and Dirichlet Processes [15]. Such models represent text content of documents in an unstructured way as a bag of words arising out of a mixture of latent topics; the latent topics are fitted to a collection of documents and represent themes of discussion occurring over the collection. Basic topic models represent text content alone, whereas recent work on text mining ([14, 18] and others) has attempted author modeling for text analysis, however, most such works are not applicable to documents with observed authors in a deep hierarchical tree such as Fig. 1, where documents (with yellow icons) can appear under any section (with blue icon) at any hierarchy level. The column at left denotes the pool of authors $A_1, \ldots, A_n$, where multiple authors can contribute to each thread and each author can contribute to threads at different nodes across all hierarchy levels (illustrative examples shown as purple arrows). We review related work in Sect. 2.

We give a solution for the challenge of effectively taking hierarchical structure of data collections and author information into account in such modeling. We introduce the *Author Tree-structured Hierarchical Dirichlet Process* (ATHDP), a new model which identifies latent topics of each section in a hierarchy and their association with authors. ATHDP is a generative model for the documents and their authors, which can model documents with multiple authors occurring at all nodes of a multi-level hierarchy. Our contributions are as follows: **1.** We develop a new nonparametric hierarchical topic model to model forum threads where multiple authors can contribute to documents, and documents and their authors can occur at any position of the section hierarchy. **2.** We develop a Gibbs sampling algorithm that extracts topics and their usage across threads and hierarchical sections. **3.** In experiments, our model outperforms the nearest state-of-the-art baseline models in terms of perplexity of held-out documents and in terms of accuracy in an author prediction task.

We point out that the latter task we consider, author prediction based on text content and section of the venue, can have many uses in online discussion venues. Authors who usually post while logged in may sometimes post with a guest username for convenience; author prediction can help associate such posts to the correct author. Moreover, when authors use different accounts on different forums, author prediction can help associate posts from the other forum to authors in the forum of interest. Author prediction models could also be applied to author similarity modeling: if posts of a known author match well (having a high classification probability) to another author, such two authors are similar and could for example be recommended as followers of each other, or could be served similar ads or other content. Such tasks assume the correct author is available in the set of candidates; in principle documents that do not match any author well could be detected simply from poor perplexity scores for all author candidates, but in this paper we do not consider such outlier detection scenarios.

The rest of the paper is structured as follows. In Sect. 2 we discuss related previous work. In Sect. 3 we introduce our new model, and in Sect. 4 we derive Gibbs sampling based Bayesian inference equations for the model. In Sect. 5 we carry out experiments on six data sets arising from two kinds of data, online forum data and online reviews data. Lastly, in Sect. 6 we draw conclusions.

2   Related Work

A topic model [5] is a parametric Bayesian model for count data such as bag-of-words representations of text documents. Several variations expand the basic topic model setting. One of the pioneering works is the Author Topic Model (ATM) [14], which explores relationships between authors, documents, topics and words. Jiang et al. [8] recommend points of interest using ATM. However, ATM models documents arising from a uniform mixture of a group of authors, and cannot model different proportions of authors, and cannot take into account hierarchical organization of documents in a venue. Yang et al. [17] proposed a model that explores asker-answerer networks between users topics for question answering applications, however no hierarchical organization of documents is considered. Another model variant considers modeling sentiment with topics jointly [3]. Author-aware Aspect Topic Sentiment Model (AATSM) [13] explores relationship between authors and sentiment to retrieve supporting opinions from reviews; again no hierarchical document organization is considered. In Link-LDA [6] occurrences of words and entities (such as authors) are not paired. It only models that the document contains a set words and a set of entities, but not which word associated with which entity. The Entity topic model (ETM) [9] models the influence of entities on word content of topics, but does not model the influence of entities on which topics are active in the first place. Thus, it cannot not model influence of sections on active topics. Moreover, all the above models are parametric models and require the number of topics to be predefined.

Teh et al. [15] proposed the Hierarchical Dirichlet Process, a nonparametric model where the number of topics does not need to be pre-specified. However, HDP does not consider author information in the model. There are several parametric/nonparametric models that consider author information. HDPauthor [18] generates documents by a group of authors, and Junyu et al. [16] proposed an infinite author topic model based on mixed Gamma-Negative Binomial Process. However, in these models it is not known which words come from which authors. Moreover, each author always has the same topic distribution regardless of where the topics occur. The only thing that then differentiates the topic proportions of different documents is the proportions of participating authors. Thus, such models cannot properly model the influence of discussion venue sections on document content, and furthermore, these models are not readily applicable to a scenario where documents could arise at any node in a deep hierarchy of sections (as shown in Fig. 1), where not only modeling the influence of sections is important, but also modeling the relationships of content among sections.

Ahmed et al. [2] create a time-dependent topic cluster model based on a recurrent Chinese restaurant process, so that content is grouped at three levels of organization such as high-level topics, individual stories, and entities over time. In PAM [11], a document is modeled as a distribution over the topics at the leaves of the topic hierarchy. In the nested Chinese restaurant process [4], a document is modeled as a distribution over a single path from the root to the leaf node. In TS-SB [1], a document is modeled by a single node of the tree. In the recursive Chinese restaurant process [10], a document has a distribution over all of the nodes of the hierarchy. In the above models, HDP is used to learn a tree structure; the difference is that in ATHDP we do not need to perform any learning on the structure of the data, our model is based on a known hierarchy which is fixed during inference. Instead we focus on modeling authors and the given hierarchy as the model structure, where documents can occur under any node in the hierarchy.

3   Author Tree-Structured Hierarchical Dirichlet Process

ATHDP is a generative model for documents arising from multiple authors at different nodes of a multilevel hierarchy of sections. Each document is represented as a bag of (*word, author*) tuples, arising out of a latent mixture of topics. Topic mixtures in the model are drawn from Dirichlet process priors: the Dirichlet process is a nonparametric prior over topic distributions that requires only a base distribution and a concentration parameter, and does not require pre-specifying the number of topics; the inference of the resulting ATHDP model will learn the number of topics from the data of documents and their authors over the hierarchy.

In the following, we describe ATHDP first as a top-down generative process from its associated graphical model shown in Fig. 2. We then introduce a restaurant-related metaphor called Fine Chocolates Banquet (FCB) for the model which provides useful terminology and intuition; such food-related metaphors are commonly used in Dirichlet process based modeling, as the Dirichlet process itself is often also described as a Chinese restaurant process. The FCB metaphor will be used in the next section to describe inference for ATHDP.

Generative process. Consider a given tree hierarchy which, in a top-down fashion, can be described as a root node (root section) connected to a set of child nodes, those in turn connected to grandchild nodes, and so on. Documents can be observed under any node, not only under leaf nodes. ATHDP is a nonparametric topic model that generates a Dirichlet process prior into each node of the tree and into each document. From that prior the topic distribution of the document is drawn, and each topic generates (*word, author*) tuples as content for the documents.

The generative process first draws a global distribution G^0_{root} from a Dirichlet process with base distribution H and concentration parameter α^0 for the

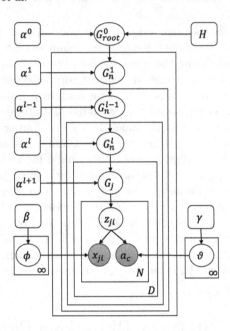

Fig. 2. Graphical model of the Author Tree-structured Hierarchical Dirichlet Process

root node of a given tree, denoted as $G_{root}^0 \sim DP(H, \alpha^0)$. A node can contain child nodes and/or documents. We index a node with v and a document with j. Therefore, for each child section v of the root node in the tree, a discrete distribution G_v^1 is generated from a Dirichlet process with base distribution G_{root}^0 and concentration parameter α^1, denoted as $G_v^1 \sim DP(G_{root}^0, \alpha^1)$. The process is repeated recursively for every child node to generate its grandchild sections, so that a node v at level l in the hierarchy (l steps down from the root) is generated a discrete distribution G_v^l by drawing it from a Dirichlet process with base distribution $G_{pa(v)}^{l-1}$ and concentration parameter α^l, where $pa(v)$ is the parent node of v, denoted as $G_v^l \sim DP(G_{pa(v)}^{l-1}, \alpha^l)$. The Dirichlet process priors describe which topics are active in each node; in order to generate topic content which are $(word, author)$ tuples, for each topic z two distributions are drawn, a distribution ϕ_z over the vocabulary V of possible words from a Dirichlet prior, denoted as $\phi_z \sim Dirichlet(\beta)$, and a distribution ϑ_z over the pool of possible authors A from another Dirichlet prior, denoted as $\vartheta_z \sim Dirichlet(\gamma)$, where β and γ are hyperparameters.

A document, or several documents, can arise under any node. Therefore, to generate a document j under a node v at level l, the model draws G_j from a Dirichlet process with base distribution G_v^l and concentration parameter α^{l+1}, denoted as $G_j \sim DP(G_v^l, \alpha^{l+1})$. From G_j, a topic index z_{ji} is drawn. Based on the topic index the model samples a word x_{ji} and an author a_{ji} from the distribution of words in that topic and the distribution of authors in that topic, respectively. Figure 2 shows the plate representation graphical model of ATHDP. In summary, the full generative process of the ATHDP model is as follows:

- For each topic $z = 1, 2, \ldots$,
    1. Sample a distribution over words, $\phi_z \sim Dirichlet(\beta)$.
    2. Sample a distribution over authors, $\vartheta_z \sim Dirichlet(\gamma)$.
- For the root, $G_{root}^0 \sim DP(\alpha^0, H)$.
- For each section v at level l from the root, $G_v^l \sim DP(\alpha^l, G_{pa(v)}^{l-1})$.
- For each document j in section v, $G_j \sim DP(\alpha^{l+1}, G_v^l)$.
- For each word x_{ji} and author a_{ji} in a document j, $z_{ji} \sim G_j$, $x_{ji} \sim \phi_z$, $a_{ji} \sim \vartheta_z$

Food-based metaphor. The formal generative process above can also be described implicitly as an iterative process where documents are filled one observed (*word*, *author*) tuple at a time. We describe the process by the FCB metaphor; the mathematical details are then provided in the next section as Gibbs sampling based inference equations.

In the FCB metaphor, a chocolate-tasting banquet, where dishes are assortment boxes of fine chocolates prepared by famous chocolatiers, is arranged in a multilevel palace: each level has several *food-delivery stations*, each of which serves several *restaurants* (dining rooms) at that level. Each topic in ATHDP is a dish, that is, a chocolate-assortment box containing a particular mixture of chocolate candies (words) created by a team of chocolatiers (authors). A customer chooses which assortment they want to eat from, and then takes a chocolate from the assortment box: each chocolate is provided in a wrapper signed by the chocolatier, thus when a customer takes a chocolate from the box they will observe a tuple of the candy itself (word) and the identity of the chocolatier (author). Attendees (i.e., customers) visit the chocolate restaurants to eat from popular dishes (popular chocolate assortments): each restaurant has tables for customers, and there is a responsible *waiter* at every table who brings a dish (chocolate-assortment box) to the table, fetching it from a table in a food-delivery station. At food-delivery stations, each table contains a pile of a particular dish (boxes of a particular chocolate-assortment), and each table also has a responsible waiter who brings the dish to the table from an upper-level delivery station, recursively. At the topmost level there is a kitchen where the chocolatiers work to create the different types of dishes (assortments). Each time a customer/waiter chooses a table, they prefer popular tables that other customers/waiters have also picked, but can also pick a new table; this property enables the FCB to make available as many dishes as are needed without specifying the number beforehand. In practice, although a potentially infinite number of dishes are available, inference yields a finite number of dishes suitable for modeling the data set.

We illustrate FCB in Fig. 3. Yellow boxes are restaurants (documents), and orange circles denote customers that each pick a chocolate representing a (*word*, *author*) tuple (x, a) from their table. Each table serves a dish (chocolate assortment box) which represents a topic, having a distribution ϕ over words

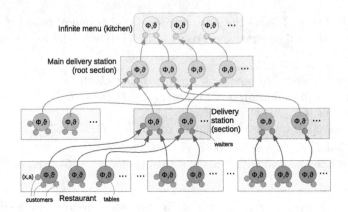

Fig. 3. An illustration of a Fine Chocolates Banquet.

and a distribution ϑ over authors. Each dish is brought to the table by a waiter from an upper-level delivery station (blue boxes), where each waiter chooses one table in the delivery station. Ultimately the dishes are created in the upper-most level (kitchen) which hosts an infinite menu of chocolate-assortment dishes. The content of the available dishes and their prevalences across restaurants and delivery stations are not observed, and will be inferred from the observed data as described in Sect. 4.

4   Inference

We introduce a Gibbs sampling scheme for ATHDP, derived based on the FCB representation. We sample tables, pointers to ancestor tables, and dishes for tables. Let $f_k^{-x_{ji},a_{ji}}(x_{ji}, a_{ji})$ denote the conditional density or likelihood of (x_{ji}, a_{ji}) given all data items except (x_{ji}, a_{ji}), where k is the dish at the table of (x_{ji}, a_{ji}). We have for a pre-existing dish and for a brand-new dish

$$f_k^{-x_{ji},a_{ji}}(x_{ji}, a_{ji}) \propto \frac{n_{kw}^{-ji}}{n_{k.}^{-ji}} \times \frac{n_{ka}^{-ji}}{n_{k.}^{-ji}} \quad \text{and} \quad f_{k_{new}}^{-x_{ji},a_{ji}}(x_{ji}, a_{ji}) \propto \frac{1}{V \times A}$$

respectively, where w is the word index of x_{ji}, a is the author index of a_{ji}, n_{kw}^{-ji} is the number of occurrences of w from dish k (other than x_{ji}), n_{ka}^{-ji} is the number of occurrences of a from dish k (other than a_{ji}), and $n_{k.}^{-ji}$ is the sum over different word indices; note that since words and authors occur in tuples, the sum over word indices is the same as the sum of n_{ka}^{-ji} over author indices. We denote

$$f_k^{-x_{jt},a_{jt}}(x_{jt}, a_{jt}) = \frac{\prod_w (\beta + n_{kw} - 1)...(\beta + n_{kw}^{-jt})}{(V\beta + n_{kw} - 1)...(V\beta + n_{k.}^{-jt})} \frac{\prod_a (\gamma + n_{ka} - 1)...(\gamma + n_{ka}^{-jt})}{(A\gamma + n_{ka} - 1)...(A\gamma + n_{k.}^{-jt})}$$

as the conditional density of (x_{jt}, a_{jt}) given all data items associated with mixture component k leaving out (x_{jt}, a_{jt}), where β and γ are hyperparameters.

Part 1. Sampling table t for a customer x_{ji} at a restaurant: For an individual customer the likelihood for a new table $t_{ji} = t^{new}$ can be calculated by integrating out the possible values of the new dish $k_{jt^{new}}$:

$$p(x_{ji}, a_{ji} | t_{-ji}, t_{ji} = t^{new}; k) = \sum_{k=1}^{K} Q_k f_{k_{jt}}^{-x_{ji}, a_{ji}}(x_{ji}, a_{ji}) + Q_{k^{new}} f_{k_{jt}^{new}}^{-x_{ji}, a_{ji}}(x_{ji}, a_{ji})$$

where t_{-ji} denotes table choices of all words other than t_{ji} and k denotes dish choices of all tables, and Q_k or $Q_{k^{new}}$ denote dish probabilities that are computed recursively, traveling from a leaf node to all the way up to the root node by summing the number of tables in each node that are assigned to a topic.

$$Q_k(v) = \frac{m_{\cdot k}^v}{m_{\cdot\cdot}^v + \alpha^l} + \frac{\alpha^l}{m_{\cdot\cdot}^v + \alpha^l} Q_k(pa(v)),$$

where $m_{\cdot k}^v$ is the number of tables assigned to topic k in node v, and $m_{\cdot\cdot}^v$ is the number of tables in node v, and l is the level of the node. Therefore, at a restaurant the conditional distribution of t_{ji} is:

$$p(t_{ji} = t) \propto n_{jt\cdot}^{-ji} \times f_k^{-x_{ji}, a_{ji}}(x_{ji}, a_{ji})$$
$$p(t_{ji} = t^{new}) \propto \alpha^{l+1} p(x_{ji}, a_{ji} | t_{-ji}, t_{ji} = t^{new}; k) \tag{1}$$

Part 2. Sampling a table t from delivery-station v for a new waiter with first customer x_{ji}: The likelihood for $t_{jt} = t^{new}$ can be calculated as follows:

$$p(t_{jt} | t_{-jt}, t_{jt} = t^{new}; k) = \sum_{k=1}^{K} \frac{c_{vt\cdot}}{c_{v\cdot\cdot} + \alpha^l} f_{k_{jt}}^{-x_{ji}, a_{ji}}(x_{ji}, a_{ji})$$
$$+ \frac{\alpha^l}{c_{v\cdot\cdot} + \alpha^l} f_{k_{jt}^{new}}^{-x_{ji}, a_{ji}}(x_{ji}, a_{ji})$$

where $c_{v\cdot k}$ is the number of tables assigned to k in node v, $c_{vt\cdot}$ is the number of tables point to table t in node v and $c_{v\cdot\cdot}$ is the number of tables point to tables in node v. Therefore, the conditional distribution of t_{jt} (with a customer at a restaurant) is

$$p(t_{jt} = t) \propto \frac{c_{vt\cdot}^{-jt}}{c_{v\cdot\cdot} + \alpha_j} f_{k_{tj}}^{-x_{ji}, a_{ji}}(x_{ji}, a_{ji})$$
$$p(t_{jt} = t^{new}) \propto \frac{\alpha_j}{c_{v\cdot\cdot} + \alpha_j} p(t_{jt} | t_{-jt}, t_{jt} = t^{new}; k) \tag{2}$$

Part 3. Sampling a delivery-station table t for a waiter with several existing customers: The likelihood for $t_{jt} = t^{new}$ for many customers in a table can be calculated as follows:

$$p(t_{jt} | t_{-jt}, t_{jt} = t^{new}; k) = \sum_{k=1}^{K} \frac{c_{vt\cdot}}{c_{v\cdot\cdot} + \alpha^l} f_k^{-x_{jt}, a_{jt}}(x_{jt}, a_{jt})$$
$$+ \frac{\alpha^l}{c_{v\cdot\cdot} + \alpha^l} f_{k_{new}}^{-x_{jt}, a_{jt}}(x_{jt}, a_{jt})$$

Algorithm 1 Gibbs Sampling for ATHDP

Input: words **w** in documents **d**, # topics K, # iterations I
Output: Topic assignments **z**
1 **for** i in I **do**
2   **for** w in **d do**
3   SampleTable(node)
4   **for** t in **d do**
5   SampleParentTable(node)

6 **Procedure** SampleTable(node)	16 **Procedure** SampleParentTable(node)
7 **if** node == document **then**	17 **if** node.parent == root.node **then**
8 $table \leftarrow$ Sample a table by Eq.(1)	18 $topic \leftarrow$ Sample a topic by Eq. (3)
9 **else**	19 **else**
10 $table \leftarrow$ Sample a table using Eq. (2)	20 node $\leftarrow$ node.parent
11 **if** $table == t_{new}$ **then**	21 $table \leftarrow$ Sample a table by Eq. (3)
12 **if** node.parent == root.node **then**	22 **if** $table == t_{new}$ **then**
13 $topic \leftarrow$ Sample a topic by Eq. (3)	23 node $\leftarrow$ node.parent
14 **else**	24 SampleParentTable(node)
15 SampleTable(node.parent)	

Therefore, the conditional distribution of t_{jt}, given all customers in the table, is

$$p(t_{jt} = t) \propto \frac{c_{vt.}^{-jt}}{c_{v..} + \alpha^l} f_k^{-\boldsymbol{x}_{jt}, \boldsymbol{a}_{jt}}(\boldsymbol{x}_{jt}, \boldsymbol{a}_{jt}), \tag{3}$$

$$p(t_{jt} = t^{new}) \propto \frac{\alpha^l}{c_{v..} + \alpha^l} p(t_{jt}|\boldsymbol{t}_{-jt}, t_{jt} = t^{new}; \boldsymbol{k})$$

If the upper level is the root level, a dish or topic is sampled from the kitchen instead of a table pointer, and the dish is propagated to all descendants of the waiter.

We summarize the Gibbs sampling algorithm for ATHDP inference in Algorithm 1. We sample a table assignment for each $(word, author)$ tuple in a document with a recursive procedure in line 3. For a $(word, author)$ tuple, we sample a table using Eq. (1), and we sample a parent table using Eq. (2) from delivery stations. If it's a new table, then we move to the parent node to sample a table from the parent node in line 15. The process is repeated until a parent table is selected or the root node is reached. If the root node is reached a topic selected using Eq. (3). After that, we update the topic of all tables in the descendant's nodes of the table in the root. We maintain a data structure to keep track of topics of all tables. For simplicity, we do not include them in the algorithm. Similarly, for each table (i.e., a group of words associated with a table) in a document, we sample a parent table, i.e., a table from the parent using Eq. (3). We repeat the process until the root is reached and eventually sample a topic for the root table using Eq. (3).

Table 1. Data sets. Total document counts at different levels from the root given in the 4th column.

	Thre-shold	# Authors	# nodes	#docs at different level of the tree	# Train docs	# Test docs
Amazon Sports		97	599	100, 57, 434, 1655, 3775, 645, 8		
	50				5965	709
Amazon Food	100	26	40	3166, 2, 56, 42, 22		
					2948	340
Amazon Home	100	37	567	25, 85, 227, 2015, 2179, 394, 80		
					4489	516
Amazon Health	100	67	484	107, 137, 550, 5501, 2912, 209		
					8444	972
S24 Relationship	100	50	25	0, 12148	13425	1514
S24 Health		71	64	0, 464, 1882		
	20				2509	661

5   Experimental Results

We evaluate ATHDP's performance by two performance measures: (1) held-out perplexity, representing ability to model unseen documents and (2) author prediction of unseen documents. Since the methods described in the related work are all not directly applicable to our case, we take the Hierarchical Dirichlet Process [15] as a baseline that would be readily available to the practitioner, and we use it in two ways to take author information into account, as described in the *Quantitative comparison* paragraph below. We used Gibbs sampling to train the models and took a sample at 100th iterations. We first describe the data sets, summarized in Table 1. We begin by a qualitative analysis of ATHDP results, and then present quantitative comparisons between ATHDP and comparison methods.

We used two different data sources, *Suomi24* (s24) and Amazon for our experiments. S24 has in total 2434 sections in the hierarchy. The data source [1] is publicly available in original and lemmatized forms. From this source, we created several datasets by taking thematic branches of the hierarchy, such as s24 relationship, s24 health for our experiments. The second data source is *reviews on Amazon.com*, one of the top shopping sites in the world with hundreds of shopping sections. We select thematic branches corresponding to several top categories (or department) such as Sports and Outdoors (Sports), Home and Kitchen (Home), Health and Personal Care (Health), Clothing Shoes and Jewelry (Clothes), Grocery and Gourmet Food (Food). Under each top category the site contains many sections. For example, there are 1933 sections under sports [7]. We select reviewers that have more than 50 or 100 reviews in each category. For each reviewer we randomly select 90% reviews for training and 10% reviews for testing. The numbers of train and test reviews for each category along with the number of reviews in different levels of the hierarchy are given Table 1. In Amazon data sets each product is considered as a document, and in s24 data sets

---

[1] https://www.kielipankki.fi/corpora/.

Table 2. Sample ATHDP topic proportion for three sections in the food data set

Section id	Section name	Top 3 topic proportions for each section
44	Peanut Butter	66:0.46, 86:0.31, 37:0.23
3	Coffee Substitutes	9:0.247, 28:0.24, 37:0.24
292	Jams & Preserves Gifts	22:0.62, 9:0.36, 48:0.002

Table 3. ATHDP topics for Amazon food data set, sections where they are active, and top words

Topic	Top 3 author ids	Stemmed top words of the topic
9	10, 19, 7	Cup recommend good coffe flavor tast drink pack highli brew keurig energi tea brewer make bold free star larg
22	18, 20, 23	Make enjoy tast nice flavor bit good eat work meal ad cup star love give lot packag mix protein morn
28	23, 0, 1	Coffe tast flavor great recommend highli make cup love chocol tea good stuff bit year product buy thing amaz awesom
37	14, 1, 0	Good tast coffe flavor great love make product sweet tea time chocol stuff perfect snack work cup soup free chicken
48	7, 5, 24	Clean top cook grape flavor work nice red kit simpl week conveni good great allergi bag expect sodium basic water
66	17, 7, 19	Sugar calori product protein flavor ingredi bar tast fiber fat high oil organ food wheat natur time sweet make raisin
86	24, 17, 10	Bar protein tast flavor calori bit eat snack fiber good meal cinnamon sugar textur fill chocol nice ingredi diet raisin

each thread is considered as a document. A document consists of many reviews or comments from many authors.

Qualitative analysis of ATHDP results. We examine how ATHDP topics covered themes within and across sections. For brevity we present the analysis of the food dataset only. We present latent topics and their proportions for three sample section, as described in Table 2. We see that sections are mixed of latent topics with different proportions. For example, the top 3 topics of the Peanut Butter section are 66, 86, and 37. The top words of each latent topic of the sample sections are presented in Table 3. We observe that extracted latent topics covered many themes including section themes. For example, top words of topics 66, 86, and 37 include many words related to peanut butter including fat, oil, protein, sugar, calori and so on. We observe that topic 66 is directly related to peanut butters and can be regarded as discussion of *bread spreads*. There is some overlap between top words of topics 66 and 86. By looking at distinct words we observed that people are discussing diet, snack, meal, cinnamon, chocolate etc. in the peanut butter section, which refers to *how good peanut butter is as a diet*. Topic 37 is about having coffee or tea, which is an occasion where peanut butter based breads might also be enjoyed, hence it is a discussion of a *use scenario of peanut butter*. Table 3 also shows top 3 author ids for each topic. Overall, ATHDP extracts reasonable meaningful word-topic, section-topic and

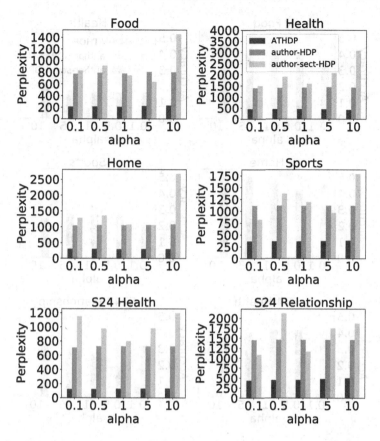

Fig. 4. Perplexity on held-out data sets with different alpha values. Smaller values are better; ATHDP outperforms author-HDP and author-sect-HDP.

topic-author distributions. We also verified that the results regarding datasets other than food were similar.

Quantitative comparison. We compare ATHDP to two baseline models in two tasks, modeling of previously unseen documents and in author prediction. We use HDP as a baseline, which takes the hierarchy into account in two simple ways – model all documents belonging to the same author (author-HDP), and all documents belonging to the same author-section pair (author-sect-HDP). In author-HDP, for example, the sports dataset consists of 97 authors, therefore we train 97 HDP models. In author-sect-HDP we train as many HDP models as there are author-section pairs. For ATHDP, we train a single model for each dataset. We run ATHDP and the author-HDP and author-sect-HDP baselines for all data sets with different values of the concentration hyperparameter alpha. We use the two HDP-based baselines as we found no related work that could fully take into account author information and the hierarchical structure of our data where documents arise in multiple places in the hierarchy, and the aim is to

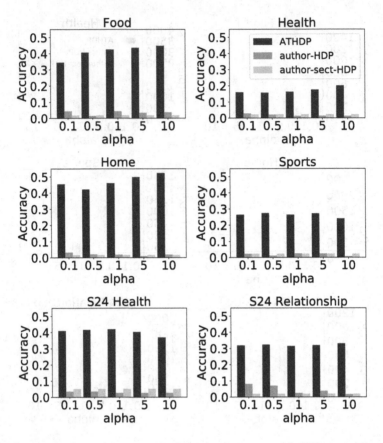

Fig. 5. Author prediction accuracy in different data sets with different alpha values. Larger values are better; ATHDP outperforms author-HDP and author-sect-HDP.

take the known document hierarchy into account. The baselines thus represent a natural way to run the existing HDP model with known divisions of data based on authors or based on authors and sections.

Modeling of previously unseen documents. We evaluate the ability of the proposed model to represent new incoming documents, by computing perplexity of held-out documents, a standard metric in information retrieval literature. We compute perplexity on held-out test documents as described in Table 1 as follows: $perplexity(D_{test}) = \frac{1}{M} \sum_{d=1}^{M} \exp\left(-\frac{log P(w_d)}{N_d}\right)$. We compute perplexity for different α values. We use the same alpha values for all levels in ATHDP. The results are shown in Fig. 4. Lower perplexity indicates the better model. We observe that ATHDP outperforms author-HDP and author-sect-HDP in perplexity for all the data sets and alpha values. The overall difference between ATHDP and the author-HDP, and between ATHDP and author-sect-HDP, is statistically significant at the $p = 0.05$ level: for both comparisons we have $p = 0.03125$ from the exact binomial test over the six data sets. Note that since ATHDP outperforms the alternatives regardless of alpha value, the choice of alpha value used to represent each method does not affect the result of the test.

Author prediction. We compare the ability of different models to predict the author of a previously unseen document, that is, to classify new documents to correct authors. To predict the author for each test document, we compute perplexity for the test document under the model for each author, and assign the document to the author that yields the lowest perplexity. We report the author prediction accuracy results in Fig. 5. We observe that ATHDP outperforms author-HDP and author-sect-HDP by a large margin. The overall difference between ATHDP and the author-HDP, and between ATHDP and author-sect-HDP, is again statistically significant at the $p = 0.05$ level: for both comparisons we have $p = 0.03125$ from the exact binomial test over the six data sets. As ATHDP outperforms the alternatives regardless of alpha value, the choice of alpha value used to represent each method again does not affect the result of the test.

The author prediction accuracies achieved by ATHDP are good, especially considering the large number of potential candidate authors. ATHDP accuracy results are up to about 50% accuracy, which although not a flawless score is practically usable for attribution (note that when there are numerous potential authors random guessing yields far worse accuracies than 50%). In contrast, the alternative systems perform poorly; a possible explanation is that the author-HDP model is unable to take hierarchical section-based variation of the authors' interests properly into account, whereas the author-sect-HDP model does not make full use of the hierarchical relationships between sections and hence has too little data per author-section combination to learn good models of authors' interests in each section. In contrast, ATHDP learns the topics and their variation across the hierarchy together, allowing successful modeling of author interests.

6   Conclusions

We introduced the Author Tree-structured Hierarchical Dirichlet process (ATHDP), a nonparametric probabilistic model of documents and their authors in a deep tree-structured hierarchical discussion venue where documents can arise at any tree node. ATHDP can to extract topics across the documents and sections in the hierarchy, and automatically computes the number of topics required to model the authors and text content across the hierarchical sections. ATHDP does not restrict content of topics to strictly match predefined sections, but infers them in a data driven way to describe users' interests. In experiments, ATHDP outperformed HDP based alternative models in modeling unseen documents (measured by perplexity), and author prediction of unseen documents (measured by accuracy).

In this first work ATHDP already proved a very well-performing and flexible model. In future work, its performance could be evaluated by a larger set of different measures, and the flexibility of the model could be further increased by, for example, modeling within-topic correlations between authors and word content, or by other such extensions. We also plan to integrate ATHDP in systems that can make use of the topic models, i.e., utilizing ATHDP topics in different applications such as recommendation [8,19], and interactive exploratory search [12].

References

1. Adams, R., Ghahramani, Z., Jordan, M.: Tree-structured stick breaking for hierarchical data. In: Proceedings of NIPS, pp. 19–27. Curran Associates Inc. (2010)
2. Ahmed, A., Ho, Q., Teo, C.H., Eisenstein, J., Smola, A.J., Xing, E.P.: Online inference for the infinite topic-cluster model: Storylines from streaming text. In: Proceedings of AISTATS, pp. 101–109 (2011)
3. Alam, M.H., Ryu, W.J., Lee, S.: Joint multi-grain topic sentiment. Inf. Sci. **339**(C), 206–223 (2016)
4. Blei, D., Griffiths, T., Jordan, M.: The nested Chinese restaurant process and Bayesian nonparametric inference of topic hierarchies. J. ACM **57**, 7:1–7:30 (2010)
5. Blei, D., Ng, A., Jordan, M.: Latent Dirichlet allocation. J. Mach. Learn. Res. **3**, 993–1022 (2003)
6. Erosheva, E., Fienberg, S., Lafferty, J.: Mixed-membership models of scientific publications. Proc. Natl. Acad. Sci. **101**(suppl 1), 5220–5227 (2004)
7. He, R., McAuley, J.: Ups and downs: Modeling the visual evolution of fashion trends with one-class collaborative filtering. In: Proceedings of WWW, pp. 507–517 (2016)
8. Jiang, S., Qian, X., Shen, J., Fu, Y., Mei, T.: Author topic model-based collaborative filtering for personalized poi recommendations. IEEE Trans. Multimed. **17**(6), 907–918 (2015)
9. Kim, H., Sun, Y., Hockenmaier, J., Han, J.: ETM: entity topic models for mining documents associated with entities. In: Proceedings of ICDM, pp. 349–358. IEEE Computer Society (2012)
10. Kim, J., Kim, D., Kim, S., Oh, A.: Modeling topic hierarchies with the recursive Chinese restaurant process. In: Proceedings of CIKM, pp. 783–792. ACM (2012)
11. Li, W., McCallum, A.: Pachinko allocation: DAG-structured mixture models of topic correlations. In: Proceedings of ICML, pp. 577–584. ACM (2006)
12. Peltonen, J., Belorustceva, K., Ruotsalo, T.: Topic-relevance map: visualization for improving search result comprehension. In: Proceedings of IUI. pp. 611–622. ACM (2017)
13. Poddar, L., Hsu, W., Lee, M.L.: Author-aware aspect topic sentiment model to retrieve supporting opinions from reviews. In: Proceedings of the 2017 Conference on Empirical Methods in Natural Language Processing, pp. 472–481. Association for Computational Linguistics (2017)
14. Rosen-Zvi, M., Griffiths, T., Steyvers, M., Smyth, P.: The author-topic model for authors and documents. In: Proceedings of UAI, pp. 487–494. AUAI Press (2004)
15. Teh, Y., Jordan, M., Beal, M., Blei, D.: Hierarchical Dirichlet processes. J. Am. Stat. Assoc. **101**, 1566–1581 (2006)
16. Xuan, J., Lu, J., Zhang, G., Xu, R.Y., Luo, X.: A Bayesian nonparametric model for multi-label learning. Mach. Learn. **106**(11), 1787–1815 (2017). Nov
17. Yang, L., et al.: CQArank: jointly model topics and expertise in community question answering. In: Proceedings of CIKM, pp. 99–108. ACM (2013)

18. Yang, M., Hsu, W.H.: HDPauthor: a new hybrid author-topic model using latent Dirichlet allocation and hierarchical Dirichlet processes. In: Proceedings of WWW, pp. 619–624. ACM (2016)

19. Zhang, S., Zhang, S., Yen, N.Y., Zhu, G.: The recommendation system of micro-blog topic based on user clustering. Mob. Netw. Appl. **22**(2), 228–239 (2017). Apr

k-NN Embedding Stability for word2vec Hyper-Parametrisation in Scientific Text

Amna Dridi[✉], Mohamed Medhat Gaber, R. Muhammad Atif Azad,
and Jagdev Bhogal

School of Computing and Digital Technology, Birmingham City University,
Millennium Point, Birmingham B4 7XG, United Kingdom
dridiamna@gmail.com

Abstract. Word embeddings are increasingly attracting the attention of researchers dealing with semantic similarity and analogy tasks. However, finding the optimal hyper-parameters remains an important challenge due to the resulting impact on the revealed analogies mainly for domain-specific corpora. While analogies are highly used for hypotheses synthesis, it is crucial to optimise word embedding hyper-parameters for precise hypothesis synthesis. Therefore, we propose, in this paper, a methodological approach for tuning word embedding hyper-parameters by using the stability of k-nearest neighbors of word vectors within scientific corpora and more specifically Computer Science corpora with Machine learning adopted as a case study. This approach is tested on a dataset created from NIPS (Conference on Neural Information Processing Systems) publications, and evaluated with a curated ACM hierarchy and Wikipedia Machine Learning outline as the gold standard. Our quantitative and qualitative analysis indicate that our approach not only reliably captures interesting patterns like *"unsupervised_learning is to kmeans as supervised_learning is to knn"*, but also captures the analogical hierarchy structure of Machine Learning and consistently outperforms the 61% sate-of-the-art embeddings on syntactic accuracy with 68%.

Keywords: Word embedding · Word2vec · Skip-gram
Hyper-parameters · k-NN stability · ACM hierarchy
Wikipedia outline · NIPS

1 Introduction

Word embeddings (WEs) are a class of natural language processing techniques that represent individual words as real-valued vectors in a predefined vector space. They were first introduced in the 1990s using statistical approaches [2, 8] with vectors computed as rows of lexical co-occurrence [8] through matrix factorization [2].

However, interest in WEs has recently skyrocketed and has found many applications. The surge in interest is due both to popularity of neural networks [1]

© Springer Nature Switzerland AG 2018
L. Soldatova et al. (Eds.): DS 2018, LNAI 11198, pp. 328–343, 2018.
https://doi.org/10.1007/978-3-030-01771-2_21

which exploit WEs for NLP tasks, and the success of low-dimensional embeddings like *word2vec* [12] and *GloVe* [16]. Due to their ability to detect semantics and meanings of words, WEs have been used as features in a variety of applications, such as document clustering [5] and classification [22], Linear Discriminant Analysis (LDA) [17], information retrieval [10], named entity recognition [20], sentiment analysis [19], and semantic discovery [21].

Typically, the reported research work that used WEs as features computed their vector representations with a default or arbitrary choice of embedding hyper-parameters. Examples of these hyper-parameters include *vocabulary size* and type (single words or phrases), *vector dimensionality*, that is, the length of the vector representations of words, and *context size* which is the span of words in the text that is taken into account - both backwards and forwards - when iterating through the words during model training. However, as this paper will show, these hyper-parameters of word embedding vectors are crucial to the prediction performance as they directly affect the accuracy of the generated analogies. Given that analogies can be used in hypotheses synthesis, consequently an accurate analogy will led to a precise hypothesis. For example, *"decision tree"* is a component of *"ensemble"* and *"decision tree"* is a *"classifier"*. So, by analogy any classifier should be a component of ensemble.

This work concerns then hyper-parametrisation of WEs in a domain-specific context with varying vocabulary sizes, and represents a gap in knowledge on present practice of WEs. The considered topic is a key practical issue while learning WEs, and is so chosen because not only is the literature on learning embedding hyper-parameters rather limited [6,14], it does not offer a method to efficiently set these hyper-parameter values. Stimulated by this shortcoming, the work we present in this paper lies within the context of word embedding hyper-parametrisation for domain-specific use. The studied problem domain is *scientific literature* and more specifically *Machine Learning* literature, which is a subcategory of literature on *Computer Science*. The choice of a scientific domain is motivated by an increasing interest in knowledge extraction from scholarly data and scientific text to understand research dynamics and forecast research trends. Since WEs have proved their ability to capture relational similarities and identify relationships in textual data without any prior domain knowledge, this work does not need to justify the use of WEs for knowledge extraction from scientific literature; instead, the presented work is a methodical approach to setting the hyper-parameters of WEs for scientific knowledge extraction.

The motivation here is to deeply understand the embedding behavior within scientific corpora which is quite different to other corpora in terms of word distributions and contexts. For instance, the term *"learning"* appears obviously in the context of education in newspapers corpora; however, *"learning"* appears in a completely different context within *Computer science*. Therefore, WEs for scientific text are worth investigating.

There have been some efforts to integrate WEs in the scientific domain [3,7, 22]; however, these efforts do not study learning the hyper-parameters suitable for a scientific text, and instead use either arbitrary or default settings (Mikolov's settings [11]).

In this research work, we aim to fill this gap. We hypothesise that by devising an approach for setting hyper-parameters of WEs in the scientific domain, this study adds a deep understanding of the sensitivity of embeddings to hyper-parametrisation. To make our point, we propose using the stability of k-nearest neighbors of word vectors as a measure to set the hyper-parameters – mainly vector dimensionality and context size – of word vector embeddings; moreover, we propose using common-sense knowledge from the *ACM hierarchy*[1] and *Wikipedia outline of Machine learning*[2]. As a result, this work adds breadth to the debate on the strengths of using WEs for knowledge extraction from scientific text. To the best of our knowledge, the proposed work represents the first attempt to methodically set WEs hyper-parameters in a scientific domain.

We list the major contributions of this work as follows: (i) we propose the stability of k-nearest neighbors of word vectors as an objective to measure while learning word2vec hyper-parameters, (ii) we enhance the standard skip-gram model by bigrams using *word2phrase* – that attempts to learn phrases by progressively joining adjacent pairs of words with a '_' character – as a method for corpus augmentation, (iii) we create an analogy dataset for the *Machine Learning* by manually curating ACM hierarchy and Wikipedia outline of *Machine Learning*, and (iv) we evaluate our work quantitatively and qualitatively on a dataset comprising of abstracts published in the NIPS conference. Our embedding detected interesting semantic relations in *Machine Learning* such as "unsupervised_learning is to kmeans as supervised_learning is to knn". The obtained results are therefore both promising and insightful.

The rest of the paper is organised as follows. Section 2 summarises the existing approaches on word embedding hyper-parametrisation and gives an overview on work that attempted to integrate word embedding in scientific domains. Section 3 presents our methodology and how we employ stability of k-nearest neighbors to optimise word2vec hyper-parameters. Section 4 describes the NIPS dataset we have used, the analogy dataset we have created from ACM hierarchy and Wikipedia as gold standard, presents and discusses results. Finally, in Sect. 5 we conclude and draw future directions.

2   Related Work

Word embedding methods depend on several hyper-parameters that have crucial impact on the quality of embeddings. For this reason, Mikolov et al. [12,13] and Pennington et al. [16] –the inventors of the popular low-dimensional embedding word2vec and GloVe, respectively – have deeply studied the optimisation of the embedding parameters, mainly the vector dimension and the context size. The performance of the embeddings has been measured based on *word* similarity that

---

[1] https://dl.acm.org/ccs/ccs_flat.cfm.
[2] https://en.wikipedia.org/wiki/Outline_of_machine_learning.

uses cosine distance between pairs of word vectors to evaluate the intrinsic quality of such word representations, and *w*ord analogies that capture fine-grained semantic and syntactic regularities using vector arithmetic. The optimal parameters have been obtained through training on large Wikipedia and Google News corpora. But, no evidence was given for generalisation of these parameters to any other corpus with a general or specific topic and guarantee the performance of embeddings. However, most of the work using WEs relies on these parameters as the default ones.

Unlike work that uses default settings, literature on learning embedding hyper-parameters is relatively short [6,14]. Levy and Goldberg [6] followed Mikolov et al. [11,12] and Pennington et al. [16] and trained their embeddings on general topic using Wikipedia corpus. They basically tested their model with different vector dimensions and different window sizes aiming to study the impact of syntactic contexts – that are derived from automatically produced dependency parse-trees – on detecting functional similarities of cohyponym nature. While Miñarro-Giménez et al. [14] trained their word embeddings on a domain-specific corpus of medical data in order to study the ability of word embeddings (word2vec) to capture linguistic regularities on the medical corpora. Similar to the previous work, Miñarro-Giménez et al. trained their word2vec embeddings with different parameter settings, *i.e.*, dimensionality of vector space, context size, and different model architectures, *i.e.*, continuous bag-of-words (CBOW) and Skip-gram (SG) [11], and simultaneously compared the relationships identified by word2vec with manually curated information from National Drug File – Reference Terminology ontology as a gold standard using word similarity and word analogies in order to evaluate the effectiveness of word2vec in identifying properties of pharmaceuticals and medical relationships. The obtained results (49% accuracy) revealed the unsuitability of word2vec for applications requiring high precision like medical applications. While this research work seems interesting mainly with its appeal to setting hyper-parameters for domain-specific word embeddings, it does not bring a defined method to efficiently set these parameters.

Leading on from the aforementioned observation, the work we present in this paper lies within the context of word embedding hyper-parametrisation for domain-specific use. The proposed domain to investigate is the *s*cientific domain and more specifically *C*omputer Science with *M*achine learning, as a case study.

There have been some efforts to integrate word embeddings in the scientific domain [3,7,22] for clustering scientific documents based on their functional structures [7] or for identifying problem-solving patterns in scientific text [3] or for paper-reviewer recommendation [22]. All the previous research work integrated word embeddings as features for their learning algorithms using either arbitrary or default settings (Mikolov's settings [11]). However, none of them has focused on training the embeddings and methodologically setting the hyper-parameters suitable for scientific text.

To the best of our knowledge, the proposed work represents the first attempt to methodologically set word embeddings hyper-parameters in the scientific domain.

3   Methodology

This study focuses on word2vec hyper-parameter optimisation applied to scientific publications, i.e., how to tune the hyper-parameters that have the largest impact in the prediction performance and what are the adoptable techniques to test the potential of word embeddings for identifying relationships from unstructured scientific text. Accordingly, the *k-NN algorithmic stability* is adopted to investigate the marginal importance of hyper-parameters of *Skip-Gram* architecture in a scientific setting. This allows us to identify three hyper-parameters, namely *vocabulary subsampling, vector dimensionality* and *context size* which can significantly affect the embedding performance. In this study, we use the popular variant word2vec architecture *Skip-Gram* as it is consistently yielded superior results comparing to *CBOW* architecture [11].

3.1   The Skip-Gram model

Previous results reported in the literature have shown that *Skip-Gram* [11] model does not only produce useful word representations, but it is also efficient to train. For this reason, we focus on it to build our embeddings for scientific text in this study. The main idea of *Skip-Gram* is to predict the *context* c given a word w. Note that the *context* is a window around w of maximum size L. More formally, each word $w \in W$ and each context $c \in C$ are represented as vectors $\vec{w} \in \mathbb{R}^d$ and $\vec{c} \in \mathbb{R}^d$ respectively, where $W = \{w_1, \cdots, w_V\}$ is the words vocabulary, C is the context vocabulary, and d is the embedding dimensionality. Recall that the vectors parameters are latent and need to be learned by maximising a function of products $\vec{w} \cdot \vec{c}$.

More specifically, given the word sequence W resulted from the scientific corpus, the objective of *Skip-Gram* model is to maximise the average log probability: $L(W) = \frac{1}{V} \sum_{i=1}^{V} \sum_{-l \leq c \leq l, c \neq 0} log Prob(w_{i+c}|w_i)$ where l is the context size of a target word. *Skip-Gram* formulates the probability $Prob(w_c|w_i)$ using a softmax function as follows: $Prob(w_c|w_i) = \frac{\exp(\vec{w_c} \cdot \vec{w_i})}{\sum_{w_i \in W} \exp(\vec{w_c} \cdot \vec{w_i})}$ where $\vec{w_i}$ and $\vec{w_c}$ are respectively the vector representations of target word w_i and context word w_c, and W is the word vocabulary. In order to make the model efficient for learning, the hierarchical softmax and negative sampling techniques are used following Mikolov et al. [11].

Word embedding vectors learned with *Skip-Gram* can be used for computing word similarities. The similarity of two words w_i and w_j can simply be measured with the inner product of their word vectors, namely $similarity(w_i, w_j) = \vec{w_i} \cdot \vec{w_j}$. Recall that *cosine distance* is the measure used to calculate the similarity between embedding vectors $\vec{w_i}$ and $\vec{w_j}$ as following:

$$similarity(w_i, w_j) = cosineDistance(\vec{w_i}, \vec{w_j}) = \frac{\vec{w_i} \cdot \vec{w_j}}{\|\vec{w_i}\| \cdot \|\vec{w_j}\|} \qquad (1)$$

As discussed in the Introduction section, we aim to evaluate the representation capability of WEs within scientific text using word similarities as a pivot to stabilise the embedding hyper-parameters.

Skip-Gram model uses a target word w to predict the surrounding window of context words. It weights nearby context words more heavily than more distant context words [11,12]. Results of word2vec training are sensitive to parametrisation. To this end, the aim of hyper-parameter optimisation is to find a tuple of hyper-parameters that yields an optimal model minimising the *loss function* for negative samples $(w, \bar{c})$ where $\bar{c}$ does not necessarily appear in the context of w. This *loss function* $\mathscr{L}$ is defined as follows: $\mathscr{L} = -log(\sigma(\vec{w} \cdot \vec{c})) - \sum_{k=1}^{n} log(\sigma(-\vec{w} \cdot \vec{c_k}))$ where σ is the sigmoid function. For each pair (w, c), the *Skip-Gram* model forms n negative pairs $(w, \bar{c_k})_{k \in \{1, \cdots, n\}}$ by sampling words that are more frequent than some threshold θ with a probability: $Prob(c) = \frac{freq(c) - \theta}{freq(c)} - \sqrt{\frac{\theta}{freq(c)}}$ where $freq(c)$ represents the frequency of the word c.

word2vec has different hyper-parameters, but *sub-sampling* that automatically affects the *corpus size, vector dimensionality* and *context window* are described by the developers of word2vec [11,12] as the most important ones for achieving good results. Consequently, in this study we focus of these hyper-parameters to produce a distributed representation of words in scientific text and evaluate the quality of embeddings in a domain-specific vocabulary.

Sub-sampling: vocabulary size. It has been proved in the literature [11,12, 16] that word2vec embedding quality increases as the corpus size increases. This is expected as longer corpus typically produce better statistics. Following on from this premise, we aim to investigate the role of vocabulary size in generating accurate embeddings for scientific text.

Unlike previous work that intuitively increments the vocabulary size by combining corpus, we propose to use the same corpus trained in two different ways that led to different vocabulary sizes. First, we train word2vec with unigrams. Second, we train the model with bigrams by using *word2phrase* – defined by Mikolov et al. [12] – that learns phrases by progressively joining adjacent pairs of words with an '_' character. Additionally, we sub-sample the frequent words on two steps which result into two different vocabulary sizes. Firstly, we remove all stop words and highly frequent academic words appearing in all publications. Secondly, we restrict the vocabulary to words that occur at least 10 times in the scientific corpus. According to Mikolov et al. [11], this sampling has proved to work well in practice. It accelerates learning and significantly improves the accuracy of the learnt embedding vectors, as it will be shown in Sect. 4.

Vector dimensionality and context window. The optimisation of *vector dimensionality* and *context window* parameters is supposed to be very crucial to achieve accurate results. The quality of embeddings increases with higher dimensionality under the assumption that it increases together with the amount of training data. But after reaching some point, the marginal gain will diminish [11].

The window size hyper-parameter corresponds to the span of words in the text that is taken into account, backwards and forwards when iterating through the words during model training. Similarly to the vector dimensionality hyper-parameter, the larger window size results in more topicality. Nevertheless, after a certain point, the marginal gain decreases.

Due to the sensitivity of these hyper-parameters and since hyper- parametrisation is generally known to be data and task dependent [4], we expect optimal hyper-parameter setting to be different for scientific text. Thus, we propose to study the marginal importance of word2vec hyper-parameters defined above using the *stability of k-nearest neighbors* of word vectors based on word similarities computed with *cosine distance* (Equation (1)) between embedding vectors.

k-NN stability for word2vec hyper-parametrisation. Stability is an important aspect of a learning algorithm. It has been widely used in clustering problems [18] to assess the quality of a clustering algorithm. Also, it has been applied in high-dimensional regression [15] for training parameter selection. Analogously and considering that word embedding presents high-dimensional word representations that led to word clusters, we propose to apply the *k-nearest neighbors* to tune the hyper-parameters of word2vec. k-NN is used to cluster similar words based on their cosine similarities.

The basic idea of word embedding stability is the following: embedding quality inevitably depends on tuning hyper-parameters defined previously, namely *vector dimensionality* and *context window*. If we choose accurate values of the tuning hyper-parameters, then we expect that the k similar words to a target word w from different embeddings should be similar. Specifically, we propose to fix one hyper-parameter, tune the second one by trying different values and training the model for each value. After each training, word similarities are computed and k-nearest neighbors words are defined. The *k-NN stability* is defined as a simple overlap rate of similar words resulted from two embeddings with different settings.

$$stability = \frac{\mathcal{S}_{E_h}^w \cap \mathcal{S}_{E_{h'}}^w}{k} \times 100 \qquad (2)$$

where $\mathcal{S}_{E_h}$ and $\mathcal{S}_{E_{h'}}$ are two sets of similar words to a target word w resulted respectively from two embeddings E_h and $E_{h'}$ with different hyper-parameter values. k is the number of nearest neighbors to w given by the cosine similarity. In this study, k is set to 5. This choice is motivated by our aim to keep the word similarities as fine-grained as possible in order to evaluate the quality of word2vec within scientific text.

3.2   Scientific Linguistic Regularities and Analogies

word2vec embeddings gain their success from their ability to capture syntactic and semantic language regularities. Surprisingly, they characterise each relationship by a relation-specific vector offset [13]. For example, the famous analogy *"king is to queen as man is to woman"* is encoded in the vector space by the vector arithmetic *"king - man + woman = queen"*. More specifically, the word analogy task aims at answering the question *"man is to woman as king is to — ?"* given the two pairs of words that share a relation ("man:woman", "king:queen") where the identity of the fourth word ("queen") is hidden.

Motivated by this ability of word2vec to identify relationships and capture analogies in textual data without any prior domain knowledge, we evaluate this ability in a domain-specific corpus, namely, scientific publications. Our aim is to assess as to what extent word2vec is able to correctly answer analogical questions in scientific text given the complexity of scientific language comparing to natural language.

The scientific word analogy we adopt is to query for scientific regularities captured in the vector model through simple vector subtraction and addition. More formally, given two pairs of words $(a : b')$ and $(b : b')$, our aim is to answer the question *(a is to a' as b is to —?)*. Thus, the vector of the hidden word b' will be the vector $(a' - a + b)$, suggesting that the analogy question can be solved by optimising:

$$\arg \max_{b' \in W}(similarity(b', a' - a + b)) \tag{3}$$

where W is the vocabulary and *similarity* is the cosine similarity measure defined in Equation (1).

This task is challenging for scientific language as no gold standard is available to evaluate the efficiency of word2vec in identifying linguistic regularities on unstructured scientific text, unlike existing work that use either the gold standard defined by Mikolov *et al.* [13] for general natural language tasks or pre-defined ontologies like NDF-RT ontology[3] for medical domain. To overcome this problem, we manually curate relationships related to *machine learning* research area from the *ACM hierarchy* and the *Wikipedia Machine Learning outline*, and we define a test set of analogy questions as *semantic questions* following the relation described above. The semantic questions are formed based on the hierarchical tree structure of both the ACM and Wikipedia outline that led to different *"Parents-Children"* relationships. For example, *"supervised_learning"* and *"unsupervised_learning"* are considered two parents for the two children *"classification"* and *"clustering"* respectively. Accordingly, the analogical question should be *"classification to supervised_learning is as clustering to —?"* To correctly answer the question, the model should identify the missing term with a correspondence counted as a correct match by finding the word *"unsupervised_learning"* whose vector representation is closest to the vector *("supervised_learning" - "classification" + "clustering")* according to the cosine similarity. Recall that for the

---

[3] National Drug File -Reference Terminology.

specificity and complexity of scientific language and respecting the interchangeability of scientific terms, instead of using the exact correspondence as the correct match, we adopt an approximate correspondence that considers an answer as correct if it belongs to the 10 nearest words given by cosine similarity in order to guarantee the applicability of our embeddings in scientific text. This is applied only for *semantic questions*. However, for *syntactic questions*, we adopt an exact correspondence. For example, the syntactic question *"classifier to classifiers is as forest to —?"* is considered correctly answered if and only if the word *"forests"* is the closest to the vector *("classifiers" - "classifier" + "forest")* according to the cosine similarity.

In addition to the *semantic questions* manually curated from ACM and Wikipedia, we define *syntactic questions* which are typically analogies about verb tenses/forms and singular/plural forms of nouns, in order to test the ability of word2vec to capture the syntactic regularities of scientific language.

4 Experimental Evaluation

4.1 NIPS Dataset: Description and Vocabulary Setup

To evaluate word embedding for scientific language, we used a subset of 2789 papers in the area of Machine Learning, published in NIPS (Neural Information Processing Systems) between 2012 and 2017. The dataset is publicly available on Kaggle[4] and contains information about papers, authors and the relation papers-authors. We used the papers database that defines six features for each paper: the *id*, the *title*, *the event type*, i.e., poster, oral or spotlight presentation, the *PDF name*, the *abstract* and the *paper text*.

The dataset needs to be pre-processed before being used for training the embedding model, since word2vec is very sensitive to vocabulary granularities like punctuation, lowercase, stop words, *etc.* which have a direct impact on the quality of generated word embeddings. After removing all punctuations and lowercasing the corpus, the pre-processing has the following steps:

(i) We removed stop-words using Stanford NLP stop word list[5] enriched by a list of 170 academic stop words that we defined from common academic vocabulary like *"introduction, abstract, table, figure, etc."*, (ii) We constructed bag of words where words are either *unigrams* used for standard word2vec training or *bigrams* used for *word2phrase* learning. The two settings resulted into different vocabulary sizes $|W_{unigrams}| = 35k$ and $|W_{bigrams}| = 96.7k$, and (iii) We discarded less frequent words that appear less than 10 times in the vocabulary in order to accelerate learning. This led to a different vocabulary size $|W_{downsampled}| = 57k$.

---

[4] https://www.kaggle.com/benhamner/nips-papers/data.

[5] http://github.com/stanfordnlp/CoreNLP/blob/master/data/edu/stanford/nlp/patterns/surface/stopwords.txt.

4.2   word2vec Training Details: Hyper-Parameters Optimisation

As described in Sect. 3, *k-NN stability* was used to optimise the word2vec hyper-parameters, namely, *vector dimensionality* and *context window size*.

Vector dimensionality. *k-NN stability*, with $k = 5$, was used to evaluate the influence of the vector dimensionality hyper-parameter using vector models generated with 20, 30, 50, 100, 150, 200, 300 and 500 dimensions, *skip-gram* architecture and three different vocabulary sizes as described in Sect. 4.1.

In Table 1, we show the results of *k-NN stability* values that vary vector length and vocabulary size. word2vec model was initially learned with 20-vector dimension. This trained model was used as a seed setting to start computing *k-NN stability*. More specifically, *k-NN stability* at 30-vector dimension was computed based on the 20-vector dimension following Equation (3) and respectively each *k*-NN stability value is computed based on the results generated by the previous dimensionality setting. The reported results correspond to the stability average of the top 100 frequent words (unigrams and bigrams) in the vocabulary.

It has been clearly seen from the three vocabulary sizes that the stability increases considerably as the dimensionality increases. But after reaching some point, it diminishes or becomes slightly invariant. For instance, for the *unigram vocabulary*, *k*-NN stability reached 67% with 100-dimension vector performing good results comparing to 30 and 50 dimensions. However, it remains basically steady with a slight increase of 1% at 200-dimension. This increase is not remarkable enough to consider 200-dimension better than 100-dimension since a higher dimension of the vectors implies a bigger size of the resulting vector model and more training time. Then, we notice that the stability decreases with larger dimensions (300 and 500). Consequently, these results suggest that 100-dimension vector consistently yielded better stability with unigrams vocabulary.

Table 1. *k*-NN stability for vector dimensionality optimisation

	D30	D50	D100	D150	D200	D300	D500
Unigrams	42%	53%	67%	67%	68%	66%	65%
Bigrams	51%	47%	56%	64%	68%	70%	71%
Downsampled bigrams	58%	61%	65%	73%	81%	n/a	n/a

Similarly, *bigrams vocabulary* shows a substantial improvement in *k*-NN stability from 30-dimension to 200-dimension with 68%. Then, it increases slightly with 300 and 500 dimensions with a 1% gain. Hence, for this vocabulary, we can fix the optimal dimensionality value to 200. Interestingly, the stability results of the *unigram vocabulary* and the *bigram vocabulary* confirm the hypothesis that vector dimensionality and the amount of training data should be increased together to have better results. As a matter of fact, 100 has shown to be the better vector length for *unigram vocabulary* of $35k$ size, while 200 is better for *bigram vocabulary* of $96.7k$ size. On the other hand, by looking at the stability values at high dimensions (300 and 500), we noticed the excess in stability of *bigrams vocabulary* comparing with *unigrams vocabulary*. This is comprehensibly justified by three facts: *(i)* this confirms the hypothesis that word2vec model quality increases as corpus size increases [14], *(ii)* this proves that n-gram enhanced skip-gram model performed better than regular skip-gram based only on unigrams, *(iii)* this confirms the specificity of scientific language and mainly the *Computer Science* area that contains an important number of bigrams like *"machine-learning"*, *"artificial-intelligence"*, etc.

Based on these findings, mainly *(i)* and *(ii)*, we ignored the 300 and 500 dimensions for training the *downsampled vocabulary* which is resulted from downsampling the *bigram vocabulary* as the vocabulary size is obviously smaller ($57k$). It is worthy to note that this downsampling improved the training speed and most importantly made the k-NN stability values more important with 81% at 200-dimension while it was 68% with *bigram vocabulary* at the same dimension. This was expected as downsampling makes the word representations significantly more accurate [12].

Overall, the k-NN stability results obtained through vector dimensionality optimisation show that bigram enhanced skip-gram model performs better with scientific language, 200 is the optimal vector length for the used dataset and the *downsampled bigram vocabulary* significantly outperforms the two other vocabularies in term of k-NN stability and computation time. Note that for all word2vec training rounds with different vocabularies and different vector dimensionalities, the hyper-parameter *window context* was set to 5, the default window size value provided by *gensim*[6].

Window context. Similarly to the setting followed to optimise vector- dimensionality, k-NN stability was adopted to find the optimal window size for the used scientific corpus in this study. Building on previous results, the trained vocabulary used is the *downsampled vocabulary* and the *vector dimensionality* is 200. word2vec embeddings were generated with skip-gram model and 7 different window sizes ranging from 2 to 8. word2vec was initially trained with a context window of size 2 as a starting point. Then k-NN stability was computed respectively based on the previous embedding results. Figure 1 presents the values of k-NN stability that vary context window size. It is clearly seen from the figure that the optimal window size is 6 with a stability of 70% for the used scientific corpus. Our results confirm the fact that larger window size results in more topicality and accordingly better accuracy of word representations. However, the marginal gain decreases after a certain point. Overall, our findings show that the combination of 200-vector dimension with context window of size 6 and downsampled bigram vocabulary proved to be the best configuration of skip-gram word2vec model. Additionally, the proposed *k-NN stability* – based on

---

[6] https://radimrehurek.com/gensim/.

word similarity as embedding properties, that we adopted in this study to optimise the word2vec hyper-parameters for scientific text – confirms all hypotheses
related to word embeddings supported in the literature and even goes beyond
them by giving a standard way to be sure about the stability of results.

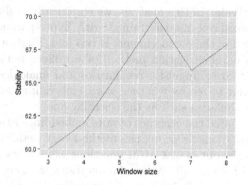

Fig. 1. k-NN stability for context window size optimisation

4.3   Analogy Evaluation

As described in Sect. 3.2, the word analogy task attempts to query for scientific regularities captured in the embedding model – trained with the previously
optimised hyper-parameters – through simple vector subtraction and addition.

The analogy dataset we created contains 1991 analogical questions, divided
into 1871 *semantic questions* and 120 *syntactic questions*. The *semantic questions*
were manually curated from ACM hierarchy (406 questions) and Wikipedia outline of *Machine Learning* (1465 question). The number of relationships generated
from Wikipedia are by far greater than the ACM counterpart. This justified by
the fact that ACM is more coarse-grained as it covers all the *Computer Science*
area, while the Wikipedia outline is a fine-grained hierarchy generated specifically for *Machine Learning* with very detailed algorithms and applications of
the area. We remove from the analogy dataset all questions that contain words
that do not exist in our vocabulary in order to fairly evaluate embedding analogies. This resulted into 1573 questions (322 ACM questions and 1251 Wikipedia
questions). Similarly to *semantic questions*, *syntactic questions* were a manually
generated subset that we created from the scientific text using typical analogies
about verb tenses/forms and singular/plural forms of nouns, in order to test the
ability of word2vec to capture the syntactic regularities of scientific language.
The number of questions is relatively small due to our aim to only preliminarily
test the word2vec ability to cover syntactic scientific regularities that do not
differ from natural language, while the *semantic questions* do. That is why we

focus more on these latter. Our analogy dataset is available online for more reproducibility and any further use by researchers[7].

For evaluating our embeddings in capturing linguistic regularities and analogies, we performed both quantitative and qualitative analysis.

Quantitative analysis. In this analysis, we empirically evaluate our proposed bigram-enhanced word2vec model trained with hyper-parameters experimentally tuned. Our goal of these experiments is two-fold. First, we aim to evaluate whether our hyper-parametrisation method of word2vec is useful for resulting embeddings able to cover linguistic regularities and analogies within scientific text. Second, we aim to assess whether word embeddings are worth using in domain-specific vocabularies such as the scientific one.

To do so, we computed the *accuracy* of word embeddings to answer the semantic and syntactic questions following the methodology detailed in Sect. 3.2. For semantic questions, 50 out of 322 ACM questions were correctly answered with an accuracy of 15.52% while 75 Wikipedia questions were correct out of a subset of 413 questions from the 1251 questions in the dataset, with an accuracy of 18%. The difference in accuracy between ACM and Wikipedia questions was expected as Wikipedia relationships were more detailed and covered *Machine Learning* names of algorithms and applications that widely occur in the vocabulary, while ACM was more coarse-grained. Although, the accuracy of both of them is very low. This could be justified by three different reasons. First, the corpus size we used is relatively small with only $57k$ while it has been shown that word2vec quality increases as corpus size increases. For instance, Mikolov et al. [12] trained their model on a corpus of $1B$ and obtained a semantic accuracy of 61%. Second, the used NIPS dataset is about very recent publications (between 2012 and 2017). So that, the vocabulary is more probably about recent topics and accordingly recent *Machine Learning* vocabulary, i.e., names of algorithms and applications might gain more frequencies in the text than the old ones, which in turn would highly affect the word representations, at the time when ACM hierarchy or Wikipedia outline are time-independent and contain generic *Machine Learning* vocabulary. Third, the scientific language is complex and does not contain explicit and accurate relationships as natural languages does. For instance, 'accuracy' and 'error rate' in the machine learning literature are used in similar contexts, despite having opposite semantics.

For all these reasons, the semantic accuracy of word embedding within the used scientific corpus is considered modest. But, it is promising as it is interpretable and improvable on one hand. On the other hand, it reveals challenges about scientific word embedding. More specifically, it is worth investigating the convergence and divergence of some *Machine Learning* algorithms and applications over time which consistently affects the word representations. Interestingly, it is challenging to find a suitable way to train and evaluate word embeddings in such dynamic vocabularies.

---

[7] https://github.com/AmnaKRDB/Machine-Learning-Analogies.

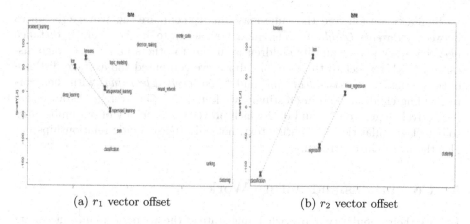

(a) r_1 vector offset   (b) r_2 vector offset

Fig. 2. Vector offsets examples of Machine Learning semantic relationships

For syntactic questions, we computed the accuracy across the 120 questions we defined. Interestingly, we found 82 questions out of 120 correctly answered with an accuracy of 68%. This result is interesting despite the small size of our vocabulary. It outperforms the syntactic accuracies of Mikolov et al. [12] which reached 61% with $1B$ vocabulary and 300-dimension vector.

Qualitative analysis. The embedding we learned revealed interesting patterns in *Machine Learning* vocabulary through relation-specific vector offsets. For instance, it captured different semantic relationships mapping *Machine Learning* techniques and related algorithms such as r_1 "*unsupervised_learning is to kmeans as supervised_learning is to knn*", and r_1 "*classification is to knn as regression is to linear_regression*". We illustrate these patterns by plotting word vector representations with *t-distributed stochastic neighbor embedding (t-SNE)* [9] as a qualitative way to evaluate our embeddings following Yao et al. [21]. Figure 2a and Fig. 2b show the t-SNE representations of r_1 and r_2 respectively.

In addition to the t-SNE visualisation used to qualitatively evaluate the accuracy of our embeddings to detect interesting patterns in the scientific text, we suggested to evaluate the capability of our model to capture the hierarchy structure "*Parent-Children*". To do so, we computed and compared similarities between every word "*parent*" and the corresponding words "*children*". The model is considered accurate if the distances are approximately equal. For instance, the distances between the parent "*supervised_learning*" and its children { "*classification*", "*regression*", "*ranking*", "*cost_sensitive*"} are approximately equal with slight differences as presented here respectively (0.369; 0.241; 0.173; 0.223) similarly to the parent "*unsupervised_learning*" and its children { "*clustering*", "*dimensionality_reduction*", "*topic_modeling*", "*anomaly_detection*", "*mixture_modeling*", "*source_separation*"} with approximately similar distances (0.259; 0.307; 0.237; 0.145; 0.145; 0.135; 0.253).

Similarly, we followed the same reasoning to compare the average distances between *"Parents-Children"*. The model is accurate if the average distance between every parent and its children is similar to others parents' average distances. With respect to the example above, we computed the average distance of the parents *"supervised_learning"* and *"unsupervised_learning"* with their corresponding children. And interestingly, we found that the average distances are respectively equal to 0.25 and 0.22 which proves the accuracy of our embedding to detect granularities of scientific text, not only the semantic relationships but also the hierarchical structure.

5 Conclusions and Future Work

Despite their popularity in overwhelming state of the art performance in semantic similarity and analogy tasks, word embeddings are still treated as black boxes and uniformly use the hyper-parameters without a methodological setting. From this perspective and aiming to provide a precise hypotheses synthesis, this work addressed word embedding hyper-parametrisation for domain-specific use, namely the scientific domain. By proposing the stability of k-nearest neighbors of word vectors, we were able to methodologically set the hyper-parameters suitable for scientific text. Our method has been validated quantitatively and qualitatively on semantic and syntactic analogies curated from ACM and Wikipedia as gold standard and has proved its effectiveness.

As a short term objective, we plan to apply our method on larger scientific vocabulary, then generalise it on different research areas. For long term objectives, we plan to investigate more settings for word embeddings within the scientific area, aiming to detect trendy and evolving patterns by performing time-aware vocabulary augmentation and sliding windows.

References

1. Bengio, Y., Ducharme, R., Vincent, P., Janvin, C.: A neural probabilistic language model. J. Mach. Learn. Res. **3**, 1137–1155 (2003)
2. Deerwester, S., Dumais, S., Furnas, G., Landauer, T., Harshman, R.: Indexing by latent semantic analysis. J. Am. Soc. Inf. Sci. **41**(6), 391–407 (1990)
3. Heffernan, K., Teufel, S.: Identifying problems and solutions in scientific text. Scientometrics (2018)
4. Hutter, F., Hoos, H., Leyton-Brown, K.: An efficient approach for assessing hyper-parameter importance. In: 31st International Conference on Machine Learning, pp. 754–762 (2014)
5. Kusner, M.J., Sun, Y., Kolkin, N.I., Weinberger, K.Q.: From word embeddings to document distances. In: 32Nd International Conference on Machine Learning, pp. 957–966 (2015)
6. Levy, O., Goldberg, Y.: Dependency-based word embeddings. In: 52nd Annual Meeting of the Association for Computational Linguistics, pp. 302–308 (2014)
7. Lu, W., Huang, Y., Bu, Y., Cheng, Q.: Functional structure identification of scientific documents in computer science. Scientometrics **115**(1), 463–486 (2018). Apr

8. Lund, K., Burgess, C.: Producing high-dimensional semantic spaces from lexical co-occurrence. Behav. Res. Methods Instrum. Comput. **28**(2), 203–208 (1996)
9. van der Maaten, L., Hinton, G.: Visualizing data using t-SNE. J. Mach. Learn. Res. **9**, 2579–2605 (2008)
10. Manning, C.D., Raghavan, P., Schütze, H.: Introduction to Information Retrieval. Cambridge University Press, New York (2008)
11. Mikolov, T., Chen, K., Corrado, G., Dean, J.: Efficient estimation of word representations in vector space. CoRR abs/ arXiv:1301.3781 (2013)
12. Mikolov, T., Sutskever, I., Chen, K., Corrado, G., Dean, J.: Distributed representations of words and phrases and their compositionality. In: 26th International Conference on Neural Information Processing Systems, pp. 3111–3119 (2013)
13. Mikolov, T., Yih, W.t., Zweig, G.: Linguistic regularities in continuous space word representations. In: HLT-NAACL, pp. 746–751 (2013)
14. Miñarro-Giménez, J.A., Marín-Alonso, O., Samwald, M.: Applying deep learning techniques on medical corpora from the world wide web: a prototypical system and evaluation. CoRR abs/ arXiv:1502.03682 (2015)
15. Meinshausen, Nicolai: Peter Bhlmann: Stability selection. J. R. Stat. Soc. **72**(4), 417–473 (2010)
16. Pennington, J., Socher, R., Manning, C.D.: Glove: global vectors for word representation. EMNLP **14**, 1532–1543 (2014)
17. Petterson, J., Buntine, W., Narayanamurthy, S.M., Caetano, T.S., Smola, A.J.: Word features for latent dirichlet allocation. In: Advances in Neural Information Processing Systems, pp. 1921–1929 (2010)
18. Rinaldo, A., Singh, A., Nugent, R., Wasserman, L.: Stability of density-based clustering. J. Mach. Learn. Res. **13**(1), 905–948 (2012). Apr
19. dos Santos, C.N., Gatti, M.: Deep convolutional neural networks for sentiment analysis of short texts. In: COLING, pp. 69–78 (2014)
20. Turian, J., Ratinov, L., Bengio, Y.: Word representations: A simple and general method for semi-supervised learning. In: 48th Annual Meeting of the Association for Computational Linguistics, pp. 384–394 (2010)
21. Yao, Z., Sun, Y., Ding, W., Rao, N., Xiong, H.: Dynamic word embeddings for evolving semantic discovery. In: 11th ACM International Conference on Web Search and Data Mining, pp. 673–681 (2018)
22. Zhao, S., Zhang, D., Duan, Z., Chen, J., Zhang, Y.p., Tang, J.: A novel classification method for paper-reviewer recommendation. Scientometrics, pp. 1–21 (Mar 2018)

Hierarchical Expert Profiling Using Heterogeneous Information Networks

Jorge Silva[1,2]([✉]) [iD], Pedro Ribeiro[1,2] [iD], and Fernando Silva[1,2] [iD]

[1] CRACS & INESC TEC, Porto, Portugal
[2] Departamento de Ciência de Computadores - Faculdade de Ciências,
Universidade do Porto, Porto, Portugal
jmbs@inesctec.pt;{pribeiro, fds}@dcc.fc.up.pt

Abstract. Linking an expert to his knowledge areas is still a challenging research problem. The task is usually divided into two steps: identifying the knowledge areas/topics in the text corpus and assign them to the experts. Common approaches for the expert profiling task are based on the Latent Dirichlet Allocation (LDA) algorithm. As a result, they require pre-defining the number of topics to be identified which is not ideal in most cases. Furthermore, LDA generates a list of independent topics without any kind of relationship between them. Expert profiles created using this kind of flat topic lists have been reported as highly redundant and many times either too specific or too general.

In this paper we propose a methodology that addresses these limitations by creating hierarchical expert profiles, where the knowledge areas of a researcher are mapped along different granularity levels, from broad areas to more specific ones. For the purpose, we explore the rich structure and semantics of Heterogeneous Information Networks (HINs). Our strategy is divided into two parts. First, we introduce a novel algorithm that can fully use the rich content of an HIN to create a topical hierarchy, by discovering overlapping communities and ranking the nodes inside each community. We then present a strategy to map the knowledge areas of an expert along all the levels of the hierarchy, exploiting the information we have about the expert to obtain an hierarchical profile of topics.

To test our proposed methodology, we used a computer science bibliographical dataset to create a star-schema HIN containing publications as star-nodes and authors, keywords and ISI fields as attribute-nodes. We use heterogeneous pointwise mutual information to demonstrate the quality and coherence of our created hierarchies. Furthermore, we use manually labelled data to serve as ground truth to evaluate our hierarchical expert profiles, showcasing how our strategy is capable of building accurate profiles.

Keywords: Expert profiling · Topic modelling · Information networks

© Springer Nature Switzerland AG 2018
L. Soldatova et al. (Eds.): DS 2018, LNAI 11198, pp. 344–360, 2018.
https://doi.org/10.1007/978-3-030-01771-2_22

1   Introduction

With the current exponential growth in web-documents, the problem of linking persons to knowledge areas and vice-versa has gained a lot of attention. This problem is known as expertise retrieval [1] and it is divided into two sub-problems: expert profiling and expert finding. The former identifies the areas of expertise of a person, while the latter finds experts in a certain topic. In literature, the expert finding task has been receiving considerably more attention than the expert profiling one. In this paper we focus on the expert profiling task. Creating accurate knowledge profiles of a person has several important applications such as [2]: categorizing personal according to their skills, identifying possible collaborations, and tracking individual or group evolution of expertise. Furthermore, the profiles generated could be used as sources of information in the expertise finding task [6,11].

In most cases, the expert profiling problem does not have a pre-defined set of knowledge areas for the persons. Instead, they are identified in a data-driven fashion using a topic modelling approach. The Latent Dirichlet Allocation (LDA) [3] model is the most widely used strategy to define the knowledge areas/topics in text. Due to its potential, the LDA algorithm was adapted to output the distribution of authors over the discovered topics [15]. This discovery fostered the development of a group of algorithms named Author-Topic models that, not only identify topics in documents, but also profile the author's expertise. Since then, several other Author-Topic models have been proposed [7,10,12].

The core of the Author-Topic models is the LDA algorithm and despite it being widely used, there are some known flaws in it [8]: lacks an intrinsic methodology to choose the number of topics, contains several hyper-parameters that can cause overfitting, and it is incompatible with properties of text such as Zipf's law for the frequency of words. In order to avoid these flaws, we propose a different strategy to the topic modelling part. The vast number of Author-Topic models that exist in literature, indicate that adding external sources of information besides text, improves the quality of expert profiles. Therefore, we use documents' meta-data to model their inter-relations in a Heterogeneous Information Network (HIN), and we uncover hidden structures in the linked data that represent topics/knowledge areas which can be used to categorize a person's knowledge. An advantage of this process when compared to LDA is that it does not require defining the number of topics to be discovered.

With respect to the expert profiling task, experts have reported that the profiles assigned to them are redundant, and either too general or too specific [2]. This occurs because the expert profiles are generated from a flat list of topics without any relation between them. A solution to the problem is to create an hierarchy of topics with "sub-topic of" relations. Unfortunately, automatically creating these structures and mapping experts into them is not trivial [16,21]. In this work, we take advantage of the HIN to organize the topics discovered in an hierarchy and to map the experts into the topics. As a result, we are capable of creating hierarchical profiles that on top represent broad knowledge areas and on the bottom more specific ones. Figure 1 illustrates the differences between a flat and an hierarchical profile.

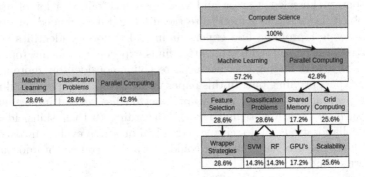

Fig. 1. Example of flat versus hierarchical organization of topics of expert profiling.

This paper is structured as follows. In Sect. 2 we discuss the related word for the topic modelling and expert profiling tasks. In Sect. 3 we formalize the task of creating an hierarchical expert profile from an heterogeneous information network. In Sect. 4 we describe our model and in Sect. 5 we evaluate the topic modelling and the expert profiles constructed. Finally, in Sect. 6 we present the conclusions and address future work.

2   Related Work

In the expert profiling domain, the Author-Topic models are widely used for the task. These models are inspired by the Latent Dirichlet Allocation (LDA) algorithm [3] which represents topics as a multinomial mixture over words, and documents as a multinomial distribution over topics. In 2004, Rosen-Zvi et al. [15] added the authors distribution of documents to the LDA model, thus creating the first Author-Topic model and fostering the motivation to several other ones. Tang et al. [20] unveiled the importance of adding the conference distribution to the author-topic models. Later, Wang et al. [22] proposed the Author-Conference-Topic-Connection model which besides adding the conference distribution, also adds the subjects of the conferences. In 2012, Daud [5] added the documents timestamps and proposed the Temporal-Author-Topic which models the topic distribution of an author over time. Later, Jeong et al. [10] proposed the Author-Topic-Flow which allows each author to directly have a temporal pattern of expertise. Duan et al. [7] explored the community information in networks, and proposed the Mutual Enhanced Infinite Community-Topic model which finds communities and the topics they discuss in text-augmented social networks. This work was the pioneer in simultaneously integrating community discovery with topic modelling, while considering communities and topics as different latent variables (i.e. a community may be interested in several topics).

There are some works in literature that rely on information networks to avoid the problems of the LDA model. Gerlach et al. [8] represented a word-document matrix as a bipartite network, and reformulated the problem of topic modelling as the task of finding communities in such network. The authors proposed the hierarchical Stochastic Block Model (hSBM) which is a probabilistic inference approach that is capable of handling the possibility of higher-order structures. Consequently, the algorithm is capable of generating an hierarchy of topics. Some different approaches that focus on topic modelling using HINs have been proposed. Rankclus [18] was a pioneer algorithm that simultaneous clusters and ranks nodes in a HIN using a generative model that operates on bipartite topologies. Netclus [19] emerged later with the intent to extend the Rankclus to HINs with a star-topology. More recently, CATHYHIN [21] extended the previous algorithms to support the following features: ranked list of attributes for each type along with a ranked list of phrases, any HIN topology, soft-clustering of all the nodes, and developing an hierarchy of topics. With respect to this work, CATHYHIN produces a similar output to our algorithm (i.e. an hierarchy of topics where each topic consists of multiple node types, see Fig. 2 for an illustration.). However there are two main differences in our work. To start CATHYHIN uses a generative model to discover the communities while we use modularity optimization. Additionally, CATHYHIN focus on discovering topics in an HIN. We extend this goal and we define strategies to map the experts into the discovered topics to create their hierarchical expert profile.

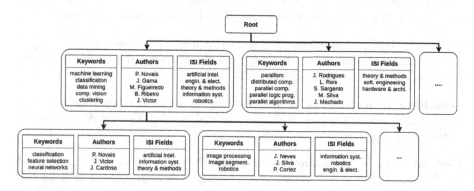

Fig. 2. Sample of the hierarchy of topics obtained from our algorithm.

In literature, there are a few works that create a expertise profile with hierarchical properties. Bin et al. [9] uses explicit feedback from persons and their bookmarks information to extract keywords that reflect their expertise. Afterwards, these keywords are mapped into a pre-defined ontology. Thus constructing an hierarchical profile. Rybak et al. [16] uses publication's meta-data to maps authors into the ACM computation classification system.[1] Since this is organized

---

[1] https://www.acm.org/publications/class-2012.

in hierarchies, the expert profile is also hierarchical. An important aspect of both strategies is the fact that they use a manually created hierarchy which requires a lot of human effort. Moreover, these structures are dynamic. As a result this is not a one time task [21]. Additionally, there is the problem of mapping the expert's knowledge into the hierarchy. In [9] the authors have to restrict the keywords to the ones that are on the ontology. On the other hand, Rybak [16] restricts the author's publication to the ones published in ACM conferences. Both strategies potentially leave out details that may be relevant to characterize the experts' knowledge. In this work we automatically create the topological hierarchy, and since our topics consist of multiple entities, we are capable of mapping experts directly (the author is part of the topics) and indirectly (author is represented by other meta-data attributes) into the hierarchy.

3　Problem Description

We formalize the problem of creating hierarchical profiles for experts as the task of receiving an HIN, generating a topical hierarchy and the mapping expert's knowledge into that structure.

Definition 1. *An information network is defined as a directed graph $G = (N, L)$ with an object type mapping $\psi : N \to A$ and a link type mapping $\varphi : L \to R$. Each node $n \in N$ belongs to an object type $a : \psi(n) \in A$. Furthermore, each link $l \in L$ belongs to a relation type $r : \varphi(l) \in R$. If two links share the same relation type, they both start at a node with type a' and end at a node with type a''.*

An HIN is a type of information network where $|A| > 1$ and/or $|R| > 1$. For a better understanding of the object types and relations, HINs have a meta-level description named network-schema [17].

Definition 2. *We define a topical hierarchy as a tree T where each node is a topic. Each topic t contains $|A'|$ lists of ranked attributes where $A' \subseteq A$ and A is the set of object types in the HIN.*

Definition 3. *An hierarchical expert profile P is a tree such that $P \subset T$. Each $t \in P$ contains a q indicating the percentage of knowledge of the expert on that topic. Additionally, $\forall l \in L$, $\sum_{t \in P_l} t_q = 1$, where L is the number of levels in the tree and P_l is the set of topics at level l.*

Our proposed model is divided into two parts. The first consists in defining a function θ such that $\theta(G) = T$. Then, we introduce two strategies to create a function λ such that $\lambda(T, e) = P_e$, where e is an expert and P_e his hierarchical expert profile. We address the construction of both functions in the next section.

4   Hierarchical Expert Profile

4.1   Network Construction

The model proposed in this work can be applied to any HIN. However, to ease understanding we present the discussion and evaluation of its components in the context of bibliographic databases. More concretely, we use data from Authenticus[2] which is a bibliographic database for the Portuguese researchers. To construct the HIN we select a set of publications and, for each one, we query the database for the following meta-data: authors, keywords and ISI fields[3]. Then, the HIN is constructed following a star-schema topology where publications are the star-nodes, and authors, keywords and ISI fields are the attribute-nodes (see Fig. 3 for an illustration). There are three different types of relations: publication-author, publication-keyword and publication-ISI field. Each relation has a different W_x that represents the importance of objects of type x in the network. The W_x values are normalized with respect to the number of attributes x connected to the star-nodes (in this case publications). For example, considering that W_a is the publication-author's weight, all the n authors of a certain publication p have a link weight of $\frac{1}{n}W_a$.

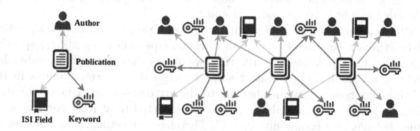

Fig. 3. Network scheme of our proposed bibliographic HIN.

4.2   Topic Modelling

Once we have an HIN we apply a modularity optimization algorithm to unveil communities on the network structure. We assume that the communities represent topics/knowledge areas for the expert profiling task. Given a network community c, modularity [14] estimates the fraction of links within c minus the expected fraction if links were randomly distributed. The value of modularity ranges between -1 and 1. Positive values indicate that the number of links in c, exceeds the number of expected ones at random. A modularity based community detection algorithm aims to maximize the global modularity of the communities in the network. However, due to the time complexity of the task, algorithms

[2] https://www.authenticus.pt.
[3] Research areas created by the Institute for Scientific Information.

must use some heuristics in order to decrease its computational cost. In this work we use Louvain algorithm [4] which is a greedy optimization method with expected runtime $O(n \log(n))$, where n is the number of nodes in the network.

With respect to our overall goal of topic modelling in HINs, using Louvain algorithm presents some drawbacks: does not account for nodes and links heterogeneity, ignores network-schema, and produces non-overlapping communities. The first two points lead to a loss of information in the HIN. The latter produces the undesired effect of hard-clustering attribute-nodes (by intuition, some authors/keywords should be part of more than one community). In order to tackle these problems, before applying the Louvain algorithm to detect communities we adapt our HIN to a similarity graph of star-nodes $G' = (N', L')$. In case of our bibliographic HIN, all the nodes in G' are publications and the links represent how related two publications are.

The process to construct G' starts with the selection of all the star-nodes from the HIN. Each one represents a different node in G'. The edge weights between every pair of nodes $(p1, p2) \in L'$ are defined by the following formula:

$$l_{p1,p2} \in L' = \sum_{n \in K} l_{p1,n} + \sum_{n \in K} l_{p2,n} \tag{1}$$

where K is the set of nodes that are adjacent to $p1$ and $p2$ in the HIN, and $l_{x1,x2}$ is the edge weight between nodes $x1$ and $x2$.

After the construction of the similarity graph we apply the Louvain algorithm which returns a community partition C that maps nodes into their respective community. Extrapolating C to the HIN, we obtain the community membership of all the star-nodes. On the next step, we expand these communities in the HIN to assign community membership to the attribute-nodes. Due to our star-schema topology, every attribute-node a is connected to at least one star-node p, that belongs to a community $c_j \in C$. Therefore, we estimate the community membership of attribute-nodes as the fraction of their link weights connected to different communities. For example, if a_i is linked to star-nodes $p1, p2$ and $p3$, and $p1$ and $p2$ are members of community c_1 and $p3$ is member of community c_2, then the community membership of a is 67% in c_1 and 33% in c_2.[4]

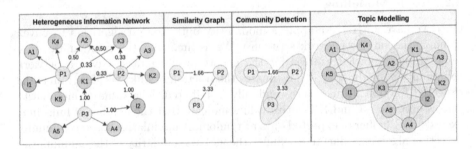

Fig. 4. Topic modelling in HINs using modularity-based community detection.

---
[4] For simplicity consider that the links have the same weight.

In the end of the whole process, all the nodes in the HIN are assigned to one or more communities. In the context of the bibliographic data of this work, we aim that our topics consist of three ranked lists of attributes: authors, keywords and ISI fields.[5] Therefore, to rank the attributes within a community, we remove the star-nodes on the network and generate a new HIN with a different network-schema. Figure 4 illustrates the different phases of topic modelling in a HIN.

4.3   Ranking Attributes Within a Topic

With respect to the information network, a topic consists of a sub-network of nodes of three attributes types. In order to better understand the topics discovered, we rank the nodes within each topic according to their importance and type. For the purpose we used several network centrality metrics: node's degree, PageRank, betweenness, closeness and eigenvector. Through experimentation we determined that PageRank seems to be the best metric for our purposes. In this work we use the node's ranking within a topic, to facilitate human interpretation of what a topic represents. However, in the case of extending our expertise profiles to other tasks such as the expert finding one, the rankings could be used to determine who is the best expert in a certain domain.

4.4   Hierarchical Topics

The topic modelling strategy presented in Sect. 4.2 creates a flat list of topics for a HIN. In this section we summarize the steps necessary to create an hierarchy of topics with a pre-defined number of l levels:

1. Start with HIN $G = (N, L)$
2. Convert the HIN into a similarity graph G' of star-nodes.
3. Apply the *Louvain* community detection algorithm such that $Louvain(G') = C$ where $C = C_1, C_2, ..., C_k$ and each C_i represents a community of star-nodes.
4. Transfer the communities information into the HIN and estimate the community membership of all the attribute nodes.
5. For each $C_i \in C$:
   (a) Create subgraph $G_{Ci} = (N', L')$ where N' is the set of the nodes in community C_i and L' the links between those nodes in G.
   (b) Rank all the attribute nodes according to their importance and object type.
   (c) If the current level is lower than l, set $G = G_{Ci}$ and go back to step 1.

4.5   Mapping Experts into the Hierarchical Topics

One of the problems of using an hierarchy of topics on the expert profiling task is that most of the times, mapping the experts into the hierarchy is either not trivial, or it requires discarding information [9,16]. In our strategy, we generate

---

[5] As illustrated by Fig. 2.

topics that consist of multiple attributes. As a result we can use them to map the experts into the topical hierarchy and create expertise profiles. In cases where the expert is represented by a node in the HIN, there is a direct mapping into the hierarchy. Otherwise, the expert can be mapped indirectly using attributes that characterize his expertise and are represented in the HIN.

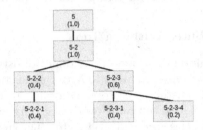

Fig. 5. Example of an hierarchical expert profile.

To create the expert profile of an expert e that is part of the HIN, we transverse the topical hierarchy T and consider all the topics he is part of. For example, let us consider that e at the lowest level of T is 40% in topic "*5-2-2-1*", 40% in "*5-2-3-1*", and 20% in "*5-2-3-4*".[6] Then, its expert profile p_e considering the complete hierarchy, would be:

- 1st level: 1.0 in topic "*5*"
- 2nd level: 1.0 in topic "*5-2*"
- 3rd level: 0.4 and 0.6 in topics "*5-2-2*" and "*5-2-3*"
- 4th level: 0.4, 0.4 and 0.2 in topics "*5-2-2-1*", "*5-2-3-1*" and "*5-2-3-4*"

Figure 5 illustrates e's expert profile. In cases where e is not represented in T, we obtain his profile by considering the set of keywords K that he has used in his publications. For each $k_i \in K$ we match it with a keyword node in the HIN by selecting the one with highest Word2Vec similarity [13] to k_i, and obtain its topical profile r_i (similar to the one illustrated in Fig. 5.) Then, we sum all the topical profiles into a single one, considering the times the expert used each keyword. For each topic in the merged profile M_p, we estimate its value (V_t) using the following formula:

$$V_t = \sum_{k \in K} \chi(r_i, t) \tag{2}$$

where χ is a function that given a topical profile r_i, extracts the value associated to topic t. On the final step, we normalize the topics' values per hierarchy level in order to make them comparable to profiles extracted directly from T. In this work we are interested in expert profiles, however using the indirect mapping we are capable of creating knowledge profiles for other entities. For example, we can create the profile for a research institution using its authors, or for a conference using the keywords used in it.

---

[6] For clarification, an '-' symbol refers to a different level on the hierarchy.

5   Experimental Evaluation

In this section we test the efficiency of the discovered topics and the quality of the profiles created using them. For the purpose, we constructed a dataset using all the computer science related publications from the Authenticus database. Our dataset consists of 8587 publications, 2715 authors, 19662 keywords and 120 ISI fields. With this data, we constructed 8 Heterogeneous Information Networks (HIN) changing the weights assigned to each type of relation. For each HIN we applied our model to create a topical hierarchy setting the number of levels to 4.[7] Table 1 shows the relational weights used and the number of topics discovered per hierarchical level.

Table 1. Relational weights and number of topics discovered for the constructed HINs. P-K: publication-keyword. P-A: publication-author and P-I: publication-ISI field.

HIN	Relation weights uniform?			Number of topics per level				
		P-K	P-A	P-I	level 0	Level 1	Level 2	Level 3	Total
CS_1	Yes	1.0	1.0	1.0	4	9	10	10	33
CS_2	No	1.0	1.0	1.0	4	55	122	200	381
CS_3	No	2.0	1.0	0.5	4	85	352	684	1125
CS_4	No	2.0	0.5	1.0	4	72	253	479	808
CS_5	No	1.0	2.0	0.5	4	51	235	563	853
CS_6	No	0.5	2.0	1.0	4	22	54	94	174
CS_7	No	1.0	0.5	2.0	4	14	30	49	97
CS_8	No	0.5	1.0	2.0	4	9	19	21	53

To evaluate the importance of normalizing the relation weights per publication, we constructed a HIN (CS_1) where the weights are uniform. From the results we observe that the relational weights have a huge impact on the number of topics discovered. Increasing the importance of the publication-keyword relation generates the most topics. On the other hand, decreasing this relation while increasing the publication-ISI field one, generates the least among the HINs with no uniform weights. The uniform HIN generated the fewest number of topics by a high margin.

5.1   Topic Evaluation

In literature, there are several metrics to evaluate the quality of topics modelled. However, they assume that the topics consists only of words, and that they were obtained using statistical inference on text. Our task of constructing an hierarchy

---

[7] Through experimentation we determined that 4 was the number of levels that achieved the most comprehensible topical hierarchy.

of topics, where each topic consists of multiple attributes has only been evaluated by the work of Wang et al. [21]. Therefore, we used the heterogeneous pointwise mutual information (HPMI) metric proposed by the authors to evaluate our topics. HPMI is an extension of the point mutual information metric which is commonly used in topic modelling. For each discovered topic, HPMI calculates the average relatedness of each pair of attributes ranked at top-k:

$$
HMPI(v^x, v^y) = \left\{ \begin{array}{ll} \frac{2}{k(k-1)} \sum_{1 \leq i < j \leq k} log(\frac{p(v_i^x, v_j^y)}{p(v_i^x)p(v_j^y)}) & x = y \\ \frac{1}{k^2} \sum_{1 \leq i, j \leq k} log(\frac{p(v_i^x, v_j^y)}{p(v_i^x)p(v_j^y)}) & x \neq y \end{array} \right\} \tag{3}
$$

where v^x is a node of type x, ranked among the top-k attributes of type x in a certain topic. The higher the HPMI is, the more coherent the topics are. We calculated the HPMI for the 8 constructed HINs using $k = 20$ and $k = 40$.[8]

Table 2. HPMI results for all the HINS. K: keywords. A: authors and I: ISI fields. Highlighted values indicate the highest score for each k.

HIN	#Topic	K-K	K-A	K-I	A-A	A-I	I-I	Overall
k = 20								
CS_1	33	−1.847	−0.960	−0.726	−1.910	−0.764	−1.056	−1.211
CS_2	381	0.204	1.420	0.222	3.164	0.439	0.057	0.918
CS_3	**1125**	**1.392**	**2.355**	**0.467**	**5.780**	**0.692**	**0.223**	**1.818**
CS_4	808	0.855	1.932	0.347	4.807	0.559	0.144	1.441
CS_5	853	1.025	1.425	0.263	2.735	0.425	0.032	0.984
CS_6	174	0.557	0.479	−0.030	−0.382	0.009	−0.209	0.071
CS_7	97	−1.040	0.492	−0.218	−0.955	−0.135	−0.270	−0.354
CS_8	53	−1.816	−0.946	−0.645	−1.899	−0.671	−0.561	−1.090
k = 40								
CS_1	33	−1.791	−0.966	−0.755	−1.912	−0.757	−1.056	−1.206
CS_2	381	0.289	1.395	0.213	3.171	0.435	0.057	0.927
CS_3	**1125**	**1.443**	**2.349**	**0.467**	**5.777**	**0.691**	**0.223**	**1.825**
CS_4	808	0.902	1.938	0.345	4.808	0.559	0.144	1.449
CS_5	853	1.082	1.423	0.269	2.739	0.422	0.032	0.995
CS_6	174	0.588	0.479	−0.018	−0.394	0.003	−0.209	0.075
CS_7	97	−0.972	0.472	−0.205	−0.969	−0.130	−0.270	−0.346
CS_8	53	−1.730	−0.944	−0.636	−1.922	−0.645	−0.561	−1.073

Table 2 shows the scores obtained. Each column represents the average relatedness of a pair of object types (x, y) for all the topics discovered. The *Overall*

---

[8] Following the idea of [21], we setted $k = 5$ for ISI fields since there are only 120 of them in the HIN. In these cases, the part $\frac{1}{k^2}$ of the formula changes to $\frac{1}{5k}$.

column is the average of the values of the 6 possible relations. The results demonstrated that the scores are very similar for k=20 and k=40. Additionally, in 5 out of 8 HINs our strategy was capable of obtaining a positive overall HPMI. Focusing on the best result (CS_3), the topics are highly coherent, specially on the author-author and keyword-author relations. With respect to the HIN construction we observed the importance of using normalization in the relation weights. The only HIN with uniform weights (CS_1) scored the worst. Regarding the non-uniform HINs, CS_7 and CS_8, the only two that assign higher importance to the publication-ISI field relation, are the only ones that achieved an overall negative HPMI. In general we discovered that in order to generate more coherent topics, we must assign an higher importance to the publication-keywords relation. The best results were obtained when doubling the importance of this relation while decreasing the weight of the publication-ISI field one (CS_3).

5.2   Profiles Evaluation

To evaluate the expert profiles created, we selected 12 authors that are computer science professors at the University of Porto. For each one, we crawled their Google Scholar page[9] and collected the research interests that they manually assigned to themselves. In this test, we assume that the research interests of an author reflect his knowledge areas. Table 3 summarizes the name of the authors, the number of publications they have (in the Authenticus database) and their research interests.

Table 3. Author's number of publications and google scholar interests.

Name	#Pubs	Google Scholar Interests
Alipio Jorge	133	Data mining; machine learning; text mining; recommender systems; artificial intelligence machine learning
Antonio Porto	30	Logic programming; coordination; artificial intelligence
Fernando Silva	91	Parallel and distributed computing; logic programming; information mining; algorithms; complex networks
Luis Torgo	90	Data mining; machine learning
Nelma Moreira	89	Automata theory; descriptional complexity; formal verification of software
Pedro Ribeiro	37	Complex networks; algorithms and data structures; parallel and distributed computing; computer science education; artificial int
Pedro Brandao	31	Communication networks; body area networks; ehealth; distributed systems
Ricardo Rocha	90	Logic programming; tabling; parallelism; language implementation
Rita Ribeiro	25	Data mining; machine learning
Rogerio Reis	81	Formal languages; automata theory; combinatorics
Sergio Crisostomo	16	Computer networks; communications; computer science
Veronica Orvalho	40	Computer graphics

---

[9] https://scholar.google.com/.

For each author we created his hierarchical expert profile using the HIN CS_3 which was the one that yielded the best HPMI results. Then, we compared the profiles against each other to obtain their similarity per hierarchy level. To compute the similarity between authors a_1 and a_2 at a certain level l, we obtain the topical distribution of each authors at l and we sum the topical intersection. The similarity value ranges from 0.0 to 1.0, where 1.0 indicates a perfect match, while 0.0 describes no match between the authors. The total similarity represents the sum of the similarities obtained for all the hierarchy levels.

The aim of this test is to use research interests as evidence to evaluate whether two authors should have high or low similarity profiles. In total we have 132 comparisons. In order to filter some cases on this analysis we divided the results into two groups considering the total similarity between authors. On the first group the total similarity is higher or equal to 2.0, while it is lower or equal to 1.0 on the second one. Tables 4 and 5 show the similarity values and the number of topics shared for both groups.

Table 4. Comparison results for total similarity ≥ 2.0

Author 1	Author 2	Level 0		Level 1		Level 2		Level 3		Total	
		Sim	#To	Sim	#To	Sim	#To	Sim	#To	Sim	#To
Nelma Moreira	Rogerio Reis	1.00	2	1.00	3	1.00	4	1.00	4	4.00	13
Fernando Silva	Pedro Ribeiro	0.73	3	0.73	3	0.60	3	0.60	3	2.66	12
Pedro Brandao	Sergio Crisostomo	0.66	2	0.66	2	0.66	2	0.66	2	2.64	8
Alipio Jorge	Luis Torgo	0.85	3	0.59	4	0.56	4	0.56	4	2.56	15
Fernando Silva	Ricardo Rocha	0.77	3	0.62	4	0.56	4	0.28	3	2.23	14
Pedro Ribeiro	Ricardo Rocha	0.76	3	0.57	3	0.42	3	0.26	2	2.01	11
Luis Torgo	Rita Ribeiro	0.67	2	0.67	2	0.34	2	0.32	2	2.01	8

Only 7 out of 132 comparisons scored a total similarity equal or higher than 2.00. This is expected due to the fact that we have a broad range of interests from the Google scholar, and the lower hierarchical levels refer to very specific topics. Thus, making it more difficult to find similar researchers at those levels. The highest similarity score (Nelma Moreira and Rogerio Reis) represent a perfect profile match at all hierarchical levels. Although their Google scholar interests are very similar, we further looked into this case due to the fact that it represents a wide gap score wise to the other cases. A co-authorship analysis on the network revealed that the two authors are co-authors in 66 publications (81.5% of Rogerio Reis's publications). Therefore, the perfect match is expected. Regarding the other cases, we observe high similarity between pairs of knowledge areas such as: machine learning (Alipio Jorge, Luis Torgo, and Rita Ribeiro), parallel programming (Fernando Silva, Pedro Ribeiro and Ricardo Rocha), and communication networks (Pedro Brandao and Sergio Crisostomo).

An interesting fact is to note that two authors, Veronica Orvalho and Antonio Porto, are not similar enough with any other author. In the case of Veronica

Orvalho, this is anticipated due to the fact that her interest on computer graphics is not shared by any other author. However, in the case of Antonio Porto, since his interests refer to areas shared by other authors an higher comparison was expected. A further look into his profile revealed that it is scattered by several topics. As a result, his intersections with other authors are not significant enough.

Table 5. Comparison results for total similarity ≤ 1.0

Author 1	Author 2	Level 0		Level 1		Level 2		Level 3		Total	
		Sim	#To	Sim	#To	Sim	#To	Sim	#To	Sim	#To
Nelma Moreira	Veronica Orvalho	0.25	1	0.25	1	0.25	1	0.25	1	1.00	4
Rogerio Reis	Veronica Orvalho	0.25	1	0.25	1	0.25	1	0.25	1	1.00	4
Antonio Porto	Pedro Ribeiro	0.66	2	0.33	1	0.00	1	0.00	1	0.99	5
Antonio Porto	Pedro Brandao	0.66	2	0.33	1	0.00	1	0.00	1	0.99	5
Fernando Silva	Luis Torgo	0.37	2	0.37	2	0.17	1	0.00	1	0.91	6
Nelma Moreira	Pedro Ribeiro	0.33	1	0.33	1	0.25	1	0.00	1	0.91	4
Nelma Moreira	Pedro Brandao	0.58	2	0.33	1	0.00	1	0.00	1	0.91	5
Nelma Moreira	Sergio Crisostomo	0.58	2	0.33	1	0.00	1	0.00	1	0.91	5
Pedro Ribeiro	Rogerio Reis	0.33	1	0.33	1	0.25	1	0.00	1	0.91	4
Pedro Brandao	Rogerio Reis	0.58	2	0.33	1	0.00	1	0.00	1	0.91	5
Rogerio Reis	Sergio Crisostomo	0.58	2	0.33	1	0.00	1	0.00	1	0.91	5
Alipio Jorge	Pedro Brandao	0.61	3	0.28	2	0.00	1	0.00	1	0.89	7
Alipio Jorge	Antonio Porto	0.64	2	0.14	1	0.00	1	0.00	1	0.78	5
Fernando Silva	Sergio Crisostomo	0.33	1	0.33	1	0.00	1	0.00	1	0.66	4
Pedro Ribeiro	Sergio Crisostomo	0.33	1	0.33	1	0.00	1	0.00	1	0.66	4
Rita Ribeiro	Sergio Crisostomo	0.33	1	0.33	1	0.00	1	0.00	1	0.66	4
Ricardo Rocha	Sergio Crisostomo	0.47	2	0.14	1	0.00	1	0.00	1	0.61	5
Luis Torgo	Sergio Crisostomo	0.34	2	0.17	1	0.00	1	0.00	1	0.51	5
Alipio Jorge	Sergio Crisostomo	0.28	2	0.14	1	0.00	1	0.00	1	0.42	5
Antonio Porto	Sergio Crisostomo	0.33	1	0.00	1	0.00	1	0.00	1	0.33	4
Sergio Crisostomo	Veronica Orvalho	0.25	1	0.00	1	0.00	1	0.00	1	0.25	4

With respect to the least similar results, in general they complement the observations from the top results that some areas (machine learning, parallel programming and communication networks) do not merge into highly similar profiles. In most of the cases we observe that there is a similarity in the level 0 of the hierarchy (i.e. on the broader topics), however as the topics get more specific the intersections between authors fade. An interesting case to highlight is the author Sergio Crisostomo, that matches on the first two levels with almost every other author, but with none (exception to Pedro Brandao, who shares a high similar profile with him) on the last two levels of the hierarchy. This indicates that from the level 2 of the topical hierarchy, there is a clear distinction of the communication network topics (his most specific google scholar interests).

Another case worth to note is the fact that although Veronica's interests are further away in comparison to the others, she still has some comparisons with total similarity higher than 1.0. A further look into her profile revealed that she is scattered through several topics however she is never a highly ranked author of the topic. In our dataset, the computer graphics area does not have as many publications as other areas such as machine learning and parallel programming. As a result, our strategy fails to model the topic correctly and scatters its information among other more predominant topics.

6   Conclusions

In this paper we addressed the problems of topic modelling and expert profiling. We avoided the problems of the LDA-based approaches by using modularity optimization in HINs to discover multi-typed topics. Additionally, we proposed a strategy to use the modelled topics to profile experts whether they are represented in the HIN or not. In order to tackle the current literature problems of constructing profiles that are redundant and either too specific or too broad, we organized the topics into an hierarchy. As a result, we create an hierarchical profile which starts with describing the expert's most broad areas, and it moves to the most specific ones. We evaluated our model with respect to the topics discovered using a state of the art metric (HPMI). This test revealed that the topics generated are coherent. Furthermore, in order to maximize topic coherency we have to assign the highest importance to the publication-keyword relation in the HIN. In another test, we used Google scholar data to evaluate the quality of the hierarchical profiles constructed. Our test revealed that we are capable of generating high similarity profiles for experts that have common research interests, while generating low similarity profiles for the ones that do not. This test also demonstrated that we need to improve our strategy to model topics that are under represented in the data.

Regarding future work, in the domain of topic discovery, we plan to test other community detection algorithms, specially the ones that not require transforming the HIN into a similarity graph. In the domain of the expert profiling, we aim to take a further look into the rankings of the nodes inside a topic and how they can be used in the profiling step. We also aim at creating an automatic summarization of the topics in such a way that we can construct a visualization of the expert's profile. Furthermore, we will look into considering the timestamps of the expert's meta-data in order to create time-sensible profiles.

Acknowledgements. This work is funded by the ERDF through the COMPETE 2020 Programme within project POCI-01-0145-FEDER-006961, and by National Funds through the FCT as part of project UID/EEA/50014/2013. Jorge Silva is also supported by a FCT/MAP-i PhD research grant (PD/BD/128157/2016).

References

1. Balog, K., Fang, Y., de Rijke, M., Serdyukov, P., Si, L.: Expertise retrieval. Found. Trends® Inf. Retriev. **6**(2–3), 127–256 (2012)
2. Berendsen, R., Rijke, M., Balog, K., Bogers, T., Bosch, A.: On the assessment of expertise profiles. J. Assoc. Inf. Sci. Technol. **64**(10), 2024–2044 (2013)
3. Blei, D.M., Ng, A.Y., Jordan, M.I.: Latent dirichlet allocation. Journal of machine Learn. Res. **3**(Jan), 993–1022 (2003)
4. Blondel, V.D., Guillaume, J.L., Lambiotte, R., Lefebvre, E.: Fast unfolding of communities in large networks. J. Stat. Mech.: Theory Exp. **2008**(10), P10008 (2008)
5. Daud, A.: Using time topic modeling for semantics-based dynamic research interest finding. Knowl.Based Syst. **26**, 154–163 (2012)
6. De Campos, L.M., Fernández-Luna, J.M., Huete, J.F.: Committee-based profiles for politician finding. Int. J. Uncertain. Fuzziness Knowl.-Based Syst. **25**(Suppl. 2), 21–36 (2017)
7. Duan, D., Li, Y., Li, R., Lu, Z., Wen, A.: Mei: Mutual enhanced infinite community-topic model for analyzing text-augmented social networks. Comput. J. **56**(3), 336–354 (2012)
8. Gerlach, M., Peixoto, T.P., Altmann, E.G.: A network approach to topic models. arXiv preprint arXiv:1708.01677 (2017)
9. bin Jamaludin, N.A., Annamalai, M., Jamil, N., Bakar, Z.A.: A model for keyword profile creation using extracted keywords and terminological ontology. In: 2013 IEEE Conference on e-Learning, e-Management and e-Services (IC3e), pp. 136–141. IEEE (2013)
10. Jeong, Y.S., Lee, S.H., Gweon, G.: Discovery of research interests of authors over time using a topic model. In: 2016 International Conference on Big Data and Smart Computing (BigComp), pp. 24–31. IEEE (2016)
11. Karimzadehgan, M., White, R.W., Richardson, M.: Enhancing expert finding using organizational hierarchies. In: European Conference on Information Retrieval, pp. 177–188. Springer (2009)
12. Li, C., Cheung, W.K., Ye, Y., Zhang, X., Chu, D., Li, X.: The author-topic-community model for author interest profiling and community discovery. Knowl. Inf. Syst. **44**(2), 359–383 (2015)
13. Mikolov, T., Chen, K., Corrado, G., Dean, J.: Efficient estimation of word representations in vector space. arXiv preprint arXiv:1301.3781 (2013)
14. Newman, M.E.: Modularity and community structure in networks. Proc. Natl Acad. Sci. **103**(23), 8577–8582 (2006)
15. Rosen-Zvi, M., Griffiths, T., Steyvers, M., Smyth, P.: The author-topic model for authors and documents. In: Proceedings of the 20th Conference on Uncertainty in Artificial Intelligence, pp. 487–494. AUAI Press (2004)
16. Rybak, Jan, Balog, Krisztian, Nørvåg, Kjetil: Temporal expertise profiling. In: de Rijke, Maarten, Kenter, Tom, de Vries, Arjen P., Zhai, ChengXiang, de Jong, Franciska, Radinsky, Kira, Hofmann, Katja (eds.) ECIR 2014. LNCS, vol. 8416, pp. 540–546. Springer, Cham (2014). https://doi.org/10.1007/978-3-319-06028-6_54
17. Shi, C., Li, Y., Zhang, J., Sun, Y., Philip, S.Y.: A survey of heterogeneous information network analysis. IEEE Trans. Knowl. Data Eng. **29**(1), 17–37 (2017)
18. Sun, Y., Han, J., Zhao, P., Yin, Z., Cheng, H., Wu, T.: Rankclus: integrating clustering with ranking for heterogeneous information network analysis. In: Proceedings of the 12th International Conference on Extending Database Technology: Advances in Database Technology, pp. 565–576. ACM (2009)

19. Sun, Y., Yu, Y., Han, J.: Ranking-based clustering of heterogeneous information networks with star network schema. In: Proceedings of the 15th ACM SIGKDD International Conference on Knowledge Discovery and Data Mining, pp. 797–806. ACM (2009)
20. Tang, J., Jin, R., Zhang, J.: A topic modeling approach and its integration into the random walk framework for academic search. In: Eighth IEEE International Conference on Data Mining, 2008 ICDM 2008, pp. 1055–1060. IEEE (2008)
21. Wang, C., Liu, J., Desai, N., Danilevsky, M., Han, J.: Constructing topical hierarchies in heterogeneous information networks. Knowl. Inf. Syst. **44**(3), 529–558 (2015)
22. Wang, J., Hu, X., Tu, X., He, T.: Author-conference topic-connection model for academic network search. In: Proceedings of the 21st ACM International Conference on Information and Knowledge Management, pp. 2179–2183. ACM (2012)

Filtering Documents for Plagiarism Detection

Kensuke Baba(✉)

Fujitsu Laboratories, Kawasaki 211-8581, Japan
baba.kensuke@fujitsu.com

Abstract. Efficient methods are required for plagiarism detection. This paper proposes a fast and scalable method for detecting "copy and paste"-type plagiarism in documents. Implementing detection methods for this type of plagiarism requires a long processing time or a large database for comprehensive matching of ordered word occurrences. The author improved the scalability of an existing fast method based on fast Fourier transform using the idea of the frequency domain filtering. He evaluated the effect of the improvement on accuracy of the plagiarism detection method, and achieved an effective trade-off between the accuracy and the required size of database.

Keywords: Plagiarism detection · Text processing
Vector representation of words · Fast Fourier transform · Filtering

1 Introduction

Efficient methods are required for plagiarism detection. A huge amount of documents became available on-line, which encourages plagiarism from copyrighted contents and academic documents. Our objective is to develop a fast and scalable method of plagiarism detection. An action "plagiarism" of humans can be formalized using features of documents in a variety of ways. In this paper, we treat "copy and paste" which can simply formalize a kind of straightforward plagiarism rather than plagiarism of ideas or rough structures of documents.

A difficulty in implementing fast detection methods for "copy and paste"-type plagiarism is the scalability. The comprehensive matching of ordered word occurrences in documents requires a long processing time or a large database. We model a plagiarism as a contiguous sequence which is approximately common in two documents, and address the problem to detect plagiarisms in a suspicious document (called a *query* document) and N documents which can be a source (called *object* documents), where the length of each document is n. The naive method needs $O(Nn^2)$ comparisons of words. An approach is using similarity based on a vector representation of documents. Similar documents on a vector space can be found in $O(\log N)$ time using a suitable data structure. The bag-of-words model [16] is an example of the methods on this approach. A number of plagiarism detection methods based on the vector space model exist [15,18,19].

© Springer Nature Switzerland AG 2018
L. Soldatova et al. (Eds.): DS 2018, LNAI 11198, pp. 361–372, 2018.
https://doi.org/10.1007/978-3-030-01771-2_23

However, this approach does not take account of the order of word occurrences, which decreases accuracy of plagiarism detection. If we use a method on this approach as a screening, then $O(n^2)$ computations are still required after the screening. Although a number of plagiarism detection methods [13,21] take accounts of the order using the edit distance [20,22], the processing time is $O(n^2)$.

We improved the scalability of an existing method of fast plagiarism detection. Baba [8] proposed a fast plagiarism detection method based on the score vector [12] between documents. The score vector between two sequences is the numbers of matches between the elements aligned with every possible gap of starting positions of the sequences. If each element of the sequences is a number, the vector is computed in $O(n \log n)$ time for the length n of each sequence using fast Fourier transform (FFT) [11], such as the Cooley–Tukey type algorithm [9]. In the plagiarism detection method, each document was converted into numerical matrix using a function from words to numerical vectors for the use of FFT. Although the method detects plagiarisms fast by storing the frequency components of object documents, the size of the data is large. We reduced the data size using an idea which corresponds to the frequency domain filtering for time series data or images; we deleted parts of the frequency components for each object document instead of reducing the number of object documents by a typical "filtering" in document processing.

We evaluated the effect of our improvement on accuracy of the FFT-based plagiarism detection method. We investigated the relation between the accuracy and the size of the frequency components of object documents required for the implementation. We applied the improved method to a data set of documents that contain randomly generated "copy and paste"-type plagiarisms. As a result of the evaluation, we achieved a better trade-off between the accuracy and the size of the data than that obtained by the original method.

The rest of this paper is organized as follows. Section 2 describes the main idea of our improvement. Section 3 introduces the FFT-based plagiarism detection method and our improvement to the method, and describes the experimental methods to evaluate the improvement. Section 4 reports the experimental results. Section 5 gives considerations on the results and future directions of our study.

2   Main Idea

This section describes the main idea of our improvement on the size of the data stored for the FFT-based plagiarism detection method.

Our idea is to find plagiarisms in documents using only their "frequency components" in a restricted range. The idea corresponds to the fact that a rough shape of a wave form or an image consists of its frequency components in a certain range. The "frequency components" of documents are obtained using a vector representation of words, a function from words to numerical vectors. A straightforward method is using binary vectors whose elements correspond to words. A number of vector representations of words with a small dimensionality that can represent a word similarity are obtained from actual document data using statistical computations, such as the recent study [17] in neural networks.

Position		0	1	2	3	4	5	6	7
Document				to	be	or	not	to	be
Document	not	to	be						
		not	**to**	**be**					

... (centered)

					not	**to**	**be**		
						not	to	be	
							not	to	be

| Score vector | | 0 | 2 | 0 | 0 | 0 | 3 | 0 | 0 |

Fig. 1. The score vector between two documents "to be or not to be" and "not to be".

$$
\begin{matrix} to \\ be \\ or \\ not \\ to \\ be \end{matrix}
\begin{pmatrix} 1\,0\,0\,0 \\ 0\,1\,0\,0 \\ 0\,0\,1\,0 \\ 0\,0\,0\,1 \\ 1\,0\,0\,0 \\ 0\,1\,0\,0 \end{pmatrix}
\quad
\begin{matrix} not \\ to \\ be \end{matrix}
\begin{pmatrix} 0\,0\,0\,1 \\ 1\,0\,0\,0 \\ 0\,1\,0\,0 \end{pmatrix}
\underset{\substack{\text{column-wise} \\ \text{convolutions}}}{\Longrightarrow}
\begin{pmatrix} 0\,0\,0\,0 \\ 1\,1\,0\,0 \\ 0\,0\,0\,0 \\ 0\,0\,0\,0 \\ 0\,0\,0\,0 \\ 1\,1\,0\,1 \\ 0\,0\,0\,0 \\ 0\,0\,0\,0 \end{pmatrix}
\underset{\substack{\text{row-wise} \\ \text{additions}}}{\Rightarrow}
\begin{pmatrix} 0 \\ 2 \\ 0 \\ 0 \\ 0 \\ 3 \\ 0 \\ 0 \end{pmatrix}
$$

Fig. 2. An example of the numerical representations of documents "to be or not to be" and "not to be", and the computation of the score vector.

The occurrences of common word sequences in two documents can be computed fast if we convert documents into numerical matrices. The score vector between documents computes the comprehensive occurrences. Figure 1 shows an example of the score vector of two documents. The comparisons for the possible combinations of words correspond to the computation of the convolution of two vectors. Therefore, using the convolution theorem [10] and FFT, the score vector for two sequences of length n is computed in $O(n \log n)$ time instead of $O(n^2)$ time. To apply the technique for numerical vectors to documents, we need to convert documents to numerical matrices using a function from words to numerical vectors. Figure 2 shows an example of the numerical representations of documents and the computation of the score vector from them, where the strict definition of the convolutions will be defined in Sect. 3. Figure 3 shows the outline of this method to compute the score vector between two documents in the situation that a document of length n is converted to an $n \times d$ matrix using a vector representation of words of dimensionality d.

Baba [8] proposed a fast but not scalable plagiarism detection method based on the score vector. This method detects an occurrence of "copy and paste"-type plagiarism as a peak in the score vector. The left part of Fig. 4 shows examples of the score vectors between two documents with and without plagiarism. The method repeats the computation of the score vector for every object documents. The upper part of Fig. 5 illustrates the outline of this method. A third of the FFT

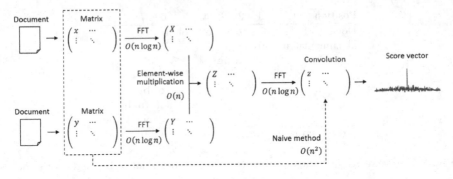

Fig. 3. Outline of the computation of the score vector between two documents using the convolution theorem and an FFT, where n is the size of each document.

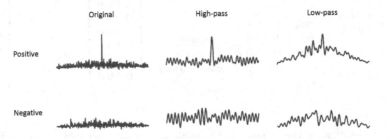

Fig. 4. Examples of the score vectors between two documents that contain a plagiarism and no plagiarism, and their filtered wave forms using high- and low-pass filters.

computations (1 in the figure) is completed before the input of a query document. Another third (2) is reduced to one Nth by using the same vector representation of words for any score vector. The other third (3) can be reduced to one dth [4], or completely omitted for rough detection [6,7]. The problem is that the size of the data stored for matching is large; the size of the data, the frequency components of object documents, is approximately $4d$-fold of that of the original documents for the dimensionality d of a vector representation of words [8]. For computing the exact score vectors, d needs to be at least the vocabulary size minus 1 [3] which is usually not practical. Although d can be reduced by an approximation of score vectors using a random vector representation of words [5], the size is still large. In the experiment conducted in [8], the vocabulary size of 4,000 abstracts of scholarly papers is approximately 42,000, and the dimensionality of the random vector representation of words that could keep plagiarism detection accuracy is several tens.

We reduced the size of the stored data using the idea of the frequency domain filtering. We masked several rows of the matrix obtained as the frequency components from each object document by FFT with 0's. The lower part of Fig. 5 illustrates the idea of our improvement. Although the output vector is approximated by this modification, the peak of the score vector remains dimly. Masking the frequency components corresponds to the process of the high- or low-pass

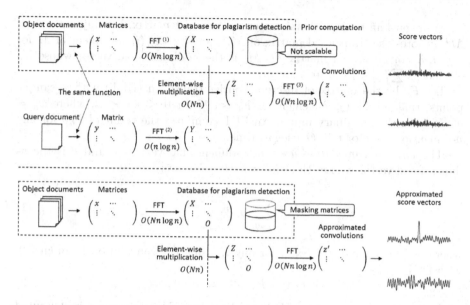

Fig. 5. Outline of the plagiarism detection method based on the score vector between documents (upper) and an improvement on the size of the data stored for detection (lower), where n is the size of each document and N is the number of object documents.

filter in image processing [14]. The middle and right parts of Fig. 4 show examples of filtered wave forms, where we used the 6.25% rows for each filter. As shown in Fig. 5, the masking is applied to the result of (the inverse of) FFT for (the matrix representation of) the score vector. Therefore, the approximated score vector is a filtered wave form of the exact score vector which is expected to represent the presence of plagiarism using fewer data.

3   Methods

This section introduces the FFT-based algorithm that computes the score vector and a plagiarism detection method based on the algorithm. This section also proposes an improvement to the method and describes how the effects of the improvement were evaluated.

3.1   Preliminaries

Let W be a set of words, called a *vocabulary*, and σ the vocabulary size $|W|$. A *document* is a list of words. For an integer $n > 0$, W^n is the set of the documents of length n over W. For a document p of length n, p_i for $0 \le i < n$ is the ith word of p. For documents p and q, pq is the document obtained by concatenating p and q. For a word w and an integer $n > 0$, w^n is the document of n w's. Let $x \notin W$ be the *never-match word* and δ a function from $(W \cup \{x\}) \times (W \cup \{x\})$ to $\{0, 1\}$ such that $\delta(w, v)$ is 1 if $w, v \in W$ and $w = v$, and 0 otherwise.

We regard an n-dimensional vector as the $n \times 1$ matrix. For any matrix M, M^T denotes the transposed matrix of M, and $M_{i,j}$ the (i,j)-element of M for $0 \le i, j$. For an $m \times n$ matrix M, M_c is the m-dimensional vector whose ith element is $\sum_{j=0}^{n-1} M_{i,j}$ for $0 \le i < m$.

Let F_n be the matrix of discrete Fourier transform (DFT) with n sample points, that is, the (j,k)-element of F_n is ω_n^{jk} for $0 \le j, k < n$, where $\omega_n = e^{-2\pi i/n}$ for the imaginary unit i. An FFT computes the result of $F_n v$ for any n-dimensional vector v in $O(n \log n)$ time.

The *circular convolution* $u * v$ of n-dimensional vectors u and v is the n-dimensional vector whose ith element for $0 \le i < n$ is

$$\sum_{j=0}^{n-1} u_j \cdot v_{i-j}, \tag{1}$$

where $u_i = u_{i+n}$ and $v_i = v_{i+n}$ for any i. Using the convolution theorem [10] with DFT,

$$u * v = F_n^{-1} \left(F_n u \circ F_n v \right), \tag{2}$$

where $\circ$ is the operator of the Hadamard product. Therefore, $u * v$ is computed in $O(n \log n)$ time using three $O(n \log n)$ computations of FFT and $O(n)$ multiplications.

3.2   The FFT-Based Algorithm

We introduce an algorithm that computes the score vector between two documents of length n over W in $O(\sigma n \log n)$ time. This algorithm can be extended to documents of different lengths by dividing the longer document in the same way as the technique used in [2].

The *score vector* $C(p,q)$ between $p \in W^m$ and $q \in W^n$ is defined to be the $(m + n - 1)$-dimensional vector whose ith element for $0 \le i < m + n - 1$ is

$$c_i = \sum_{j=0}^{m-1} \delta(p_j, q'_{i+j}), \tag{3}$$

where $q' = x^{m-1} q x^{m-1}$.

First, we extend the idea of the circular convolution of vectors to matrices. For two $n \times d$ matrices M and N, $M * N$ is defined to be the $n \times d$ matrix whose (i,j)-element for $0 \le i < n$ and $0 \le j < d$ is

$$(M * N)_{i,j} = \sum_{k=0}^{n-1} M_{k,j} \cdot N_{i-k,j}, \tag{4}$$

where $M_{i,j} = M_{i+n,j}$ and $N_{i,j} = N_{i+n,j}$ for any i and j. Then, using Eq. 2,

$$M * N = F_n^{-1} \left(F_n M \circ F_n N \right). \tag{5}$$

Therefore, $M * N$ is computed from M and N in $O(dn \log n)$ time using an FFT.

The $O(\sigma n \log n)$ algorithm is obtained from the fact that the score vector between two documents in W^n is

$$C(p, q) = (P * Q)_c, \tag{6}$$

where P and Q are $(2n - 1) \times \sigma$ matrices

$$P = \begin{pmatrix} \phi(p_{n-1})^T \\ \phi(p_{n-2})^T \\ \vdots \\ \phi(p_0)^T \\ O \end{pmatrix} \quad \text{and} \quad Q = \begin{pmatrix} \phi(q_0)^T \\ \phi(q_1)^T \\ \vdots \\ \phi(q_{n-1})^T \\ O \end{pmatrix} \tag{7}$$

for $\phi : W \to \{0, 1\}^\sigma$ such that the ith element of $\phi(w)$ for $0 \le i < \sigma$ and $w \in W$ is 1 if $\varphi(w) = i$, and 0 otherwise, for a bijection $\varphi : W \to \{0, 1, \ldots, \sigma - 1\}$. A precise proof is described in [4].

The algorithm is summarized as follows:

1. Convert p and q to P and Q using Eq. 7,
2. Compute $P * Q$ from P and Q using Eq. 5,
3. Compute $C(p, q)$ from $P * Q$ using Eq. 6.

The processing time is $O(\sigma n \log n)$: the first process requires $O(\sigma n)$ time even by a naive method; the second process requires $O(\sigma n \log n)$; and the last process consists of $O(\sigma n)$ additions. More precisely, the score vector can be computed using the set $W_{p,q}$ of the words that appear in both p and q instead of the total vocabulary W. In this case, the processing time is bound by $O(\sigma' n \log n)$ for $\sigma' = |W_{p,q}|$.

3.3 Plagiarism Detection Method

We introduce the plagiarism detection method proposed in [8] which uses the FFT-based algorithm introduced in Sect. 3.2 for a query document and N object documents.

Plagiarism detection is to predict either "positive" or "negative" for instances of plagiarism between a pair of documents. The *accuracy* of a plagiarism detection method is defined to be the ratio of the number of correct predictions to the number of total predictions.

The plagiarism detection method in this paper repeats the following process for every object documents. For input documents,

1. Calculate the score vector using the FFT-based algorithm, and
2. Predict "positive" or "negative" using the obtained vector and a threshold.

In the first process, we used a random vector representation ϕ_r of words instead of ϕ used in Sect. 3.2, which approximates the score vector using vectors of a small dimensionality for representing words. Let ϕ_r be a function from $W \cup \{x\}$ to $\{-1, 0, 1\}^d$ such that $\phi_r(x)$ is the d-dimensional zero-vector and $\phi_r(w)$ for $w \in W$ is a vector chosen randomly from $\{-1, 1\}^d$. In the second process, we determined the threshold from training data by applying a support vector machine with a linear kernel to 3-tuples of

- The peak value of the score vector, where the *peak value* of a vector v is the minimum element in v'' and $v'_i = (vv)_{i+1} - v_i$ for $0 \leq i < |v|$,
- The average of the elements in the score vector, and
- The length of the shorter document

for the pairs in the training data. The computation for detecting the peak value of an n-dimensional vector needs $O(n)$ time; therefore, the resulting processing time of the method is mainly due to the $O(dn \log n)$ computation for the score vector.

Most of the $O(dn \log n)$ computations in the method can be completed before the input of a query document. Using Eqs. 5, 6, and 7, the score vector is computed as

$$C(p, q) = \left(F_\ell^{-1} \left(F_\ell P \circ F_\ell Q \right) \right)_c \qquad (8)$$

for $\ell = 2n - 1$. On the assumption that the object documents are given in advance, we can compute and store the frequency components $F_\ell P$'s of the N object documents p's. The number of FFT computations in this process is one third of the total number required in the method, and another third is reduced to a Nth by using ϕ_r for any conversion (Fig. 5).

3.4   Improvement

We proposed an improvement to the plagiarism detection method introduced in Sect. 3.3.

The problem is that the size of the frequency components $F_\ell P$'s for the object documents p's is large. For this problem, we used an approximated score vector defined to be

$$C'(p, q) = \left(F_\ell^{-1} \left(A_{x,k}(F_\ell P) \circ F_\ell Q \right) \right)_c \qquad (9)$$

for $x \in \{h, l\}$ and $0 \leq k \leq \ell$, where $A_{x,k}$ is an $\ell \times \ell$ matrix such that

$$A_{h,k} = \begin{pmatrix} E_k & O \\ O & O \end{pmatrix} \quad \text{and} \quad A_{l,k} = \begin{pmatrix} O & O \\ O & E_k \end{pmatrix} \qquad (10)$$

for the identity matrix E_k of size k. Applying $A_{x,k}$ to $F_\ell P$'s is masking of the $F_\ell P$'s which is the modification from the upper part to the lower part of Fig. 5.

The size of practical data for storing $A_{x,k} M$ for an $\ell \times d$ matrix M is approximately k/ℓ of that for M. We call $\rho = k/\ell$ the *reduction rate* of applying $A_{x,k}$ to M.

Additionally, we modified the definition of the peak value of a vector v used for detecting plagiarisms to be the minimum element in v'' and $v'_i = (vv)_{i+w} - v_i$ for $w = 1/\rho$ and $0 \leq i < |v|$.

3.5   Evaluation

We applied the plagiarism detection method introduced in Sect. 3.3 with the improvement proposed in Sect. 3.4 to a dataset and investigated the accuracy and the data size required for the implementation.

The dataset used for our experiment was a set of document pairs with randomly generated plagiarisms. We used 5,000 pairs of abstracts of articles published in Nature [1] from 1975 to 2017. The pairs were chosen from 28,146 abstracts so that

- An abstract of each pair is the nearest neighbor of the other abstract based on the similarity of the bag-of-words model in the total set, and
- The length of each abstract is longer than 100 words and shorter than 300 words.

We generated plagiarisms for 2,500 pairs chosen randomly from the 5,000 pairs by inserting a word sequence in an abstract into the other on the condition that

- The word sequence is chosen randomly from the first abstract,
- The word sequence is inserted into a randomly chosen position of the second abstract, and
- The word sequence is longer than 10% of the second abstract.

The positive and negative pairs were divided equally into 4,000 pairs for training and 1,000 pairs for test in validation.

We aimed to clarify the relation between the accuracy and the size of the data used in the plagiarism detection method, which is affected by the dimensionality d of ϕ_r and the reduction rate ρ of $A_{x,k}$'s. We investigated

- The accuracy of the original method against $d = 2^i$ for $1 \leq i \leq 5$ and
- The accuracy of the improved method against $\rho = 2^{-j}$ for $0 \leq j \leq i$ and for $2 \leq i \leq 5$.

4   Results

Figure 6 shows the relation between the accuracy of the plagiarism detection method and the data size required for storing the frequency components of the objective documents. In each graph, the horizontal axis means the ratio of the data size to that of the method with no filtering in the case where the dimensionality of the vector representation ϕ_r is $d = 1$. The dotted line represents the accuracy of the method with no filtering in which the data size was changed by d. The other lines represent the accuracy generated using high-pass $A_{h,k}$ or low-pass $A_{l,k}$ filtering in which the data size was changed by the reduction rate ρ. In the right graph, the accuracy of the method with the low-pass filtering were better than the dotted line. The data size could be reduced to a half in exchange for a slight decrease of the accuracy.

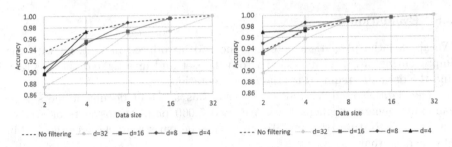

Fig. 6. Accuracy of the plagiarism detection algorithm against the data size with (left) high- and (right) low-pass filtering.

The base plagiarism detection method is more scalable and accurate than straightforward methods. The accuracy of the method with the exact score vectors was 0.997 which is approximately equal to that 0.999 in the case of no filtering with $d = 32$. However, the total vocabulary size was $\sigma = 75336$ which needs a large data size. The accuracy of another plagiarism detection using the Jaccard index of the word sets and an optimized threshold was 0.823 for the training and test data.

5   Discussion

5.1   Major Conclusion

We achieved a fast and scalable method of plagiarism detection. As shown in Fig. 6, our improvement could reduce the size of the data required for detection in exchange for a slight decrease of the accuracy. In the experiment, we could reduce the size by half with a small decrease of the accuracy in some conditions, which means that we can implement the fast plagiarism detection method using a smaller space than the original method.

5.2   Key Findings

As shown in Fig. 6, the accuracy of the plagiarism detection method with the low-pass filter was better than that with the high-pass filter. One of the reasons is supposed that the wave forms modified by the high- and the low-pass filters represent global and local changes of the original wave form, respectively, and the computation for finding a peak defined in Sect. 3.4 was suitable for the low-pass filter.

5.3   Future Directions

We are interested in what a filtered document is as a document. In this paper, we treated the filtering for matrix representations of documents. The filtered "documents" can be defined formally by using the inverse of the vector representation of words. However, we could find no meaning in the filtered documents like sharpening and blurring for the high- and low-pass filter in image processing. We expect that there exists vector representation of words which can give a meaning to filtered documents.

6   Conclusion

We proposed a fast and scalable method for plagiarism detection. We improved the scalability of an existing method of fast plagiarism detection; we reduced the size of data prepared for the method by applying the idea of the frequency domain filtering into documents. We evaluated the effect of the improvement by conducting experiments with document data that included plagiarisms. As a result, we achieved an effective trade-off between the accuracy and the required size of the data. In the experiment, we could reduce the size to less than a half with a small loss of the accuracy. Thus, we can implement the fast plagiarism detection method using a small space.

References

1. Nature. http://www.nature.com/nature/. Accessed 15 Jan 2018
2. Atallah, M.J., Chyzak, F., Dumas, P.: A randomized algorithm for approximate string matching. Algorithmica **29**(3), 468–486 (2001)
3. Baba, K.: String matching with mismatches by real-valued FFT. In: Taniar, D., Gervasi, O., Murgante, B., Pardede, E., Apduhan, B.O. (eds.) ICCSA 2010. LNCS, vol. 6019, pp. 273–283. Springer, Heidelberg (2010). https://doi.org/10.1007/978-3-642-12189-0_24
4. Baba, K.: An acceleration of FFT-based algorithms for the match-count problem. Inf. Process. Lett. **125**, 1–4 (2017)
5. Baba, K.: An extension of the FFT-based algorithm for the match-count problem to weighted scores. IEEJ Trans. Electr. Electron. Eng. **12**(S5), 97–100 (2017)
6. Baba, K.: A fast algorithm for plagiarism detection in large-scale data. J. Digit. Inf. Manag. **15**(6), 331–338 (2017)
7. Baba, K.: Fast plagiarism detection based on simple document similarity. In: Proceedings of the Twelfth International Conference on Digital Information Management, pp. 49–53. IEEE (2017)
8. Baba, K.: Fast plagiarism detection using approximate string matching and vector representation of words. In: Wong, R., Chi, C.-H., Hung, P.C.K. (eds.) Behavior Engineering and Applications. ISCEMT, pp. 67–79. Springer, Cham (2018). https://doi.org/10.1007/978-3-319-76430-6_3
9. Cooley, J.W., Tukey, J.W.: An algorithm for the machine calculation of complex fourier series. Math. Comput. **19**(90), 297–301 (1965)
10. Cormen, T.H., Stein, C., Rivest, R.L., Leiserson, C.E.: Introduction to Algorithms, 2nd edn. McGraw-Hill Higher Education, Boston (2001)
11. Fischer, M.J., Paterson, M.S.: String-matching and other products. In: Complexity of Computation (Proceedings of the SIAM-AMS Applied Mathematics Symposium, New York, 1973), pp. 113–125 (1974)

12. Gusfield, D.: Algorithms on Strings, Trees, and Sequences: Computer Science and Computational Biology. Cambridge University Press, New York (1997)
13. Irving, R.W.: Plagiarism and collusion detection using the smith-waterman algorithm. Technical report (2004)
14. Jain, A.K.: Fundamentals of Digital Image Processing. Prentice-Hall Inc., Upper Saddle River (1989)
15. Lin, W.-Y., Peng, N., Yen, C.-C., Lin, S.-D.: Online plagiarism detection through exploiting lexical, syntactic, and semantic information. In: Proceedings of the ACL 2012 System Demonstrations, ACL 2012, Stroudsburg, PA, USA, pp. 145–150. Association for Computational Linguistics (2012)
16. Manning, C.D., Raghavan, P., Schütze, H.: Introduction to Information Retrieval. Cambridge University Press, Cambridge (2008)
17. Mikolov, T., Sutskever, I. Chen, K., Corrado, G.S., Dean, J.: Distributed representations of words and phrases and their compositionality. In Burges, C.J.C., Bottou, L., Welling, M., Ghahramani, Z., Weinberger, K.Q. (eds.) Advances in Neural Information Processing Systems 26, pp. 3111–3119. Curran Associates Inc. (2013)
18. Misra, H., Cappé, O., Yvon, F.: Using LDA to detect semantically incoherent documents. In: Proceedings of the Twelfth Conference on Computational Natural Language Learning, CoNLL 2008, Stroudsburg, PA, USA, pp. 41–48. Association for Computational Linguistics (2008)
19. Řehůřek, R.: Plagiarism detection through vector space models applied to a digital library. In: RASLAN 2008, Brno, pp. 75–83. Masarykova Univerzita (2008)
20. Smith, T.F., Waterman, M.S.: Identification of common molecular subsequences. J. Mol. Biol. **147**, 195–197 (1981)
21. Su, Z., Ahn, B.-R., Eom, K.-Y., Kang, M.-K., Kim, J.-P., Kim, M.-K.: Plagiarism detection using the Levenshtein distance and Smith-Waterman algorithm. In: Innovative Computing Information and Control, p. 569 (2008)
22. Wagner, R.A., Fischer, M.J.: The string-to-string correction problem. J. ACM **21**(1), 168–173 (1974)

Most Important First – Keyphrase Scoring for Improved Ranking in Settings With Limited Keyphrases

Nils Witt[1(✉)], Tobias Milz[2(✉)], and Christin Seifert[3(✉)]

[1] ZBW - Leibniz Information Centre for Economics, Kiel, Germany
n.witt@zbw.eu
[2] University of Passau, Passau, Germany
tobias.milz@uni-passau.de
[3] University of Twente, Enschede, The Netherlands
c.seifert@utwente.nl

Abstract. Automatic keyphrase extraction attempts to capture keywords that accurately and extensively describe the document while being comprehensive at the same time. Unsupervised algorithms for extractive keyphrase extraction, i.e. those that filter the keyphrases from the text without external knowledge, generally suffer from low precision and low recall. In this paper, we propose a scoring of the extracted keyphrases as post-processing to rerank the list of extracted phrases in order to improve precision and recall particularly for the top phrases. The approach is based on the tf-idf score of the keyphrases and is agnostic of the underlying method used for the initial extraction of the keyphrases. Experiments show an increase of up to 14% at 5 keyphrases in the F1-metric on the most difficult corpus out of 4 corpora. We also show that this increase is mostly due to an increase on documents with very low F1-scores. Thus, our scoring and aggregation approach seems to be a promising way for robust, unsupervised keyphrase extraction with a special focus on the most important keyphrases.

1 Introduction

Automatic text summarization is applied in Natural Language Processing and Information Retrieval to provide a quick overview of longer texts. More specifically, automatic keyphrase extraction methods are employed to allow human readers to quickly assess relevant concepts in the text. As was pointed out by Miller on average people can hold 7 (± 2) items in their short-term memory [19]. This indicates that people who read keyphrase lists that exceed 5 to 9 keyphrases forget the first items of the list as they reach the end. Thus 5 keyphrases per document is an optimal number regarding human perception and it is worth optimizing keyphrase extraction methods towards this threshold. In this paper, we investigate a method-agnostic approach that enhances the important first part of a keyphrase list. That is, given a long (e.g. 20 keyphrases) ranked list

L. Soldatova et al. (Eds.): DS 2018, LNAI 11198, pp. 373–385, 2018.
https://doi.org/10.1007/978-3-030-01771-2_24

of keyphrases extracted by some keyphrase extraction method, our tf-idf-based approach reorganizes that list such that more suitable keywords are at the top of the list. The tf-idf value has been shown to be an informative feature for keywords [9]. Thus, we apply tf-idf-based scoring on extracted keyphrases a-posteriori, assuming that words with a high tf-idf value are more likely to be high quality keywords or part of high quality keyphrases (i.e. a short sequence of words)[1]. Concretely, the contributions of this paper are the following:

1. We propose a tf-idf-based scoring and re-ranking of keyphrases which is agnostic to the underlying keyphrase extraction method.
2. In experiments on four different corpora, we show that tf-idf-based scoring can enhance the precision and recall of well-known keyphrase extraction algorithms.

The source code and the data that were used to conduct the experiments are publicly available[2]. After reviewing related work we explain the details on keyphrase scoring. Then we report on the experimental setup (Sect. 4) and results (Sect. 5). Finally, we discuss and conclude our work in Sect. 6.

2   Related Work

Due to the rapid growth of available information, the ability to automatically generate summarized short texts has become a valuable tool for many Natural Language Processing tasks. Summarization approaches aim to generate sentences, keyphrases or keywords that condense the information provided by a document. Summaries that are extracted directly from the document and abstractive summaries that are created based on the content of the document with words not necessarily appearing in the document, are two main concepts of these approaches [15]. This paper focuses on extractive summaries and in particular *Rake* [21] and *TextRank* [18]. *TextRank* similarly to Wan and Xiao [24] and Liu et al. [11] searches for POS tag combinations in the document to identify possible keyphrase candidates. Other systems use different NLP methods and heuristics such as the removal of stop words [14], finding matching n-grams in Wikipedia articles [7] or extracting n-grams with specific syntactic patterns [10,16,26]. As these methods often produce too many and poor candidates for long documents, a second step is required to separate those candidates that are more likely keyphrases. Previous approaches [6,22,26] applied supervised binary classification techniques to select the keyphrases from the candidates. Binary classification, however, yields the problem that a candidate is simply deemed as either worthy or not worthy and their relative importance is not compared to the other candidates. As a result, other approaches adapted a ranking based system such as the unsupervised graph-based approach implemented by *TextRank*, *CollabRank* [23] and *TopicRank* [2]. Here, each candidate is represented as a node in

---

[1] Throughout the document we will use the unifying term *keyphrase* to refer to keywords as well as keyphrases as defined in the Introduction.

[2] https://doi.org/10.5281/zenodo.1435518.

a graph and its importance is recursively computed based on the number of connections the node has and how important the connected candidates are. Among these systems *TextRank* has established itself as the most popular graph-based ranking system and was also adapted in topic-based clustering concepts such as *TopicalPageRank* [13] and *CommunityCluster* [7]. Both systems apply *TextRank* multiple times (once for each topic in the document) and add to the importance of the topic to the computation. More recently, topic-based keyphrase extraction was done using topic modeling to find clusters of co-occurring words, which were used to construct candidate keyphrases [5]. Those candidates were then ranked according to several different properties, with the best-performing one being *purity* which prefers keyphrases consisting of words that are frequent in a given topic and rare in other topics. Sequence-to-sequence models based on recurrent neural networks have shown to perform very well not just on the task of keyphrase extraction but also on the more challenging task of keyphrase prediction, which includes finding keyphrases that do not appear in the text [17]. Similar results were achieved using convolutional neural networks [27], but due to the concurrent nature of convolutional neural networks the training time could be reduced by a factor of 5–6.

For our experiments we focus on fast, unsupervised methods with solid implementations as they are not constrained to extract only those keyphrases they saw during training. Since these algorithms do not rely on training data, they also have a larger domain of application. We also include the tf-idf baseline as still it is still a comparative baseline, despite its simplicity. Textrank remains a very important method and is still used by the community [8,12,17]. Rake was chosen as is it was able to outperform Textrank while scaling much better on longer documents (see Fig. 1.7 in [21]).

3 Approach

In this section, we explain how we assign scores to keyphrases irrespective of the algorithm that extracted it. An overview of the approach is shown in Fig. 1.

3.1 Keyphrase Scoring

We follow the idea described in Sect. 1 to create ranked keyphrase lists per document. The keyphrases in those lists can come from one or more keyphrase extracting algorithms. The lists are supposed to have the property that, on average, the highest ranked keyphrases are the "best" keyphrases, followed by the second highest ranked keyphrase, etc. Moreover, we act on the assumption that the gold standard keyphrases are "good", which allows us to formulate our expectations towards the ranked lists more formally: For any given document, the probability of a higher ranked keyphrase of being in the set of gold standard keyphrases is higher than the probability of a keyphrase with a lower rank:

$$rank(kp) \propto P(kp \in GS), \tag{1}$$

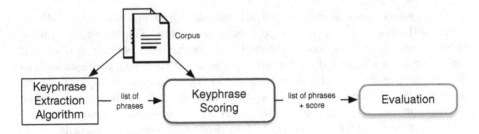

Fig. 1. Overview of keyphrase scoring approach and its evaluation. Standard evaluation measures precision, recall and F1-measure are compared before and after scoring and reranking.

where $P(kp \in GS)$ is the probability of keyphrase kp being in the set of gold standard keyphrases GS and $rank(kp)$ is the rank of the keyphrase. Since tf-idf offers only scores for single words rather than phrases we need a mechanism by which the score of a phrase can be computed in order to rank a set of keyphrases. We use a simple weighting approach:

$$score(kp) = \left(\sum_{i=1}^{|t|} ts_i \right) \cdot (1 - \alpha \cdot |t|), \tag{2}$$

where $|t|$ is the number of tokens in the keyphrase and ts_i is the token score (i.e. the tf-idf value of the token) of the token at position i. The parameter α determines how long phrases are penalized. In our experiments, we set $\alpha = \frac{1}{10}$ meaning that keyphrases with 10 tokens are always assigned a keyphrase score of 0.0 and keyphrases with more than 10 tokens get a negative score. This property might seem undesirable, but is reasonable as most gold standard keyphrases in the corpora used have less than 6 words (see Table 1). Therefore it is reasonable to penalize long extracted keyphrases in this scenario. Different keyphrase extraction algorithms can be used to extract keyphrases for a corpus. The quality of those extractions can be evaluated using gold standard keyphrases GS obtaining precision, recall and F1. For keyphrase scoring the score values (in our case tf-idf) are calculated on the corpus and used to rank or rerank the output of the keyphrase extraction step. The reranked lists are then evaluated in a similar fashion against the gold standard GS.

3.2   Keyphrase Extraction Ensembles

The introduced keyphrase scoring provides a unified, comparable score for all phrases, independent of the respective extraction algorithm. Thus, this score can be used to combine the output from different keyphrases extraction methods (as depicted in Fig. 2), similarly to the idea of bagging in machine learning [3]. Therefore we also measure the performance of multiple keyphrase extraction methods combined to see whether the overall performance can be enhanced

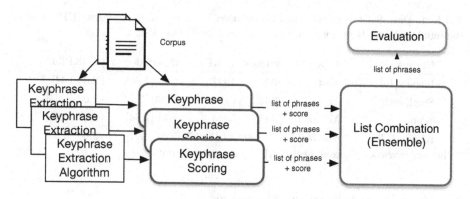

Fig. 2. Overview of the ensemble approach and its evaluation. Multiple keyphrase extraction methods combined using our keyphrase scoring approach.

in comparison to the individual methods. The keyphrase extraction ensemble works as follows: Given a document, k different keyphrase extraction methods can be applied, resulting in k result lists. Each of the keyphrases in the list might or might not have an algorithm-specific score, depending on the extraction method used. We remove duplicates from those lists, score each of the keyphrases as described in the previous section and create a unified result list containing keyphrases ordered by descending score.

4   Experimental Setup

In this section, we describe the data sets and the base keyphrase algorithms as well as the evaluation methodology for the experiments.

4.1   Data Sets

To evaluate our approach we used four corpora containing abstracts of scientific publications. Although the datasets are homogeneous as they all contain abstracts of scientific publications, the corresponding keyphrases exhibit vastly different characteristics. They not only differ in the average number of keyphrases per document but also in the average number of words per keyphrase (see Table 1). SemEval for example contains keyphrases that are as long as a whole sentence (e.g. *controllable size reduction with high resolution towards the observation of size- and quantum effects*), also there are unsuitable keywords (e.g. *defense simulation platform is discussed in*) which makes this corpus very challenging. The Scopus corpus is challenging for another reason. There are documents whose keyphrases mostly or only consist of abbreviations (e.g. *ARPA, CSPs, ERP, IaaS, NIST, PaaS, SaaS*) that are not mentioned in the text, leading to zero scores on all performance metrics for that document. KP20k stands out as it contains much more abstracts than the other corpora of which we only used 100,000 randomly sampled documents due to computational constraints.

Table 1. Data Set Overview. The abbreviation KP refers to keyphrases. |KP|denotes the number of words in a keyphrase, σ the standard deviation.

Corpus	Type	# Docs	# KP	$\varnothing$ KP/doc $[\sigma]$	$\varnothing$ |KP|$[\sigma]$
Inspec [10]	Abstracts	2000	19275	9.64 [4.80]	2.3 [0.44]
SemEval2017 [1]	Abstracts	493	5846	11.90 [7.44]	3.03 [1.31]
Scopus[a]	Abstracts	745	3385	4.54 [1.34]	2.16 [0.65]
KP20k [17]	Abstracts	570,809	3,017,637	5.29 [3.77]	2.05 [0.63]

[a]https://www.kaggle.com/neelshah18/scopusjournal/data

4.2   Keyphrase Extraction Algorithms

In this section we briefly describe the three keyphrase extraction algorithms that are used for our experiments.

The **tf-idf** keyphrase extraction is based on POS tags, following Wen and Xiao [25]. We determined the 12 most common POS tags in gold standard keyphrases among all corpora using the NLTK POS tagger[3]. In order to generate candidate keyphrases we determine the POS tag of each word in a given text and extract word sequences where all POS tags are "good". A sequence like *[bad, good, bad, good, good, bad]* generates two keyphrase candidates. One of length 1 corresponding to the word at position two and one of length 2 corresponding to the words at positions three and four. Finally, the candidates are ranked by the mean of their tf-idf values of the individual words.

Rake is based on the observation that keyphrases rarely contain stop words and punctuation. Therefore all sequences in a text not containing stop words or punctuation are identified and treated as candidate keyphrases. Then a matrix is constructed where the co-occurrence of words within a keyphrase is counted. Finally each keyphrase is scored based on the co-occurrence scores of its individual words. The phrases with the highest scores are the keyphrases of the document. We used the stopword list introduced in [20] and the implementation provided by NLTK[4].

Textrank builds a graph of lexical units (e.g. words). Only words passing a syntactic filter (e.g. nouns and adjectives only) are added to the graph. These words are connected based on co-occurrence in a sliding window. Once the graph is built PageRank [4] is used to determine the importance of each node in the graph. In a post-processing step sequences of adjacent keywords are merged into keyphrases and their scores are added for the final ranking. We used the stopword list introduced in [20] and the jgtextrank[5].

---

[3] https://www.nltk.org/.
[4] https://pypi.org/project/rake-nltk/.
[5] https://github.com/jerrygaoLondon/jgtextrank.

Table 2. The effect of keyphrase scoring on one example. The top cell contains an original abstract. The cell below contains the expert-assigned ground-truth keyphrases. The bottom row contains the extracted keyphrases from multiple algorithms. Entries written in italic indicate a match to a ground-truth keyphrase. Note: This example was chosen as it clearly shows the positive effect of our scoring approach. But there are also examples where the scoring has no effect.

Abstract

A comparison theorem for the iterative method with the preconditioner (I + S/sub max/) A.D. Gunawardena et al. (1991) have reported the modified Gauss-Seidel method with a preconditioner (I + S). In this article, we propose to use a preconditioner (I + S/sub max/) instead of (I + S). Here, S/sub max/ is constructed by only the largest element at each row of the upper triangular part of A. By using the lemma established by M. Neumann and R.J. Plemmons (1987), we get the comparison theorem for the proposed method. Simple numerical examples are also given.

Ground-truth keyphrases

iterative method, preconditioner, modified Gauss-Seidel method, comparison theorem

Extracted keyphrases			
Rake	**Rake$_s$**	**Textrank**	**Textrank$_s$**
1. upper triangular part	*1. preconditioner*	*1. iterative method*	*1. preconditioner*
2. simple numerical examples	*2. comparison theorem*	2. modified Gauss-Seidel method	*2. comparison theorem*
3. Seidel method	*3. iterative method*	3. method	*3. iterative method*
4. proposed method	4. upper triangular part	4. R.J. Plemmons	*4. modified Gauss-Seidel method*
5. modified Gauss	5. lemma established	5. M. Neumann	5. upper triangular part

4.3 Evaluation Method

In user-facing applications the quality of the complete keyphrase list is more important than the quality of the individual keyphrases. Therefore, we evaluate the quality of keyphrase lists, similar to evaluations in previous work [9]. To simplify further discussion, we introduce the term *n-sublist*, which is a list of the first n elements of a larger list. For example the 2-sublist of the list $(1, 2, 3)$ is $(1, 2)$. We expected that the 1-sublists exhibit the highest precision scores at a low recall score (because in scenarios where multiple gold standard keyphrases are given, single keyphrases cannot reach high recall scores). When assessing longer keyphrase sublists (e.g. the 2-sublists, 3-sublists etc.) the precision is expected to decline, due to the lower precision of lower ranked keyphrases, while the recall increases, as more extracted keyphrases match the gold standard keyphrases. For each document and each algorithm, we then compute precision, recall and F1 for each n-sublist ($n \in [1, 20]$ in the experiments). Measures are macro-averaged, that means, we calculate the measure for each document and then average over the total number of documents.

5 Results

In this section, we provide results on the influence of the keyphrase scoring and show the effect of combining the scored output of different extraction algorithms.

Table 2 shows keyphrases extracted from an example document. We can see that Rake initially does not find a ground-truth keyphrase, but after scoring there are three matches. Similarly, Textrank initially finds two ground-truth

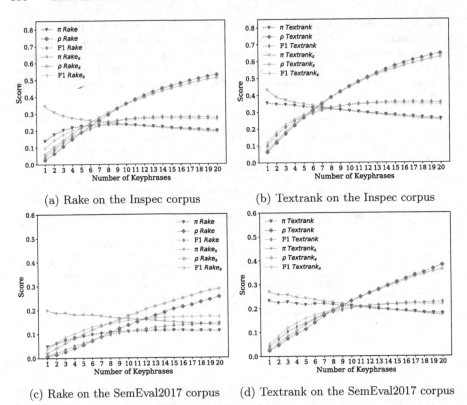

(a) Rake on the Inspec corpus   (b) Textrank on the Inspec corpus

(c) Rake on the SemEval2017 corpus   (d) Textrank on the SemEval2017 corpus

Fig. 3. Precision (π), Recall (ρ) and F1-score of Textrank and Rake compared to the tf-idf-based reranking on the Inspec and the SemEval2017 corpora.

keyphrases but after scoring it finds all four ground-truth keyphrases. Figure 3 compares precision (π), recall (ρ) and $F1$ for the scored and unscored versions of Rake and TextRank on the Inspec corpus and on the SemEval2017 corpus. The tf-idf-based scoring increases the precision of Rake significantly for up to five keywords. The precision of Textrank is also enhanced but the effect is significantly smaller.

For instance, for only one keyphrase on the Inspec corpus, Rake is below the baseline (baseline 0.24 π, Rake 0.14 π) but the scored version of Rake is considerably better than the baseline ($Rake_s$ 0.35 π). The already better-than-baseline performance (0.36 π) of Textrank is enhanced (0.43 π). At 5 keyphrases the situation is similar. But here the performance of Rake is above the baseline and the performance gain due to the scoring is smaller (Rake +0.04 π, Textrank +0.02 π). Figure 4 shows the ranked F1-scores on the Inspec and SemEval2017. There we can see that our method increases the performance mostly on documents with mediocre to low scores. The performance on documents where to score is already good is not affected as much.

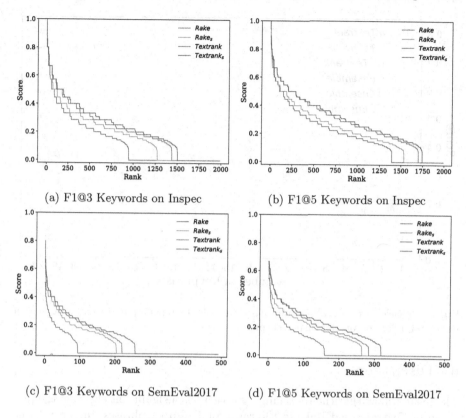

(a) F1@3 Keywords on Inspec

(b) F1@5 Keywords on Inspec

(c) F1@3 Keywords on SemEval2017

(d) F1@5 Keywords on SemEval2017

Fig. 4. Individual results ordered by F1-scores.

In general we can say that as the number of keyphrases increases the positive effect of the reranking diminishes due to the fact that scoring the list of keywords has no influence anymore if the whole list is used. This behaviour is also observable for the other corpora. Figures are omitted here due to space constraints, but the snapshots of performance curves at 1, 5, 10 and 20 keyphrases are provided in Table 3. In this table, it can also be seen, that Textrank always outperforms Rake.

Table 3 also shows the performance of the ensemble method. Performance of the ensemble is consistently better than $Rake_s$ but worse than $Textrank_s$. This means the ensemble method is not able to incorporate the additional keyphrases provided by Rake to enhance the performance of Textrank. Figure 5 depicts the performance of the ensemble versus the best performing algorithm $Textrank_s$ on the Inspec corpus.

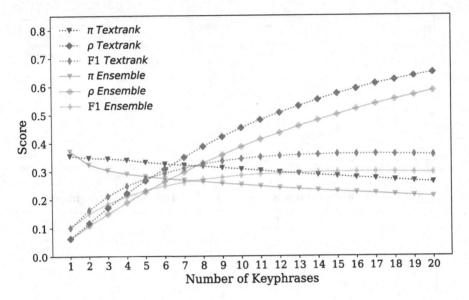

Fig. 5. Precision (π), Recall (ρ) and F1-score of textrank compared to an ensemble of Rake and Textrank on the Inspec corpus.

6   Discussion

The results show that the reranking approach has a significant effect on the precision of Textrank and Rake in the range of 1 to 5 keyphrases. In general, the effect size is strongest when only a single keyphrase is extracted and declines as the number of extracted keyphrases increases. Similarly, recall benefits from the reranking but the absolute effect size is much smaller. The experiments also show that the ensemble is not able to retain the performance of the strongest individual algorithm. Instead it consistently performs better than the weaker algorithm (Rake) and worse than the stronger algorithm (Textrank). Also it must be noted that from these results one cannot conclude that *in general* keyphrases with a higher tf-idf value are better keyphrases than keyphrases with lower tf-idf values. Instead, one can only state that the probability of being a gold standard keyphrase is proportional to tf-idf value. Moreover, the way keyphrase-function may differ depends on the scenario. We chose a simple linear method (as shown in Eq. 2) which favors keyphrases with 2–5 tokens. Preferences for longer or shorter keyphrases can be steered with the parameter α in Eq. 2, which was set to $\alpha = \frac{1}{10}$ in our experiments. However, its influence on the quality of the result list would need to be investigated with a parameter study in the future.

7   Summary

We presented a framework that allows to rank a list or a set of keyphrases based on the tf-idf values of their individual tokens. Moreover, the framework

Table 3. Performance of base keyphrase extraction algorithms, their scored version and the ensemble-based keyphrase extractor.

Algo		1 phrase			5 phrases			10 phrase			20 phrases		
		π	ρ	F1	π	ρ	F1	π	ρ	F1	π	ρ	F1
Inspec	tf-idf	0.24	0.04	0.06	0.15	0.11	0.12	0.12	0.17	0.13	0.09	0.22	0.12
	RK	0.14	0.03	0.04	0.23	0.19	0.19	0.23	0.36	0.27	0.20	0.53	0.27
	TR	0.36	0.06	0.10	0.33	0.27	0.27	**0.31**	**0.45**	**0.34**	**0.26**	**0.65**	**0.36**
	RK_s	0.35	0.06	0.10	0.27	0.21	0.22	0.24	0.35	0.26	0.19	0.51	0.26
	TR_s	**0.43**	**0.07**	**0.12**	**0.35**	**0.28**	**0.29**	**0.31**	**0.45**	**0.34**	0.25	0.63	0.34
	ENS_s	0.37	0.06	0.10	0.28	0.23	0.23	0.25	0.39	0.29	0.21	0.58	0.29
SemEval17	tf-idf	0.24	0.02	0.04	0.16	0.08	0.10	0.12	0.12	0.11	0.08	0.17	0.10
	RK	0.05	0.01	0.01	0.09	0.05	0.06	0.11	0.14	0.11	0.11	0.25	0.14
	TR	0.23	0.02	0.04	0.22	0.11	0.14	**0.21**	**0.23**	**0.20**	**0.18**	**0.38**	**0.23**
	RK_s	0.20	0.02	0.04	0.18	0.10	0.12	0.17	0.18	0.16	0.14	0.29	0.17
	TR_s	**0.27**	**0.03**	**0.05**	**0.24**	**0.13**	**0.16**	**0.21**	**0.23**	**0.20**	0.17	0.36	0.22
	ENS_s	0.24	**0.03**	0.04	0.19	0.10	0.13	0.18	0.20	0.17	0.15	0.32	0.19
Scopus	tf-idf	0.11	0.04	0.06	0.05	0.09	0.06	0.03	0.13	0.05	0.02	0.16	0.04
	RK	0.02	0.01	0.01	0.04	0.08	0.05	0.05	0.20	0.07	0.05	0.36	0.08
	TR	0.08	0.03	0.04	0.09	0.18	0.11	**0.08**	**0.33**	**0.13**	**0.07**	**0.49**	**0.11**
	RK_s	0.10	0.04	0.06	0.09	0.17	0.11	0.07	0.26	0.10	0.05	0.36	0.08
	TR_s	**0.15**	**0.06**	**0.08**	**0.11**	**0.22**	**0.14**	**0.08**	0.32	**0.13**	0.06	0.44	0.10
	ENS_s	0.11	0.05	0.06	0.09	0.18	0.11	0.07	0.27	0.11	0.05	0.39	0.09
KP20k	tf-idf	0.10	0.04	0.05	0.05	0.09	0.06	0.03	0.12	0.05	0.02	0.15	0.04
	RK	0.01	0.01	0.01	0.03	0.07	0.04	0.04	0.17	0.06	0.04	0.31	0.07
	TR	0.09	0.04	0.05	0.08	0.17	0.10	0.07	0.29	0.11	**0.06**	**0.43**	**0.10**
	RK_s	0.09	0.04	0.05	0.08	0.16	0.10	0.06	0.24	0.09	0.05	0.33	0.08
	TR_s	**0.12**	**0.05**	**0.07**	**0.10**	**0.20**	**0.12**	**0.08**	**0.30**	**0.12**	**0.06**	0.41	**0.10**
	ENS_s	0.10	0.04	0.06	0.08	0.16	0.10	0.06	0.26	0.10	0.05	0.37	0.08

is agnostic to the method applied to extract the keyphrases. In fact, it is also able to deal with keyphrases extracted by multiple methods, regardless whether these methods rank the keyphrases they extract or not. This property provides a normalized, common score for all keyphrases and thus allows to combine results from different algorithms. For two keyphrase extraction algorithms, we showed that the keyphrases with high tf-idf values are more likely to be gold standard keyphrases. Thus, they are – on average – more informative keyphrases for end users. The results could be reproduced on four different corpora. We also showed a method to merge multiple keyphrase extraction algorithms into a single one, although it failed to achieve the top performance of the best individual method. Future work includes finding and investigating other keyphrase scoring functions and more extensive experiments with more keyphrase extraction algorithms to aggregate.

References

1. Augenstein, I., Das, M., Riedel, S., Vikraman, L., McCallum, A.: Semeval 2017 task 10: Scienceie-extracting keyphrases and relations from scientific publications. arXiv preprint arXiv:1704.02853 (2017)
2. Bougouin, A., Boudin, F., Daille, B.: Topicrank: graph-based topic ranking for keyphrase extraction, pp. 543–551, Oct 2013
3. Breiman, L.: Bagging predictors. Mach. Learn. **24**(2), 123–140 (1996)
4. Brin, S., Page, L.: The anatomy of a large-scale hypertextual web search engine. Computer. Netw. ISDN Syst. **30**(1–7), 107–117 (1998)
5. Danilevsky, M., Wang, C., Desai, N., Ren, X., Guo, J., Han, J.: Automatic construction and ranking of topical keyphrases on collections of short documents. In: Proceedings of the 2014 SIAM International Conference on Data Mining, pp. 398–406. SIAM (2014)
6. Frank, E., Paynter, G.W., Witten, I.H., Gutwin, C., Nevill-Manning, C.G.: Domain-specific keyphrase extraction. In: Proceedings of the Sixteenth International Joint Conference on Artificial Intelligence, IJCAI 1999, pp. 668–673. Morgan Kaufmann Publishers Inc., San Francisco, CA, USA (1999)
7. Grineva, M., Grinev, M., Lizorkin, D.: Extracting key terms from noisy and multi-theme documents. In: Proceedings of the 18th International Conference on World Wide Web, WWW 2009, pp. 661–670. ACM, New York, NY, USA (2009)
8. Hasan, K.S., Ng, V.: Conundrums in unsupervised keyphrase extraction: making sense of the state-of-the-art. In: Proceedings of the 23rd International Conference on Computational Linguistics: Posters, COLING 2010, pp. 365–373. Association for Computational Linguistics, Stroudsburg, PA, USA (2010)
9. Hasan, K.S., Ng, V.: Automatic keyphrase extraction: a survey of the state of the art. In: Proceedings of the 52nd Annual Meeting of the Association for Computational Linguistics (Volume 1: Long Papers), vol. 1, pp. 1262–1273 (2014)
10. Hulth, A.: Improved automatic keyword extraction given more linguistic knowledge. In: Proceedings of the 2003 Conference on Empirical Methods in Natural Language Processing, EMNLP 2003, pp. 216–223. Association for Computational Linguistics, Stroudsburg, PA, USA (2003)
11. Liu, F., Pennell, D., Liu, F., Liu, Y.: Unsupervised approaches for automatic keyword extraction using meeting transcripts. In: Proceedings of Human Language Technologies: The 2009 Annual Conference of the North American Chapter of the Association for Computational Linguistics, NAACL 2009, pp. 620–628. Association for Computational Linguistics, Stroudsburg, PA, USA (2009)
12. Liu, Z., Chen, X., Zheng, Y., Sun, M.: Automatic keyphrase extraction by bridging vocabulary gap. In: Proceedings of the Fifteenth Conference on Computational Natural Language Learning, pp. 135–144. Association for Computational Linguistics (2011)
13. Liu, Z., Huang, W., Zheng, Y., Sun, M.: Automatic keyphrase extraction via topic decomposition. In: Proceedings of the conference on empirical methods in natural language processing, pp. 366–376. Association for Computational Linguistics (2010)
14. Liu, Z., Li, P., Zheng, Y., Sun, M.: Clustering to find exemplar terms for keyphrase extraction. In: Proceedings of the Conference on Empirical Methods in Natural Language Processing: Volume 1 - Volume 1, EMNLP 2009, pp. 257–266. Association for Computational Linguistics, Stroudsburg, PA, USA (2009)

15. Mani, I.: Advances in Automatic Text Summarization. MIT Press, Cambridge (1999)
16. Medelyan, O., Frank, E., Witten, I.H.: Human-competitive tagging using automatic keyphrase extraction. In: Proceedings of the Conference on Empirical Methods in Natural Language Processing: Volume 3 - Volume 3, EMNLP 2009, pp. 1318–1327. Association for Computational Linguistics, Stroudsburg, PA, USA (2009)
17. Meng, R., Zhao, S., Han, S., He, D., Brusilovsky, P., Chi, Y.: Deep keyphrase generation. arXiv preprint arXiv:1704.06879 (2017)
18. Mihalcea, R., Tarau, P.: Textrank: bringing order into texts. In: Proceedings of Conference on Empirical Methods in Natural Language Processing, Barcelona, Spain (2004)
19. Miller, G.A.: The magical number seven, plus or minus two: Some limits on our capacity for processing information. Psychol. Rev. **63**(2), 81 (1956)
20. Ren, X., El-Kishky, A., Wang, C., Tao, F., Voss, C.R., Han, J.: Clustype: effective entity recognition and typing by relation phrase-based clustering. In: Proceedings of the 21th ACM SIGKDD International Conference on Knowledge Discovery and Data Mining, pp. 995–1004. ACM (2015)
21. Rose, S., Engel, D., Cramer, N., Cowley, W.: Automatic keyword extraction from individual documents, pp. 1–20. Wiley, Chichester (2010)
22. Turney, P.: Learning to extract keyphrases from text, Jan 1999
23. Wan, X., Xiao, J.: Collabrank: towards a collaborative approach to single- document keyphrase extraction. In: Proceedings of the 22nd International Conference on Computational Linguistics (Coling 2008), pp. 969–976. Coling 2008 Organizing Committee, Manchester, UK, August 2008
24. Wan, X., Xiao, J.: Single document keyphrase extraction using neighborhood knowledge. In: Proceedings of the 23rd National Conference on Artificial Intelligence - Volume 2, AAAI 2008, pp. 855–860. AAAI Press (2008)
25. Wan, X., Xiao, J.: Single document keyphrase extraction using neighborhood knowledge. AAAI **8**, 855–860 (2008)
26. Witten, I.H., Paynter, G.W., Frank, E., Gutwin, C., Nevill-Manning, C.G.: Kea: practical automatic keyphrase extraction. In: Proceedings of the Fourth ACM Conference on Digital Libraries, pp. 254–255. ACM, New York, NY, USA (1999)
27. Zhang, Y., Fang, Y., Weidong, X.: Deep keyphrase generation with a convolutional sequence to sequence model. In: 2017 4th International Conference on Systems and Informatics (ICSAI), pp. 1477–1485. IEEE (2017)

WS4ABSA: An NMF-Based Weakly-Supervised Approach for Aspect-Based Sentiment Analysis with Application to Online Reviews

Alberto Purpura[✉], Chiara Masiero[✉], and Gian Antonio Susto[✉]

University of Padova, Padova 35122, Italy
{purpuraa,masieroc,gianantonio.susto}@dei.unipd.it

Abstract. The goal of Aspect-Based Sentiment Analysis is to identify opinions regarding specific targets and the corresponding sentiment polarity in a document. The proposed approach is designed for real-world scenarios, where the amount of available information and annotated data is often too limited to train supervised models. We focus on the two core tasks of Aspect-Based Sentiment Analysis: aspect and sentiment polarity classification. The first task – which consists in the identification of the opinion targets in a document – is tackled by means of a weakly-supervised technique based on Non-negative Matrix Factorization. This strategy allows users to easily embed some a priori domain knowledge by means of short seed terms lists. Experimental results on publicly available data sets related to online reviews suggest that the proposed approach is very flexible and can be easily adapted to different languages and domains.

Keywords: Aspect-based sentiment analysis
Non-negative matrix factorization · Text mining
Weakly-supervised learning

1 Introduction

Sentiment Analysis (SA) [11] is a growing area of research in Natural Language Processing. While SA aims at inferring the overall opinion of the writer in a document, Aspect-Based Sentiment Analysis (ABSA) is concerned with fine-grained polarity analysis, and its purpose is two-fold:

1. extracting relevant aspects - For instance, in a context of on-line reviews on restaurants, relevant aspects could be food, service, location, etc;
2. evaluating the sentiment polarity of each aspect separately.

In the context of real-world applications, there is a clear need for ABSA solutions that are interpretable and flexible, ie. that can be adapted to different

© Springer Nature Switzerland AG 2018
L. Soldatova et al. (Eds.): DS 2018, LNAI 11198, pp. 386–401, 2018.
https://doi.org/10.1007/978-3-030-01771-2_25

languages and domains. For example, ABSA is particularly suitable for Business to Consumer (B2C) companies to improve and develop their products, services or marketing strategies based on the feedback provided by customers in the form of online reviews; however, such online reviews may come from different countries and may be related to different product categories (e.g. laptops and smartphones), making ABSA a difficult task. Moreover, we remark that data annotation in this situation is time consuming and expensive, also because the subjectivity of this task is generally tackled by employing a panel of human annotators; thus, unsupervised or weakly-supervised approaches are often preferred to supervised ones. Furthermore, some domain knowledge is generally available, even if limited and partial. For instance some keywords used to describe aspects are generally known in advance and the same holds true for opinion words. We remark that the sentiment associated with opinion words is aspect-specific; for example, the opinion word 'cheap' conveys a positive sentiment with respect to a 'value for money' aspect in online reviews, while it conveys a negative sentiment on a 'quality' and an 'appearance' in case these aspects are considered in the ABSA task. With this scenario in mind, we propose Weakly-Supervised Approach for ABSA (WS4ABSA), a technique that is able to accomplish the core tasks of ABSA, and can be easily adapted to deal with different domains and languages. WS4ABSA tackles ABSA in two steps: (i) aspect classification and (ii) sentiment polarity classification. As for (i), we present a novel approach based on a well-known Topic Modeling technique, called Non-negative Matrix Factorization (NMF). The proposed approach allows the user to include some domain knowledge – in a weakly-supervised way – in order to link each topic discovered by NMF to the aspects which are referred to in a document. Indeed, WS4ABSA allows the user to embed a list of seed words to guide the algorithm towards more significant topic definitions. Regarding (ii), WS4ABSA employs another weakly-supervised framework based on the definition of a positive and a negative seed list with a few sentiment terms for each topic. These lists are then extended using Word2Vec [13] and used to assign a polarity to each aspect identified in step (i). Our system distinguishes itself from the ones in literature (detailed in Sect. 2) because it does not rely on any auxiliary resources or annotated data sets, but only on the aforementioned lists and some simple grammar rules to deal with negations. Therefore, WS4ABSA can be applied easily for the analysis of documents from any domain and in any language with the advantages of being easily interpretable and implementable. Moreover, WS4ABSA allows the user to include his prior knowledge on the problem and to iteratively improve the results thanks to the reformulation of the NMF problem objective function proposed here. This additional information can be used to steer the classification in a precise and predictable manner. The rest of the paper is organized as follows: WS4ABSA is presented in Sects. 3.1 and 3.2, by illustrating the procedure for aspect extraction and sentiment polarity classification, respectively. In Sect. 4, we test this approach on publicly available data sets, while final remarks and future research directions are reported in Sect. 5.

2 Related Work

Available approaches to ABSA can be divided into supervised [18], semi-supervised [23], weakly-supervised [5] or unsupervised [2] techniques. Support Vector Machines [12], Naive Bayes classifiers [20] and Maximum Entropy classifiers [26] are the most common approaches among the supervised machine learning methods to detect the aspects in a sentence or for sentiment classification. In the realm of supervised approaches, Neural Networks (NN) have also received an increasing interest in recent years. Convolutional NN, for example, have been successfully applied to ABSA such as in [18]. Unfortunately, supervised methods – particularly neural networks – require large annotated corpora to perform well. As said, this is an issue, especially for low-resource languages and specific application domains. Thus, unsupervised approaches are frequently adopted. In general, this type of techniques need labeled data only to test and validate the model. Most of the approaches that fall under this category use topic models to extract aspect and sentiment terms. The most adopted topic modeling techniques are Latent Dirichlet Allocation (LDA) [1] and NMF [14], mainly because their results are easy to interpret thanks to positivity constraints. NMF has two main advantages when compared to LDA: first, it allows for an easier tuning and manipulation of its internal parameters [22]; second, there are efficient and completely deterministic algorithms for computing a reliable approximate solution. W2VLDA [5] uses LDA to detect aspects and extracts the corresponding polarity based on very simple lists of seed words in input. The main drawback of W2VLDA is that it requires a language model trained on a large specific domain corpus to embed domain knowledge in aspect classification. Another topic modeling-based approach is UTOPIAN [3] that provides an interactive topic modeling system based on NMF that allows users to steer the results by embedding their domain knowledge. Finally, also [10] uses NMF to identify general sentiment linguistic indicators from one domain and then gauge sentiment around documents in a new target domain. Another example of the versatility of NMF can be seen in [24], where the authors attempt to learn topics from short documents using term correlation data rather than the usual high-dimensional and sparse term occurrence information in documents.

3 Weakly-Supervised Approach for ABSA (WS4ABSA)

Given a collection of documents, WS4ABSA tackles ABSA in two steps:

(i) based on a list of seed words for each aspect, WS4ABSA performs aspect extraction by means of NMF;
(ii) using a set of sentiment seed words for each aspect, for each document WS4ABSA assigns a sentiment polarity to each of the detected aspects.

3.1 Aspect Classification

In WS4ABSA, aspect classification is achieved by means of NMF. This technique aims at solving the following problem: given a non-negative $m \times n$ matrix

A (i.e. a matrix where each element $A_{ij} \geq 0, \forall i, j$), find non-negative matrix factors $W \in \mathbb{R}_+^{m \times k}$ (*term-topic matrix*) and $H \in \mathbb{R}_+^{k \times n}$ (*topic-document matrix*), for a given number of aspects $k \in \mathbb{N}_+$, such that $A \approx WH$. In our formulation of the problem, A represents the collection of n documents we want to analyze, for example using the Term Frequency-Inverse Document Frequency (TF-IDF) weighting scheme [17] with respect to the m distinct terms contained in the collection. W represents the associations between the terms contained in the collection and the k considered aspects, and H represents the associations between the aspects and each document in the indexed collection. Among the different problem formulations for NMF [8], here we consider the factorization problem based on the Frobenius Norm:

$$\min_{W \geq 0, H \geq 0} f(W, H) = ||A - WH||_F^2, \tag{1}$$

where, with $W, H \geq 0$ we impose the constraint on each element of the matrices to be non-negative, and with $|| \cdot ||_F$ we indicate the Frobenius norm. Although NMF is an NP-hard problem [21], one can still hope to find a local minimum as an approximation. In this work we will focus on the Block Coordinate Descent (BCD) method that is an algorithmic framework to optimize the above objective function. BCD divides variables into several disjoint subgroups and iteratively minimizes the objective function with respect to the variables of each subgroup at a time. Under mild assumptions, it is possible to prove that BCD converges to stationary points [7]. Multiplicative Updating (MU) is another popular framework for solving NMF [9], however it has slow convergence and may lead to inferior quality solutions [7]. In Sect. 3.1 we introduce a novel NMF problem formulation that is particularly suitable to embed domain knowledge, while Sects. 3.1 and 3.1 provide additional implementation details.

Proposed NMF Resolution Method. To solve the NMF problem with the BCD approach, we referred to a method called Hierarchical Alternating Least Squares (HALS) [4]. Let us partition the matrices W and H into $2k$ blocks (k blocks each that are respectively the columns of W and the rows of H), in this case we can see the problem in the objective function in Eq. (1) as

$$||A - WH||_F^2 = ||A - \sum_{i=1}^k w_{\cdot i} h_{i \cdot}||_F^2. \tag{2}$$

To minimize each block of the matrices we solve

$$\min_{w_{\cdot i} \geq 0} ||h_{i \cdot}^T w_{\cdot i}^T - R_i^T||_F^2, \quad \min_{h_{i \cdot} \geq 0} ||w_{\cdot i} h_{i \cdot} - R_i||_F^2, \tag{3}$$

where $R_i = A - \sum_{\tilde{i}=1, \tilde{i} \neq i}^k w_{\cdot \tilde{i}} h_{\tilde{i} \cdot}$. The promising aspect of this $2k$-block partitioning is that each subproblem in (3) has a closed-form solution using Theorem 2 from [7]. The convergence of the algorithm is guaranteed if the blocks of W and H remain nonzero throughout all the iterations and the minima of (3) are attained at each step [7]. Finally, we include in Eq. 1 a regularization factor for H to induce sparse solutions, so that each document is modeled as a mixture of just a few topics. At the same time, we also add a regularization term on W to prevent its entries from growing too much, and add a prior to exploit available knowledge of the user. In particular, for each of the topics, s/he can identify some terms as relevant, or decide to exclude others. As a result, we obtain the following expression:

$$\min_{W,H \geq 0} ||A - WH||_F^2 + \phi(\alpha_p, W, P) + \psi(H), \tag{4}$$

where $\psi(H) = \beta \sum_{i=1}^{n} ||h_{i\cdot}||_1^2$ and $\phi(\alpha_p, W, P) = \sum_{i=1}^{m} \sum_{j=1}^{k} \alpha_{p_{ij}} (w_{ij} - p_{ij})^2 + \alpha ||W||_F^2$. The notation $h_{i\cdot}$ is used to represent the i-th row of H, the l_1 term promotes sparsity on the rows of H, while the Frobenius norm, (equivalent to l_2 regularization on the columns of W) prevents values in W from growing too large. The prior term P is an $m \times k$ matrix in which entries are either 1 for terms that, according to the available domain knowledge, should be assigned to a certain aspect and 0 otherwise. For example, if the prior terms list of the aspect Food contains the terms 'curry' and 'chicken', the values in the columns corresponding to that aspect in the rows corresponding to these terms are set to 1. Since these seed lists are expected to be very short, matrix P will be a sparse matrix. The values in α_p serve as normalizing factors for the element-wise difference between W and P and to activate/deactivate the prior on a specific term for any of the k aspects. In other words, for what concerns matrix P, if $p_{ij} = 1$ we are suggesting to assign i-th term to aspect j. On the contrary, if $p_{ij} = 0$ we want the i-th term not to be assigned that aspect. This allows us, by manipulating the values in α_p and P, to choose to what extent we want to influence the link between certain topics and the seed terms. To the best of our knowledge, the regularization term based on P is novel and distinguishes our approach from other similar techniques such as [3]. In the same vein as [6], the new update formula for matrices W and H can be obtained in closed form as

$$w_{\cdot k} \leftarrow \frac{[\nu]_+}{\alpha_{p \cdot k} + (HH^T)_{kk}}, \quad h_{i\cdot}^T \leftarrow \left[h_{i\cdot}^T + \xi \right]_+,$$

$$\nu = (AH^T)_{\cdot k} - (WHH^T)_{\cdot k} + W_{\cdot k}(HH^T)_{kk}\alpha_{p \cdot k} \odot P_{\cdot k} - 0.5\alpha 1_m, \tag{5}$$

$$\xi = \frac{(A^T W)_{\cdot i} - H^T((W^T W)_{\cdot i} + \beta 1_k)}{(W^T W)_{ii} + \beta}.$$

Here, $\odot$ indicates an element-wise product, $[x]_+ = \max(0, x)$, 1_t indicates a vector of ones of length t and the division in the fraction is element-wise. After we performed the factorization of matrix A into the factors W and H, we normalize each column of matrix H, then, in order to identify the set of relevant topics in a document, we set an Aspect Detection Threshold (ADT) to an appropriate value – according to the total number of considered topics in the data set. Finally, we associate a document to a topic if the corresponding weight in matrix H is greater or equal to the chosen threshold.

Indexing of the Collection. The initialization of the term-document matrix A is a crucial aspect that is often overlooked in the related literature on NMF.

Whenever prior knowledge on the topics is available, we should make sure that all relevant elements with respect to these topics appear in A. Hence, we propose a novel method based on short seed lists of terms, the same used by the user to influence the aspect classification task. In particular, we select the set of terms $\mathcal{D}$ used to index the collection by means of three steps:

1. We add all seed words that appear at least once in our collection to $\mathcal{D}$;
2. Since we do not assume to have complete domain knowledge, the seed lists are extended automatically by means of Word2Vec and included in $\mathcal{D}$;
3. We compute the TF-IDF weight for each term in the collection and add the top few hundreds to $\mathcal{D}$.

As for the Word2Vec model used in the second step, no additional source of information is used because the model is trained on the same data set that has to be analyzed. Even if the latter is not very rich, it turns out that Word2Vec is still capable of mapping seed words close to other words coming from the same topic. More details are provided in Sect. 4.1.

W and H Matrices Initialization. There are different approaches to initialize the matrices W and/or H in the BCD framework and their initialization deeply impacts on the achieved solution. In the context of topic modeling, since we assume that we have some a priori knowledge on which terms should be associated to a specific topic/aspect in the form of words lists, we initialize the matrices according to this knowledge. We begin by extending these lists using a Word2Vec model trained on the same collection of documents that we have to analyze. Specifically, for each term in each list, we add the two closest terms in the Word2Vec model. Then, we perform a search of the few terms from the extended lists corresponding to each topic in each sentence of the collection, and set the corresponding elements of matrix H to 1 or 0 if the sentence contains one of them or not, respectively. We apply the same procedure for the term-topic matrix W. The only pre-processing step involved prior to the training of the Word2Vec model is stopwords removal.

3.2   Sentiment Polarity Classification

After the identification of the relevant aspects in each sentence, we compute the polarity for each of them, classifying the corresponding opinions as positive or negative. Again, we propose a weakly-supervised approach articulated as follows:

- we manually compile two lists of seed terms, one for the positive and one for the negative sentiment terms for each aspect;
- we extend the previously created lists using a Word2Vec model, as we did for document indexing (see Sect. 3.1);
- for each document in our data set, we run a pre-processing step that involves stemming and stopwords removal[1] and we do the same on each extended sentiment terms list;

---

[1] In this work we employed the stopwords and stemmers provided in Python nltk 3.2.5, https://www.nltk.org.

– finally, we look for these sentiment terms in each document, considering only the sentiment terms relative to the most relevant topics identified in it.

In order to compute the polarity for each aspect, we average over the number of positive and negative terms found in relation to that topic, weighting 1 the positive terms and −1 the negative ones. The final label assigned to the opinion will depend on the sign of this score. We found this approach to be the best performing to extend the seed lists available. We also implemented a simple negation detection system that employs a short manually compiled list of negation terms that can flip the polarity of a word[2]. If we detect one of these terms in a range of 3 tokens before a sentiment term, we flip its polarity. For other languages which follow similar strategies to indicate a negation, this rule can be easily modified to comply with the new structure. In [25], it is shown that the use of negation in these terms can be easily transferred to other languages.

4   Experimental Results

We evaluate WS4ABSA on public data sets[3]:

– 2016 Track 5 Subtask 1 [15], training data set of Restaurant reviews, in English [Rest-EN] (1152 documents, 18779 tokens, 1.18 labels on average for each document);
– 2016 Track 5 Subtask 1 [15], training data set of Restaurant reviews in Spanish [Rest-ES] (1047 documents, 21552 tokens, 1.35 labels on average for each document);
– 2015 Track 12 [16], test data set of Hotel reviews in English [Hotels] (86 documents, 1316 tokens, 1.10 labels on average for each document).

For the aspect classification task, we focused firstly on Restaurant reviews [Rest-EN, Rest-ES] and considered the following aspects:

– Ambiance: the atmosphere or the environment of the restaurant's interior or exterior space;
– Food: the food in general or specific dishes;
– Service: the customer/kitchen/counter service or the promptness and quality of the restaurant's service in general.

The corresponding entities in SemEval data sets are shown in Table 1.

---

[2] In this work we will consider corpora in English and Spanish (see Sect. 4); the lists of negation terms for English (16 terms) and Spanish (12 terms) are included in our code repository https://gitlab.dei.unipd.it/dl_dei/ws4absa.

[3] The code for deploying and evaluating WS4ABSA is available on https://gitlab.dei.unipd.it/dl_dei/ws4absa.

Table 1. Aspects definitions for the aspect classification task.

Data	Aspect	Labels in SemEval data sets
Rest.	Ambiance	AMBIENCE#GENERAL
	Food	FOOD#PRICES, FOOD#QUALITY, FOOD#STYLE
	Service	SERVICE#GENERAL
Hotel	Ambiance	FACILITIES#DESIGN_FEATURES, ROOMS#DESIGN_FEATURES, HOTEL#DESIGN_FEATURES
	Food	FOOD_DRINKS#PRICES, FOOD_DRINKS#QUALITY, FOOD_DRINKS#STYLE_OPTIONS
	Service	SERVICE#GENERAL

4.1   Training of the Word2Vec Model

Before diving into the experimental results, we report here how we use the available prior knowledge. The word lists provided in input by a user are extended employing a Word2Vec model trained on the data set we are currently analyzing. Even if this model is not an accurate representation of the relations between terms in the considered language in general, we found it good enough for our goal of adding terms related or used in the same context of the available seed terms. We employ these extended lists of terms for document indexing – together with the features selected with TF-IDF – and document classification, with the assumption that words used in the same context are relevant for the same aspect. An example of the resulting word lists obtained with this technique is reported in Table 2. The model has been trained[4] on each collection using the Continuous Bag-Of-Words (CBOW) training algorithm for 10 epochs, generating word embeddings of size 300.

4.2   Evaluation of Aspect Classification

Initially we tackle aspect classification for English reviews. The hyperparameters used in the NMF optimization are listed in Table 3, while the seed lists used to perform aspect classification on the [Rest-EN] data set are reported in Table 4.

Since LDA-based methods are the main alternative to NMF-based ones for the unsupervised document classification task, we choose to compare our approach to two other weakly-supervised methods, LocLDA [2] and ME-LDA [26]. These were developed under the assumption that each sentence is assigned to a single aspect. Thus, we compare WS4ABSA with them on the subset of [Rest-EN] sentences with a single aspect label (972 sentences). The comparison is reported in Table 5. While LocLDA and ME-LDA outperform WS4ABSA in most of the cases, we remark that, differently from WS4ABSA, these approaches

---

[4] Word2Vec implementation from https://radimrehurek.com/gensim/models/word2vec.html.

Table 2. [Rest-EN]: A few of the terms obtained by extending the English seed lists for aspect classification (from Table 4) with Word2Vec.

Aspect	Seeds included with Word2Vec
Ambiance	Cheap, classic, clean, describe, interesting, Italy, looks
Food	Drip, dumplings, oil, pay, perfect, price, sausages, starter, vegetarian
Service	Happy, help, hookah, hours, personality, professional, recommend, service

Table 3. Hyperparameters used in NMF for the aspect classification task, the ADT, and the number of terms selected with TF-IDF weights for the document indexing (TDI), for each of the considered test data sets. These parameters have been obtained with a grid search over a portion of the dataset, used for validation.

Data set	α	β	α_p	ADT	TDI
[Rest-EN]	1.00	0.10	1.00	0.13	200
[Rest-ES]	0.01	1^{-16}	0.10	0.16	300
[Hotels]	1^{-3}	1.00	1.00	0.19	200

Table 4. Seed lists employed for the aspect classification task in the [Rest-EN] and [Hotels] data sets.

Aspect	Seeds
Ambiance	Bad, beautiful, big, ceilings, chic, concept, cool, cozy, cramped, dark, decor, elegant, expensive, interior, lightning, loud, modern, nice, noisy, setting, trendy, uninspired, vibe, wall
Food	Beef, chewy, chicken, crispy, curry, drenched, dry, egg, groat, moist, onions, over-cooked, pizza, pork, red, roasted, seared, shrimp, smoked, soggy, sushi, tender, tuna, undercooked
Service	Attentive, chefs, efficient, employees, helpful, hostess, inattentive, knowledgeable, making, manager, owner, packed, polite, prompt, rude, staff, unfriendly, wearing, workers

heavily rely on additional resources and can be used only in a single-label context. In particular, in [2] and [26] the authors compute a topic model with 14 topics first, then they examine each of them manually and assign a label to them according to the aspects provided in input. Thus, whenever a new dataset is considered, human inspection of the topic modeling results is required to choose the correct number of topics to use. In addition, in LocLDA and ME-LDA, the discovered topics have to be manually linked to the aspects under examination, while this is not necessary in WS4ABSA, where seed words define the aspects. Moreover, the methods in [2] and [26] both employ some language-dependent resources such as Part-Of-Speech (POS) taggers to identify adjectives in sentences and improve the identification of aspects. Furthermore, in ME-LDA, the authors also employ an annotated dataset to train a Maximum Entropy (ME) classifier. On the contrary, WS4ABSA requires no additional resources beyond the dataset but a list of seed words based on available domain knowledge. Moreover, it is also suitable to deal with the more general multi-label assumption for aspect extraction (indeed, as mentioned above, the average number of labels per sentence is always greater that one in our dataset).

4.3 Evaluation of Sentiment Polarity Classification

For the sentiment polarity classification task, we first perform our experiments on [Rest-EN] data set. We formalize sentiment polarity classification as a single-label multi-class classification problem and used the seed words listed in Table 7. Table 8 describes the results of the sentiment polarity classification task, obtained on the Restaurants data sets. We computed these results considering only the opinions which were classified correctly in the previous aspect classification stage. Our performance results in this task are aligned with other state of the art approaches [16] but stand out for the independence from external resources and for the high language flexibility. As expected, negative polarity is the most challenging to detect. However, we highlight that the we rely on extremely simple rules, described in Sect. 3.2, that may be further enriched to achieve better performance.

Table 5. Aspect classification performance on [Rest-EN] data set, considering the documents with a single relevant aspect in the performance evaluation. We remark that the amount of resources used in our approach is lower than the other methods included in the comparison. Indeed, we only required a short list of seed words defining aspects, while the other two methods are based on language-specific POS tagging, additional annotated data sets and manual topic inspection to retrieve aspects.

	WS4ABSA	LocLDA	ME-LDA
Ambiance: Precision	0.21	0.60	**0.77**
Recall	**0.79**	0.68	0.56
F1 score	0.33	0.64	**0.65**
Food: Precision	0.79	**0.90**	0.87
Recall	0.53	0.65	**0.79**
F1 score	0.64	0.75	**0.83**
Service: Precision	**0.88**	0.80	0.78
Recall	0.39	**0.59**	0.54
F1 score	0.54	**0.68**	0.64
Overall: Precision	0.74	0.77	**0.81**
Recall	0.52	**0.64**	0.63
F1 score	0.56	0.69	**0.70**

Table 6. Terms not present in the seed lists assigned by NMF to the chosen aspects from the Hotels data set.

Term	Ambiance	Food	Service
Curtain	1	≈ 0	≈ 0
Pool	1	≈ 0	≈ 0
Breakfast	≈ 0	1	≈ 0
Buffet	0.03	0.93	0.03
Response	≈ 0	≈ 0	1
Management	≈ 0	≈ 0	1

4.4   Domain Flexibility Evaluation

To assess the flexibility of WS4ABSA, we use the seed lists defined on Restaurants to perform aspect classification on another data set with similar topics coming from a different domain, i.e. Hotels. The results, shown in Table 9, suggest that WS4ABSA is able to generalize the information provided by seed words to similar aspects from different domains. In this case we consider aspect classification as a multi-label classification problem [19]. This is a more general and challenging scenario. Thus, we measure accuracy in this task by means of the Jaccard index $J = \frac{1}{N} \sum_{i=1}^{N} \frac{|\hat{y}_i \wedge y_i|}{|\hat{y}_i \vee y_i|}$, where N is the total number of samples that have been evaluated (in order to compute the average), $\hat{y}_i$ is a binary vector that is 1 only in the positions corresponding to the aspects predicted for the i-th sample and y_i is another binary vector that is 1 only in the positions corresponding to the true aspects to associate to the i-th sample. These results may be explained by the fact that the method is able to leverage partial prior information, i.e. the seeds play a key role in defining final topics, but they can also be extended automatically to other terms in the collection, if this improves the quality of the factorization. Indeed, recall that we have an active penalization term on W_{ij}, related to the prior, only if domain knowledge suggests that term i is relevant for topic j. Then, we induce a penalization policy that only acts on a subset of the entries of W, denoted by $\mathcal{S} := \{(i,j) \mid i \in \bar{I}, j \in \bar{J}\}$ for $\bar{I}$ and $\bar{J}$ defined based on prior knowledge on topics. No penalization is imposed for entries $\{W_{i,j} \mid (i,j) \notin \mathcal{S}\}$. This approach differs from the penalization strategy adopted by methods, such as Utopian [3], that allows the user to include domain knowledge, but embed it in the form of a distribution over all the available terms. If we assume to set equal to zero all the elements of P corresponding to positions $(i,j) \notin \mathcal{S}$, and we impose a topic-wise penalization, such as in Utopian, i.e.

$$H, W = \underset{H \geq 0, W \geq 0}{\operatorname{argmin}} \|A - WH\|_F^2 + \|(W - P)D\|_F^2,$$

with D diagonal matrix of weights, we force the algorithm towards solutions which do not assign new words to a topic for which seed words were already provided. In fact, the results of this test led to achieve an accuracy of just 0.30 on the [Rest-EN] data set. Therefore, we infer that WS4ABSA can work well

Table 7. Seed lists employed for the sentiment polarity classification task in the [Rest-EN] and [Hotels] data sets.

Aspect	Polarity	Seeds
Ambiance	Positive	beautiful, chic, cool, cozy, elegant, modern, nice, trendy, winner
	Negative	bad, beaten, big, cramped, dark, expensive, loud, noisy, uninspired
Food	Positive	crispy, groat, moist, red, roasted, seared, smoked, tender, winner
	Negative	beaten, chewy, drenched, dry, over-cooked, soggy, undercooked
Service	Positive	attentive, efficient, helpful, knowledgeable, polite, prompt, winner
	Negative	beaten, inattentive, making, packed, rude, unfriendly, wearing

Table 8. Performance results in the sentiment classification task on the Restaurants data set for the Positive and Negative polarities.

	[Rest-EN]	[Rest-ES]
Accuracy	0.85	0.57
Precision (Pos.)	0.86	0.96
Recall (Pos.)	0.89	0.48
Precision (Neg.)	0.84	0.31
Recall (Neg.)	0.80	0.92

Table 9. Aspect classification performance in multi-label classification.

	[Rest-EN]	[Rest-ES]	[Hotels]
Accuracy	0.58	0.52	0.60
Average precision	0.52	0.48	0.58
Average recall	0.74	0.55	0.67
F1 score	0.61	0.51	0.62

with much weaker supervision than Utopian. In this regards, in Table 6, which shows a few rows of the term-topic matrix W, we can see that the algorithm includes some terms that were absent from the initial seed lists. Thus, the proposed classification generalizes well the initial information that was provided in the seed lists.

4.5 Language Flexibility

To assess the flexibility of WS4ABSA with regard to different languages, we considered [Rest-ES] data set and used as seed lists the same words used for the English data set, – see Tables 4 and 7 – translated when necessary with Google Translate. The resulting seed words are shown in in Tables 10 and 11. The results, described in Tables 8 and 9, suggest that our approach can be straightforwardly adapted to different domains and languages by translating the terms in the seed lists with a machine translation system. This is a simple method to leverage the same prior knowledge for cross-domain and cross-languages applications. As for aspect classification task, the performance on [Rest-ES] is very close to what we obtained on [Rest-EN]. As it might be expected, we notice a decrease in the average recall in this case, since some terms that might be very frequently used in a language in a specific context might not be used as frequently in other languages. We see the same effect on the recall of the positive class in Table 8. The impact of the machine translation of seed terms is lower on the recall of the negative class in the same table because we employ a set of terms to recognize negations which was compiled manually based on basic Spanish grammar rules. In fact, these terms would not have been easy to obtain by automatically translating the ones in the list we employed for the datasets in English. We expect that fine tuning of seed words and negation rules could further improve the performance of WS4ABSA. Yet, experiments suggest that an almost automatic adaptation to a different language achieves acceptable performance.

Table 10. Seed lists employed for the aspect classification task in the [Rest-ES] data set. These have been obtained by translating the ones in Table 4.

Aspect	Seeds
Ambiance	Acogedor, ambiente, apretado, bonito, caro, chic, concepto, decoración, elegante, escenario, fuerte, genial, grande, hermoso, interior, malo, moderno, no inspirado, oscuro, pared, relámpago, ruidoso, techos
Food	Ahumado, atún, camarón, carne de res, cauterizado, cebolla, cocido, crujiente, curry, empapado, groat, huevo, húmedo, masticable, mojado, pizza, pollo, puerco, rojo, seco, sobre cocinado, sushi, tierno, tostado
Service	Anfitriona, antipático, atento, cocineros, conocedor, cortés, eficiente, embalado, empleados, falta de atención, gerente, grosero, personal, propietario, rápido, servicial, trabajadores, usar

4.6 Impact of Initialization Policy

We also analyzed how the initialization of matrices W and H affects the results of aspect classification. In particular, we compared the policy described in Sect. 3.1 with 50 random initializations of the matrices W and H in order to evaluate the improvement of our initialization strategy in the classification task on the [Rest-EN] data set. With a random initialization, we obtained an average accuracy in the multi-label classification problem of 0.36, while the proposed initialization approach leads to an accuracy of 0.58, with an improvement of 38%[5] compared to a random initialization of the matrices on average. Furthermore, we also tested our method for feature extraction, i.e. for the document indexing process and

---

[5] The difference was computed as: difference ÷ other value * 100.

Table 11. Seed lists employed for the sentiment polarity classification task in the [Rest-ES] data set. These have been obtained by translating the ones in Table 7.

Aspect	Polarity	Seeds
Ambiance	Positive	Acogedor, agradable, chic, de moda, elegante, ganador, genial, hermoso, moderno
	Negative	apretado, caro, fuerte, golpeado, grande, malo, oscuro, ruidoso, sin inspiración
Food	Positive	Ahumado, cauterizado, crujiente, ganador, groat, húmedo, rojo, tierno, tostado
	Negative	batido, demasiado cocido, masticable, mojado, poco cocido, seco
Service	Positive	Atento, conocedor, educado, eficiente, ganador, rápido, útil
	Negative	Antipático, desgastado, embalado, fabricación, falta de atención, golpeado, grosero

the creation of the A matrix. In particular, we compared the results obtained by following the initialization procedure described in Sect. 3.1 with simple TF-IDF initialization on the same [Rest-EN] data set. As a result, we obtained an accuracy of 0.38. Hence noticing a performance improvement with our new feature selection approach of 34%.

5   Conclusions and Future Directions

We propose Weakly-Supervised Approach for ABSA (WS4ABSA), a weakly-supervised approach for ABSA based on NMF that allows users to include domain knowledge in a straightforward fashion by means of short seed lists. Thus, we address one of the drawbacks of most of the available topic modeling strategies, i.e. the fact that the beneficiary of the results is not able to improve them. WS4ABSA can be easily adapted to other domains or languages, as suggested by tests performed on publicly available data sets, and achieves performance comparable with other weakly and semi-supervised approaches in the literature, even though it relies on less external resources. Future research directions include deeper investigations on the effect of the prior on W and possibly H, and also the release of simple rules to deal with negations more effectively in different languages. It might also be useful to implement an on-line version of the NMF classification algorithm, so that it can receive a feedback from the user, and recompute the output on-the-fly more efficiently, i.e. without running again from scratch.

References

1. Blei, D.M., Ng, A.Y., Jordan, M.I.: Latent dirichlet allocation. Journal Mach. Learn. Res. **3**, 993–1022 (2003)
2. Brody, S., Elhadad, N.: An unsupervised aspect-sentiment model for online reviews. In: Human Language Technologies: The 2010 Annual Conference of the North

American Chapter of the Association for Computational Linguistics, pp. 804–812. Association for Computational Linguistics (2010)

3. Choo, J., Lee, C., Reddy, C.K., Park, H.: Utopian: User-driven topic modeling based on interactive nonnegative matrix factorization. IEEE Trans. Vis. Comput. Graph. **19**(12), 1992–2001 (2013)

4. Cichocki, A., Phan, A.H.: Fast local algorithms for large scale nonnegative matrix and tensor factorizations. IEICE Trans. Fundam. Electron., Commun. Comput. Sci. **92**(3), 708–721 (2009)

5. García-Pablos, A., Cuadros, M., Rigau, G.: W2vlda: almost unsupervised system for aspect based sentiment analysis. Expert. Syst. Appl. **91**, 127–137 (2018)

6. Kim, H., Park, H.: Nonnegative matrix factorization based on alternating nonnegativity constrained least squares and active set method. SIAM J. Matrix Anal. Appl. **30**(2), 713–730 (2008)

7. Kim, J., He, Y., Park, H.: Algorithms for nonnegative matrix and tensor factorizations: A unified view based on block coordinate descent framework. J. Glob. Optim. **58**(2), 285–319 (2014)

8. Kuang, D., Choo, J., Park, H.: Nonnegative matrix factorization for interactive topic modeling and document clustering. In: Celebi, M.E. (ed.) Partitional Clustering Algorithms, pp. 215–243. Springer, Cham (2015). https://doi.org/10.1007/978-3-319-09259-1_7

9. Lawson, C.L., Hanson, R.J.: Solving Least Squares Problems, vol. 15. SIAM, Philadelphia (1995)

10. Li, T., Sindhwani, V., Ding, C., Zhang, Y.: Bridging domains with words: Opinion analysis with matrix tri-factorizations. In: Proceedings of the 2010 SIAM International Conference on Data Mining, pp. 293–302. SIAM (2010)

11. Liu, B.: Sentiment analysis and opinion mining. Synth. Lect. Hum. Lang. Technol. **5**(1), 1–167 (2012)

12. Maas, A.L., Daly, R.E., Pham, P.T., Huang, D., Ng, A.Y., Potts, C.: Learning word vectors for sentiment analysis. In: Proceedings of the 49th annual meeting of the association for computational linguistics: Human language technologies-vol. 1, pp. 142–150. Association for Computational Linguistics (2011)

13. Mikolov, T., Sutskever, I., Chen, K., Corrado, G.S., Dean, J.: Distributed representations of words and phrases and their compositionality. In: Advances in Neural Information Processing Systems, pp. 3111–3119 (2013)

14. Paatero, P., Tapper, U.: Positive matrix factorization: A non-negative factor model with optimal utilization of error estimates of data values. Environmetrics **5**(2), 111–126 (1994)

15. Pontiki, M., et al.: Semeval-2016 task 5: aspect based sentiment analysis. In: Proceedings of the 10th International Workshop on Semantic Evaluation (SemEval-2016), pp. 19–30 (2016)

16. Pontiki, M., Galanis, D., Papageorgiou, H., Manandhar, S., Androutsopoulos, I.: Semeval-2015 task 12: aspect based sentiment analysis. In: Proceedings of the 9th International Workshop on Semantic Evaluation (SemEval 2015), pp. 486–495 (2015)

17. Salton, G., Buckley, C.: Term-weighting approaches in automatic text retrieval. Inf. Process. Manag. **24**(5), 513–523 (1988)

18. Toh, Z., Su, J.: Nlangp at semeval-2016 task 5: improving aspect based sentiment analysis using neural network features. In: Proceedings of the 10th International Workshop on Semantic Evaluation (SemEval-2016), pp. 282–288 (2016)

19. Tsoumakas, G., Katakis, I.: Multi-label classification: an overview. Int. J. Data Warehous. Min. **3**(3), 1–13 (2006)

20. Varghese, R., Jayasree, M.: Aspect based sentiment analysis using support vector machine classifier. In: 2013 International Conference on Advances in Computing, Communications and Informatics (ICACCI), pp. 1581–1586. IEEE (2013)
21. Vavasis, S.A.: On the complexity of nonnegative matrix factorization. SIAM J. Optim. **20**(3), 1364–1377 (2009)
22. Wang, F., Li, T., Zhang, C.: Semi-supervised clustering via matrix factorization. In: Proceedings of the 2008 SIAM International Conference on Data Mining, pp. 1–12. SIAM (2008)
23. Xiang, B., Zhou, L.: Improving twitter sentiment analysis with topic-based mixture modeling and semi-supervised training. In: Proceedings of the 52nd Annual Meeting of the Association for Computational Linguistics (Volume 2: Short Papers), vol. 2, pp. 434–439 (2014)
24. Yan, X., Guo, J., Liu, S., Cheng, X., Wang, Y.: Learning topics in short texts by non-negative matrix factorization on term correlation matrix. In: Proceedings of the 2013 SIAM International Conference on Data Mining, pp. 749–757. SIAM (2013)
25. Zagibalov, T., Carroll, J.: Automatic seed word selection for unsupervised sentiment classification of chinese text. In: Proceedings of the 22nd International Conference on Computational Linguistics-Volume 1, pp. 1073–1080. Association for Computational Linguistics (2008)
26. Zhao, W.X., Jiang, J., Yan, H., Li, X.: Jointly modeling aspects and opinions with a maxent-lda hybrid. In: Proceedings of the 2010 Conference on Empirical Methods in Natural Language Processing, pp. 56–65. Association for Computational Linguistics (2010)

Applications

Finding Topic-Specific Trends and Influential Users in Social Networks

Eleni Koutrouli(✉), Christos Daskalakis, and Aphrodite Tsalgatidou

Department of Informatics & Telecommunications, National & Kapodistrian
University of Athens, Panepistimiopolis, 157 84 Ilisia, Athens, Greece
{ekou, cdaslaka, atsalga}@di.uoa.gr

Abstract. Social networks (SNs) have become an integral part of contemporary life, as they are increasingly used as a basic means for communication with friends, sharing of opinions and staying up to date with news and current events. The general increase in the usage and popularity of social media has led to an explosion of available data, which creates opportunities for various kinds of utilization, such as predicting, finding or even creating trends. We are thus interested in exploring the following questions: (a) Which are the most influential - popular internet publications posted in SNs, for a specific topic? (b) Which members of SNs are experts or influential regarding a specific topic? Our approach towards answering the above questions is based on the functionality of hashtags, which we use as topic indicators for posts, and on the assumption that a specific topic is represented by multiple hashtags. We present a neighborhood-based recommender system, which we have implemented using collaborative filtering algorithms in order to (a) identify hashtags, urls and users related with a specific topic, and (b) combine them with SN-based metrics in order to address the aforementioned questions in Twitter. The recommender system is built on top of Apache Spark framework in order to achieve optimal scaling and efficiency. For the verification of our system we have used data sets mined from Twitter and tested the extracted results for influential users and urls concerning specific topics in comparison with the influence scores produced by a state of the art influence estimation tool for SNs. Finally, we present and discuss the results regarding two distinct topics and also discuss the offered and potential utility of our system.

Keywords: Influence · Social networks · Recommender systems

1 Introduction

E-communities necessitate mechanisms for the identification of credible entities which can be trusted and used in a particular context. Various reputation systems have been proposed to effectively address this need [19], based on the evaluation of individual transactions. In social networks (SNs), where there is an abundance of information, rich social activity of many people and dynamic relationships and interactions of various forms, apart from the need to find credible entities, new requirements and new possibilities arise, which are related to the identification of influential entities, i.e. entities which attract the interest of users and can provoke actions [20]. A vast amount of

L. Soldatova et al. (Eds.): DS 2018, LNAI 11198, pp. 405–420, 2018.
https://doi.org/10.1007/978-3-030-01771-2_26

research works have focused in the exploration of the concept of influence in social networks, its estimation and its use [1–3, 14]. These works are usually based on the various relationships and social actions of entities and focus on different aspects such as identification of influential entities and content [1, 14] influence propagation [27], influence maximization [7], combination of influence and trust [25]. Collaborative filtering mechanisms have also been widely studied and used to identify similarities and to produce recommendations in SN-based applications [9, 15, 26].

SN applications include also the useful "hashtag" functionality, i.e. the possibility to tag content using hashtags, which is a way of mapping content to specific topics. This functionality, when combined with the vast amount of social network activity information, creates the opportunity to explore influence in a more specialized context. Useful examples of specialized influence estimation include finding influential news and influential users regarding the specific topic. In this paper we present our work towards answering the following questions: (a) Which are the most influential - popular internet publications posted in SNs for a specific topic? (b) Which members of SNs are experts or influential regarding a specific topic? Answers to these questions are vital in various areas such as marketing, politics, social media analysis, and generally in all fields which need to quickly understand and respond to current trends.

Our approach towards answering the aforementioned questions combines influence estimation techniques and collaborative filtering mechanisms. More specifically, it is based on (a) the hashtag functionality and the assumption that a topic is represented by one or more hashtags, (b) collaborative filtering techniques for finding similar hashtags based on their common usage and the links assigned to them, and (c) analysis of social network-based actions. The contribution of this work is thus a solution for finding topic-specific trends both for content and for users: collaborative filtering is used for identifying a set of similar hashtags which represent a topic, which are then used for filtering user activity in order to find the topic-specific influential users and content. This solution is, to the best of our knowledge, novel with respect to related works which examine influence from different perspectives.

In the following section we present related work which focuses on influence estimation in SNs. This is followed by an overview of our approach and a description of its steps. In the fourth section we present the implemented system and its evaluation using various scenarios and in relation to a benchmark influence estimation tool and we discuss the produced results. Our conclusive remarks follow in the last section.

2   Related Work

Various works have focused on estimating influence of entities and content in social networks. Influence is dealt with in various ways: e.g. as an indirect reputation concept [1, 14], i.e. an indication of how much trust or popularity can be assigned to an entity or content based on indirect information rather than on direct ratings, or as an indication of action propagation [11], or from a social analysis perspective [25], etc.. The various approaches use data related with the social activity in the SN, i.e. the actions of users towards other users or content in the SN. Most specifically, these approaches use algorithms which combine (a) entity-centered characteristics related with social actions, e.g. the number of likes of a post or the number of followers of a user [6, 11] and (b) social action-related characteristics of entity pairs, e.g. the number of likes user A assigns to posts of user B [25]. Trust relationships between two entities are also incorporated in the latter case, while a usual representation of a SN in the context of influence estimation is a graph, where the nodes represent individual entities and the edges represent the links between two entities, accompanied with one or more weights (each one for different kinds or relationships).

Along with the different kinds of information used, related works differentiate according to: (a) influence estimation technique and (b) requirements stemming from the kind of social network where influence is estimated. Various influence estimation techniques are found in the literature, such as probabilistic [2], deterministic [14], graph-based [6, 25], and machine-learning-based techniques (for influence prediction) [11]. According to the specific kind of social network, different requirements for influence estimation occur. For example, in microblogging SNs, such as Twitter [24], influence is related with a number of factors, such as recognition and preference [1, 14], which are attributed to social network activity of users (numbers of shares, likes, followers, the followers social activity, etc.). In other works, such as in review SNs, e.g. Epinions [10] or [8], a combination of social activity information with trust relationships is used [25].

In the following, we briefly present some works related with influence estimation in SNs with a focus on the influence estimation technique and the data they use. Agarwal et al. [1] deal with estimating the influence of bloggers in individual blogs. Four factors are considered as vital for defining influence: recognition, activity generation, novelty and eloquence. These properties are defined according to specific post characteristics and the social activities of bloggers, and are then combined for assessing the user's influence. Anger et al. [3] measure influence of both users and content in Twitter. They take into consideration various Twitter statistics, such as the numbers of followers, tweets, retweets and comments, and they estimate two measures on the content and on the action logs. Similarly to this, the work in [14] presents an influence estimation system both for hashtags and users in Twitter, based on various social activity-based data. Further performance indicators for Twitter are presented in [3], whereas [18] analyzes Twitter influence tools, including Klout [13], which had been a widely accepted influence estimation tool for SNs until May 2018. A distinct approach for finding the most important urls regarding a specific topic in Twitter is proposed by Yazdanfar et al. [26], who reason about the importance of url recommendation in Twitter and implement such recommendations using collaborative filtering techniques and a three-dimensional matrix of users, urls and hashtags.

Ahmed et al. [2] integrate the concept of trust in their approach for estimating influence probabilities. Their suggested algorithm discovers the influential nodes based on trust relationships and action logs of users. Varlamis et al. [25] integrate also trust relationships in their influence estimation mechanism, which uses both social network analysis metrics and collaborative rating scores, where the latter take into account both the direct and the indirect relationships and actions between two users. In [11] various influence models are constructed for a number of different time models and various algorithms used in the literature are analyzed and discussed. Bento [6] implements various social network analysis algorithms for finding influential nodes in location-based SNs and in static SNs.

Our approach focuses on topic-specific influence, but unlike topic-specific recommendation systems for SNs, such as [6], it is not restricted to collaborative filtering techniques. Furthermore, it is not restricted to SN activity–based influence estimation, which is adapted in [1, 2, 14]. It comprises rather a specialized influence estimation which combines the user activity characteristics responsible for influence estimation with collaborative filtering techniques for the topic-specific filtering of posts.

3 Estimating Influence on a Specific Topic

The goal of the proposed approach is to estimate influence of users and urls regarding a specific topic, and to find the most influential ones among them. The idea is that we first choose a hashtag which is representative of the topic of interest and then find a set of hashtags which are similar to the initial hashtag. We then aggregate social network metrics for the tweets which have used these hashtags in order to estimate influence scores for the users which have posted these tweets and for the urls which have been used in them.

For the purposes of this paper we focused on micro blogging systems like Twitter [24]; however, the proposed approach can be generalized due to the fact that its elements (e.g. hashtags, numbers of likes and followers) are common to most social networks.

Here is a step-by-step description of the approach we follow:

- Step 1: Given a specific hashtag h_i, we first find the N-top similar hashtags based on collaborative filtering techniques which take into consideration the level of usage of hashtags by users and the usage of common urls together with hashtags, as explained in Sect. 3.1. We define **H** as the set containing h_i and the most similar hashtags to h_i.
- Step 2: We collect the sets of tweets, users and urls which have used at least one hashtag belonging to **H**, as analytically presented in Sect. 3.2.
- Step 3: Based on the above sets, we find the most influential users. The criteria for estimating a user's influence are based on social activity-based metrics related to tweets which contain the specific url. This step is presented in Sect. 3.3.
- Step 4: In a similar way, we use the above sets to find the most influential urls, using various social activity-based criteria, as described in Sect. 3.4.

3.1 Finding Similar Hashtags for Topic Representation

In order to identify a set of hashtags which represent a topic, we use an initial hashtag "h" and try to find hashtags which are similar to h, using two criteria for assessing similarity: (a) the number of common links that two distinct hashtags have (if two hashtags have the same number of references to a link, this link is related to the same level to these hashtags) and (b) the level of their usage by users who have used them in common (if two users have used them with similar frequency this means that these hashtags are of the same level of interest for the users, and are considered similar in this context). We thus define two similarity measures for hashtags according to the two criteria and combine them in one. Specifically, we use the similarity measures (1) and (2) that appear below, in order to estimate the Euclidean distance of two hashtags regarding the two criteria.

$$sim_{euclidean}\left(h_i, h_j\right)_{url} = 1 \Big/ \left(\sqrt{\sum_{l=1}^{L} \left(r_{h_{i,l}} - r_{h_{j,l}}\right)^2} \right) \tag{1}$$

where

- h_i, h_j are two distinct hashtags,
- L is the set of the links (urls) which have been used in at least one tweet of each of the hashtags h_i, h_j
- l is a url belonging to L,
- $r_{hi,l}$, $r_{hj,l}$ are the numbers of tweets which have used the link l and have also used the hashtag h_i and h_j, and
- $sim_{euclidean}\left(h_i, h_j\right)_{url}$ is the similarity of hashtags h_i, h_j regarding their usage of common links.

$$sim_{euclidean}\left(h_i, h_j\right)_{user} = 1 \Big/ \left(1 + \sqrt{\sum_{u=1}^{U} \left(r_{h_{i,u}} - r_{h_{j,u}}\right)^2} \right) \tag{2}$$

where

- hi, hj are two distinct hashtags,
- U is the set of the users which have used the two hashtags in their tweets,
- u is a user belonging in U,
- $r_{hi,u}$, $r_{hj,u}$ are the numbers of tweets of user u which have used the hastag h_i and h_j, and
- $sim_{euclidean}\left(h_i, h_j\right)_{user}$ is the similarity of hashtags h_i, h_j regarding their common usage by users.

We also use the cosine similarity measure presented in (3) to estimate the similarity between two hashtags according to the criteria of the level of interest and the number of commonly used urls (sim_{cosine} (h_i,h_j)$_{user}$, and sim_{cosine} (h_i,h_j)$_{urlr}$ accordingly):

$$sim_{cosine}\left(h_i, h_j\right)_{url/user} = \frac{\sum_{k=1}^{n}(r_{ik}) * (r_{jk})}{\sqrt{\left[\sum_{k=1}^{n}(r_{ik})^2\right]} * \sqrt{\left[\sum_{k=1}^{n}(r_{jk})^2\right]}} \tag{3}$$

where

- $sim_{cosine}(h_i, h_j)_{url}$ is the cosine similarity of hashtags h_i, h_j regarding their common urls,
- $sim_{cosine}(h_i, h_j)_{user}$ is the cosine similarity of hashtags h_i, h_j regarding their common usage by users,
- h_i, h_j are two distinct hashtags,
- n is the number of users which have both used the two hashtags (when $sim_{cosine}(h_i, h_j)_{user}$ is estimated) or the number of the common urls used by the two hashtags (when $sim_{cosine}(h_i, h_j)_{url}$ is estimated),
- k is a user (when $sim_{cosine}(h_i, h_j)_{user}$ is estimated) or a url (when $sim_{cosine}(h_i, h_j)_{url}$ is estimated),
- r_{ik}, r_{jk} are the numbers of times a user k has used the hashtags h_i, h_j respectivley (when $sim_{cosine}(h_i, h_j)_{user}$ is estimated) or the numbers of times a url k has been used in hashtags h_i and h_j respectively (when $sim_{cosine}(h_i, h_j)_{url}$ is estimated)

We use either one of the above similarity measures (Cosine or Euclidean distance-based similarity), or average measures to define final similarity metrics for user-based similarity ($sim(h_i,h_j)_{user}$) and url-based similarity ($sim(h_i,h_j)_{url}$). We then combine the two similarity measures using a weighted average to estimate the similarity between two hashtags.

$$sim(h_i, h_j) = w_{simuser} * sim(h_i, h_j)_{user} + w_{simurl} * sim(h_i, h_j)_{url} \qquad (4)$$

where

- $w_{simuser}$, w_{simurl} are the weights we use for the two kinds of similarity, and
- $w_{simuser} + w_{simurl} = 1$

We thus find a list of top-N hashtags which have the highest similarity with the original hashtag h_i. We define as **H**, the set which contains h_i and the N hashtags of this list. The original set of hashtags which are examined for the extraction of **H** can be obtained by various ways e.g. Twitter Streaming API [17], Twitter Rest API [23] Twitter widgets [22]. The selection of the original hashtag h_i from the available hashtags can be done either based on personalized criteria, e.g. one can select a hashtag which she believes as representative of a topic, or by searching available hashtags with text similarity criteria.

3.2   Collection of Data

Having acquired the set **H** of hashtags which represent a topic, we collect the following data, which are needed for finding the influential users and urls in the context of a specific topic:

1. The set T_H of all the tweets which have at least one hashtag belonging to the set H.
2. The set U_H of all users which have tweeted at least one tweet belonging to the set T_H, i.e. users which have used hashtags belonging to H.
3. The set L_H of all urls which have been attributed to one or more tweets belonging to the set T_H (or equivalently to one or more hashtags belonging to the set H).

In the following sections we describe the ways we use to extract lists of influential users and influential urls regarding a specific topic.

3.3   Finding Influential Users on a Specific Topic

For each one of the users belonging to in U_H we estimate her influence score regarding each hashtag belonging to H, based on the triple: *(weighted number of likes of related tweets, weighted number of retweets of related tweets, absolute number of user's followers)*. The triple is used to represent a number of criteria which we consider as important for determining influence. These criteria are presented in the rest of this section, along with the related metrics – formulae used for the estimation of a user's influence regarding a hashtag.

Estimating a User's Influence on a Specific Hashtag. The criteria for estimating a user's influence regarding a specific hashtag are described below, together with the metrics which represent them.

Adaptation: The total number of retweets a user has got for a specific topic (hashtag) shows the interest of other users to adapt or share the user's posts. We are interested in the adaptation level of u_i's tweets containing h_i, compared to the general level of adaptation that tweets containing h_i generate. We thus use the following adaptation metric:

$A(u_i, h_i)$ = the ratio of the number of retweets of the posts of a specific user u_i containing a specific hashtag h_i, to the total number of retweets which contain h_i. This metric shows the relative interest of users to share u_i's tweets compared to the total amount of interest that related tweets generate.

$$A(u_i, h_i) = \frac{number\ of\ retweets\ of\ u_i's\ tweets\ which\ contain\ h_i}{number\ of\ retweets\ of\ all\ tweets\ containing\ h_i}$$

Preference: A user's influence can be measured by the number of her followers; the more friends a user has got, the more she is trusted / preferred.

$$P(u_i) = number\ of\ followers\ of\ u_i$$

Endorsement: (concerning a specific topic expressed by a hashtag h_i): In today's social networks every post of a user can be endorsed by other users. In Facebook you can endorse the post of a user by reacting to it (like, Wow, etc.), in Twitter you can declare you like it. The more users endorse a post, the more influence this post has over users. This gives us an insight of how valuable is the user's opinion on some topic. We are interested in the value of the user's opinion is on a topic, in relation to the value of other users' opinions on that topic. For the endorsement metric we have thus used the following formula:

$E(u_i, h_i)$ = the ratio of the number of favorites that u_i's tweets containing a hashtag h_i have been assigned, to the total number of favorites assigned to tweets which contain h_i. This metric shows the relative endorsement in user's tweets compared to the total endorsement for tweets containing h_i.

$$E(u_i, h_i) = \frac{number\ of\ favorites\ of\ u_i's\ tweets\ containing\ h_i}{number\ of\ favorites\ of\ all\ tweets\ containing\ h_i}$$

We are using a weighted mean to estimate the influence score of a user u_i concerning a specific hashtag h_j, as a combination of the result scores of the above three influence factors:

$$InfUserHashtag(u_i, h_j) = w_A * A(u_i, h_j) + w_E * E(u_i, h_j) + w_P * P(u_i) \qquad (5)$$

where

- w_A, w_P, w_F are the weights we assign to the factors described above, and
- $w_A + w_P + w_F = 1$.

The choise of the values of these weights should be done according to the importance we want to give to each criterion. Machine learning methods can also been used to find the most appropriate values for the weights.

Estimating a User's Influence on a Topic. Having estimated the individual influence score of users regarding each (Top-N similar) hashtag of the hashtag set H which is representative of a topic, according to the previous section, we estimate the total influence score of every user using the following formula:

$$Inf(u_i) = \sum_{j=1}^{N} w_j * InfUserHashtag(u_i, h_j) \qquad (6)$$

where

- $Inf(u_i)$ is the influence score of a user u_i,
- $InfUserHashtag(u_i, h_j)$ is the influence score of a user u_i concerning a specific hashtag h_j, where h_j belongs to the set of the top N similar hashtags H and has the jth order in similarity with the initial hashtatg), and
- w_j is the weight assigned to the score of each hashtag h_j,

We have adjusted the values of weighs w_j which we assign to the various InfUserHashtag scores, depending on the similarity of the specific hashtag to the initial hashtag according to (4). Specifically, considering that $InfUserHashtag(u_i, h_1)$, $InfUserHashtag(u_i, h_2)$, ..., $InfUserHashtag(u_i, h_N)$ are ordered according to their similarity with the initial hashtag, then the first score will refer to the initial hashtag itself (h_1) and its weight w_1 will be estimated according to (7) and will be used as reference for estimating the other weights, in a way that $w_{i+1} = (w_i/2)$.

$$w_1 = 1/(1 + \sum_{i=1}^{k} \frac{1}{2^i})$$ (7)

where k: $2^k <= N * 2$ and $2^{k+1} > N * 2$

Using this way of weight estimation, we achieve $\sum_{i=1}^{N} w_i \approx 1$ [1] and we give higher weights to the influence scores of the most similar hashtags.

We finally extract the most influential users regarding the topic based on the estimated influence of users, according to formula (6).

3.4   Finding Influential Urls on a Specific Topic

As explained in Sect. 3.2, for each one of the hashtags belonging to the set **H** of similar hashtags, we find the links which have been used with them (set **L_H**) and the related tweets (set **T_H**). Then, for each link l of **L_H** we find the subset **T_l** of **T**, which contain the tweets of **T** which have used this link. We also find the numbers of likes and retweets of the tweets belonging in **T_l**.

We extract thus a list with recommended links, i.e. the ones with the highest influence score, which depends on the number of likes and retweets (of the relative tweets that contain them) and the number of tweets that contain these links, according to formula (8).

$$Inf(l) = w_{likes} * \left(\frac{TlNoLikes}{TlNoTweets}\right) + w_{retweets} * \left(\frac{TlNoRetweets}{TlNoTweets}\right)$$ (8)

where

- *TlNoLikes* is the number of likes on tweets belonging to **T_l**,
- *TlNoRetweets* is the number of retweets of tweets belonging to **T_l**,
- *TlNoTweets* is the number of tweets belonging to **T_l**
- w_{likes}, $w_{retweets}$ are the weights of the numbers of likes and retweets respectively, related to the total number of topic-related tweets.

We consider these links as the most influential ones for a specific topic, since they are the ones which are mapped to the most related hashtags to the topic and also are attributed to the highest social activity.

---

[1] When $2 * N = 2^k$ for an integer number k, then $\sum_{i=1}^{N} w_i = 1$.

4 Implementation of the Proposed System

We have developed a system that implements the proposed approach, as a data analysis platform for Twitter, which features three core functionalities in Twitter:

- Finding the most similar topics (hashtags) using collaborative filtering techniques based on the posts gathered from a social network.
- Finding the most influential links in terms of popular internet publications, using the information extracted from the above step.
- Finding the most influential users for a specific toping, exploiting the set of similar topics from the first step.

In the following subsections we present the technology used, the collection of real social network data and the evaluation tests.

4.1 Technology and Datasets

For the implementation of this big data analysis platform we have created an application that leverages the possibilities provided by the Apache Spark framework [4], as well as the Java programming language. Apache Spark is used in order to achieve optimal scaling and the possibility to process large volumes of data faster and more efficiently.

For the purposes of this paper we used the Twitter Rest API [23] in order to collect the last 1–3.300 tweets for different groups of Twitter users. The first group consisted of the most followed Twitter accounts. The second one had all the users that are considered the most influential ones for the Twitter platform according to [21] and the third one consisted of random Twitter users with a count of total users equal to 52. We used users from these three selected groups for our tests in order to have a wide range of users with different levels of popularity and influence in the social network. The tweets collected where filtered in order to contain at least one hashtag and were stored in the following format:

- UserId|TweetId|CreatedDate|Lang|text|FavCount|ShareCount|#|...|#|url|...|url|
  For each one of the aforementioned users, we stored the following data concerning them
- UserId|FollowersCount|FriendsCount|StatusesCount

We note that we intend to expand our experimental evaluation in larger datasets, leveraging the scalability possibilities offered by Apache Spark, in order (a) to extract results for influential urls and users in various scenarios involving different topics and user groups, and (b) to evaluate the performance of the system, e.g. running times in big data volumes.

4.2 Finding Topic-Specific Hashtags and Influential Users and Urls

From the 72.029 collected tweets we were able to extract 40.924 pairs of hashtags that (a) were used by at least two users in their tweets and (b) had at least one url in common. For each pair we were then able to define their similarity scores according to the formulae (1)–(4).

For estimating the total similarity between two hashtags, we chose the Cosine similarity metric, and used formula (4), with the following weight values: $w_{simuser} = 0.4$ and $w_{simurl} = 0.6$, considering that the similarity score of two hashtags regarding their common urls of greater importance than their similarity concerning their common level of usage by users. We note that the reason we chose to use the Cosine similarity metric for estimating both similarity components, is its suitability when we are interested in the cosine of the rating vectors [5], i.e. the similarity between their trends (in our case the trends of the url rating vectors and the trends of the user rating vectors).

Influential Users based on a Specific Topic. We have examined the influence scores of users for various topics and have compared our results with the scores provided by the user influence estimation tool InfluenceTracker [13] for the same users. We have chosen InfluenceTracker as a benchmark for our comparison, as it is included in the state of the art of influencer discovery and Twitter [18]. In the rest of this section we present preliminary evaluation results for the topics "marketing" and "bigdata" considering they are represented by the initial hashtags #marketing and #bigdata respectively. We have thus first extracted from our dataset the top most similar hashtags for #marketing and present them in Fig. 1, along with their similarity scores based on (4).

Figure 2 presents the influence scores of the top ten most influential users regarding the topic "marketing", considering this topic is represented by the hashtag #marketing and its most similar hashtags presented above. For the same users we have also estimated the scores produced by InfluenceTracker [16, 12] and present them in Fig. 3. The InfluenceTracker score combines the numbers of a user's followers, followees and tweets. Its lowest value is "0", while the highest has no upper limit. The higher this value is, the more impact has an account on the social network [12].

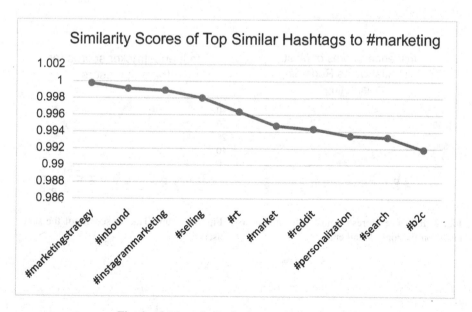

Fig. 1 10 Most similar hashtags with #marketing

Figures 2 and 3 give different insights on the users, as InfluenceTracker takes into consideration all the tweets of a user, whereas the proposed systems is based on topic-related tweets. Differences are also due to the different time periods of the tweets that were used in the two metrics, as InfluenceTracker takes into consideration the last 100 tweets of a user, whereas our metric is based on the collected dataset.

Influential Links based on a Specific Topic. In order to extract the most influential links for a specific topic we have implemented the algorithm described in Sect. 3.4. We have used as an example the topic "marketing" and the urls with the highest influence scores are shown in Fig. 4. These urls redirect to articles written about marketing, and content marketing in social media. We have used as a second example the hashtag #bigdata assigning to the topic "big data". We searched for hashtags similar to #bigdata and have identified the similar hashtags which are shown in Fig. 5. We have then estimated the influence scores of the users identified in the previous example as the most influential users regarding the topic "marketing". The results of these users' influence scores are presented in Fig. 6. It is evident that influence scores of the users differ according to the topic examined. We can see that for the "big data" topic the same users which were first examined for the topic "marketing" have different influencer scores. For example user uid1 that was the most influencing user in the marketing topic now has one of the lowest scores. Furthermore, users uid3, uid5 and uid7 have zero influence score in this specific topic, whereas they were considerably influencing regarding the topic "marketing". We also found and present in Fig. 7 the most influential urls regarding "big data". These urls redirect to articles related to the hashtags (#smartcities, #healthtech, etc.) that our platform identified as mostly related with the used topic.

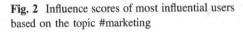

Fig. 2 Influence scores of most influential users based on the topic #marketing

Fig. 3 InflunceTracker scores for the same users

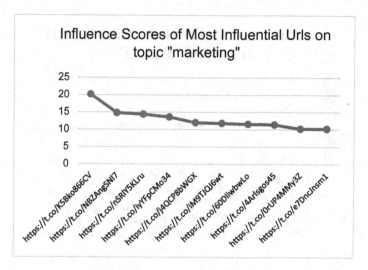

Fig. 4 Influence scores of most influential urls based on the topic "marketing"

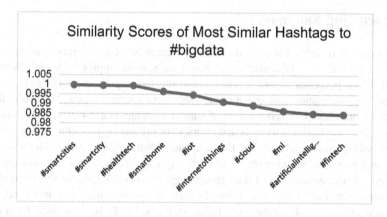

Fig. 5 Most similar hashtags to #bigdata

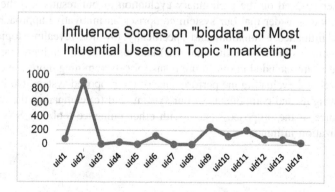

Fig. 6 "Big data"-based Influence scores of most influential users on topic "marketing"

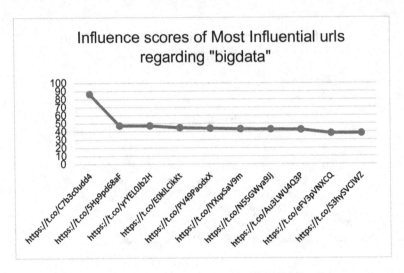

Fig. 7 Influence scores of urls for topic "bigdata"

5 Concluding Summary

In this paper we describe the design and implementation of a recommendation system for influential urls and influential users based on specific topics in Social Networks (SNs). For the purposes of our work, we use the hashtag functionality of SNs and, based on the assumption that the hashtags represent specific topics, we use collaborative techniques for identifying a pool of similar hashtags that correspond to a specific topic. We then use social activity based metrics for estimating the influence of urls and users by taking into consideration the identified hashtags, so as to achieve specialization on topics. For the implementation, the Apache Spark framework was used for achieving scalable searches and data processing. Results for specific topics based on the datasets we have extracted from Twitter were presented and analyzed. The benchmark influence estimation tools InfluenceTracker [12, 16] was used for a comparison of our results with the influence values this system estimates for the most influential users. Based on the preliminary evaluation of our results and the study of related work, we consider that our system comprises an innovative approach towards topic-specific influence estimation, and specifically towards revealing topic-specific trends in content (as the most influential urls) and topic-specific influencers. We note that further tests are included in our future plans for (a) examining more use cases with bigger data sets and evaluating the performance of the proposed system, (b) fine tuning the various weights and components of our system, and (c) comparing the results and the performance of the proposed system with other influence tools for SNs and with topic specialization in mind.

References

1. Agarwal, N., Liu, H., Tang, L., Yu, P.S.: Identifying the influential bloggers in a community. In: Proceedings of the International Conference on Web Search and Web Data Mining (WSDM '08), pp. 207–218. ACM (2008)
2. Ahmed, S., Ezeife, C.I.: Discovering influential nodes from trust network. In: Proceedings of the 28th Annual ACM Symposium on Applied Computing (SAC '13), pp. 121–128. ACM (2013)
3. Anger, I., Kittl, C.: Measuring influence on Twitter. In: Proceedings of the 11th International Conference on Knowledge Management and Knowledge Technologies (i-KNOW '11), ACM (2011)
4. Apache, https://spark.apache.org/. Accessed 28 June 2018
5. Bagchi, S.: Performance and quality assessment of similarity measures in collaborative filtering using mahout. Procedia Comput. Sci. **50**, 229–234 (2015)
6. Bento, C.: Finding influencers in social networks. Dissertation, Instituto Superior Technico, Universidate Tecnica de Lisboa (2012)
7. Chen, W., Wang, Y., Yang, S.: Efficient influence maximization in social networks. In: Proceedings of the 15th ACM SIGKDD International Conference on Knowledge Discovery and Data Mining (KDD '09), pp. 199–208. ACM (2009)
8. De Meo, P., Ferrara, E., Fiumara, G., Provetti, A.: Improving recommendation quality by merging collaborative filtering and social relationships. In: Proceedings of the 11th International Conference on Intelligent Systems Design and Applications (ISDA 2011), pp. 587–592. IEEE (2011)
9. Ekstrand, M.D., Riedl, J.T., Konstan, J.A.: Collaborative filtering recommender systems. Found. Trends Hum.-Comput. Interact. **4**(2), 81–173 (2011)
10. Epinions, http://www.epinions.com/. Accessed 28 June 2018
11. Goyal, A., Bonchi, F., Lakshmanan, L.: Learning influence probabilities in social networks. In: Proceedings of the 3rd ACM International Conference on Web Search and Data Mining (WSDM '10), pp. 241–250. ACM (2010)
12. InfluenceTracker, http://www.influencetracker.com/. Accessed 28 June 2018
13. Klout, https://klout.com/, https://www.lithium.com/products/klout. Accessed 28 June 2018
14. Koutrouli, E., Kanellopoulos, G., Tsalgatidou, A.: Reputation mechanisms in on-line social networks: the case of an influence estimation system in Twitter. In Proceedings of the SouthEast European Design Automation, Computer Engineering, Computer Networks and Social Media Conference, pp. 98–105. ACM (2016)
15. Prawesh, S., Padmanabhan, B.: The, "most popular news" recommender: count amplification and manipulation resistance. Inf. Syst. Res. **25**(3), 569–589 (2014)
16. Razis, G., Anagnostopoulos, I. InfluenceTracker: Rating the impact of a Twitter account. In: 10th IFIP International Conference on Artificial Intelligence Applications and Innovations, pp. 184–195. Springer, Berlin (2014)
17. Realtime Tweets, https://developer.twitter.com/en/docs/tweets/filter-realtime/overview. Accessed 28 June 2018
18. Riquelme, F., González-Cantergiani, P.: Measuring user influence on Twitter. Inf. Process. Manag. **52**(5), 949–975 (2016)
19. Ruohomaa, S., Kutvonen, L. Koutrouli, E.: Reputation management survey. In: Proceedings of the 2nd International Conference on Availability, Reliability and Security (ARES), pp. 103–111 (2007)
20. Trusov, M., Bodapati, A.V., Bucklin, R.: Determining influential users in internet social networks. J. Mark. Res. **47**(4), 643–658 (2010)

21. Twitter Counter, https://twittercounter.com/pages/100. Accessed 28 June 2018
22. Twitter embedded timelines, https://dev.twitter.com/web/embedded-timelines. Accessed 28 June 2018
23. Twitter Rest API, https://dev.twitter.com/twitterkit/android/access-rest-api. Accessed 28 June 2018
24. Twitter, https://twitter.com. Accessed 28 June 2018
25. Varlamis, I., Eirinaki, M., Louta, M.: Application of social network metrics to a trust-aware collaborative model for generating personalized user recommendations. In: The Influence of Technology on Social Network Analysis and Mining, vol 6. Springer, Berlin (2013)
26. Yazdanfar, N., Thomo, A.: Link recommender: collaborative-filtering for recommending urls to Twitter Users. Procedia Comput. Sci. **19**, 412–419 (2013)
27. Ye, S., Wu, S.F.: Measuring message propagation and social influence on Twitter.com. In: Proceedings of the 2nd International Conference on Social Informatics (SocInfo'10), Springer, pp. 216–231 (2010)

Identifying Control Parameters in Cheese Fabrication Process Using Precedence Constraints

Melanie Munch[1]([✉]), Pierre-Henri Wuillemin[2], Juliette Dibie[1],
Cristina Manfredotti[1], Thomas Allard[3], Solange Buchin[4],
and Elisabeth Guichard[3]

[1] UMR MIA-Paris, AgroParisTech, INRA, Université Paris-Saclay,
75005 Paris, France
melanie.munch@agroparistech.fr
[2] Sorbonne University, UPMC, Univ Paris 06, CNRS UMR 7606, LIP6,
75005 Paris, France
[3] CSGA, AgroSupDijon, CNRS, INRA, Université Bourgogne Franche-Comté,
21000 Dijon, France
[4] UR342 INRA, 39800 Poligny, France

Abstract. Modeling cheese fabrication process helps experts to check their assumption on the domain such as finding which parameters (denoted as control parameters) can explain the final products and its properties. This modeling is however complex as it involves various parameters and a reasoning over different steps. Our previous work presents a method to learn a probabilistic relational model in order to check a user's (an expert on the considered domain) assumption on a transformation process domain, using a knowledge base of this domain and his expert knowledge. However this method did not include temporal information, and thus the learned model is not enough to reason on the cheese fabrication process. In this article we present an extension of our previous work that allows a user to integrate causal and temporal information represented by precedence constraints in order to model a cheese fabrication process. This allows the user to check his assumption to identify the transformation process control parameters.

Keywords: Ontology · Probabilistic relational model · Temporality

1 Introduction

Cheese processing is a complex domain involving many different variables. Their combination leads to final products that can differ in quality which can be assessed by different criteria (i.e. sensory, nutritional). Parameters that are enough to explain all these criteria are denoted as **control parameters**. In order to help experts assess and check their assumptions (e.g. identifying control parameters), tools and methods are needed to analyze data. In a previous

© Springer Nature Switzerland AG 2018
L. Soldatova et al. (Eds.): DS 2018, LNAI 11198, pp. 421–434, 2018.
https://doi.org/10.1007/978-3-030-01771-2_27

work [7], we have defined a method helping an expert to check an assumption about possible causal relations between variables by combining a knowledge base and a probabilistic relational model. However cheese processing being composed of a succession of different steps also includes temporal information, that was not considered in our method. In this article, we therefore propose a generalized version including temporality.

This work has been applied on a real application about cheese processing using data from the TrueFood project. The goal of the TrueFood project is to investigate to what extent the impact of some combinations of thermophile lactic bacteria (i.e. Streptococcus thermophilus, Lactobacillus helveticus LH with 2 distinct levels and Lactobacillus delbrueckii LD with 2 distinct levels) on the characteristics of hard cooked cheese is affected by the use of milks with various compositions and by the use of different technological conditions (such as distinct temperature for the heating of the milk in the vat). Our study focuses on 24 hard cooked cheese of 10kg each manufactured during three weeks in January 2008, and made using 100 liters vats. Three kinds of milk, differing in their protein content and their production conditions, were used for the cheese making. During the cheese making, three different temperatures (53 °C, 55 °C and 57 °C) were applied for the milk heating. During this study various parameters were monitored, such as different measures of proteolysis. In particular, the potentially bioactive peptides content of the cheeses were measured at several steps of the cheese ripening. Their sensory properties were also assessed at the end of the ripening step: texture and flavor were evaluated by 11 panelists on a 10 points scale.

The influence of milk heating and of combination of lactic bacteria during cheese manufacture on the formation of peptides has already been observed in the literature [12]. Moreover the impact of the type of milk used for the cheese manufacture (especially the influence of the cows feeding system) on the organoleptic properties of hard cheeses has been shown in [9].

In our study, the experts make the assumption that the three factors of variation of the experiments (i.e. type of milk used for the cheese making, combination of thermophile lactic bacteria added to it, and the milk temperature) are the control parameters for the potentially bioactive peptide content of the cheese and its sensory properties. Our aim is to check this assumption using our method extended to take into account both causal and temporal informations.

This paper is structured as follows. Sect. 2 presents the background on probabilistic relational models and related works. Section 3 presents the state of the art on dealing with temporal information in probabilistic models. Section 4 presents our improved method to help the experts check their assumption using causal and temporal constraints. Section 5 presents our study on the data of the True-Food project. Section 6 concludes this article.

2   Background

2.1   BNs and PRMs

Probabilistic relational models (PRMs) extend Bayesian networks (BNs) with the notion of class of relational databases. A BN is the representation of a joint probability over a set of random variables that uses a Directed Acyclic Graph (DAG) to encode probabilistic relations between variables. However, in the case of numerous random variables with repetitive patterns, it cannot efficiently represent every probabilistic relations.

PRMs extend the BN representation with a relational structure between potentially repeated fragments of BN called classes [14]. They define the high-level, qualitative description of the structure of the domain and the quantitative information given by the probability distribution over the different **attributes** [3], where the attributes represent the different possible values for the variables. In the following, we consider attributes as the objects we want to reason with: we want to assess whether a specific attribute's value can explain another attribute's value. A **class** is defined as a DAG over a set of attributes. These can be inner attributes or attributes from other classes referenced by so-called **reference slots**. The high level structure of a PRM (i.e. its **relational schema**) describes a set of classes C, associated with attributes $A(C)$ and reference slots $R(C)$. A slot chain is defined as a sequence of reference slots that allows one to put in relation attributes of objects that are indirectly related. The probabilistic models are defined on the low level structure (i.e. at the class level) over the set of inner attributes, conditionally to the set of outer attributes and represent generic probabilistic relations inside the classes. This is the **relational model** of the PRM. Classes can be instantiated for each specific situation. A **system** of a PRM provides a probability distribution over a set of instances of a relational schema [15] and, once instantiated, is equivalent to a BN.

2.2   Essential Graph

An instantiated system of a PRM is equivalent to a BN. As a consequence, alongside the construction of the PRM, we also learn an Essential Graph (EG). An EG is a semi-directed graph associated to a BN and composed of edges and oriented arcs. They both share the same skeleton, but the orientation of the EG's edges can vary. If the orientation of an edge is the same for all the Markov equivalent graphs of the BN, this edge is also oriented in the EG; if not, the edge remains unoriented. All directed edges in the EG are called essential arcs [5]. An example of a EG and its two possible interpretations is given by Fig. 1.

The EG expresses whether the orientation of an arc between two nodes can be reversed without modifying the probabilistic relations encoded in the graph. It is useful when presenting results to the user as it can help him visualizing the causal relations learned: when a model has been learned with causal constraints, if an edge is oriented in the EG, it could mean that there is a causal dependence.

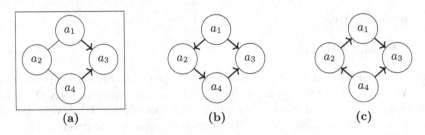

Fig. 1. Example of an essential graph (a) and two BNs (a) and (b) representing possible interpretations.

2.3   Causal Relation Discovery Driven by an Ontology

In a previous work, we have proposed a method to learn a PRM to discover causal relations in a knowledge base $\mathcal{KB}$ relying on a user's assumption, the user being an expert of the studied domain. We consider, in the following, a knowledge base $\mathcal{KB} = (\mathcal{O}, \mathcal{F})$ where the ontology $\mathcal{O}$ is represented in OWL[1] and the data $\mathcal{F}$ in RDF[2]. The user's assumption about possible causal relations between data is of the form "$E_1, ..., E_n$ *have a causal influence on* $C_1, ..., C_p$", with E_i attributes the user has determined as **explaining** and C_j attributes the user has determined as **consequence**. From the assumption and the knowledge base $\mathcal{KB}$, a database B is created and, afterwards used for the learning. It is composed of the explaining and consequence attributes as well as other inferred attributes as presented in [7].

Given the distinction between explaining and consequence attributes, we introduce some constraints in the learning. In particular, explaining attributes may have an influence over consequence attributes but the inverse is not possible. As a result, if during the learning a relation is found between explaining and consequence attributes, then it has to be oriented from explaining to consequence. These causal constraints guide the probabilistic model construction: indeed, learning using constraints that reflect causality results in a model including causal information and allows the validation of the user's assumption.

Our method gives the user the possibility to check his assumption about possible causal relations between data of a knowledge base. The integration of explaining and consequence attributes helps him express his own knowledge of the domain, and guide the learning towards a coherent causal model. This model however does not take into account possible temporal relations between data and the fact that explaining attributes at one time step can become consequence attributes at the next time step. We denote by **event** a group of attributes that happen at the same time. When dealing with temporal information, it is possible that the consequence attributes of an event e_t at time t become the explaining ones of another attribute of another event e_{t+1} at time $t + 1$, which would be hardly represented by our previous explaining and consequence attributes.

---

[1] https://www.w3.org/OWL/.
[2] https://www.w3.org/RDF/.

Moreover, we can suppose that all the attributes from an event can have an influence over all the attributes of the following events.

We propose, in this paper, an extension of our method dealing with both causality and temporality constraints.

3   State of the Art

Both causality and temporality impose a direction to the relation between attributes. For this reason, we have to consider how to take into account constraints while learning PRMs. We first study works on precedence constraints and then how the temporality have been addressed in previous works.

3.1   Learning Under Constraints

Related works have established that using constraints while learning BNs brings more efficient and accurate results. Parameters learning can be improved by allowing users to specify their knowledge through constraints estimations and priors [10]. In [2] an exact structure learning algorithm that uses data and expert's knowledge constraint is presented by defining two types of constraints. In particular one of those identifies where arcs may or may not be included. In [4] it is argued that combining analogical generalization and structure mapping with statistical machine learning methods allows state-of-the-art performances on standards tasks.

In the K2 algorithm [1], a complete ordering of the attributes is required before learning a BN. In this way, the authors introduce **precedence constraints** between the attributes. If, in this order, an attribute A is before an attribute B, then a precedence constraints is applied between those two from A to B, meaning that during the learning we do not consider the possibility of a relation from B to A. If a relation is found between A and B, then the direction of this relation has to be from A to B. However K2 requires a complete knowledge over all the different attributes precedences (since all attributes have to be sorted), which is not always the case as we generally don't know everything about the domain. In this paper we present a method to learn with only partial constraints.

3.2   Integrating Temporality

Temporality has been expressed in Markov models such as Markov chains (MCs, [11]) or Hidden Markov models (HMMs, [11]). MC is a stochastic model describing a sequence of possible events in which the probability of each depends only on the state of the previous one. HMM is a MC with unobserved states, meaning that some attributes' values can vary with an unknown attribute. Temporal information can be gained from both of them: following the flow of time, we can deduce that if an event happens at time t, then it can have an influence over all events that happen at time $t + i$. Moreover we can also deduce subjective independence information between events: if an event only depends on the previous one, then there is no relation between two events which are not consecutive.

However MC and HMM are limited in our case as they cannot handle our need to represent numerous attributes and their relations in time. Both can be extended by dynamic BNs (DBNs, [11]). Widely used to model sequential data, in particular time-series, DBNs introduce the notion of relation between variables over adjacent time steps. For instance, in [6] DBNs are used to model a long-term simulation of clinical complications in type 1 diabetes. They define two models, one Data-Driven only, and another designed with expert inputs. However, DBNs impose to look at the same attributes and their evolution through time, which is not our aim.

Moreover, we want to consider that every event can have an influence on the attributes of the events that happen after it. This would allow us to better study the possible influence of all attributes on the following ones, which is useful in our problem where we want to assess the relations between attributes of different events (and not only between attributes of consecutive events). This leads us to define a new kind of model, we call it **stack model** as presented below.

4   Stack Model

4.1   Determining Precedence Constraints

Using [7] where we defined explaining and consequence attributes, we propose to decompose the precedence constraints into two sub-constraints: the causal constraints and the temporal constraints. **Causal constraints** are information on the relations between attributes of the type *"The attribute A is a possible cause for the attribute B"*. **Temporal constraint** are information on the relations between attributes of the type *"The attribute A happens before the attribute B"*. These causal and temporal constraints both imply two things: (1) the value of B *can be* explained by A (but it doesn't have to); (2) B can never explain the value of A.

Causal and temporal constraints are differentiated by their nature: temporality is immediate and objective (i.e. the past can influence the future and not the contrary), while causality usually needs a supply of expert knowledge.

Temporal constraints. When possible, the temporal information is provided in the knowledge base through the time ontology[3] that helps anchoring its events in time. In some cases it is also possible to introduce temporal information from other ways (e.g. directly from experts). In all cases we suppose that attributes can be attached to a specific event in time, and as a consequence they also contain temporal information.

Causal constraints. Causal information can be brought by experts or by the ontology itself. In certain cases it is also possible to use statistical independence

---

[3] https://www.w3.org/TR/owl-time/.

tests such as χ^2 test used for the construction of causal BNs in order to guess some possible causal relations [13].

4.2   Description

The main idea of our stack model is that it is built in order to graphically represent the two kinds of precedence constraints we defined in Sect. 4.1. If an attribute is put higher in the stack then it has precedence constraint on all attributes below it; if two attributes are on the same level then they do not have precedence constraints.

It is also possible to encounter parallel events. In this case, we suppose we have enough information from the knowledge base to differentiate the events, in order to know which attribute correspond to which event. In this case, we define paths for each parallel events. Events on the same path all have parenthood links: temporal constraints can be established between them. On the contrary, events that do not share the same parent events are on two separated paths, and we suppose they cannot influence each other. As a consequence, there cannot be precedence constraints between them, neither causal nor temporal.

Starting from a user's assumption, the model construction is based on the two operations described below[4].

1. Defining temporal constraints. Groups of attributes that happen at the same time are put at the same level. If they are from a same event, they are put in a same stack; on the contrary if they are from parallel event we create different paths, each with a stack, for each parallel event.

2. Defining causal constraints. Inside a stack some attributes might have a causal influence over others. In order to express those causal constraints, we sort the attributes such as higher attributes can explain lower attributes and that attributes at the same level share no causal influence between each other.

An example of this construction is given in Fig. 2. We consider here four events: one at time t_1, two parallel at time t_2 and one at time t_3 (a). When constructing the model we first only consider temporal constraints (b): two paths are created with on one side a stack with the group of attributes A and on another side two stacks with respectively the group of attributes B and C, the first being above the second. Finally, a fourth stack is created below all the others, including the group of attributes D. Temporal constraints are defined between the different stacks: since the group B is not on the same path as A, no temporal constraint is drawn between them. In the end, if needed, causal constraints are defined (c). In our example, we suppose that our expert distinguishes between explaining and consequence attributes in the group A, respectively subgroups A_1 and A_2.

In order to lighten the figure, arrows between groups of attributes inside different stacks are not represented: however, if two stacks are linked, then it

---

[4] For convenience and in order to ease the readability of the presentation we use in this article a *top-down* construction (from temporality to causality). However nothing prevents us to use the opposite *bottom-up* construction (from causality to temporality).

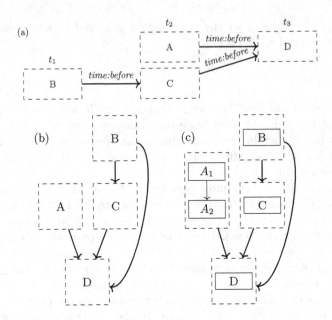

Fig. 2. (a) Example for a system with parallel events. (b) Definition of the temporal constraints. (c) Definition of the causal constraints.

means that each attribute on the higher stack have a temporal constraint over the lower.

4.3   From Stack Models to PRMs

The final model is used to construct a PRM's relational schema, which defines the classes and their attributes of the PRM. Each subgroup of attributes becomes a class, which are linked together with reference slots following the different precedence constraints. For instance in the model in Fig. 2 (c), it would lead to five classes and six reference slots.

Once the relational schema is defined, the PRM can be learned using the database B extracted from the knowledge base [8]. This PRM can then be instantiated in order to obtain a BN representing our learned model. It will include causal information as it was learned under precedence constraints; however, it is not a complete causal BN considering that the learning of dependencies between attributes inside the same group is dealt like a classical BN. In order to deal with causal information, the EG is used: if an arc is oriented in it (meaning that its orientation cannot be changed without changing the likelihood of the BN) then it can mean that there is a causal relation.

5   Experiment

5.1   Data Description

Considering the TrueFood project, the experts would like to model the different relations between the attributes in order to explain the products at the end and infer its characteristics. More particularly they want to check the formulated assumption: *"The temperature, ferments and type of milk have a causal influence on the potentially bioactive peptide content of the cheese and its sensory properties"*. Following the approach presented in [7] temperature, ferments and type of milk are the only explaining attributes of the problem, while the other are consequences. Since those three are fixed at the beginning, they correspond to the **control parameters**.

The dataset is composed of data from three different steps that are part of a cheese fabrication and tasting process: Step in the vat, Ripening and Mastication.

- **Step in the vat:** is described by three processing control parameters (Temperature, Starters and Type of milk), and two measured (hardening and clotting times).
- **Ripening:** is described by the measured value of five different concentrations in cheese: butyric acid, propionic acid, acetic acid, free amino acids and free amino groups.
- **Mastication:** In this step, a panel of 11 judges has evaluated each cheese sample on 45 different criteria (e.g. spice aroma, sugar or fat perception). Those sensory notes can be divided into two categories, texture of the cheese (10 attributes) and flavor (35 attributes). The scores ranged from 0 to 10.

The times measured during the step in the vat are a pre-requisite to **study bioactive peptide contents,** even if they do not represent their quantities. On another hand the attributes measured during the ripening and the mastication steps are useful to **evaluate the cheese sensory properties**.

5.2   Model Construction

A first descriptive analysis over the notes attributes during the mastication step shows that some have a variance $\sigma < 0.25$. Given the standard variation calculated by $\sqrt{\sigma}$, it means that for these attributes the variation over the whole samples is less than ± 0.5 points. We consider it to be too low to observe meaningful variations among the different samples, and remove them from the studied set, leaving 39 attributes (9 texture attributes and 30 flavor attributes).

In order to apply our method to this dataset, we first separate the attributes per steps. We order them in stacks, following the temporal order: first *Step in the vat*, then *Ripening* and then *Mastication*. Temporal constraints are then drawn. Once this has been done, the only causal constraints that need to be introduced are taken into account in order to separate the control attributes from the rest. Temperature, Ferments and Type of Milk are stacked above the Hardening and clotting times inside the same step.

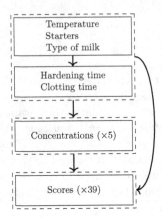

Fig. 3. Model constructed from the expert assumption.

The obtained model is presented in Fig. 3, where the different steps are underlined by the dashed squares and $\times i$ denotes the number of attributes of the given type.

5.3 Analysis

While analyzing the PRM we focus on two types of relations: the intra-step and the inter-step relations. While the analysis of the intra-step relations in general has already been tackled in our previous work (leading to causal information analysis), the inter-step gives a whole new reading of the model. It indeed helps us generate new information about the temporal aspect, in particular discovering if some steps can explain all the other attributes, or, on the contrary, if a step has no influence on the process. In our case, we would like to see in what extent the control parameters are able to explain (in)directly the other attributes. As a consequence we extend our study on inter-step relations, also including the inter-subgroup relations between the control parameters and the two attributes Solidifying time and Clotting time.

In order to illustrate our results, we consider three attributes A, B and C with A a control parameter in a step, and B and C two attributes of the step after (Fig. 4). When checking whether A can explain C, two cases are possible:

1. **A has a complete or partial control over C.** In the first case (Fig. 4 (a)) there exists a inter-step relation directly from A to C: it means that knowing the state of A will give the maximum possible information on C. In the second case (Fig. 4 (b)) the inter-step relation between A and C is intercepted by other attributes, B in our example. Since there is no direct relation between A and C, then knowing A will only give partial information on C. Moreover, knowing B makes the knowledge of A obsolete, as B alone is enough to have a complete information on C.

2. **A has no control over C.** There are also two possible cases. The first (Fig. 4
   (c)) is straight-forward: since there are no inter-step relations between A and
   B nor C, then A has no control over C. In the second (Fig. 4 (d)) a *v-structure*
   $A \rightarrow B \leftarrow C$ makes A and C independent: A and C are *d-separated* by B,
   meaning that controlling A cannot influence C, however fixing B gives partial
   information on both A and C.

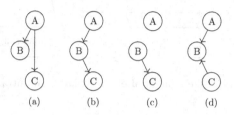

Fig. 4. Possible cases encountered during the results analysis of the assumption "*A
has an influence over C*".

It is important to note that we assume the direction of the relation between
A and B, C because we previously defined A as a control parameter. However,
since B and C are not control parameters, there is no precedence constraint
that indicates whether B has an influence on C or otherwise. In order to do this
analysis we, therefore, use the EG, that indicates whether the direction of the
relation is sure (i.e. changing this relation direction would modify the likelihood
of the learned BN). If a relation is oriented in the EG, we can suppose that this
orientation is due to the precedence constraints used during the learning that
brought causal information and, therefore, this relation orientation might be
causal. This is not however automatic, and relations orientation in the EG can
sometimes reflect other problem such as learning artifact or missing attributes.
If, finally, a relation is not oriented in the EG then we cannot assume causal
information from it. For the following we assume that, when talking about relations orientation, we are using the EG informations.

5.4   Results

The vast majority of the observed inter-steps relations found **confirms the
experts assumption**: "*The temperature, ferments and type of milk have a
causal influence on the potentially bioactive peptide content of the cheese and
its sensory properties*". Some of them are directly explained, while others are
linked to attributes of the same group that are explained by the control parameters. Only three sensory notes are not linked at all to any parameter. Those
results and the number of found relations are summarized by Fig. 5.

While the study of the times during processing and concentrations is pretty
straight-forward, all being completely or partially explained by the control

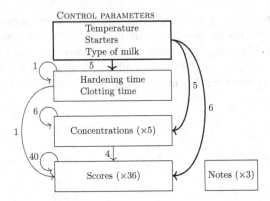

Fig. 5. Summary of the number of observed inter and intra step relations.

parameters, an interesting trend in the sensory notes attributes can be observed while looking at the EG. Indeed we can notice that a large group of 21 flavor attributes (over the 30) is d-separated from the control parameters by another sensory attribute, meaning that this part is in fact equally independent from the control parameters despite being part of the network. More generally we observe in the EG a difference between flavor and texture attributes. Figure 6 shows an excerpt of the learned EG where texture attributes are denoted by T_i ($i \in [1, 9]$) and flavor attributes are separated into two groups denoted by F_j ($i \in [1, 30]$). This choice has been made in order to ease the reading: however one must keep in mind that two relations between F_j and different T_i do not involve the same attribute of F_j.

Flavor attributes in F_1 are d-separated from the control parameters, and the other in F_2 are partially explained by them. Moreover when looking at F_1 we observe a large number of intra-step relations between them. This leads us to assume that (1) they are highly correlated with each other and (2) their relations are not causal due to the high number of attributes learned together without any precedence constraints. On the contrary texture attributes are mostly directly explained by the control parameters (especially the type of milk), and some are partially explained by them. Most of them do not have intra-step relations with each other as seen in Fig. 6. The two texture attributes not represented in the network are related to time and concentration attributes and not directly linked to the control parameters. These observations are validated by the experts: considering the milk differences in terms of production conditions and composition, milk on the cheese texture was expected. In addition, flavor attributes are indeed more likely to be correlated with each other.

Since nearly all flavor attributes are linked together, it could be interesting to profile the cheeses with their different flavor values. This way, instead of reasoning with all the numerous flavor attributes, we could directly check the influence of the control parameter on the cheese type.

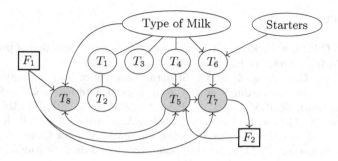

Fig. 6. Excerpt of the EG learned with T_i texture attributes and F_i groups of flavor attributes. Grey attributes d-separates F_1 from the control parameters.

6 Conclusion

In this article we present a new method able to help experts to study a domain represented by a knowledge base. Our aim is to provide to the experts a method to check the assumptions they can formulate on that domain by learning a PRM that presents the different probabilistic relations between its attributes. In order to guide this learning we also allow the experts to integrate causal and temporal informations by defining precedence constraints. Considering the TrueFood project we use our method to check the following assumption: *"The temperature, ferments and type of milk have a causal influence on the potentially bioactive peptide content of the cheese and its sensory properties"*.

To integrate precedence constraints, we extend our previous work, that already included causal information, in order to include temporal information. To do so, we define a new model, denoted by the **stack model**, where attributes are organized so that higher ones can have a precedence constraint over the ones below (i.e. they can be their cause). Using this model, we learn a PRM which, once instantiated, gives us a BN we can use to check the assumption.

The learned BN gives us two ways of analysis. First, using its EG (a graph that shows arcs whose direction is sure considering the BN structure), we can check the assumption. Considering that our control parameters are fixed at the beginning of the process, if a relation is found between them and an attribute, then we can conclude that the parameters may control this attribute. Second, once the model has been validated by the experts, it can be used to predict results. For instance if we want to control the cheese texture scoresin order to keep them in a certain range, we identify the control parameters we have to act on.

In future work we want to study in more detail the validation and introspection of the learned model in order to improve our help to the user.

References

1. Cooper, G.F., Herskovits, E.: A bayesian method for the induction of probabilistic networks from data. Mach. Learn. **9**(4), 309–347 (1992)
2. de Campos, C.P., Zeng, Z., Ji, Q.: Structure learning of bayesian networks using constraints. In: Proceedings of the 26th Annual International Conference on Machine Learning, ICML 2009, pp. 113–120. ACM, New York, NY, USA (2009)
3. Friedman, N., Getoor, L., Koller, D., Pfeffer, A.: Learning probabilistic relational models. In: Proceedings of the Sixteenth International Joint Conference on Artificial Intelligence, IJCAI 1999, Stockholm, Sweden, July 31 - August 6, 1999. 2 Volumes, 1450 pages, pp. 1300–1309 (1999)
4. Liang, C., Forbus, K.D.: Learning plausible inferences from semantic web knowledge by combining analogical generalization with structured logistic regression. In: Proceedings of the Twenty-Ninth AAAI Conference on Artificial Intelligence, AAAI 2015, pp. 551–557. AAAI Press (2015)
5. Madigan, D., Andersson, S.A., Perlman, M.D., Volinsky, C.T.: Bayesian model averaging and model selection for markov equivalence classes of acyclic digraphs. Commun. Stat.-Theory Methods **25**(11), 2493–2519 (1996)
6. Marini, S., et al.: A dynamic bayesian network model for long-term simulation of clinical complications in type 1 diabetes. J. Biomed. Inform. **57**, 369–376 (2015)
7. Munch, M., Wuillemin, P.-H., Manfredotti, C.E., Dibie, J.: Towards interactive causal relation discovery driven by an ontology. Technical report (2018). https://hal.archives-ouvertes.fr/hal-01823862v1
8. Munch, M., Wuillemin, P.-H., Manfredotti, C.E., Dibie, J., Dervaux,S.: Learning probabilistic relational models using an ontology of transformation processes. In: On the Move to Meaningful Internet Systems. OTM 2017 Conferences - Confederated International Conferences: CoopIS, C&TC, and ODBASE 2017, Rhodes, Greece, 23–27 October 2017, Proceedings, Part II, pp. 198–2105 (2017)
9. O'Callaghan, T.F., et al.: Effect of pasture versus indoor feeding systems on quality characteristics, nutritional composition, and sensory and volatile properties of full-fat cheddar cheese. J. Dairy Sci. **100**(8), 6053–6073 (2017)
10. de Campos, C.P., Ji, Q.: Improving bayesian network parameter learning using constraints, Jan 2009
11. Murphy, K.P.: Dynamic bayesian networks: representation, inference and learning, Jan 2002
12. Santiago-López, L., Aguilar-Toalá, J.E., Hernández-Mendoza, A., Vallejo-Cordoba, B., Liceaga, A.M., González-Córdova, A.F.: Invited review: bioactive compounds produced during cheese ripening and health effects associated with aged cheese consumption. J. Dairy Sci. **101**(5), 3742–3757 (2018)
13. Spirtes, P., Glymour, C., Scheines, R.: Causation, Prediction, and Search, 2nd edn. MIT press, Cambridge (2000)
14. Torti, L., Wuillemin, P.-H., Gonzales, C.: Reinforcing the object-oriented aspect of probabilistic relational models. In: PGM 2010 - The Fifth European Workshop on Probabilistic Graphical Models, Helsinki, Finland, pp. 273–280, Sept 2010
15. Wuillemin, P.-H., Torti, L.: Structured probabilistic inference. Int. J. Approx. Reason. **53**(7), 946–968 (2012)

Less is More: Univariate Modelling to Detect Early Parkinson's Disease from Keystroke Dynamics

Antony Milne[1]([✉]), Katayoun Farrahi[2], and Mihalis A. Nicolaou[1,3]

[1] Department of Computing, Goldsmiths, University of London, London, UK
antony.milne@gmail.com
[2] Electronics and Computer Science, University of Southampton, Southampton, UK
[3] The Cyprus Institute, Nicosia, Cyprus

Abstract. We analyse keystroke hold times from typing logs to detect early signs of Parkinson's disease. We develop a feature that captures the dynamic variation between consecutive keystrokes and demonstrate that it can be be used in a univariate model to perform classification with AUC = 0.85 from only a few hundred keystrokes. This is a substantial improvement on the current baseline. We argue that previously proposed methods are based on overcomplicated models—our simpler method is not only more elegant and transparent but also more effective.

1 Introduction

After Alzheimer's, Parkinson's disease (PD) is the world's second most prevalent neurodegenerative disease [1]. Currently, diagnosis is based on a specialist's interpretation of neurological tests completed by the patient at a clinic [2]. This procedure is time-consuming, expensive, subjective, and rather inaccurate (especially for identifying early stages of PD) [3].

Giancardo et al. [4] suggest that early PD can be detected through the analysis of typing logs, studying data obtained from 85 subjects (42 Parkinson's, 43 control) each transcribing text for around 15 min. Subsequent analysis of the keystroke dynamics focusses on the length of time between pressing and releasing each key (hold time), a measure believed to be outside a subject's conscious control and independent of typing skills. The so-called neuroQWERTY index (nQi) method is developed to classify a typing session as that of a Parkinson's sufferer or a control subject.

We regard this as a valuable line of research that demonstrates promising results for detecting early PD. In this paper, however, we present results indicating that the analysis presented in Ref. [4] is opaque and overly complicated for the problem at hand. Following the philosophy that 'less is more', we find that the classification performance of nQi can be equalled, and even surpassed, by a far simpler and more easily reproducible methodology.

We begin in Sect. 2 by outlining the nQi formalism and results. This is followed by an exploration of the basic features of the hold time data (Sect. 3),

© Springer Nature Switzerland AG 2018
L. Soldatova et al. (Eds.): DS 2018, LNAI 11198, pp. 435–446, 2018.
https://doi.org/10.1007/978-3-030-01771-2_28

and a demonstration that a univariate model can be used to straightforwardly achieve classification performance equal to nQi (Sect. 4). In Sects. 5 and 6, we develop more sophisticated dynamic features of the data that can be used, again in a univariate model, to substantially outperform nQi. Section 7 discusses a recent contribution to the literature [5], which also suffers from significant overengineering and, more importantly, reports results we believe to be invalid.

2   Classification with neuroQWERTY Index

Let us briefly describe nQi and the datasets involved. These are labelled early PD (those within five years of confirmed diagnosis: 18 Parkinson's, 13 control) and de novo PD (newly diagnosed and untreated: 24 Parkinson's, 30 control). Each dataset consists of a set of typing sessions. We use h_n to denote the hold time of the nth keystroke during a typing session, which has N keystrokes in total. Both nQi and our proposed classification methods are concerned with the one-dimensional time series $\boldsymbol{h}$.

Reference [4] begins by partitioning each time series into non-overlapping windows of length 90 s. We write $\boldsymbol{h} \equiv (\boldsymbol{h}^1, \boldsymbol{h}^2, \ldots, \boldsymbol{h}^I)$ to indicate this partitioning, where I gives the total number of windows. Any $\boldsymbol{h}^i$ with fewer than 30 elements is removed. Then, for each window i, a 7-dimensional feature vector $\boldsymbol{x}^i$ is calculated for $\boldsymbol{h}^i$. Let q_j^i be the jth quartile of the elements of $\boldsymbol{h}^i$, and denote the interquartile range as $\Delta q^i \equiv q_3^i - q_1^i$. Then $\boldsymbol{x}^i$ consists of the following features:

- The proportion of elements that are outliers, defined as $h_n^i < q_1^i - \frac{3}{2}\Delta q^i$ or $h_n^i > q_3^i + \frac{3}{2}\Delta q^i$.
- The skewness, given by $(q_2^i - q_1^i)/\Delta q^i$.
- The flight time between consecutive keystrokes.[1]
- The proportion of elements in $\boldsymbol{h}^i$ that are in each of four equally-spaced bins between 0 and 500 ms.

Training is performed with an ensemble of 200 Linear ϵ-Support Vector Regression models, where hyperparameters are selected using a grid search approach on an external dataset. During testing, a value of nQi for each $\boldsymbol{x}^i$ is calculated by applying all 200 regression models to $\boldsymbol{x}^i$ and then finding the median score, nQi^i. To arrive at a single nQi score for the typing session, these median scores are then averaged over the I windows: $\text{nQi} = \frac{1}{I}\sum_{i=1}^{I} \text{nQi}^i$.

To evaluate nQi, Giancardo et al. [4] perform cross-validation by training on the early PD dataset and testing on the de novo PD dataset, and then vice-versa. This yields a single prediction of nQi for each of the 85 subjects in the combined

---

[1] As given, this will yield a number for each keystroke; it is not explained in Ref. [4] how this measure is then aggregated over the window. Moreover, we note that, contrary to the principles promoted by Giancardo et al., this measure appears to use more than purely hold time data.

dataset.[2] Area under the Receiving Operating Characteristic curve (AUC) is used to evaluate the binary classification of each subject as either Parkinson's sufferer or control subject.

In our work, we follow precisely the same evaluation strategy, so that our classification results can be directly compared with those given in Ref. [4]. We are able to reproduce AUC = 0.81 reported by Giancardo et al. for classification using nQi.

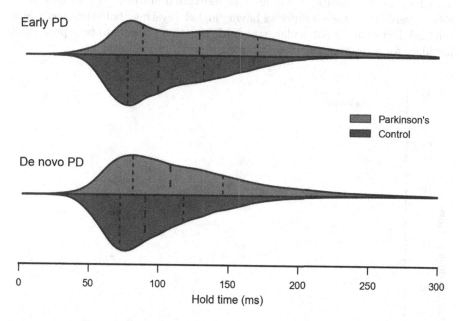

Fig. 1. The distribution of hold times for each of the two datasets used, distinguishing Parkinson's sufferers from control subjects. Each half of the violins are normalised to the same area. Dashed lines indicate the position of the lower quartile, median and upper quartile. Hold times above 300 ms are not shown here (corresponding to about 0.85% of the total data, and overwhelmingly from Parkinson's sufferers).

3   Exploratory Analysis

We begin by performing initial analysis of the early PD and de novo PD datasets, something that has not previously been presented in the literature. Figure 1 shows the distribution of all the hold times in each dataset, split between Parkinson's and control subjects. Unsurprisingly, there is a clear shift towards longer

---

[2] In fact, each subject in the early PD dataset produced two typing sessions. While training or testing, each typing session is handled independently. If a subject has produced multiple typing sessions then the average nQi is computed to produce a single score.

hold times for Parkinson's sufferers, especially for the early PD dataset. The plots also suggest that there is a greater variance in hold time for Parkinson's sufferers compared to control.

However, we are interested in classifying individual subjects rather than groups as a whole. To probe the difference in distributions suggested by Fig. 1, we calculate the hold time mean $\langle h \rangle \equiv \frac{1}{N} \sum_{n=1}^{N} h_n$ and standard deviation $\sigma(h) \equiv \sqrt{\langle h^2 \rangle - \langle h \rangle^2}$ for each subject. Figure 2 suggests that these statistics could be used to classify at the level of individual subjects. There is a clear trend towards Parkinson's sufferers having higher keystroke hold time mean and standard deviation. In particular, standard deviation appears to be a promising candidate for a discriminatory statistic.

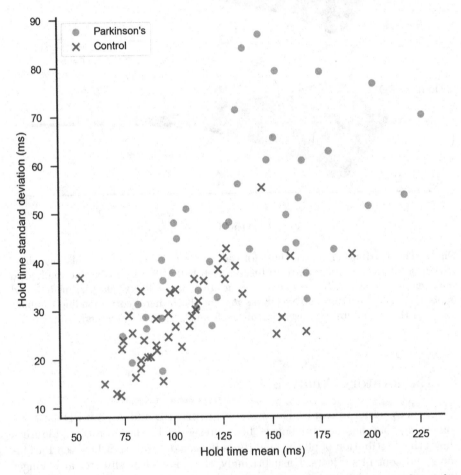

Fig. 2. The mean hold time $\langle h \rangle$ and standard deviation $\sigma(h)$ for all users in the study. Data from the early PD and de novo PD datasets are shown the same way. The average (std) of $\sigma(h)$ is 47 (18) for Parkinson's and 29 (9) for control, suggesting the power of this statistic as a discriminatory feature.

4   Classification with Elementary Statistics

One might well wonder whether these basic statistics alone are sufficient to effectively discriminate between Parkinson's and control subjects. We perform Logistic Regression using the features $\langle h \rangle$ and $\sigma(h)$ for each subject using scikit-learn's default parameters [6] and immediately obtain a classification performance comparable to nQi. In fact, we obtain AUC = 0.82 using standard deviation alone as a single feature (compared to AUC = 0.81 for nQi).[3] Figure 3 and Table 1 show the performance of this univariate method with standard deviation feature, which we refer to as the Stdev model, along with the performance of nQi and two other models which will be discussed in later sections. The classification performance achieved using a single elementary statistical feature is very similar to that obtained using nQi.

It is for this reason that we believe nQi is a contrived method for performing the classification task. Let us highlight the differences between nQi and our Stdev model:

- nQi splits the time series h for each user into several windows, calculates features for each window separately, and then recombines statistics at the end; we use a feature that uses the time series as a whole.
- nQi uses seven features that capture, in various ways, properties of the distribution of hold times;[4] we use one feature. Furthermore, standard deviation is an extremely well-known and transparent statistic.
- nQi uses an ensemble of 200 classifiers, with hyperparameters optimised using an external dataset; we use a single Logistic Regression algorithm with no optimisation of hyperparameters required.

Clearly the seven features of nQI capture more of the typing behaviour, and these features could be used to paint a more complete picture of a subject. However, for the purposes of classification on the datasets provided, there is no evidence to suggest that nQi outperforms the considerably simpler and more elegant Stdev method. One can achieve strong classification performance without the need to engineer particular statistical features, use anything beyond hold times, or perform carefully optimised ensemble models. We emphasise that the method we propose here has been evaluated using exactly the same cross-validation strategy on the same data as nQi (as are all models discussed in this paper).

Of course, this is not to say that performing a Logistic Regression with default hyperparameters on a single feature is the best possible method. Indeed, we will later formulate a method which substantially outperforms both the Stdev model and nQi. We present the Stdev model in order to show that one may immediately

---

[3] This classification performance is very similar to that obtained using using both $\langle h \rangle$ and $\sigma(h)$ as features, whilst the performance using just $\langle h \rangle$ as a feature is substantially lower.

[4] We note again that, unlike the Stdev method, nQi actually appears to use information about the flight time in addition to purely hold time data.

and very straightforwardly obtain a baseline classification performance that is comparable to the convoluted methods of nQi. We note that in a related paper on smartphone typing data [7], a univariate model using an elementary statistical feature (sum of covariances) was in fact found to outperform all of the more complicated multivariate methods studied.

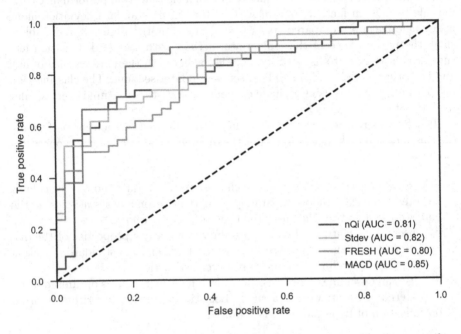

Fig. 3. The ROC curves for all the models evaluated in this paper. nQi values are taken from Ref. [4] (and reproduced by us). The other three methods use a Logistic Regression algorithm with different features. Stdev and MACD correspond to the univariate models with features $\sigma(h)$ and $\langle|\Delta|\rangle$ respectively. FRESH refers to the multivariate model with the five most relevant time series features automatically extracted from each training set. All models were evaluated using the same cross-validation strategy as that used in Ref. [4] (training on the early PD dataset and testing on the de novo PD dataset, and then vice-versa).

5   Feature Extraction

We now consider what features might be the most relevant for detecting early PD. We have already seen that using a univariate method based on the standard deviation yields strong classification performance, but can we do better by using more sophisticated features and a multivariate model?

Recall that the data we are working with is a one-dimensional set h, whose elements h_n ($n = 1, 2, \ldots, N$) are ordered according to the order of keystrokes

Table 1. The performance of all the models evaluated in this paper, labelled as in Fig. 3. We follow the same evaluation strategy as Ref. [4] by reporting values of the confusion matrix and accuracy at the cut-off point determined by maximising Youden's J Statistic [8].

Model	TP	FN	TN	FP	Accuracy	AUC
nQi	30	12	36	7	0.76	0.81
Stdev	27	15	37	6	0.75	0.82
FRESH	36	6	26	17	0.73	0.80
MACD	34	8	35	8	0.81	0.85

recorded. Simple statistical measures such as standard deviation discard information encoded in the ordering of the elements h_n; typing behaviour might be captured more effectively by measures that take into account the actual dynamics of h.

There are countless features that one could extract from a time series, but not all will be relevant for identifying discriminatory behaviour. We use the Feature Extraction based on Scalable Hypothesis (FRESH) algorithm and associated library `tsfresh` [9,10]. This characterises time series using a comprehensive set of well-established features, including those that are 'static' (e.g. standard deviation) and truly 'dynamic' (e.g. Fourier transform coefficients). The relevance of each feature is evaluated by quantifying its significance for predicting the target label (for us, Parkinson's or control).

We perform a classification of the time series data with FRESH using the following procedure. The training data is analysed to find the m most relevant features for predicting whether the user has PD. These m features are then extracted on the test data and used to perform classification using Logistic Regression. Features are standardised by scaling to vanishing mean and unit variance. By running this model on $m = 1, 2, \ldots, 10$, we find that the best performance is achieved by $m = 5$. The AUC for this is again comparable to nQi and our univariate standard deviation method (see Fig. 3 for the ROC curve and Table 1 for evaluation metrics).

Let us look at the features extracted by FRESH on the time series h. We perform cross-validation based on two datasets (early PD and de novo PD), and hence two different sets of $m = 5$ features are found as being the most relevant during training. These are given in full in Table 2.

For both the early PD and the de novo PD datasets, FRESH finds that several features given by the function `change_quantiles` are highly relevant. This function aggregates consecutive differences between elements of h. More precisely, we fix a corridor set by the quantiles `ql` and `qh` and take only those elements for which both $ql \leq h_n \leq qh$ and $ql \leq h_{n+1} \leq qh$. Define $\Delta_n \equiv h_{n+1} - h_n$; then the feature found by `change_quantiles` is given by the aggregator function `f_agg` applied to the set of all Δ_n ($|\Delta_n|$ when `isabs` is set). In other words, we are analysing (a subset of) the differences in hold time between consecu-

Table 2. The five most relevant features found by FRESH on the early PD and de novo PD datasets. Features are given by the functions and parameters used to calculate them with the `tsfresh` package [10].

Early PD
`change_quantiles(ql=0.8,qh=1.0,isabs=True,f_agg=mean)`
`change_quantiles(ql=0.0,qh=1.0,isabs=True,f_agg=var)`
`spkt_welch_density(coeff=5)`
`variance`
`standard_deviation`
De novo PD
`change_quantiles(ql=0.6,qh=0.8,isabs=True,f_agg=var)`
`change_quantiles(ql=0.4,qh=1.0,isabs=True,f_agg=mean)`
`change_quantiles(ql=0.6,qh=1.0,isabs=True,f_agg=mean)`
`change_quantiles(ql=0.6,qh=0.8,isabs=False,f_agg=var)`
`max_langevin_fixed_point(r=30, m=3)`

tive keystrokes. This captures a more complex element of variance that 'static' measures such as standard deviation do not (although it is worth noting that standard deviation is in fact identified as a highly relevant feature for at least the early PD dataset).

Given the thoroughness of the FRESH algorithm, which extracts several hundred features, it is perhaps at first surprising that this multivariate method does not significantly outperform the univariate method using standard deviation. However, note that none of the most relevant features are common between the two datasets. We are effectively suffering from overfitting: FRESH identifies some rather obscure features that fit the training data very well but do not generalise to the test data. Take, for example, the feature discovered using `spkt_welch_density`, which is present in the early PD but not the de novo PD dataset. This corresponds to the cross power spectral density at a particular frequency after h has been transformed to the frequency domain. This is a feature that happens to correlate strongly with the binary classification targets on the early PD data, but that should clearly not be taken as a feature that truly captures a genuine difference between the typing behaviours of Parkinson's sufferers compared to control subjects.

6   Classification with Mean Absolute Consecutive Difference

Using the analysis produced by FRESH, we believe that features based on `change_quantiles` are suitable for capturing the intricate dynamic behaviour of our time series without overfitting. In particular, we take $ql = 0.6$ and $qh = 1.0$ to mark the corridor of hold times, i.e. we take only the elements of h for which both h_n and h_{n+1} are in the 60th percentile. We then take the mean of the absolute difference in hold time between these consecutive keystrokes to give the feature $\langle |\Delta| \rangle \equiv \frac{1}{N} \sum_{n=1}^{N} |\Delta_n|$, where we recall that $\Delta_n \equiv h_{n+1} - h_n$. We refer to this as the mean absolute consecutive difference (MACD).

Reference [4] notes that in order to identify Parkinson's sufferers effectively, it is necessary to capture transient bradykinesia effects that prevent the subject from lifting their fingers from keys in a consistent manner. However, static features that describe the distribution of hold times do not yield such information. In contrast, MACD captures precisely the dynamic variation in hold time between one keystroke and the next. We restrict MACD to analysing hold times in the 60th percentile as typing patterns involving longer hold times appear to be particularly discriminatory.

Using MACD as a univariate feature and classifying with Logistic Regression, we obtain the ROC curve and evaluation scores shown in Fig. 3 and Table 1. Crucially, we find AUC = 0.85, significantly outperforming all the models previously considered. In fact, using MACD, one can obtain effective classification without needing to analyse every element of the hold time series h. In Fig. 4 we demonstrate how classification performance depends on the number of keystrokes analysed. We truncate h after a certain number of elements and perform classification according to the same scheme outlined above, using the MACD model. Figure 4 demonstrates that one may achieve very good performance (AUC > 0.80) from analysing only 200 keystrokes in a typing session.

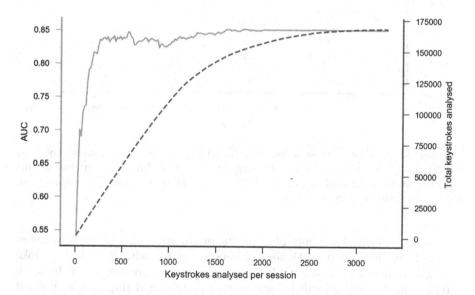

Fig. 4. The dependence of classification performance on the number of keystrokes analysed. The x axis gives the length of the truncated time series h. In red (left y axis) we show the AUC achieved by the MACD model operating on the truncated time series; in blue (right y axis) we show the total number of keystrokes that are analysed across all typing sessions in the whole dataset of 85 users.

7 Tappy Study

Finally, we make some important remarks regarding the 'Tappy' dataset and associated analysis performed in a recent study by Adams [5]. Some concern peculiarities with the data; some concern the methods used during the analysis; and some concern the validity of the results. Although we believe that Adams' work should be of considerable interest to researchers, we were not able to replicate the perfect evaluation results claimed. Other researchers have similarly struggled to achieve the performance claimed by Adams [11]. Here we suggest where there may be flaws in the analysis presented in Ref. [5]. Moreover, we see once again the use of severely overcomplicated methods.

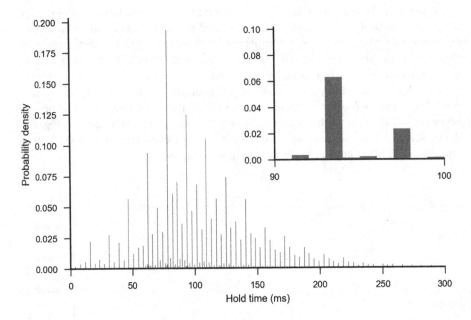

Fig. 5. Hold times for every keystroke used in the Tappy study, with a bin size of 1 ms, indicating a peculiar form of noise affecting the data. Hold times greater than 300 ms are not shown (corresponding to about 0.25% of the data). The inset plot zooms in on hold times between 90 ms and 100 ms.

Again, we begin by simply plotting the distribution of hold times analysed in the study (Fig. 5). As with the datasets associated with Ref. [4], keystroke timing is recorded to an accuracy of 3 ms. However, there appears to be some artefact affecting the recorded times, so that certain hold times are very much more likely than others. For example, a hold time of precisely 78.1 ms accounts for 9.5% of all the hold times recorded; overall, the 13 most common hold times recorded account for more than 50% of the data. Adams uses features that should not be unduly affected by the unnatural spikiness of the hold time distribution;

we highlight these peculiarities for two reasons: firstly, to demonstrate the value of performing data exploration, and secondly, as a caution to researchers that future studies on similar problems may benefit from smoothing the data prior to analysis.

Reference [5] performs the classification task of distinguishing Parkinson's sufferers from control based on both hold time and latency (the interval between pressing one key and the next). These are analysed using elementary statistical features describing the distributions, e.g. mean, standard deviation, skewness and kurtosis, giving a total of 9 features for hold time and 18 for latency. As Adams notes, given the dataset of 53 subjects (20 Parkinson's, 33 control), this large selection of features could easily lead to overfitting. As such, Linear Discriminant Analysis (LDA) is performed on each set of features as a means of dimensionality reduction to produce a single combined feature for hold time and a single combined feature for latency. Each single combined feature is then classified using an ensemble of eight separate models (Support Vector Machine, Decision Tree Classifier, K-Nearest Neighbours, etc.), the results of which are aggregated using a weighted average to produce an overall classification prediction.

We believe that, much like Ref. [4], this is an overengineered approach. The space produced by LDA is limited to one dimension (as constrained by the rank of the between-classes scatter matrix in a binary classification problem). Therefore the optimal decision criterion requires a single threshold value to be established. The use of ensemble techniques to perform such a task is unnecessary and overcomplicated.

Most importantly, however, we believe that the classification results of the study are not reproducible. Adams reports a perfect cross-validated performance, with every subject correctly classified as Parkinson's or control (AUC = 1.00). Based on our efforts to replicate the results, we find this to be wholly implausible and suspect it is an error resulting from flaws in the data acquisition or analysis. In particular, we speculate that the claimed perfect performance is the result of erroneously performing the supervised dimensionality reduction method of LDA on both the training and test data. This flaw is suggested by the description of the pre-processing stage given in Ref. [5]. If this is indeed the case then it would lead to gross overfitting of the data and hence an exaggerated AUC score for the classification task.

8   Conclusion

We have presented a critical analysis of methods proposed in Refs. [4,5] for detecting early signs of Parkinson's disease from typing data. Whilst we believe that such work offers exciting possibilities for improved healthcare, we find the proposed methods to be overengineered and opaque. Moreover, the complexity of the neuroQWERTY index model [4] is demonstrably unnecessary: we achieve equal classification performance (AUC = 0.82) using the standard deviation as the single feature in a Logistic Regression. By performing a thorough investigation of more sophisticated time series features, we formulate the concept of mean absolute consecutive difference (MACD), which can be used as a single feature to classify the data with AUC = 0.85. Importantly, we demonstrate that such performance can be obtained from only a few hundred keystrokes, thereby achieving state of the art results while using significantly fewer samples than previous techniques. We select relevant features from a huge range of complicated time series features and find that multivariate models using up to such 10 features do not outperform the univariate model using MACD by itself—sometimes the simplest method is indeed the best.

References

1. Elbaz, A., Carcaillon, L., Kab, S., Moisan, F.: Epidemiology of Parkinson's disease. Rev. Neurol. **172**(1), 14–26 (2016)
2. Martínez-Martín, P., Gil-Nagel, A., Gracia, L., Gómez, J., Martínez-Sarriés, J., Bermejo, F.: Unified Parkinson's disease rating scale characteristics and structure. Mov. Disord. **9**(1), 76–83 (1994)
3. Pagan, F.L.: Improving outcomes through early diagnosis of Parkinson's disease. Am. J. Manag. Care **18**(7 Suppl), S176–82 (2012)
4. Giancardo, L., Sánchez-Ferro, A., Arroyo-Gallego, T., Butterworth, I., Mendoza, C.S., Montero, P., Matarazzo, M., Obeso, J.A., Gray, M.L., Estépar, R.: Computer keyboard interaction as an indicator of early Parkinson's disease. Sci. Rep. **6**, 34468 (2016)
5. Adams, W.R.: High-accuracy detection of early Parkinson's Disease using multiple characteristics of finger movement while typing. PLoS One **12**(11), e0188226 (2017)
6. Pedregosa, F., Varoquaux, G., Gramfort, A., Michel, V., Thirion, B., Grisel, O., Blondel, M., Prettenhofer, P., Weiss, R., Dubourg, V., Vanderplas, J., Passos, A., Cournapeau, D., Brucher, M., Perrot, M., Duchesnay, E.: Scikit-learn: machine learning in Python. J. Mach. Learn. Res. **12**, 2825–2830 (2011)
7. Arroyo-Gallego, T., Ledesma-Carbayo, M., Sánchez-Ferro, A., Butterworth, I., Sanchez-Mendoza, C., Matarazzo, M., Montero- Escribano, P., Lopez-Blanco, R., Puertas-Martín, V., Trincado, R., Giancardo, L.: Detection of motor impairment in Parkinson's disease via mobile touchscreen typing. IEEE Trans. Biomed. Eng. **64**(9), 1994–2002 (2017)
8. Youden, W.J.: Index for rating diagnostic tests. Cancer **3**(1), 32–35 (1950)
9. Christ, M., Kempa-Liehr, A. W., Feindt, M.: Distributed and parallel time series feature extraction for industrial big data applications. arXiv:1610.07717 (2016)
10. tsfresh, https://github.com/blue-yonder/tsfresh. Accessed 12 Feb 2018
11. Kaggle: raw data used to predict the onset of Parkinson's from typing tendencies, https://www.kaggle.com/valkling/tappy-keystroke-data-with-parkinsons-patients. Accessed 1 Aug 2018

Sky Writer: Towards an Intelligent Smart-phone Gesture Tracing and Recognition Framework

Nicholas Mitri and Mariette Awad[(✉)]

Department of Electrical and Computer Engineering, American University of
Beirut, Beirut, Lebanon
mariette.awad@aub.edu.lb

Abstract. We present Sky Writer, an intelligent smartphone gesture tracking and recognition framework for free-form gestures. The design leverages anthropo-morphic kinematics and device orientation to estimate the trajectory of complex gestures instead of employing traditional acceleration based techniques. Orientation data are transformed, using the kinematic model, to a 3D positional data stream, which is flattened, scaled down, and curve fitted to produce a gesture trace and a set of accompanying features for a support vector machine (SVM) classifier. SVM is the main classifier we adopted but for the sake of comparison, we couple our results with the hidden Markov models (HMM). In this experiment, a dataset of size 1200 is collected from 15 participants that performed 5 instances for each of 16 distinct custom developed gestures after being instructed on how to handle the device. User-dependent, user-independent, and hybrid/mixed learning scenarios are used to evaluate the proposed design. This custom developed gesture set achieved using SVM 96.55%, 96.1%, and 97.75% average recognition rates across all users for the respective learning scenarios.

Keywords: Support vector machines · Gesture recognition
Forward kinematics · Inverse kinematics
Hidden Markov models · Machine learning

1 Introduction

Gesture control [GestureTek] uses sensors that require line of sight operation which pose challenges including computational complexity, energy requirements, robust segmentation, sensitivity to light conditions, object occlusion, and line of sight (Prigge 2004) to name a few. With current smart phones typically equipped with a bevy of both hardware (accelerometer, gyroscope, proximity) and software/virtual sensors (orientation), vision-based gesture detection and motion tracking challenges could be circumvented by employing inertial sensors instead. Coupled with machine learning (ML), these sensors can enable more complex and meaningful motion control in mobile platforms beyond tilts and shakes.

Gesture recognition leveraging ML techniques such as hidden Markov models (HMM), finite state machines, dynamic time warping (DTW), data-driven template matching, or feature based statistical classifiers have reported recognition rates above 90% on average in literature as reviewed in the Sect. 2. Though the most popular,

© Springer Nature Switzerland AG 2018
L. Soldatova et al. (Eds.): DS 2018, LNAI 11198, pp. 447–465, 2018.
https://doi.org/10.1007/978-3-030-01771-2_29

HMM requires some knowledge of the dataset (specifically gesture complexity) to configure the model with adequate states (Kauppila et al. 2007). Additionally, to create more accurate probability distributions, HMM demands a higher number of training samples per gesture (Kauppila et al. 2007; Pylvänäinen 2005; Kallio et al. 2003; Khanna et al. 2015). Thus, we propose to use support vector machines (SVM) with 'Sky Writer', a smartphone gesture recognition system. Like (Zhang et al. 2011), our framework leverages a fusion of inertial sensors which allows for user arm-pose estimation when coupled with our proposed kinematics anthropomorphic model that are also partly documented in our US patent application (Mitri 2016). This is presented as an alternative to the conventional use of depth sensors for human pose estimation e.g. (Zhang et al. 2011). Additionally, we designate the end effector of the kinematic model as a virtual pen and employ Bezier curve fitting to extract control points as features like (Chan et al. 2014). This unique combination of techniques (Fujioka et al. 2006) allows us to store a parametric version of a gesture that can be used for visual feedback.

The rest of this manuscript is structured as follows: Sect. 2 exposes work related to Sky Writer while Sect. 3 details Sky Writer framework. Section 4 elaborates on the adopted methods and Sect. 5 presents the evaluation results. Finally, Sect. 6 concludes the manuscript with follow on remarks.

2   Related Work

While there is no standard library of gestures for mobile platforms, there is common ground with respect to the ML techniques employed with HMM being the most popular and DTW a close second.

The work in Kauppila et al. (2007); Pylvänäinen (2005), Kallio (2003), Awad et al. (2015), Zhang et al. (2011), Chan et al. (2014), Fujioka et al. (2006), N. Mitri et al. (2016), Amma et al. (2012), Raffa et al. (2010), S. Choi et al. (2006), Liu et al. (2009), Kratz et al. (2013), He et al. (2010), Wu et al. (2009), E. S. Choi et al. (2005), Fuccella et al. (2015), Wobbrock et al. (2007) ignores trajectory estimation. The classifiers apply their learning techniques on either raw or processed sensor data. Very few offer a reconstructed visualization of the user-made gesture as in Cho et al. (2004) due to the prevalence of accelerometer usage and accumulated drifting errors. Cho et al. (2004) presented a gesture input device, Magic Wand, for free form gestures recorded using inertial sensors. Acceleration and angular velocity were recorded and a trajectory estimation algorithm was employed to project the gesture onto a 2D plane. Zero velocity compensation was used to account for the error growth caused by double integration. A Bayesian network (BN) was used with a stroke model for recognition over predefined gesture classes and it achieved an average of 99.2 % writer independent recognition rate using a database of 15 writers, 13 gestures, and 24 samples per gesture.

Thus, gesture recognition on mobile/handheld devices has achieved good results when simple gesture sets are employed. However, the choice of motion data is traditionally acceleration. Due to the associated drifting errors, most related work avoided preprocessing and used raw sensor data for classification of gestures, making reconstruction of performed gestures infeasible. Compensation techniques that allow for reconstruction as in Cho et al. (2004) exist but it is unclear how robust they are to the variability in the scale of gestures and the time required for performing them. Additionally, acceleration based tracking requires that the user be stationary since the device is tracked with respect to a non-moving reference frame. Otherwise, tracked trajectories would have to be processed to estimate and filter out secondary motion, a process that is computationally expensive and error prone.

In this work, we explore the use of orientation data coupled with a constraint system for motion tracking. We propose Sky Writer as a framework that leverages a handheld device's orientation and combines it with a constraint model based on human kinematics to achieve "soft" trajectory estimation and gesture recognition. The framework exploits the enhanced precision in orientation sensors to account for various scales and times of gesturing. It allows the user to move while performing gestures since the device is tracked with respect to the user's shoulder, thus positioning itself better in the field of portable on-the-go human-computer interaction (HCI).

3   System Overview

With Sky Writer, components of the pipeline are designed to extract meaningful information from the orientation data of the device in such a way as to acquire a good estimate of the device's 3D position and consequently a reconstructed trajectory that can be provided to the user as visual feedback. A block diagram of the system and its components is shown in Fig. 1. The front end of the process is handled by a smart-phone (Samsung S2) which is responsible for acquiring data and transferring it to a backend server using WIFI. Sky Writer has two phases: Gesture Tracing Processing and Gesture Classification.

3.1   Gesture Tracing Processing

Data Acquisition. For gesture recognition, Skywriter leverages device orientation, specifically the rotation vector of the phone. On the Android platforms, the associated sensor is software based and is implemented using a preconfigured extended Kalman filter that fuses data from the accelerometer, gyroscope, and magnetometer sensors. The rotation vector represents the world orientation of the phone and is a combination of the axis of rotation represented by the unit vector $\hat{K} = (k_x, k_y, k_z)$ and the angle through which the device was rotated around the axis, θ. The three elements of the rotation vector recorded at each sample are:

$$k_x sin(\frac{\theta}{2}), k_y sin(\frac{\theta}{2}), k_z sin(\frac{\theta}{2}) \tag{1}$$

Our android application collects the rotation vector values from the rotation sensor sampled at approximately 90 Hz. With respect to the tracing and learning algorithms and manual gesture recording, the user is prompted to touch and hold the display to record the data. On release, the collected data are sent via transmission control protocol (TCP) to a server. Upon receiving the data packet, the data matrix is subsampled. Sample counts of 100–200 provided fast processing while retaining enough information. Next, the axis of rotation $\hat{K}$ is obtained by normalizing each rotation vector sample and is then converted to a rotation matrix using (2). We use the ${}^{A}_{B}R$ notation here to represent the matrix describing the relative rotation of a coordinate system 'B' with respect to a coordinate frame 'A'. Thus, ${}^{world}_{phone}R$ describes the rotation of the phone w.r.t the world frame.

$$
\hat{K} = \begin{bmatrix} k_x \\ k_y \\ k_z \end{bmatrix} \rightarrow R_{\hat{K}}(\theta) = {}^{world}_{phone}R
$$

$$
= \begin{bmatrix} k_x * k_x * v\theta + c\theta & k_x * k_y * v\theta - k_z * s\theta & k_z * k_x * v\theta + k_y * s\theta \\ k_x * k_y * v\theta + k_z * s\theta & k_y * k_y * v\theta + c\theta & k_z * k_y * v\theta - k_x * s\theta \\ k_x * k_z * v\theta - k_y * s\theta & k_y * k_z * v\theta + k_x * s\theta & k_z * k_z * v\theta + c\theta \end{bmatrix} \quad (2)
$$

where $c\theta = \cos\theta$, $s\theta = \sin\theta$, and $v\theta = 1 - \cos\theta$.

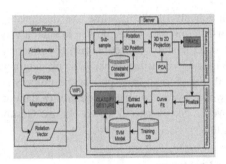

Fig. 1. System block diagram.

Kinematic Constraint Model. Orientation data alone do not provide sufficient information to allow for unique mapping to 3D position. When the object is considered as an end effector to a joint chain, a correlation that is dependent on the degrees of freedom (DOFs) associated with the chain is created. With this knowledge and the fact that the device in motion is hand-held, we propose a method for extracting positional data from orientation using a robotic model inspired by an anthropomorphic arm/joint chain with limited degrees of freedom.

Theoretical Background. A manipulator is defined in our context as a chain of joints connected by rigid links; akin to joints and bones in a skeletal frame (Fig. 3). We follow the Denavit-Hartenberg notation (Craig 2004) where every link of the chain is assigned four quantities. Two describe the link itself, while the other two describe its relation to the neighboring links. These relational parameters are also dependent on the choice of standard procedure followed in assigning frames to every joint of the mechanism. As a rule, the z-axis of the frame is aligned with the axis of the joint. The latter axis is defined by the joint type. In our proposed model, this is the axis of rotation. The x-axis is placed along the perpendicular line connecting two consecutive joints. The y-axis is the result of the cross product of the two. Various intricacies are involved in assigning frames and the assignment is not always unique. See Craig (2004) for more details. Here, we provide a brief definition of the link parameters and how they apply to our system. Let:

- a_i = distance from Z_i to Z_{i+1} measured along X_i;
- αi = angle from Z_i to Z_{i+1} measured about X_i;
- d_i = distance from X_{i-1} to X_i measured along Z_i;
- θi = angle from X_{i-1} to X_i measured about Z_i;

where i is the location of the joint in the chain starting with i = 0 being the root. Defining these parameters for every link of a mechanism allows us to determine the transformation matrix relating two consecutive joints i − 1 and i as:

$$
{}^{i-1}_{i}T = \begin{bmatrix} c\theta_i & -s\theta_i & 0 & a_{i-1} \\ s\theta_i c\alpha_{i-1} & c\theta_i c\alpha_{i-1} & -s\alpha_{i-1} & -s\alpha_{i-1}d_i \\ s\theta_i s\alpha_{i-1} & c\theta_i s\alpha_{i-1} & c\alpha_{i-1} & c\alpha_{i-1}d_i \\ 0 & 0 & 0 & 1 \end{bmatrix} \tag{3}
$$

and, the transformation matrix between any two joints of the hierarchy by multiplying the individual transformations e.g. ${}^{0}_{N}T = {}^{0}_{1}T^{1}_{2}T^{2}_{3}T\ldots{}^{N-1}_{N}T$. This allows us to define the relative position and orientation of any joint w.r.t any other joint.

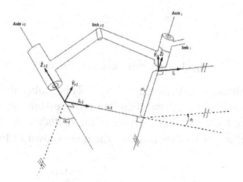

Fig. 2. Attached rigid links with their corresponding frames and parameters.

Forward kinematics (FK) is the static geometrical problem of computing the position and orientation of an end-effector given the parameters of all preceding joints and links. Inverse kinematics (IK) is the problem of calculating all possible sets of link parameters that can be used to achieve a given position and orientation of an end-effector (Craig 2004). We employ both notions in tandem here to estimate a smart phone's location in a user space given its orientation (Fig. 2).

Proposed model description. To manipulate a hand-held device in space, a user does not need to engage all possible degrees of freedom (DOF). In fact, as few as 3 DOFs can be used. The Sky Writer model is based on 4 DOFs (three in the shoulder, one in the elbow) as shown in Fig. 4 with the associated link parameters listed in Table 1. In Fig. 3, link lengths are not visualized accurately for clarity of representation. In actuality, frames 0–3 have coinciding origins. Since non-uniqueness of solution is a common challenge in IK, this restriction in movement (due to limited DOFs) allows us to reduce the solution space for position estimation. The fewer the DOFs employed, the fewer heuristics need to be enforced in order to retain a single unique solution.

Implementation. To extract positional information from the device's orientation, additional information is necessary. The readily available device's orientation with respect to the world frame is converted into the rotation matrix ${}^{world}_{phone}R$ in Eq. (2); signifying the rotation of "phone" in the "world" frame. To solve the trajectory estimation problem, we propose attaching the device to the end effector of a joint chain

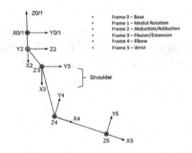

Fig. 3. Illustrative kinematic model.

rooted at the user's shoulder as depicted in Fig. 4. This applies the same constraints to the phone that the hand abides by and therefore provides information for localization. To that end, we place the phone in the frame of the root/shoulder frame (frame 0) and use the compound effect of rotational transformations shown in Eq. (3):

$$
{}^{0}_{phone}R = {}^{0}_{word}R * {}^{world}_{phone}R \tag{4}
$$

Table 1 Link Parameters.

i	α_{i-1}	a_{i-1}	d_{i-1}	θ_{i-1}
1	0	0	0	θ_1
2	−90°	0	0	90°+θ_2
3	90°	0	0	θ_3
4	0	L_1	0	θ_4
5	0	L_2	0	-

Equation (4) means the rotation of the phone with respect to the shoulder/root frame is the compound rotation of the world frame's rotation w.r.t to the root and the rotation of the phone w.r.t. to the world frame. The latter piece of information is already available while $^{0}_{word}R$ can be obtained from its transpose $^{word}_{0}R$ which relates to the users facing direction w.r.t to North. This is of course hidden to us.

To surmount this challenge, we propose defining a fixed way to handle the device in hand. The holding pose is depicted in Fig. 4. The device's pointing direction can now be used as a rough estimate of the user's facing direction. With an initial assumption of the two being perfectly aligned, an adaptive re-orientation is then performed based on a continuity metric to achieve a better estimate. If the distance between two consecutive estimated positions exceeds a pre-defined threshold, the initial assumption is deemed false and modified. This resolves the missing $^{0}_{word}R$. The fixed handling reveals the rotation of the phone w.r.t the wrist $^{5}_{phone}R$ and subsequently w.r.t the elbow $^{4}_{phone}R$ if wrist rotations are disallowed. With that, all the necessary information to derive the needed constraints for the mapping strategy become available:

$$^{4}_{phone}R \begin{bmatrix} 0 & 1 & 0 \\ -1 & 0 & 0 \\ 0 & 0 & 1 \end{bmatrix} \tag{5}$$

$$^{0}_{4}R = {}^{0}_{phone}R * {}^{4}_{phone}R^{T} = \begin{bmatrix} r_{11} & r_{12} & r_{13} \\ r_{21} & r_{22} & r_{23} \\ r_{31} & r_{32} & r_{33} \end{bmatrix} \tag{6}$$

$$^{0}_{4}T = \begin{bmatrix} & {}^{0}_{4}R & & t_{14} \\ & & & t_{24} \\ & & & t_{34} \\ 0 & 0 & 0 & 1 \end{bmatrix} \tag{7}$$

$$r_{11} = -c_1 s_2 c_3 c_4 - s_1 s_3 c_4 + c_1 s_2 s_3 s_4 - s_1 s_4 c_3 \tag{8}$$

$$r_{12} = c_1 s_2 c_3 s_4 + s_1 s_3 s_4 + c_1 s_2 s_3 c_4 - s_1 c_4 c_3 \tag{9}$$

$$r_{13} = c_1 c_2 \tag{10}$$

$$r_{21} = -s_1 s_2 c_3 c_4 + c_1 s_3 c_4 + s_1 s_2 s_3 s_4 + c_1 s_4 c_3 \tag{11}$$

$$r_{22} = s_1 s_2 c_3 c_4 - c_1 s_3 s_4 + s_1 s_2 s_3 c_4 + c_1 c_4 c_3 \tag{12}$$

$$r_{23} = s_1 c_2 \tag{13}$$

$$r_{31} = -c_2 c_3 c_4 + c_2 s_3 s_4 = -c_2 c_{3+4} \tag{14}$$

$$r_{32} = c_2 c_2 s_4 + c_2 s_3 c_4 = c_2 s_{3+4} \tag{15}$$

$$r_{33} = -s_2 \tag{16}$$

$$t_{14} = -c_1 s_2 c_3 L_1 - s_1 s_3 L_1 \tag{17}$$

$$t_{24} = -s_1 s_2 c_3 L_1 + c_1 s_3 L_1 \tag{18}$$

Fig. 4. Phone holding pose.

$$t_{34} = -c_1 c_3 L_1 \tag{19}$$

where $c_i/s_i = cos(\theta_i)/sin(\theta_i)$ and t stands for translation. The next step is to use the available information to make two passes along the joint hierarchy. The first IK pass reveals the set of joint angles necessary to produce the recorded orientation of the phone. From the previous equations, we can derive the following:

$$\theta_1 = \arctan(r_{23}, r_{13}) \tag{20}$$

$$\theta_2 = \arctan\left(-r_{33}, \frac{r_{13}}{c_1}\right) \tag{21}$$

$$\theta_3 + \theta_4 = \arctan(\frac{r_{32}}{c_2}, \frac{-r_{31}}{c_2}) \tag{22}$$

The final step in the IK pass is to solve for joints 3 and 4 whose rotation axes are parallel and therefore act as a deterrent to a unique solution to the system at hand. For that purpose, we propose forcing a coupling relation between the joints (i.e. $\theta_4 = w_c \theta_3$). w_c acts as a coupling weight determining the proportional relation between the joints. Any choice of w_c is an assumption about the degree to which a user prefers engaging his/her elbow to his/her shoulder while making a gesture. The system is tolerant to a selection of weights since the end trace is warped uniformly and preserves its shape. Values of 1 and higher are noted to be more aligned with the principles of natural motion. The system's tolerance to a range of possible configurations extends to the choices of L1 and L2, corresponding to the lengths of the upper arm and the forearm respectively. This ties back to the implementation of motion tracking that we refer to as "soft" trajectory estimation. Unlike systems like Kinect and Wii, Skywriter does not demand point to point accuracy in device tracking. Instead, it is sufficient that the general shape and readability of the gesture being performed be preserved (i.e. mild warping due to non-optimal choices are tolerated). Thus, parameters like L1 and L2 can be chosen generically instead of requiring the user to perform a laborious tuning phase to derive system parameters after specific motions are done.

Next, the FK pass allows us to use the obtained angles to position the phone with respect to the base frame attached to the shoulder. We require the translational components of the transformation matrices. Up to this point, we had utilized ^0_4T for all derivations. Since the phone and the wrist are assumed to coincide in space, all that is needed is to translate the elbow frame L2 units along its x-axis according to (23).

$$\begin{bmatrix} x \\ y \\ z \end{bmatrix} = \begin{bmatrix} t_{14} \\ t_{24} \\ t_{34} \end{bmatrix} + \begin{bmatrix} r_{11} \\ r_{21} \\ r_{31} \end{bmatrix} * L_2 \tag{23}$$

Projection. Since most gestures are both visualized and used in 2D space, principal component analysis (PCA) is used to project the trace along the axis of least variance. PCA performs orthogonal transformations from a set of interrelated variables into a new set of linearly uncorrelated Principal Components (PCs) ordered so that the first PC accounts for the most variability in the data. The last PC is responsible for the least variance and is therefore a good candidate for a projection axis. Since using all 3 PCs to reframe the trace adds undesired rotations, we derive from the PC of least variance an azimuth angle about z0 and rotate the trace accordingly before discarding its y-

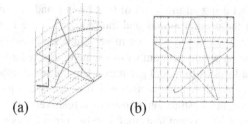

(a)   (b)

Fig. 5. (a) 3D trace extracted from orientation data with 3rd PC/projection vector shown. (b) 2D projection of trace using PCA for dimensionality reduction.

dimension. This has the effect of flattening the gesture with respect to the virtual vertical 2D drawing plane. Figure 5 shows an example of the projection method.

A less computationally expensive approach, better suited when the phone becomes responsible for all necessary computations, is to take advantage of the phone's pointing direction. With the holding pose suggested, the user handles the phone much like s/he would a pointing device with the pointing direction being that of the y-axis of the phone (x-axis of the wrist/frame 5). The advantage here is that averaging the directional vector over all collected samples provides an excellent estimate of a projection vector. Also, since this vector is already computed as part of ^0_4R, it is much less resource intensive.

Feature Extraction. With the 3D gesture trajectory projected onto a planar surface, the obtained 2D trace is processed for distinctive features. This places Sky Writer between traditional gesture recognition techniques and handwritten character recognition (HWR) techniques. We opt for a geometric feature extraction, which shies away from standard global features extraction commonly used in HWR. We resort to a parameterized Bezier curve fitting approach and make use of the obtained control points. To achieve this fit, the 2D trace obtained after projection is first scaled and centered in a square 64 × 64 pixel frame. This injects pseudo scale invariability into the system, which is crucial for the generalization and accuracy of the learning algorithm and the gesture prediction.

A Bezier curve can be used to model smooth scalable curves, commonly referred to as paths in image editing. Although it has varying polynomial degrees, m, a Bezier curve is typically cubic and generated using (24) as seen in Khan (2007):

$$q(t_i) = \sum_{k=0}^{m} \binom{m}{k} P_k (1 - t_i)^{m-k} t_i^k, 0 \le t_i \le 1 \qquad (24)$$

where m = 3 for cubic, $q(t_i)$ is the interpolated point at parameter value t_i, and P_k is the kth control points. The advantage of utilizing Bezier curves is that they fit a data curve with a smooth path defined by a small set of control points. The quality of the fit is defined by its deviation from the original curve. We calculate this deviation using least square error (Khan 2007).

Since cubic Bezier curves can only model 4-dimensional data vectors, segmentation is used to model larger data vectors. In Khan (2007), this is achieved by segmenting the input data using an initial set of breakpoints and defining an error ceiling. If the error between the original points and the modeled curve is exceeded, the input data are further segmented, thus producing more connected cubic curves. We opt for a reversed implementation where the number of segments is fixed and the error ceiling is set to infinity. This allows us to model gestures using a fixed number of control points and therefore a fixed length feature vector, which is necessary for the learning algorithm of SVM. Since the gesture set consists of relatively simple gestures, fixing empirically the segmentation count to a small number provides us with a close fit and a feature vector with manageable size to reduce unnecessary computations. Figure 6 shows an example of a gesture fitted using 10 segments. For our scenario, the fit retains the integrity of the original shape while producing a 62-dimensional feature vector made up of the x and y coordinates of the control points (31 in total) defining the cubic segments.

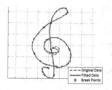

Fig. 6. Bezier curve fit using 10 segments.

3.2 Kernel-Based SVM Gesture Classifier

Since Sky Writer is designed to classify for multiple gestures, we chose the one-against-one multiclass SVM (OAO-MCSVM), a pair-wise classification method that builds $c(c-1)/2$ binary SVM, each of which is used to discriminate two of the c classes. OAO requires the evaluation of (c-1) SVM classifiers.

Our selection for SVM classifier is multifold. SVM offers a principled approach to ML problems because of its mathematical foundation in statistical learning theory. A SVM seeks to find the optimal parameters for a hyper-plane that acts as a separator between different classes in a multi-dimensional feature space. Due to the formulation of the objective function being solved, this hyper-plane is defined by a few support vectors (SV) in such a way as to provide strong generalization for the classifier while minimizing classification errors. SVM uses the kernel trick to map the data into a higher-dimensional space before solving the ML machine learning task as a convex optimization problem in which optima are found analytically. Of course, selection and settings of the kernel function are crucial for SVM optimality (Awad 2015; Saab 2014).

4 Evaluation Methods

4.1 Gesture Set

To evaluate our recognition system, we aim for a gesture set that is universal with functional gestures that are quite sophisticated. However, it is difficult to find among relevant literature such a universal gesture set. Although some attempts have been made in literature like (Ruiz et al. 2011), they still failed by design to meet our criterion. We therefore elect to use our own gesture set while maintaining a design that allows users to customize the system and supplement it with their own gestures. For our gesture set, the rule of thumb is to adopt a gesture set that is concise enough for the user to remember, and one that consists of gestures that intuitively fit the application (meaningful gestures). Figure 7 shows the proposed gesture set consisting of 10 digits and 6 additional shapes. This set enables features that range from speed dialing to simple browser control (e.g. star for favorite, arrows for navigation, 'S' for auto-scroll) to simplified kinect-like game control and others. Additionally, the inclusion of digits allows us to compare against other works that utilized a digit only set.

Fig. 7. Gesture set for digits and six gesture codes for shapes

4.2   A. Performance Evaluation of Sky Writer

Despite most smartphones being equipped with sufficient processing power to handle relatively intensive tasks, maintaining low computational demands is still necessary to conserve a device's battery life or shift Sky Writer to computationally less powerful devices such as smartwatches. In this direction, evaluating latency is important as it affects the choice of hardware and the sampling rate, which is related to the accuracy of the results obtained. To achieve that, we measure the number of operations of Sky Writer's major stages. The main stages are as follows: OrientToPosition maps the 3D orientation of the device to 3D position. ProjectTrace converts the captured gesture from 3D to 2D. Pixelize creates the 64×64 pixel frame on which the 2D gesture trace is rendered. The last preprocessing stage is Bezier curve fitting for feature extraction. Instead of detailing the number of operations required during OrientToPosition and Bezier fitting, we measure runtime in milliseconds, calculated as a function of the number of available samples as shown in Sect. 5.A.

4.3   Data Collection

Participants. Fifteen participants (all students among which are four males, with average age 25.67 and a standard deviation of 7.02) were asked to supply gesture samples. Because this project aims first at providing a proof of concept that such a system is promising, we tapped into the resources available on campus.

Procedure. Participants were given a few minutes to get acquainted with how to start recording their gestures as well as get familiar with the handling limitations. Due to the device's learning curve, not all user attempts resulted in the expected shape being traced. The samples were therefore categorized as "hits" and "misses". Misses were gestures that were either significantly clipped at the start or at the end or gestures that, due to the user's lack of experience with the device, turned out drastically misshapen to the point of affecting readability. Users were therefore asked to make as many gestures as necessary to acquire 5 hits. The 5 hits were collected into one data set while all attempts were binned in another. Specifically, a dataset of size 1200 was collected from 15 participants that performed 5 hits for each of 16 unique gestures. Finally, the miss rate associated with every user was recorded as well for further analysis.

Learning Scenarios. For Sky Writer's evaluation, we adopt multiple learning scenarios. While most of the literature considers either user-dependent or user-independent learning, we include both and offer a hybrid third option that combines both. In all scenarios, the same data set is used, but is partitioned and treated uniquely. We rely on k fold cross validation (CV) as it is widely used for system evaluation in the ML communities. To classify the gesture, a trace is compared to the contents of a training database since it is supervised learning and analyzed by the learning module of SVM before a label is assigned to the gesture. The performance of the system is calculated as the resultant error between the predicted gesture label and its true one.

User-dependent. The user is asked to provide sample gestures to train the learning model. For evaluation, each user set is partitioned to create training and testing sets and validated using 5-fold cross-validation. Sky Writer is trained on 4 samples of each gesture class and tested on the 5th. This is repeated 5 times so that each sample is used for testing once.

User-independent. In user-independent learning, the user is not expected to train the system before it can be used. The system is trained using data from 14 users and then tested it on the 15th user. Training data here are referred to as community data.

Mixed. The mixed scenario is a combination of both user-dependent and user-independent learning. Here, the user is expected to provide training data for the system to learn from. For evaluation, user data are partitioned using 5-fold cross validation. The resulting training partitions are supplemented by data collected from other users, community data. Testing is then performed using the user's testing folds.

4.4 Classifiers Used

SVM is used as the main classifier for its superior generalization ability. Additionally, we offer a glimpse into the performance of HMM. For that purpose, a 13 state HMM was coupled with an added encoding scheme. For the latter, the tangential direction of gesture segments was discretized as described in Elmezain et al. (2008) and used to create a feature vector that fed into the HMM model. Both number of states and number of discrete directions were chosen via a rough grid search. Similarly, the SVM models were optimized using a grid search for basis function and regularization parameters.

5 Evaluation Results

5.1 A. Computational Performance

For today's mobile processors, the computational demand of Skywriter is a non-issue especially when considering system on chips (SoCs) in flag ship phones. As for smartwatches, we present the Apple Watch as an example with its S1 processor running at 520 MHz and its PowerVR SGX543 GPU (Voica 2014), a combination of low power hardware with processing power rivaling phones only a few generations old. Table 2 shows estimates of the number of operations required by Sky Writer's major stages where N stands for the number of samples used.

Table 2 Computational demand per selected function

Function	Total operations
OrientToPosition	1200N+73
ProjectTrace	64+8N
Pixelize	54+22N

OrientToPosition is by far the most demanding of the major processing stages. Measured against current hardware, nonetheless, it does not pose a challenge to user experience. When tested on a galaxy S2, negligible time was observed between releasing the phone at the end of the gesture and the gesture being drawn on screen. The latency in play here involves all three processing stages. A similarly low latency was noted for recognition. This is promising to consider the hardware used in this prototype.

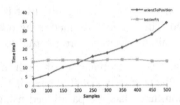

Fig. 8. Runtime (in milliseconds) vs. number of samples used.

As for the last preprocessing stage, i.e. the Bezier curve fitting, Fig. 8 shows the run time analysis for the function using a 2.0 GHz i7 PC with 16 GB of RAM. For comparison purposes, the run-time plot of the OrientToPosition function is plotted, verifying its linear growth. While the curve fitting algorithm is of order O(N), the graph reveals a constant run-time. The reason is the fact that the Pixelize function reframes the trace in a square pixel grid uses interpolation to fill the trace. This over-samples the trace to a number of points that does not vary significantly for the same gesture. Thus, the recorded times do not obey the expected growth of the algorithm. Hyper-sampling improves classification accuracy by providing the curve fitting procedure with enough points to produce break points that are quasi-uniformly spaced.

5.2   Example User Data

Figure 9 shows examples of samples provided by one of the users as plotted at the server side. This user was selected because his resulting trajectories are smooth and highly legible. Figure 10 shows an example of more complex traces (complete names) that Skywriter can generate as seen on the client side (phone screenshot).

Fig. 9. Samples of gestures traced by one user.

Fig. 10. Examples of full names generated by Skywriter.

On the other hand, when the testing device is mishandled, some of the generated traces were clipped drastically at both ends. In more problematic cases where the phone's pose limitations were not respected, significant warping of the gestures was observed as seen in Fig. 11. Such data are treated as outliers and not included into the data set.

Fig. 11. Samples of gestures categorized as "misses".

5.3 Learning Results

Accuracy results and standard deviation of using the proposed gesture set and a digit-only version of SVM for 15 users are reported for the different learning scenarios (Figs. 12, 13 14). The average over all users of the entire gesture set and of the digit-only set for SVM is referred to as "AVE" in the figures and similarly the average over all users for HMM is referred to as "HMM" in the aforementioned figures. Moreover, in Figs. 12 and 13 the accuracy results are compared to the results of digit-only gesture set of S. Choi et al. (2006) (referred to in the figures as "S. Choi").

User-dependent. Figure 12 summarizes the results for user-dependent learning. For the entire set, accuracies above 93.75% are recorded for all users with rates as high as 99.25% for users 13 and 14. With digits only, the lowest accuracy is 90% while the highest is 100%. On average, the 2 sets (gesture and digits-only sets) achieved accuracy rates of 96.55±0.33% and 96.56±0.95% respectively. As for HMM, the entire gesture set achieved an average rate of 71.66±2.58% with a minimum of 61% and a maximum of 83% while the digit-only set scored higher with 76.53±3.57% with a minimum of 65.2% and a maximum of 90.4%. Our results are shy of some of the results noted in relevant work. Namely, S. Choi et al. (2006) claimed a rate of 100% for the user-dependent case using HMM with velocity and acceleration data. We attribute this to the fact that their digit only dataset consisted of shapes that were more distinguishable than ours. This is especially relevant to our HMM results since the combination of more complex gestures and a limited training data set lead to subpar performance.

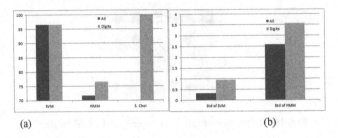

(a)   (b)

Fig. 12. Accuracy results (a) and standard deviation (b) for the user-dependent case. Results are shown per user for both full set and digits-only. A user average is compared to the results of [S. Choi et al. 2006].

User-independent. Using SVM under the user-independent setup, accuracy rates ranging between 88.75% and 100% were recorded for all gestures and between 93.6% and 100% for the digit only set as seen in Fig. 13. On average, the two sets achieved accuracy rates of 96.4±0.31% and 97.46±0.51% respectively. This is a minor drop of 0.15% for the entire set but an increase of 0.90% for digits. Both are significantly higher than the 90.2% achieved by S. Choi et al. (2006) for this scenario using DTW.

HMM benefits from the larger training data set. The entire gesture set achieved an average rate of 90.26±1.85% with a minimum of 74.75% and a maximum of 96.5% while the digit-only set achieved 90.82±2.39% with a minimum of 81.6% and a maximum of 97.2%. Both results are significantly better than the user dependent scenario with a margin of 20+%.

While not the expected result, the rates achieved here for the independent case are very promising. With sufficiently diverse independent training in the back-end, the user is not required to perform any training for the system to perform well.

Mixed/Hybrid. Figure 14 shows per user accuracy rates with the mixed strategy. With overall average rates of 97.75% and 97.66% with SVM and 93.83% and 94.66% with HMM for the entire gesture set and the digit set respectively, it's clear that this strategy provides the best performance even though some users took a minor hit in accuracy rates. This is especially true for our HMM model which achieved an increase of 22.17% and 18.3% for the respective sets in comparison to the user dependent case.

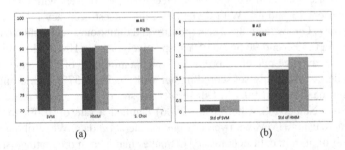

(a)   (b)

Fig. 13. Accuracy results (a) and standard deviation (b) for the user-independent case, for both full set and digits-only. A user average is compared to the results of [S. Choi et al. 2006].

Combining both "community" data and "self" data had an interesting effect here. For some users, the classifier could perform better with more information provided by the community. That information aided the classifier in creating a better model for the gesture classes. Specifically, we consider user 9 for whom accuracy took a sharp dip for the independent scenario. Despite the "community" data strongly biasing the learning model, a few training samples from the users were enough to disambiguate some problematic gestures, namely decreasing the misclassifications of 'S' and '6'. For other users, the former information acted as a minor hindrance: the user's own data did not coincide with the community norm enough. Including gestural data from others seemed to have biased the model away from the user's own data.

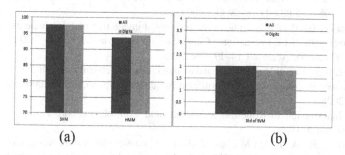

(a)   (b)

Fig. 14. Accuracy results per user for both full set and digits-only for the mixed case.

6 Conclusion

Sky Writer uses a novel combination of device orientation and kinematic inspired constraints to estimate the trajectory of the device in a virtual 2D drawing plane

With promising results proving the applicability orientation data for gesture tracing in mobile platforms, future work needs to focus on refining the workflow to ensure great user experience. This is especially important given the wide array of applications Skywriter can evolve into and inspire like digital signature generation and cursive letter tracing and recognition. The suggested workflow could also be even adapted to study its applicability to fall detection which is an active research area. A switch to a smartwatch form factor to improve usability accompanied by a comprehensive performance analysis that addresses cost, latency, power consumption, and availability of minimum hardware requirements in modern smart-devices will be the target of follow up work; especially that smartwatchs, according to recent specs (SmartWatchr 2015), are becoming more powerful in terms of performance and functionality (1.2 GHz CPU, 8 GB storage and several sophisticated sensors).

References

Android developers, "Sensor Overview" http://developer.android.com/guide/topics/sensors/sensors_overview.html (2016). Accessed 5 Jan 2016

Amma, C., Georgi, M., Schultz, T.: Airwriting: hands-free mobile text input by spotting and continuous recognition of 3d-space handwriting with inertial sensors. In: 2012 16th International Symposium on Wearable Computers (ISWC), pp. 52–59 (2012). Accessed 18–22 June 2012

Awad, M., Khanna, R.: Efficient Learning Machines: Theories, Concepts, and Applications for Engineers and System Designers. A Press (2015)

Boser, B.E., Guyon, I.M., Vapnik, V.N.: A training algorithm for optimal margin classifiers. In: Proceedings of the 5th Annual Workshop on Computational Learning Theory, pp. 144–152 (1992)

Chan, K., Koh, C., Lee, C.S.G.: A 3-D-point-cloud system for human-pose estimation. IEEE Trans. Syst. Man Cybern.: Syst. 44(11), 1486–1497 (2014)

Cho, S., et al.: Magic wand: a hand-drawn gesture input device in 3-D space with inertial sensors. In: IWFHR-9 2004. Ninth International Workshop on Frontiers in Handwriting Recognition, 2004, pp. 106–111 (2004) 26–29 Oct 2004

Choi, E.S., et al.: Beatbox music phone: gesture-based interactive mobile phone using a tri-axis accelerometer. In: IEEE International Conference on Industrial Technology, 2005. ICIT 2005, pp. 97–102 (2005). Accessed 14–17 Dec 2005

Choi, S., Lee, A.S., Lee, S.Y.: On-line handwritten character recognition with 3D accelerometer. In: International Conference on Information Acquisition, pp. 845–850 (2006). Accessed 20–23 Aug 2006

Craig, J.J.: Introduction to Robotics: Mechanics and Control, 3rd edn. Prentice Hall, Upper Saddle River (2004)

Elmezain, M., et al.: A Hidden Markov Model-based continuous gesture recognition system for hand motion trajectory. In: ICPR 2008. 19th International Conference on Pattern Recognition, 2008, pp. 1–4 (2008). Accessed 8–11 Dec 2008

Fuccella, V., Costagliola, G.: Unistroke Gesture recognition through polyline approximation and alignment. In: Proceedings of the 33rd Annual ACM Conference on Human Factors in Computing Systems, pp. 3351–3354. ACM (2015)

Fujioka, H., et al.: Constructing and reconstructing characters, words, and sentences by synthesizing writing motions. IEEE Trans. Syst. Man Cybern. Part A: Syst. Hum. 36(4), 661–670 (2006)

GestureTek mobile. http://www.gesturetekmobile.com

He, Z.: A new feature fusion method for gesture recognition based on 3D accelerometer. In: 2010 Chinese Conference on Pattern Recognition (CCPR), pp. 1–5 (2010). Accessed 21–23 Oct 2010

He, Z.: Accelerometer based gesture recognition using fusion features and SVM. J. Softw. 6(6), 1042–1049 (2011)

Kallio, S., Kela, J., Mantyjarvi, J.: Online gesture recognition system for mobile interaction. In: IEEE International Conference on Systems, Man and Cybernetics, 2003, vol. 3, pp. 2070–2076 (2003). Accessed 5–8 Oct 2003

Kratz, S., Rohs, M., Essl, G.: Combining acceleration and gyroscope data for motion gesture recognition using classifiers with dimensionality constraints. In: Proceedings of the 2013 International Conference on Intelligent user Interfaces, pp. 173–178. ACM (2013)

Liu, J., et al.: uWave: accelerometer-based personalized gesture recognition and its applications. In: PerCom 2009. IEEE International Conference on Pervasive Computing and Communications, 2009, pp. 1–9 (2009). Accessed 9–13 March 2009

Mitri, N., Wilkerson, C., Awad, M.: Recognition of free-form gestures from orientation tracking of a handheld or wearable device (2016). http://www.freepatentsonline.com/y2016/0092504.html

Prigge, E.: A positioning system with no line-of-sight restrictions for cluttered environments. PhD dissertation, Stanford University, p. 1 (2004)

Kauppila, M., et al.: Accelerometer based gestural control of browser applications. In: Proceedings of UCS, Tokyo (2007)

Khan, M.: Cubic Bezier least square fitting. Matlab Central (2007). http://www.mathworks.com/matlabcentral/fileexchange/15542-cubic-bezier-least-square-fitting

Pylvänäinen, T.: Accelerometer Based gesture recognition using continuous HMMs. In: Marques, Jorge S., Pérez de la Blanca, N., Pina, P. (eds.) IbPRIA 2005. LNCS, vol. 3522, pp. 639–646. Springer, Heidelberg (2005). https://doi.org/10.1007/11492429_77

Raffa, G., et al.: Don't slow me down: bringing energy efficiency to continuous gesture recognition. In: 2010 International Symposium on Wearable Computers (ISWC), pp. 1–8. Accessed 10–13 Oc. 2010

Rico, J., Brewster, S.: Gesture and voice prototyping for early evaluations of social acceptability in multimodal interfaces. In: Proceeding ICMI-MLMI '10 International Conference on Multimodal Interfaces and the Workshop on Machine Learning for Multimodal Interaction, Article No. 16

Ruiz, J., Li, Y., Lank, E.: User-defined motion gestures for mobile interaction. In: Proceedings of the SIGCHI Conference on Human Factors in Computing Systems, pp. 197–206. ACM (2011)

SmartWatchr: 2015 Smartwatch Specs Comparison Chart. Smartwatch News. (2015). http://www.smartwatch.me/t/2015-smartwatch-specs-comparison-chart/979

Saab, A., Mitri, N., Awad, M.: Ham or Spam? A comparative study for some content-based classification algorithms for email filtering. In: 17th IEEE Mediterranean Electrotechnical Conference - Information & Communication Systems, Beirut, Lebanon (2014). Accessed 13–16 April 2014

Timo, P.: Accelerometer based gesture recognition using continuous HMMs. In: Pattern Recognition and Image Analysis Lecture Notes in Computer Science, vol. 3522, pp. 639–646 (2005)

Vapnik, V.: The Nature of Statistic Learning Theory. Springer, New York (1995)

Voica, A.: PowerVR GX5300: the world's smallest GPU for next-generation wearables and IoT. Imagination Blog. 2014. http://blog.imgtec.com/powervr/powervr-gx5300-the-worlds-smallest-gpu-for-next-generation-wearables-and-iot. Accessed 5 Jan 2016

Wang, X., et al.: Gesture recognition using mobile phone's inertial sensors. In: Distributed Computing and Artificial Intelligence, pp. 173–184. Springer, Berlin (2012)

Wobbrock, J.O., Wilson, A.D., Li, Y.: Gestures without libraries, toolkits or training: a $1 recognizer for user interface prototypes. In: Proceedings of the 20th Annual ACM Symposium on User Interface Software and Technology, pp. 159–168. ACM (2007)

Wu, J., et al.: Gesture recognition with a 3-D accelerometer. In: UIC '09 Proceedings of the 6th International Conference on Ubiquitous Intelligence and Computing, pp. 25–38 Brisbane (2009). Accessed July 7–9 2009

Zhang, X., et al.: A framework for hand gesture recognition based on accelerometer and EMG sensors. In: IEEE Trans. Syst. Man Cybern. Part A: Syst. Hum. **41**(6), 1064–1076 (2011)

Visualization and Analysis of Parkinson's Disease Status and Therapy Patterns

Anita Valmarska[1]([⊠]), Dragana Miljkovic[1], Marko Robnik–Šikonja[3], and Nada Lavrač[1,2,4]

[1] Jožef Stefan Institute, Jamova 39, Ljubljana, Slovenia
[2] Jožef Stefan International Postgraduate School, Jamova 39, Ljubljana, Slovenia
{anita.valmarska,dragana.miljkovic,nada.lavrac}@ijs.si
[3] Faculty of Computer and Information Science, University of Ljubljana, Ljubljana, Slovenia
marko.robnik@fri.uni-lj.si
[4] University of Nova Gorica, Vipavska 13, Nova Gorica, Slovenia

Abstract. Parkinson's disease is a neurodegenerative disease affecting people worldwide. Since the reasons for Parkinson's disease are still unknown and currently there is no cure for the disease, the management of the disease is directed towards handling of the underlying symptoms with antiparkinson medications. In this paper, we present a method for visualization of the patients' overall status and their antiparkinson medications therapy. The purpose of the proposed visualization method is multi-fold: understanding the clinicians' decisions for therapy modifications, identification of the underlying guidelines for management of Parkinson's disease, as well as identifying treatment differences between groups of patients. The resulting patterns of disease progression show that there are differences between male and female patients.

Keywords: Data mining · Parkinson's disease · Disease progression
Therapy modifications · Visualization

1 Introduction

Parkinson's disease is the second most common neurodegenerative disease after Alzheimer's disease. It is connected to the decreased levels of dopamine and it affects the central nervous system. Symptoms mostly associated with Parkinson's disease include bradykinesia, tremor, rigidity, and instability. In addition to the motor symptoms, patients also experience sleeping, behavioral, and mental problems. These symptoms significantly affect the quality of life of the patients and of their families.

The cause of Parkinson's disease is still unknown. Currently, there is no cure for the disease and the treatment of Parkinson's disease patients is directed towards management of the symptoms using antiparkinson medications. These medications can be grouped into three groups: levodopa, dopamine agonists,

L. Soldatova et al. (Eds.): DS 2018, LNAI 11198, pp. 466–479, 2018.
https://doi.org/10.1007/978-3-030-01771-2_30

and MAO-B inhibitors. Their role is to help regulate the patients' dopamine levels. The prolonged use of antiparkinson medications can result in side effects, prompting the clinicians to try and find a personalized balance of different antiparkinson medications for each patient, which will offer a good trade-off between controlling the Parkinson's disease symptoms and avoiding the possible side effects.

Recent data mining research has addressed the issue of both disease progression and changes in antiparkinson medications therapies. Valmarska et al. [22] use clustering and analysis of short time series to determine patterns of disease progression and patterns of medications change. This work is followed by analyzing the symptoms' influence on disease progression and by analyzing how the patients' antiparkinson medications therapy changes as a result of the status of the analyzed symptoms [21].

Visual representations of medical data can come in various shapes and forms. Kosara and Miksch [8] reviewed visualization techniques from the perspective of the application: visualizing measured data, visualizing events or incidents, and planning actions (therapeutic steps). Medical measurements of patients condition and symptoms occurrence carry important information for finding disease causes and preparing therapies. A simple representation of the recorded measured data, named a time line [20], is by drawing a line during the occurrence of the symptom.

LifeLines [12] is an approach that develops this idea by drawing lines for different types of symptoms and incidents in order to visualize the patients' personal health histories. To the best of our knowledge, no research has been performed in the Parkinson's disease domain, that would allow to visualize and analyze the changes of the overall patient's disease status in relation with the actions the clinicians take to keep the patient's status stable as long as possible. In this work we combine the LifeLines method for visualization with additional visualization shapes in order to visualize patterns of disease progression and therapy modifications of Parkinson's disease patients. For comprehensibility reasons, due to the vast number of symptoms associated with Parkinson's disease, we decided to include only information about patients' overall status.

We address this problem by presenting a method for visualization of the changes of the patients' overall status and the corresponding therapy modifications. This study builds upon our previous research [22] where the overall status of the patients is represented by their assignment to clusters. We combine the LifeLines method for visualization with the building block information from [22] in order to showcase how the overall status of Parkinson's disease patients change. The proposed visualization can provide comprehensible insights into the clinicians' decisions for therapy modifications. The graphical representation of both the patients' status and their therapy with antiparkinson medications can reveal the causal nature of the patient's status and the changes in the prescribed medications therapy. The analysis of the potential causal interaction between the patients' condition and their therapy modifications may reveal the underlying guidelines for treatment of Parkinson's disease. However, a deeper analysis of the antiparkinson medications therapies can reveal other regularities or phenomena, given that the clinicians base their decision for patient's treatment both on the existing medical guidelines as well as on the patient's preferences, given that some patients actually request an immediate treatment of the symptoms while others are not necessarily very bothered by them. To this end, our study explicitly reveals some differences between the treatment of male and female patients, which were previously not explicitly known.

The visualization of patients' Parkinson's disease progression and the corresponding antiparkinson medications treatment has been appreciated by the clinicians, as it offers them the opportunity to visually examine how the status of a particular patient has changed in previous visits and allows them to easily find out which treatments have been suitable or unsuccessful for the stabilization of the patient's status. The later can be used to assist the clinicians when considering future therapy modifications. In future work, we will include the possibility for the users to show the severity of the chosen symptoms, thus offering even deeper insights into the patients' status.

The paper is structured as follows. In Sect. 2 we give a short overview of the research closely related to our work. Section 3 outlines the data used in the analysis. In Sect. 4 we present the proposed visualization methodology and the visualization results through an illustrative use case. Section 5 presents the analysis of disease progression patterns of male and female patients. We conclude by presenting the plans for further work in Sect. 6.

2 Background

Parkinson's disease is a neurodegenerative disease. The status of the patients will change as the disease progresses. The progression of the disease and the actual overall status of the patients through time is mostly dependent on the natural progression of the disease and the therapy with antiparkinson medications.

The issue of Parkinson's disease progression in the data mining domain is only partially explored. The exception is the work of Tsanas et al. [15–19] addressing the Parkinson's disease progression in terms of the patients' motor and overall UPDRS (Unified Parkinson's Disease Rating Scale) score. Their evaluation of Parkinson's disease progression is performed on data from non-invasive speech tests for a six months period, during which all of the patients were off their antiparkinson medications.

To the best of our knowledge, the problem of Parkinson's disease progression over a longer period of time in combination with the patients' antiparkinson medications therapy was addressed only in our own past research [22,23] that investigates the progression of Parkinson's disease by analyzing short time series data, where the patients' overall status is determined by data clustering. Clustering is performed on the so-called *merged* data set that consists of sums of symptoms severity scores assessing different aspects of the patients' life. In this research, it was shown that the patients can be divided into three patient groups, which can be ordered according to the severity of the patients' motor status. The overall status of the patients will change over time. In [22] this change is reflected by assignment of patients to clusters and the patterns of disease progression are determined by using skip-grams [7].

The issue of dosage modifications of antiparkinson medications is addressed in our past work [21, 22]: in [21] we used predictive clustering trees [1, 14] to uncover how particular symptoms affect the therapy modifications for Parkinson's disease patients, while in [22] we explored the aggregate medications dosage changes that lead to improvement or worsening of the patients' overall status.

3   Parkinson's Disease Data Set

In this study, we use the PPMI data collection [9] gathered in the observational clinical study to verify progression markers in Parkinson's disease. The PPMI data collection consists of data sets describing different aspects of the patients' daily life. Below we describe the selection of PPMI data used in this study.

3.1   Symptoms Data

The severity of patients' symptoms and the overall quality of life of Parkinson's disease patients is determined through several standardized questionnaires. The most widely used questionnaire is the Movement Disorder Society (MDS) sponsored revision of Unified Parkinson's Disease Rating Scale (MDS-UPDRS) [6]. It is a four-part questionnaire addressing 'non-motor experiences of daily living' (Part I, subpart 1 and subpart 2), 'motor experiences of daily living' (Part II), 'motor examination' (Part III), and 'motor complications' (Part IV). The MDS-UPDRS questionnaire consists of 65 questions, each addressing a particular symptom. Each question is anchored with five responses that are linked to commonly accepted clinical terms: 0 = normal (patient's condition is normal, symptom is not present), 1 = slight (symptom is present and has a slight influence on the patient's quality of life), 2 = mild, 3 = moderate, and 4 = severe (symptom is present and severely affects the normal and independent functioning of the patient, i.e. her quality of life is significantly decreased).

The Montreal Cognitive Assessment (MoCA) [2] is a rapid screening instrument for mild cognitive dysfunction. It consists of 11 questions, designed to assess different cognitive domains: attention and concentration, executive functions, memory, language, visuoconstructional skills, conceptual thinking, calculations, and orientation.

Scales for Outcomes in Parkinson's disease – Autonomic (SCOPA-AUT) is a specific scale to assess autonomic dysfunction in Parkinson's disease patients [24]. Physical Activity Scale for the Elderly (PASE) [25] is a questionnaire which is a practical and widely used approach for physical activity assessment in epidemiologic investigations.

The above data sets are periodically updated to allow the clinicians to monitor patients' disease development over time. Answers to the questions from each questionnaire form the vectors of attribute values. All of the considered questions have ordered values, and—with the exception of questions from MoCA and PASE—increased values suggest higher symptom severity and decreased quality of life.

The symptoms data used in this study are represented in a single data table and constructed by using the sums of values of attributes of the following data sets: MDS-UPDRS Part I (subpart 1 and subpart 2), Part II, Part III, MoCA, PASE, and SCOPA-AUT. Goetz et al. [5] use sums of symptoms values as an overall severity measure of a given aspect of Parkinson's disease. Similarly, we use sums of attribute values from different data sets to present the overall status of patients concerning respective aspects of their everyday living. Table 1 gives a short description of the attributes used for determining the overall status of the patients.

Table 1. Characteristics of the attributes used to determine the patients' overall status.

Attribute	Questionnaire	Value range
NP1SUM	MDS-UPDRS Part I	0–24
NP1PSUM	MDS-UPDRS Part Ip	0–28
NP2PSUM	MDS-UPDRS Part II	0–52
NP3SUM	MDS-UPDRS Part III	0–140
MCATOT	MoCA	0–30
PASESUM	PASE	7–14
SCOPASUM	SCOPA-AUT	0–63

The overall status of the patients is determined using clustering on the merged data set. Description of the merged data set and the process of determining the overall status of the patients can be found in [22].

3.2   Medications Data

The PPMI data collection offers information about all of the concomitant medications that the patients used during their involvement in the study. The medications data in the concomitant medications log are described by their name, the medical condition they are prescribed for, as well as the time when the patient started and (if) ended the medications therapy. In our research, we concentrate only on the patients' therapy with antiparkinson medications. The main families of drugs used for treating the symptoms of Parkinson's disease are levodopa, dopamine agonists, and MAO-B inhibitors [10]. Dosages of PD medications are translated into a common Levodopa Equivalent Daily Dosage (LEDD) which allows for comparison of different therapies (different medications with personalized daily plans of intake). We visualize the medications data by their World Health Organization (WHO) name, the group they belong to, the dosage in

LEDD, the date when the therapy was introduced, and the date when the therapy has stopped. In addition to the regular visits, clinicians do phone call-ups to patients in order to stay informed of their status. If necessary, they modify the therapy between visits in order to control and stabilize the status of the patients. The concomitant medications log contains medications data with the appropriate LEDD values for 380 patients.

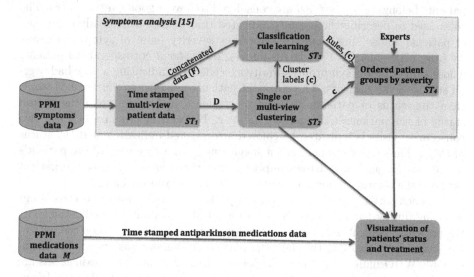

Fig. 1. Methodology for PPMI patients' overall status and medications therapy visualization.

4 Visualization: Methodology and Use Case

The proposed visualization methodology builds on the Parkinson's disease patients overall status corresponding to one of the three disease severity groups [22], which is used for automated visualization of the status and the antiparkinson medications therapies of 380 patients from the PPMI study. The visualization method is implemented in Python using the ReportLab toolkit.[1]

4.1 Methodology

The proposed methodology for visualizing the patients' status and their medications therapy is outlined in Fig. 1. The methodology is based on the patients'

---

[1] The visualization method is closely related to the PPMI data and the cluster label results from [22]. As the permission to use the PPMI data can be obtained only from www.ppmi-info.org/data, we cannot share the complete solution but the code is available upon request to the first author of the paper.

overall status data from [22], graphically presented in the upper part of the figure. The procedure consists of clustering of patients into groups with similar symptoms based on PPMI data and describing the characteristics of these groups by applying classification rule learning algorithms. The ultimate cluster validation and ordering are done by the experts.

For each patient, the patient's overall status on a particular visit to the clinician is determined by the patient's assignment to one of the three clusters. The patients belonging to *cluster 0* are considered to have a good overall status. The rules used to describe the clusters indicated that *cluster 0* is mostly composed of patients whose sum of motor symptoms severity (sum of symptoms severity from MDS-UPDRS Part III) is under 22. *Cluster 2* corresponds to patients whose sum of motor symptoms severity exceeded 42, indicating a very bad overall status of the patients. The status of patients assigned to *cluster 1* is worse than the status of patients assigned to *cluster 0* and better than the overall status of the patients assigned to *cluster 2*. The cluster crossing between two consecutive visits is indicated by colored arrows—red indicating that the status of the patients has worsened, green suggesting an improvement of the patient's overall status, and blue arrows implying that the overall status of the patient between the corresponding consecutive visits has stayed unchanged.

For each patient, we also draw the antiparkinson medications that the patient is taking during the recorded visits. In the PPMI concomitant medications log, the patients' medications therapies are described with the date when the patient has started the therapy, the date the particular therapy has ended, the medication's WHO name, and the corresponding LEDD value. The medications are arranged according to the antiparkinson medications group they belong to. In our visualization, the medications groups are indicated by color—in red we present the levodopa based medications, dopamine agonists are presented in green, while we present the MAO-B inhibitors using blue lines. The LEDD values are presented with the thickness of the lines. The start and the end of a particular therapy are indicated by the start and the end of the corresponding line. We look at whether there are differences in the disease progression patterns of male and female patients. For this purpose, we adopted the skip-gram approach presented in detail in [22].

4.2   Use Case

The visualization of patients' status and their corresponding antiparkinson medication therapies was performed for all 380 patients. Due to space restrictions and more comprehensible presentation of the visualization results, we showcase the progression of the disease and the medications therapy of a single patient.

Figure 2 presents a time line of overall status change for a particular patient. It is a male patient who was 65 years old at the time of his involvement in the PPMI study (baseline visit). The data shows that less than a year has passed between the patient's time of diagnosis and the baseline visit. The visits for which we have data for the patient's overall status are presented on the time line axis. As an additional information, we also include the actual time when the visit has occurred (month and year).

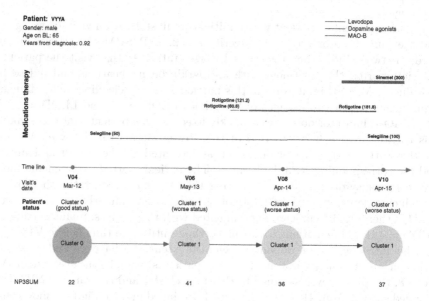

Fig. 2. Inspection of a cluster change time line of a single patient. Points on the time line present the patient's visits to the clinician. The patient's medications therapy is presented in the upper part of the figure, showing the antiparkinson medications the patient has received during his involvement in the PPMI study. The lower part of the figure shows the patient's overall status as indicated by his cluster assignment on each visit.

The patient's medications therapy is presented in the upper part of Fig. 2. The patient's medications therapy is presented by the groups of antiparkinson medications the patient has received during his involvement in the PPMI study. The color of medications therapy determines the group of antiparkinson medications—MAO-B inhibitors are presented with a green line, dopamine agonists with a blue line, and levodopa based medications with a red line on the top. The line width indicates the value of LEDD, i.e. the thicker the line the higher the value of LEDD. The line endpoints indicate the beginning and the end of treatment with a particular medication. The line endpoints are placed proportionally, according to the actual date when the therapy has started/ended. For comprehensibility reasons, we also included the WHO names of the medications. The corresponding LEDD values are written in parenthesis.

Below the visit time line we visualize what has happened to the patient's overall status as indicated by his cluster assignment on each visit. The user can also choose whether to show the sum of the patient's motor symptoms severity (NP3SUM). Higher values of NP3SUM indicate a decreased quality of life and more severe motor symptoms. The arrows between clusters indicate the change of the patient's status: a red arrow indicates that the patient's status has worsened between the two consecutive visits, the green arrow denotes the improvement of the status, and the blue arrow indicates that the patient's overall status remained unchanged.

Figure 2 presents a patient whose initial overall status—on visit 4—was good and the sum of motor symptoms severity was 22. NP3SUM on V04 was on the border between indicating a good and worse status. At that visit the patient's clinician started the treatment with antiparkinson medications and did so by introducing MAO-B inhibitors (in this particular case, Selegiline). The clinician experimented with two dosages of Selegiline (LEDD = 50 and LEDD = 100). The MAO-B inhibitor dosage was quickly fixed and the patient continued taking this therapy over a longer period of time (even after V10). Between V4 and V6 the status of the patient degraded, as indicated by the cluster assignment and the corresponding NP3SUM values. The patient started experiencing more severe motor symptoms thus prompting the clinician to start the therapy by introducing dopamine agonist (Rotigotine) after V06. Similarly to the patient's MAO-B therapy, the clinician experimented with the dosages, finally setting on LEDD = 181.8. The patient took this therapy until some time before V10.

The introduction of dopamine agonists seems to have caused a slight, although insufficient improvement of the patient's overall status—between V6 and V8, the patient was assigned to the same cluster and the value of NP3SUM dropped slightly from 41 to 36. This trend continued in V10. Figure 2 shows that there are no signs of improvement of the patient's status between V6 and V8, and V8 and V10. This can be interpreted as that the introduction of dopamine agonists did not have the desired effect on decreasing the severity of the motor symptoms, thus prompting the clinician to introduce levodopa between V8 and V10. As evident from Fig. 2, the updated medications therapy again did not influence the change of the patient's overall status (*cluster 1*, NP3SUM=37 in V10). It is expected that after V10 the clinician would increase the dosage of levodopa.

5   Analysis of Disease Progression Patterns for Different Patient Groups: A Case for Male and Female Patients

The PPMI study includes data for patients from different gender, geographical locations, age groups, etc. The clinicians' decisions for how they treat their patients are based on official guidelines for treatment of Parkinson's disease patients [3,4,11,13]. These decisions are always made in the context of the patient and their quality of life. According to our consulting clinicians, the clinician will take into account the patient's family and employment status. If it is an older patient who is not employed and has family members to support him/her during the disease, the clinician will not be very forceful with the introduction of antiparkinson medications therapy. However, for example, if the patient is a working professional, who finds that some symptoms are impeding him/her in their working environment, the clinician will be inclined towards more rigorous control of symptoms with medications.

As the clinician's therapy partly depends on the context, i.e. the patient's preferences, gender, employment, geographical location, etc., it is interesting to research whether there are differences in the therapies and patterns of disease progression based on the context. In this study, we have focused on gender analysis.

Table 2. List of most influential symptoms for female and male patients from PPMI according to our previous study [21].

Female patients	Male patients
Rigidity	Toe tapping
Sleep problems (night)	Daytime sleepiness
Finger tapping	Finger tapping
Bradykinesia	Hand movement
Toe tapping	Bradykinesia
Hand pronation/supination	Hand pronation/supination
Facial expression	Facial expression
Hand movement	Sleep problems (night)
Leg agility	Rigidity
Constancy of rest	Constancy of rest

Table 3. List of most influential symptoms for female and male patients from PPMI according to our previous study [21].

Patients	First cluster assignment ($\%c_0$, $\%c_1$, $\%c_2$)	Last cluster assignment ($\%c_0$, $\%c_1$, $\%c_2$)
Females	(47.14%, 39.29%, 13.57%)	(45.71%, 37.14%, 17.14%)
Males	(39.25%, 47.55%, 13.21%)	(30.19%, 40.38%, 29.43%)

Table 2 presents a list of 10 attributes that change most frequently as the overall status of the patients changes for both male and female patients. These indicators were obtained by adapting the method for detection of influential symptoms in [21]. The symptoms listed in Table 2 identify differences between the symptoms that most frequently change their severity as the overall status of female or male patients change. The symptoms that change most frequently as the overall status of female patients change are rigidity and problems with sleeping at night. On the other hand, the most influential symptoms to the overall status of male patients are problems with toe-tapping and daytime sleepiness.

Table 3 presents the patients' cluster assignment distribution on the first (V04) and their last recorded visit. Results show that on their first recorded visit, most of the females were assigned to *cluster 0* (47.14%), while most of the male patients already had worsened overall status, and were assigned to *cluster 1*. On the last recorded visit, females tended to stay in the initially assigned cluster, while the status of the male patients significantly worsened (*cluster 2*, 29.43%). The average number of recorded visits for male patients is 3.36 and for female patients is 3.23.

The difference between the lists of most influential symptoms for male and female patients indicate possible differences in the patterns of disease progression of male and female patients. We looked into what are the patterns of disease progression for male and female patients. Details of using skip-gram analysis for determining patterns of disease progression can be found in [22].[2]

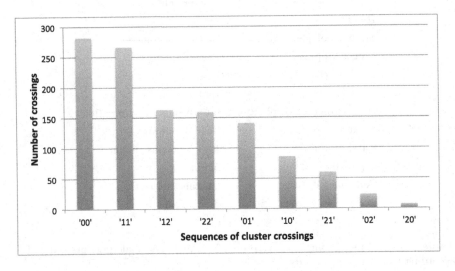

Fig. 3. Cluster crossings for male PPMI patients.

Figure 3 presents patterns of cluster sequences for male patients. The results show that the status of the patients is mostly stable. In most cases, the patients stayed in the clusters they were initially assigned to: *cluster 0* and *cluster 1*. Additionally, in many cases, the patients' status became or was very uncomfortable (*cluster 2*, cluster crossings '12' and '22').

Figure 4 presents patterns of cluster crossings for female patients. Similarly to the male patients, the status of the female patients is mostly stable and the patients stayed in the clusters they were initially assigned to, *cluster 0* and *cluster 1*. The status of the female patients mostly switched between these two

---

[2] A skip gram, e.g., a d-skip-n-gram, is a sequence of n items (disease progression phases, in our case), which are not necessarily consecutive, but gaps of up to d intermediate items are tolerated. The advantage of skip-grams over ordinary n-grams is that they are more noise tolerant and offer stronger statistical support for possibly interrupted sequence patterns.

clusters, and only on rare occasions the patients overall status significantly worsened and the patients were assigned to *cluster 2*.

Further data analysis and consultations with medical professionals is required to determine the reason for the different patterns of disease progression between male and female patients. We will look into the differences of the periods of patients' diagnosis and their introduction to the PPMI study, the period between their first symptoms and their diagnosis, as well as the differences in their medications treatment.

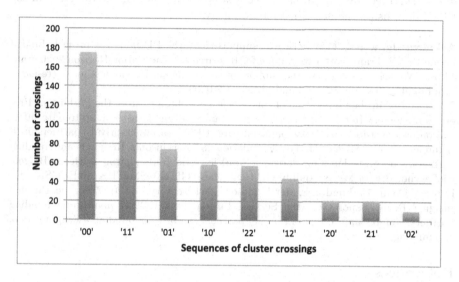

Fig. 4. Cluster crossings for female PPMI patients.

6   Conclusions

This work presents a methodology for visualization and analysis of Parkinson's disease patients' status and medications therapy patterns. The visualization method builds on previously detected data representing the patients' overall status. The simultaneous visualization of patients' overall status and their medications therapy can be a step further towards personalized treatment of Parkinson's disease patients. It can keep doctors in the loop and allows them to more readily understand why the medications therapy of a certain patient needs to be changed.

The analysis of the patients' status between consecutive visits revealed differences in the patterns of disease progression for male and female patients. The analysis reveals that male patients are more likely than female patients to experience severe overall motor status, while female patients are more likely to stay and switch between clusters indicating good or intermediate status.

In future work, we aim to analyze the potential differences in medication patterns for patients from different countries, different age groups, etc. The

PPMI study collects data from patients from several countries, including the United States, Israel, and Italy. Although the clinicians follow the official guidelines for treatment of Parkinson's disease, their decisions are also highly influenced by their previous experiences and the context, including the patients' own demand for treatment of particular symptoms. An interesting research avenue is to explore how the treatment of PPMI patients adheres to the official guidelines for treating Parkinson's disease patients. The knowledge from the official guidelines is available in the form of text. This knowledge can be transformed into more structured inputs that can be compared to the patterns extracted from the PPMI data using the described methodology.

Acknowledgments. This work was supported by the PD_manager project, funded within the EU Framework Programme for Research and Innovation Horizon 2020 grant 643706. We acknowledge also the support of the Slovenian Research Agency (research core funding program P2-0103 and project P2-0209).

Data used in the preparation of this article were obtained from the Parkinsons Progression Markers Initiative (PPMI) (www.ppmi-info.org/data). For up-to-date information on the study, visit www.ppmi-info.org. PPMI—a public-private partnership—is funded by the Michael J. Fox Foundation for Parkinson's Research and funding partners. Corporate Funding Partners: AbbVie, Avid Radiopharmaceuticals, Biogen, BioLegend, Bristol-Myers Squibb, GE Healthcare, GLAXOSMITHKLINE (GSK), Eli Lilly and Company, Lundbeck, Merck, Meso Scale Discovery (MSD), Pfizer Inc, Piramal Imaging, Roche, Sanofi Genzyme, Servier, Takeda, Teva, UCB. Philanthropic Funding Partners: Golub Capital. List of funding partners can be also found at www.ppmi-info. org/fundingpartners.

References

1. Blockeel, H., Raedt, L.D., Ramon, J.: Top-down induction of clustering trees. In: Proceedings of the 15th International Conference on Machine Learning, ICML 1998. pp. 55–63 (1998)
2. Dalrymple-Alford, J., et al.: The MoCA: well-suited screen for cognitive impairment in Parkinson disease. Neurology **75**(19), 1717–1725 (2010)
3. Ferreira, J., et al.: Summary of the recommendations of the EFNS/MDS-ES review on therapeutic management of Parkinson's disease. Eur. J. Neurol. **20**(1), 5–15 (2013)
4. Fox, S.H., et al.: The movement disorder society evidence-based medicine review update: treatments for the motor symptoms of Parkinson's disease. Mov. Disord. **26**(S3), S2–S41 (2011)
5. Goetz, C., Luo, S., Wang, L., Tilley, B., LaPelle, N., Stebbins, G.: Handling missing values in the MDS-UPDRS. Mov. Disord. **30**(12), 1632–1638 (2015)
6. Goetz, C., et al.: Movement disorder society-sponsored revision of the unified Parkinson's disease rating scale (MDS-UPDRS): scale presentation and clinimetric testing results. Mov. Disord. **23**(15), 2129–2170 (2008)
7. Guthrie, D., Allison, B., Liu, W., Guthrie, L., Wilks, Y.: A closer look at skip-gram modelling. In: Proceedings of the 5th International Conference on Language Resources and Evaluation, LREC 2006. pp. 1–4 (2006)
8. Kosara, R., Miksch, S.: Visualization methods for data analysis and planning in medical applications. Int. J. Med. Inform. **68**(1), 141–153 (2002)

9. Marek, K., et al.: The Parkinson's progression markers initiative (PPMI). Prog. Neurobiol. **95**(4), 629–635 (2011)
10. National Collaborating Centre for Chronic Conditions: Parkinson's Disease: National Clinical Guideline for Diagnosis and Management in Primary and Secondary Care. Royal College of Physicians, London (2006)
11. Olanow, W., Watts, R., Koller, W.: An algorithm (decision tree) for the management of Parkinson's disease (2001): treatment guidelines. Neurology **56**(suppl 5), S1–S88 (2001)
12. Plaisant, C., Milash, B., Rose, A., Widoff, S., Shneiderman, B.: Lifelines: visualizing personal histories. In: Proceedings of the SIGCHI Conference on Human Factors in Computing Systems, pp. 221–227. CHI 1996, ACM, New York, NY, USA (1996)
13. Seppi, K., et al.: The movement disorder society evidence-based medicine review update: treatments for the non-motor symptoms of Parkinson's disease. Mov. Disord. **26**(S3) (2011)
14. Struyf, J., Ženko, B., Blockeel, H., Vens, C., Džeroski, S.: CLUS: User's Manual (2010)
15. Tsanas, A.: Accurate telemonitoring of Parkinsons disease symptom severity using nonlinear speech signal processing and statistical machine learning. Ph.D. thesis, Oxford University, UK (2012)
16. Tsanas, A., Little, M., McSharry, P., Ramig, L.: Accurate telemonitoring of Parkinson's disease progression by noninvasive speech tests. IEEE Trans. Biomed. Eng. **57**(4), 884–893 (2010)
17. Tsanas, A., Little, M.A., McSharry, P.E., Ramig, L.O.: Accurate telemonitoring of Parkinson's disease progression by noninvasive speech tests. IEEE Trans. Biomed. Eng. **57**(4), 884–893 (2010)
18. Tsanas, A., Little, M.A., McSharry, P.E., Ramig, L.O.: Enhanced classical Dysphonia measures and sparse regression for telemonitoring of Parkinson's disease progression. In: Proceedings of the IEEE International Conference on Acoustics Speech and Signal Processing, ICASSP 2010, pp. 594–597. IEEE (2010)
19. Tsanas, A., Little, M.A., McSharry, P.E., Spielman, J., Ramig, L.O.: Novel speech signal processing algorithms for high-accuracy classification of Parkinson's disease. IEEE Trans. Biomed. Eng. **59**(5), 1264–1271 (2012)
20. Tufte, E.R.: The Visual Display of Quantitative Information. Graphics Press, Cheshire, CT, USA (2001)
21. Valmarska, A., Miljkovic, D., Konitsiotis, S., Gatsios, D., Lavrač, N., Robnik-Šikonja, M.: Symptoms and medications change patterns for Parkinson's disease patients stratification. Artif. Intell. Med. (2018). https://doi.org/10.1016/j.artmed.2018.04.010
22. Valmarska, A., Miljkovic, D., Lavrač, N., Robnik-Šikonja, M.: Analysis of medications change in Parkinson's disease progression data. J. Intell. Inf. Syst. **51**(2), 301–337 (2018)
23. Valmarska, A., Miljkovic, D., Robnik-Šikonja, M., Lavrač, N.: Multi-view approach to Parkinson's disease quality of life data analysis. In: Proceedings of the International Workshop on New Frontiers in Mining Complex Patterns, pp. 163–178. Springer (2016)
24. Visser, M., Marinus, J., Stiggelbout, A.M., Van Hilten, J.J.: Assessment of autonomic dysfunction in Parkinson's disease: the SCOPA-AUT. Mov. Disord. **19**(11), 1306–1312 (2004)
25. Washburn, R.A., Smith, K.W., Jette, A.M., Janney, C.A.: The physical activity scale for the elderly (PASE): development and evaluation. J. Clin. Epidemiol. **46**(2), 153–162 (1993)

Author Index

Printed in the United States
By Bookmasters